Graduate Texts in Mathematics 40

Denumerable Markov Chains

John G. Kemeny *J. Laurie Snell*
Anthony W. Knapp

Springer-Verlag New York Heidelberg Berlin

John G. Kemeny
President
Dartmouth College
Hanover, New Hampshire 03755

J. Laurie Snell
Dartmouth College
Department of Mathematics
Hanover, New Hampshire 03755

Anthony W. Knapp
Cornell University
Department of Mathematics
Ithaca, New York 14850

AMS Subject Classifications
60J05, 60J10

Library of Congress Cataloging in Publishing Data

Main entry under title:
Denumerable Markov Chains.
(Graduate texts in mathematics; 40)
First ed. by J. G. Kemeny, J. L. Snell, and
A. W. Knapp is entered under Kemeny, John G.
Bibliography: p. 471
Includes indexes.
1. Markov processes. I. Kemeny, John G.
II. Series.
QA274.7.K45 1976 519.2′33 76-3535
Second Edition

Originally published in the University Series in Higher Mathematics
(D. Van Nostrand Company); edited by M. H. Stone, L. Nirenberg,
and S. S. Chern.

Printed in the United States of America

ISBN 0-387-90177-9 Springer-Verlag New York Heidelberg Berlin
ISBN 3-540-90177-9 Springer-Verlag Berlin Heidelberg New York

PREFACE TO THE SECOND EDITION

With the first edition out of print, we decided to arrange for republication of *Denumerable Markov Chains* with additional bibliographic material. The new edition contains a section Additional Notes that indicates some of the developments in Markov chain theory over the last ten years. As in the first edition and for the same reasons, we have resisted the temptation to follow the theory in directions that deal with uncountable state spaces or continuous time. A section entitled Additional References complements the Additional Notes.

J. W. Pitman pointed out an error in Theorem 9-53 of the first edition, which we have corrected. More detail about the correction appears in the Additional Notes. Aside from this change, we have left intact the text of the first eleven chapters.

The second edition contains a twelfth chapter, written by David Griffeath, on Markov random fields. We are grateful to Ted Cox for his help in preparing this material. Notes for the chapter appear in the section Additional Notes.

J.G.K., J.L.S., A.W.K.
March, 1976

PREFACE TO THE FIRST EDITION

Our purpose in writing this monograph has been to provide a systematic treatment of denumerable Markov chains, covering both the foundations of the subject and some topics in potential theory and boundary theory. Much of the material included is now available only in recent research papers. The book's theme is a discussion of relations among what might be called the descriptive quantities associated with Markov chains—probabilities of events and means of random variables that give insight into the behavior of the chains.

We make no pretense of being complete. Indeed, we have omitted many results which we feel are not directly related to the main theme, especially when they are available in easily accessible sources. Thus, for example, we have only touched on independent trials processes, sums of independent random variables, and limit theorems. On the other hand, we have made an attempt to see that the book is self-contained, in order that a mathematician can read it without continually referring to outside sources. It may therefore prove useful in graduate seminars.

Denumerable Markov chains are in a peculiar position in that the methods of functional analysis which are used in handling more general chains apply only to a relatively small class of denumerable chains. Instead, another approach has been necessary, and we have chosen to use infinite matrices. They simplify the notation, shorten statements and proofs of theorems, and often suggest new results. They also enable one to exploit the duality between measures and functions to the fullest.

The monograph divides naturally into four parts, the first three consisting of three chapters each and the fourth containing the last two chapters.

Part I provides background material for the theory of Markov chains. It is included to help make the book self-contained and should facilitate the use of the book in advanced seminars. Part II contains basic results on denumerable Markov chains, and Part III deals with discrete potential theory. Part IV treats boundary theory for both transient and recurrent chains. The analytical prerequisites for the two chapters in this last part exceed those for the earlier parts of the book and are not all included in Part I. Primarily, Part IV presumes that the reader is familiar with the topology and measure theory of compact metric spaces, in addition to the contents of Part I.

Two chapters—Chapters 1 and 7—require special comments. Chapter 1 contains prerequisites from the theory of infinite matrices and some other topics in analysis. In it Sections 1 and 5 are the most important for an understanding of the later chapters. Chapter 7, entitled "Introduction to Potential Theory," is a chapter of motivation and should be read as such. Its intent is to point out why classical potential theory and Markov chains should be at all related.

The book contains 239 problems, some at the end of each chapter except Chapters 1 and 7.

For the most part, historical references do not appear in the text but are collected in one segment at the end of the book.

Some remarks about notation may be helpful. We use sparingly the word "Theorem" to indicate the most significant results of the monograph; other results are labeled "Lemma," "Proposition," and "Corollary" in accordance with common usage. The end of each proof is indicated by a blank line. Several examples of Markov chains are worked out in detail and recur at intervals; although there is normally little interdependence between distinct examples, different instances of the same example may be expected to build on one another.

A complete list of symbols used in the book appears in a list separate from the index.

We wish to thank Susan Knapp for typing and proof-reading the manuscript.

We are doubly indebted to the National Science Foundation: First, a number of original results and simplified proofs of known results were developed as part of a research project supported by the Foundation. And second, we are grateful for the support provided toward the preparation of this manuscript.

J. G. K.
J. L. S.
A. W. K.

Dartmouth College
Massachusetts Institute of Technology

RELATIONSHIPS AMONG MARKOV CHAINS

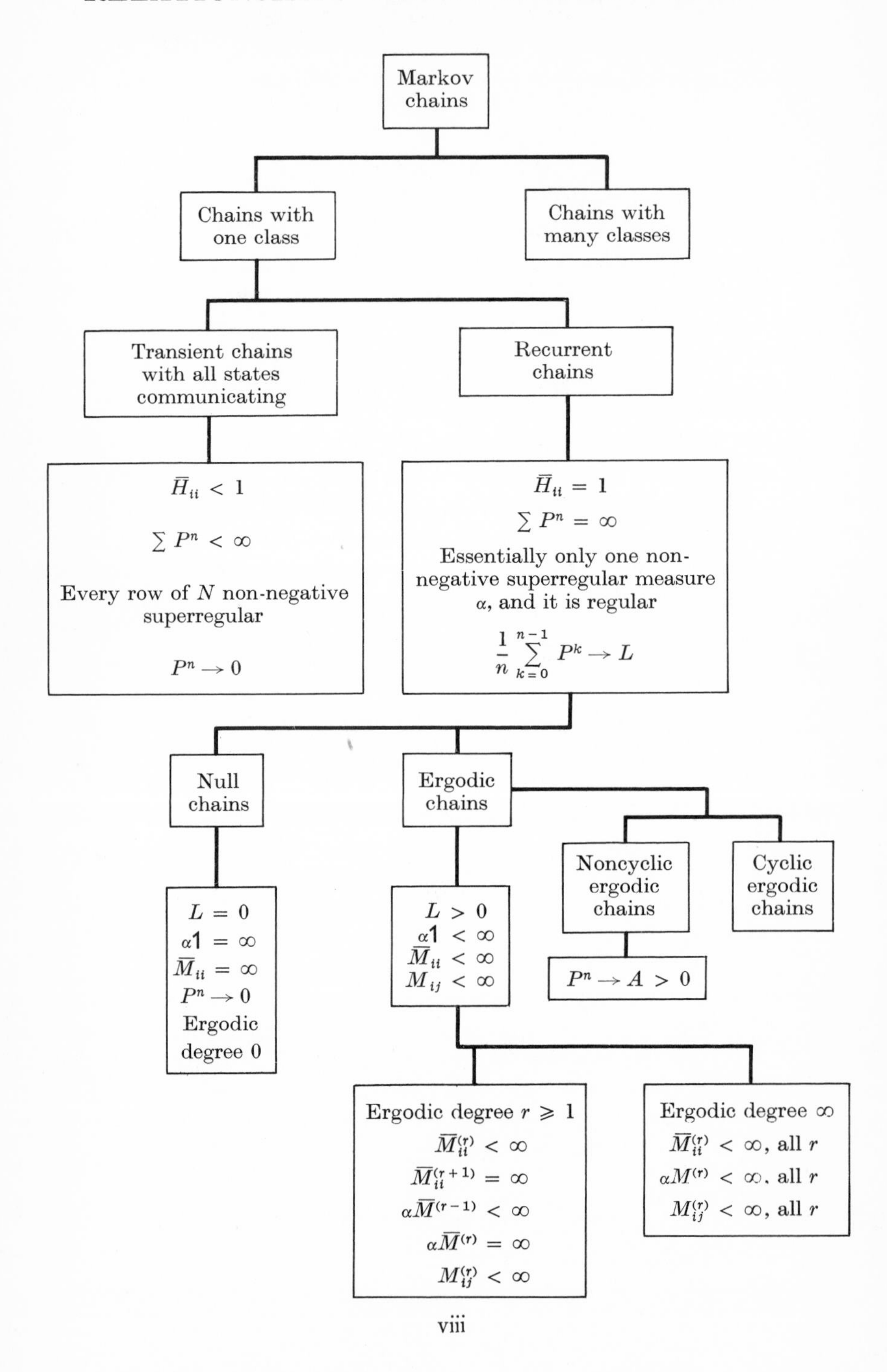

TABLE OF CONTENTS

Chapter 5: TRANSIENT CHAINS

Chapter 6: RECURRENT CHAINS

Chapter 7: INTRODUCTION TO POTENTIAL THEORY

Chapter 8: TRANSIENT POTENTIAL THEORY

CHAPTER 9: RECURRENT POTENTIAL THEORY

CHAPTER 10: TRANSIENT BOUNDARY THEORY

CHAPTER 11: RECURRENT BOUNDARY THEORY

CHAPTER 12: INTRODUCTION TO RANDOM FIELDS

CHAPTER 1

PREREQUISITES FROM ANALYSIS

1. Denumerable matrices

The word **denumerable** in the sequel means finite or countably infinite. Let M and N be two non-empty denumerable sets. A **matrix** is a function with domain the set of ordered pairs (m, n), where $m \in M$ and $n \in N$, and with range a subset of the extended real number system—the reals with $+\infty$ and $-\infty$ adjoined. We call the sets M and N **index sets**. The matrix is called a **finite matrix** if both M and N are finite sets.

To say that the m-nth **entry** of the matrix is x or is equal to x, we mean that the value of the function on the pair (m, n) is x. A matrix is said to be **non-negative** if all of its entries are non-negative, and it is said to be **positive** if all of its entries are positive. We agree to use upper-case italic letters to stand for matrices. If A is a matrix, we denote the m-nth entry of A by A_{mn}. Some examples of matrices are as follows:

(1) If all entries of a matrix are equal to zero, we say that the matrix is the **zero matrix**, denoted by 0.

(2) A matrix for which M and N are the same set is called a **square matrix**. The entries corresponding to $m = n$ are **diagonal entries**; other entries are **off-diagonal entries**.

(3) A square matrix whose off-diagonal entries all equal zero is a **diagonal matrix**. The diagonal matrix obtained from a square matrix A by setting all of its off-diagonal entries equal to zero is denoted A_{dg}.

(4) A diagonal matrix whose diagonal entries are all equal to one is called the **identity matrix**, denoted by I.

(5) A matrix whose second index set contains only one element is called a **column vector**. If we wish to distinguish a column vector from an arbitrary matrix, we shall denote the former by a lower-case italic letter.

(6) A matrix whose first index set contains only one element is called a **row vector**. If we wish to distinguish a row vector from an arbitrary matrix, we shall denote the former by a lower-case Greek letter.

(7) If A is a matrix defined on index sets M and N, define a matrix A^T, called the **transpose** of A, to have index sets N and M and to have entries given by $(A^T)_{nm} = A_{mn}$. The transpose of the transpose of A is simply A.

(8) The column vector all of whose entries are equal to one is denoted $\mathbf{1}$; the row vector with all entries one is $\mathbf{1}^T$. A matrix other than a row or column vector which has all entries equal to one is denoted by E.

(9) If A is an arbitrary matrix and c is a real number, cA is the matrix whose entries are given by $(cA)_{mn} = cA_{mn}$.

(10) The matrix $-A$ is defined to be the matrix $(-1)A$.

(11) A **constant** (column) **vector** is a vector of the form $c\mathbf{1}$ for some extended real number c.

(12) A **bounded vector** is a vector all of whose entries are less than or equal in absolute value to some finite real number c.

Two matrices A and B are equal, written $A = B$, if they have the same index sets and if $A_{mn} = B_{mn}$ for every m and n. Inequalities are defined similarly. For example, $A > B$ if A and B have the same index sets and if $A_{mn} > B_{mn}$ for every m and n. In particular, non-negative matrices are those for which $A \geq 0$, and positive matrices are those for which $A > 0$.

Addition of matrices is defined for matrices A and B having the same index sets M and N. Their **sum** $C = A + B$ has the same index sets, and addition is defined entry-by-entry:

$$C_{mn} = A_{mn} + B_{mn}.$$

The sum $C = A + B$ is **well defined** if no entry of C is given by $\infty - \infty$ or by $-\infty + \infty$. We leave the verification of the following properties of matrices with index sets M and N to the reader:

(1) $A + 0 = A$ for every A.
(2) For every A having all entries finite, $A + (-A) = 0$.
(3) For any matrices A, B, and C,

$$A + (B + C) = (A + B) + C$$

if the indicated sums on at least one side of the equality are well defined.

Up to now, we have imposed no orderings on our index sets, and in fact nothing we have done so far necessitates doing so. We shall define even matrix multiplication shortly in a way that requires no ordering.

There is, however, a standard way of representing matrices as rectangular arrays, and for this purpose one normally orders the index sets with the usual ordering on the non-negative integers. The elements of the index sets are thus numbered 0, 1, 2, ... either up to some integer r if the index set is finite or indefinitely if the index set is infinite. Under such orderings of its index sets, a matrix A is represented as

$$A = \begin{pmatrix} A_{00} & A_{01} & A_{02} & \cdots \\ A_{10} & A_{11} & A_{12} & \cdots \\ A_{20} & A_{21} & A_{22} & \cdots \\ \vdots & & & \ddots \end{pmatrix}.$$

We note that other representations are possible if at least one of the index sets is infinite; such representations come from ordering the index sets with an order type other than that of the non-negative integers. We shall meet another order type with its corresponding representation at the end of this section. We point out, however, that orderings are completely irrelevant as far as the fundamental properties of matrices are concerned, and we shall have little occasion to refer to them again.

For any real number a_m, define a_m^+ and a_m^- by

$$a_m^+ = \max(a_m, 0)$$
$$a_m^- = -\min(a_m, 0).$$

The sum of denumerably many non-negative terms $\sum_{m \in M} a_m^+$ or $\sum_{m \in M} a_m^-$ always exists independently of any ordering on M. Therefore, we say that $\sum_{m \in M} a_m = \sum_{m \in M} a_m^+ - \sum_{m \in M} a_m^-$ is **well defined** if not both of $\sum_{m \in M} a_m^+$ and $\sum_{m \in M} a_m^-$ are infinite.

Definition 1-1: Let A be a matrix with index sets K and M, and let B be a matrix with index sets M and N. Suppose the sums

$$\sum_{m \in M} A_{km} B_{mn}$$

are well defined for every k and every n. Then the **matrix product** $C = AB$ is said to be well defined; its index sets are K and N, and its entries are given by $C_{kn} = \sum_{m \in M} A_{km} B_{mn}$. Matrix multiplication is not defined unless all of these properties hold.

Most of the propositions and theorems about matrices that we shall deal with are statements of equality of matrices $A = B$. Such statements are really just assertions about the equality of corresponding entries of A and B, and a proof that A equals B need only contain an argument that an arbitrary entry of A equals the corresponding entry of B. With this understanding, we see that the proof of the additive properties of matrices is reduced to a trivial repetition of the properties of real numbers. Propositions about multiplication, however, when looked at entry-by-entry involve a new idea.

Let A be a matrix with index sets M and N and let m and n be fixed elements of M and N, respectively. The mth **row** of A is defined to be the restriction of the function A to the domain of pairs (m, s), where s runs through the set N. Similarly the nth **column** of A is defined to be the restriction of the function A to the domain of pairs (t, n), where t runs through the elements of the set M. We note that the mth row of a matrix is a row vector and that the nth column is a column vector. With these conventions matrices can be thought of as sets of rows or as sets of columns, and addition of matrices is simply addition of corresponding rows or columns of the matrices involved. Furthermore, the k-nth entry in the matrix product of A and B is the product of the kth row of A by the nth column of B and is of the form $\sum_{m \in M} \pi_m f_m$, where π is a row vector and f is a column vector. That is, propositions about matrix multiplication, when proved entry-by-entry, may sometimes be proved by considering only the product of a row vector and a column vector.

Because of the correspondence of row vectors to rows and column vectors to columns, we shall agree to call the domain of a row vector or a column vector the elements of a single index set.

Connected with any definition of multiplication are five properties which may or may not be valid for the structure being considered. All five of the properties do hold for the real numbers, and we state them in this context:

(1) Existence and uniqueness of a multiplicative identity. The real number 1 satisfies $c1 = 1c = c$ for every c.
(2) Commutativity: $ab = ba$
(3) Distributivity: $a(b + c) = ab + ac$
$(a + b)c = ac + bc$
(4) Associativity: $a(bc) = (ab)c$
(5) Existence and uniqueness of multiplicative inverses of all non-zero elements.

We can easily settle whether the first two properties hold for matrix multiplication. First, the identity matrix I plays the role of the identity, and the identity is clearly unique. Second, commutativity can be expected to fail except in special cases because it is not even necessary for the index sets of two matrices to agree properly after the order of multiplication has been reversed.

The validity of the third property, that of distributivity, is the content of the next proposition.

Proposition 1-2: If A, B, and C are matrices and if AB, AC, and $AB + AC$ are well defined, then $A(B + C) = AB + AC$. Similarly $(D + E)F = DF + EF$ if DF, EF, and $DF + EF$ are all well defined.

PROOF: We prove only the first assertion. We may assume that A is a row vector π and that B and C are column vectors f and g. Then

$$\begin{aligned}\pi f + \pi g &= \sum_{m \in M} \pi_m f_m + \sum_{m \in M} \pi_m g_m \\ &= \sum_{m \in M} (\pi_m f_m + \pi_m g_m) \\ &= \sum_{m \in M} \pi_m (f_m + g_m) \\ &= \pi(f + g).\end{aligned}$$

The fourth and fifth properties are related and nontrivial. Associativity does not always hold, but useful sufficient criteria for its validity are known. For an example of how associativity may fail, let A be a matrix whose index sets are the non-negative integers and whose entries are given by

$$A = \begin{pmatrix} 1 & -1 & 0 & 0 & 0 & \cdots \\ 0 & 1 & -1 & 0 & 0 & \cdots \\ 0 & 0 & 1 & -1 & 0 & \cdots \\ 0 & 0 & 0 & 1 & -1 & \cdots \\ 0 & 0 & 0 & 0 & 1 & \cdots \\ \vdots & & & & & \ddots \end{pmatrix}$$

Then

$$\mathbf{1}^T(A\mathbf{1}) = 0,$$

whereas

$$(\mathbf{1}^T A)\mathbf{1} = 1.$$

All the products involved are well defined, but the multiplications do not associate.

We shall not consider the problem of existence of inverses, but uniqueness rests upon associativity. For suppose $AB = BA = AC = CA = I$. Since $AC = I$, we have $B(AC) = B$, and since $BA = I$, we have $(BA)C = C$. Therefore, $B = C$ if and only if $B(AC) = (BA)C$. With this note we proceed with some sufficient conditions for associativity.

Lemma 1-3: Let b_{ij} be a sequence of real numbers nondecreasing with i and with j. Then $\lim_i \lim_j b_{ij} = \lim_j \lim_i b_{ij}$, both possibly infinite.

PROOF: In the extended sense $\lim_i b_{ij} = L_j$ exists and so does $\lim_j b_{ij} = L_i^*$. Now $\{L_j\}$ is nondecreasing, for if $L_j > L_{j+k}$, then for i sufficiently large $b_{ij} > L_{j+k} \geq b_{i,j+k}$, which is impossible. Similarly $\{L_i^*\}$ is nondecreasing, so that $\lim_j L_j = L$ and $\lim_i L_i^* = L^*$ exist in the extended sense. If $L \neq L^*$, we may assume $L^* > L$ and hence L is finite. Then there exists an i such that $L_i^* > L$. Hence

$$\begin{aligned} L_i^* &> L \\ &\geq L_j \quad \text{for all } j \\ &\geq b_{ij} \quad \text{for all } j. \end{aligned}$$

Thus b_{ij} is bounded away from its limit on j, a contradiction.

Following the example of Lemma 1-3, we agree that all limits referred to in the future are on the **extended real line**.

Proposition 1-4: Non-negative matrices associate under multiplication.

PROOF: Since we are interested in each entry separately of a triple product, we may assume that we are to show that $\pi(Af) = (\pi A)f$, where $\pi \geq 0$, $A \geq 0$, $f \geq 0$, π is a row vector, f is a column vector, and the index sets are subsets of the non-negative integers. Then

$$\pi(Af) = \sum_m \sum_n \pi_m A_{mn} f_n$$

and

$$(\pi A)f = \sum_n \sum_m \pi_m A_{mn} f_n.$$

Set $b_{ij} = \sum_{m=0}^{i} \sum_{n=0}^{j} \pi_m A_{mn} f_n = \sum_{n=0}^{j} \sum_{m=0}^{i} \pi_m A_{mn} f_n$ and apply Lemma 1-3 to complete the proof.

If A is an arbitrary matrix we define A^+ and A^- by the equations

$$\begin{aligned} (A^+)_{mn} &= \max \{A_{mn}, 0\} \\ (A^-)_{mn} &= -\min \{A_{mn}, 0\}. \end{aligned}$$

Then $A = A^+ - A^-$, $A^+ \geq 0$, and $A^- \geq 0$. For row and column vectors, the matrices π^+, π^-, f^+, and f^- are defined analogously. We note that if Af is well defined, then so are Af^+ and Af^-. Powers of matrices are defined inductively by $A^0 = I$, $A^n = A(A^{n-1})$. The absolute value of a matrix A is $|A| = A^+ + A^-$. Proposition 1-4 now gives us five corollaries.

Corollary 1-5: Matrices associate if the product of their absolute values has all finite entries.

PROOF: We are again to prove that $\pi(Af) = (\pi A)f$, and we do so by setting $\pi = \pi^+ - \pi^-$, $A = A^+ - A^-$, and $f = f^+ - f^-$, applying distributivity, and using Proposition 1-4 on the resulting non-negative matrices.

Corollary 1-6: Finite matrices with finite entries associate.

PROOF: The result follows from Corollary 1-5.

Corollary 1-7: If A and B are non-negative matrices and f is a column vector such that $A(Bf)$ and $(AB)f$ are both well defined, then $A(Bf) = (AB)f$. In particular, if C is a non-negative matrix, if $n > 0$, and if C^nf and $C(C^{n-1}f)$ are well defined, then $C^nf = C(C^{n-1}f)$.

PROOF: Consider f^+ and f^- separately and apply Proposition 1-4. For the second assertion, set $A = C$ and $B = C^{n-1}$.

Similarly one proves two final corollaries.

Corollary 1-8: If A, B, C, and D are non-negative matrices such that either

(1) ABD, AB, and BD, or
(2) ACD, AC, and CD

are finite-valued, then $(A(B - C))D = A((B - C)D)$.

Corollary 1-9: If A, B, and C are matrices such that either

(1) A has only finitely many non-zero entries in each row,
(2) C has only finitely many non-zero entries in each column, or
(3) B has only finitely many non-zero entries,

and if $(AB)C$ and $A(BC)$ are well defined, then

$$(AB)C = A(BC).$$

Some of these conditions are cumbersome to check, but there is a simple sufficient condition. Suppose that we write a general product as $\prod_{i=1}^n (A_i - B_i)$, with $A_i \geq 0$ and $B_i \geq 0$. If all the 2^n products

$$A_1A_2 \ldots A_n,\ B_1A_2 \ldots A_n, \ldots, B_1B_2 \ldots B_n$$

are finite, then we see from Proposition 1-2 and Corollary 1-5 that we may freely use distributivity and associativity.

The effect of matrix multiplication on matrix inequalities is summarized by the next proposition, whose proof is left to the reader.

Proposition 1-10: Matrix inequalities of the form $A \geq B$ or $B \leq A$ are preserved when both sides of the inequality are multiplied by a non-negative matrix. Inequalities of the form $A > B$ or $B < A$ are preserved when both sides are multiplied by a positive matrix, provided the products have all entries finite.

Next we consider the problem of "block multiplication" of matrices. The picture we have in mind is the following decomposition of the matrices involved in a product:

$$\begin{pmatrix} A_1 & A_2 \\ A_3 & A_4 \end{pmatrix}\begin{pmatrix} B_1 & B_2 \\ B_3 & B_4 \end{pmatrix} = \begin{pmatrix} C_1 & C_2 \\ C_3 & C_4 \end{pmatrix}.$$

More specifically, let K, M, and N be index sets and let K', M', and N', respectively, be non-empty subsets of the index sets. Impose orderings on K, M, and N so that the elements of K', M', and N' precede the other elements, which comprise the complementary sets $\tilde{K}'$, $\tilde{M}'$, and $\tilde{N}'$. Let A, B, and C be matrices such that

(1) A is defined on K and M,
(2) B is defined on M and N, and
(3) $AB = C$ is well defined.

Let matrices A_1, A_2, A_3, and A_4 be defined as the restriction of the function A to the sets

(1) K' and M' for A_1
(2) K' and $\tilde{M}'$ for A_2
(3) $\tilde{K}'$ and M' for A_3
(4) $\tilde{K}'$ and $\tilde{M}'$ for A_4.

Pictorially what we are doing is writing A as four submatrices with

$$A = \begin{pmatrix} A_1 & A_2 \\ A_3 & A_4 \end{pmatrix}.$$

We perform the same kind of decomposition for B and C and obtain

$$\begin{pmatrix} A_1 & A_2 \\ A_3 & A_4 \end{pmatrix}\begin{pmatrix} B_1 & B_2 \\ B_3 & B_4 \end{pmatrix} = \begin{pmatrix} C_1 & C_2 \\ C_3 & C_4 \end{pmatrix}.$$

The proposition to follow asserts that the submatrices of A, B, and C

multiply as if they were entries themselves. Its proof depends on the fact that matrix multiplication is defined independently of any ordering on the index sets.

Proposition 1-11:

$$\begin{aligned} A_1B_1 + A_2B_3 &= C_1 \\ A_1B_2 + A_2B_4 &= C_2 \\ A_3B_1 + A_4B_3 &= C_3 \\ A_3B_2 + A_4B_4 &= C_4. \end{aligned}$$

PROOF: We prove only the first identity since the others are similar.

$$\begin{aligned} (C_1)_{ij} = C_{ij} = \sum_{m \in M} A_{im}B_{mj} &= \sum_{m \in M'} A_{im}B_{mj} + \sum_{m \in \tilde{M}'} A_{im}B_{mj} \\ &= (A_1B_1)_{ij} + (A_2B_3)_{ij}. \end{aligned}$$

Notice that if the submatrix A_1 has at least one infinite index set, then the representation of A by

$$A = \begin{pmatrix} A_1 & A_2 \\ A_3 & A_4 \end{pmatrix}$$

is not the standard one

$$A = \begin{pmatrix} A_{00} & A_{01} & \cdots \\ A_{10} & A_{11} & \cdots \\ \vdots & & \ddots \end{pmatrix}.$$

The ordering on the index sets of A is not of the same type as that of the non-negative integers. We recall once more, however, that the fundamental properties of matrices are independent of any orderings on the index sets. It is only the representation of a matrix as an array which requires these orderings.

Limits of matrices play an important part in the study of denumerable Markov chains. We shall touch only briefly at this time on the problems involved.

Definition 1-12: Let $\{A^{(k)}\}$ be a sequence of matrices. We say that $A = \lim_{k\to\infty} A^{(k)}$ exists if $A_{mn} = \lim_{k\to\infty} (A^{(k)})_{mn}$ exists for every m and n.

Notice that limits of matrices are defined entry-by-entry. No uniformity of convergence to the limiting matrix is assumed.

The type of problem that arises is as follows. Let π be a row vector and let $\{f^{(k)}\}$ be a sequence of column vectors converging to a column vector f. Is it true that $\{\pi f^{(k)}\}$ necessarily converges to πf? The answer to this question is in the negative unless some additional hypothesis is added. What is being attempted is an interchange of the order of two limit operations—one from the series which defines $\{\pi f^{(k)}\}$ and the other from the limit as k tends to ∞. Such an interchange can be justified only under special circumstances, and we shall obtain later in this chapter some sufficient conditions as special cases of theorems of measure theory.

2. Measure theory

Let X be an arbitrary non-empty set of points and let $\mathscr{F}$ be a family of subsets of X. We say that $\mathscr{F}$ is a **field of sets** if

(1) the empty set $\varnothing$ is in $\mathscr{F}$,
(2) whenever A is a set of $\mathscr{F}$, the complement of A, denoted $\tilde{A}$, is in $\mathscr{F}$, and
(3) whenever A and B are sets of $\mathscr{F}$, so is their union, denoted $A \cup B$.

A field of sets $\mathscr{F}$ is called a **Borel field** if it has the additional property that whenever $A_n \in \mathscr{F}$ for $n = 1, 2, 3, \ldots$, so is $\bigcup_{n=1}^{\infty} A_n$.

The intersection of sets A and B is indicated by $A \cap B$, and the difference $A \cap \tilde{B}$ is denoted $A - B$. From the above definitions the reader can easily establish the following result.

Proposition 1-13: If $\mathscr{F}$ is a field of sets, then $\mathscr{F}$ contains $\varnothing$ and X and is closed under complementation, finite unions, finite intersections, and differences. If $\mathscr{F}$ is a Borel field, then $\mathscr{F}$ is closed under denumerable intersections.

Proposition 1-14: For any class of sets $\mathscr{C}$ of the points of a set X, there exists a unique smallest Borel field containing $\mathscr{C}$.

PROOF: The family of all subsets of X forms a Borel field containing $\mathscr{C}$. Form the intersection of all Borel fields which contain $\mathscr{C}$ and call the resulting family of sets $\mathscr{F}$. Let A be in $\mathscr{F}$; then A is in all Borel fields containing $\mathscr{C}$ and so is $\tilde{A}$. Hence $\tilde{A}$ is in $\mathscr{F}$. A similar argument applies to intersections and denumerable unions. Thus $\mathscr{F}$ is the smallest Borel field containing $\mathscr{C}$.

Definition 1-15: A function ρ from a field of sets $\mathscr{F}$ to the extended real number system is called a **set function**. If $\rho(A) \geq 0$ for every A in $\mathscr{F}$, ρ is said to be **non-negative**. If $\rho(A \cup B) = \rho(A) + \rho(B)$ whenever A and B are in $\mathscr{F}$ and $A \cap B = \varnothing$, ρ is said to be **additive**. Suppose A_n is in $\mathscr{F}$ for $n = 1, 2, 3, \ldots$, and suppose $A_i \cap A_j = \varnothing$

whenever $i \neq j$. If $\rho(\bigcup_{n=1}^{\infty} A_n) = \sum_{n=1}^{\infty} \rho(A_n)$ holds whenever $\bigcup_{n=1}^{\infty} A_n$ is a set of $\mathscr{F}$, then ρ is said to be **completely additive**. In discussing set functions, we shall assume that there are no two sets A and B in $\mathscr{F}$ such that $\rho(A) = +\infty$ and $\rho(B) = -\infty$, and we shall assume that ρ is not identically infinite.

An additive set function ρ has the properties that

(1) $\rho(\varnothing) = 0$,
(2) $\rho(\bigcup_{n=1}^{N} A_n) = \sum_{n=1}^{N} \rho(A_n)$ for disjoint sets $\{A_n\}$, and
(3) $\rho(A \cup B) + \rho(A \cap B) = \rho(A) + \rho(B)$.

If ρ is non-negative and additive and if A is contained in B, then $\rho(A) \leq \rho(B)$. To see this, set $C = B \cap \tilde{A}$ so that A and C are disjoint and $A \cup C = B$. Then $\rho(A) + \rho(C) = \rho(B)$ by additivity, and the result follows at once. We shall now establish two facts about completely additive set functions.

Proposition 1-16: Let ρ be an additive set function defined on a field of sets $\mathscr{F}$. Let $\{A_n\}$ be a sequence of sets in $\mathscr{F}$ such that $A_1 \subset A_2 \subset \cdots$, and suppose $A = \bigcup_{n=1}^{\infty} A_n$ is in $\mathscr{F}$. If ρ is completely additive, then $\lim_{n\to\infty} \rho(A_n) = \rho(A)$. Conversely, if $\lim_{n\to\infty} \rho(A_n) = \rho(A)$ for all such sequences, ρ is completely additive.

PROOF: Set $B_1 = A_1$ and $B_n = A_n \cap \tilde{A}_{n-1}$. Then $A_n = \bigcup_{k=1}^{n} B_k$ disjointly, and by additivity $\rho(A_n) = \sum_{k=1}^{n} \rho(B_k)$. But $A = \bigcup_{k=1}^{\infty} B_k$ and by complete additivity $\rho(A) = \sum_{k=1}^{\infty} \rho(B_k)$. The proof of the converse is left to the reader.

A consequence of this proposition is the following:

Corollary 1-17: Let ρ be an additive set function defined on a field of sets $\mathscr{F}$ in such a way that $\rho(A) < \infty$ for every A. Let $\{A_n\}$ be a sequence of sets in $\mathscr{F}$ such that $A_1 \supset A_2 \supset A_3 \supset \cdots$ and $\bigcap_{n=1}^{\infty} A_n = \varnothing$. If ρ is completely additive, then $\lim_{n\to\infty} \rho(A_n) = 0$. Conversely, if $\lim_{n\to\infty} \rho(A_n) = 0$ for all such sequences, then ρ is completely additive.

A non-negative completely additive set function on a field of sets $\mathscr{F}$ is called a **measure**. The set of points X with a measure defined on its field $\mathscr{F}$ is called a **measure space**. We shall usually denote measures by μ or ν. If there is no ambiguity about what measure is involved, we shall frequently refer to X by itself as the measure space.

If X is a measure space with field of sets $\mathscr{F}$ and measure μ, then X is a set in $\mathscr{F}$, and we define $\mu(X)$ to be the **total measure** of the space. A **probability space** is a measure space of total measure one.

We give four examples of measure spaces.

(1) Let X be any set, let $\mathscr{F} = \{\varnothing, X\}$, and define $\mu(\varnothing) = 0$ and $\mu(X) = a \geq 0$. Then X is the trivial measure space.

(2) Let X be Euclidean n-space, let $\mathscr{F}$ be the Lebesgue measurable sets, and let μ be Lebesgue measure (the natural generalization of length, area, or volume).

(3) Let X be the set of six possible outcomes for tossing a die. Assign weight $\frac{1}{6}$ to each of the six points in the space, and for any subset of X assign as a measure the sum of the weights of the points in the set. Then $\mathscr{F}$ is the family of all subsets of X, and X is a probability space.

(4) Let X be a denumerable index set, and let π be a non-negative row vector with X as its index set. Assign as a weight to each point of X the value of the corresponding entry of π. For any subset of X assign as a measure the sum of the weights of the points in the set. Then $\mathscr{F}$ is the family of all subsets of X, and X is a measure space with total measure $\pi\mathbf{1}$.

The sets of a field on which a measure μ is defined are called the **μ-measurable** or simply the **measurable** subsets of X. In the construction of a measure on a field, it is possible for a non-empty set A to be assigned measure zero. In example (2) above, for instance, every denumerable set and even certain uncountable sets are sets of measure zero. Suppose B is a subset of such a set A. If B is measurable, then $\mu(B) \geq 0$ since μ is a measure. But

$$\mu(B) \leq \mu(A) = 0$$

since $B \subset A$ and A is of measure zero. Thus, a measurable subset of a set of measure zero is of measure zero. But there is no reason why such a set B has to be measurable. However, one can agree to add all subsets of sets of measure zero to a field and extend the resulting family of sets to the smallest field containing the family. Such an extended field is called an **augmented field**. It consists precisely of all sets of the form $(C - D) \cup E$, where C is a set in the original field and D and E are subsets of a set of measure zero. Therefore the augmented field of a Borel field is again a Borel field. Note that in any augmented field every subset of a set of measure zero is measurable and has measure zero. In later chapters of this book all fields will be augmented.

If a statement about the points of a measure space X fails to be true only for a set of points which is a subset of a set of measure zero, we say that the statement holds for **almost all** points of X or that it is true **almost everywhere** (abbreviated **a.e.**).

Proposition 1-18: Let μ be a measure defined on a field of sets $\mathscr{F}$. If $\{A_n\}$ is a sequence of sets in $\mathscr{F}$, if A is in $\mathscr{F}$, and if $A \subset \bigcup_n A_n$, then

$$\mu(A) \leq \sum_n \mu(A_n).$$

PROOF: Write $B_n = A_n - (\bigcup_{k=1}^{n-1} A_k)$. The sets B_n are disjoint in pairs, and consequently the sets $A \cap B_n$ are also disjoint. Furthermore, $\bigcup_n B_n = \bigcup_n A_n$ so that

$$\begin{aligned} A &= A \cap \left(\bigcup_n A_n\right) \\ &= A \cap \left(\bigcup_n B_n\right) \\ &= \bigcup_n (A \cap B_n). \end{aligned}$$

By hypothesis μ is a measure. It is therefore completely additive and

$$\begin{aligned} \mu(A) &= \sum_n \mu(A \cap B_n) \\ &\leq \sum_n \mu(B_n) \quad \text{since } A \cap B_n \subset B_n \\ &\leq \sum_n \mu(A_n) \quad \text{since } B_n \subset A_n. \end{aligned}$$

To conclude this section we shall establish a result known as the Extension Theorem. The proof follows the proof of Rudin [1953].

Theorem 1-19: Let $\mathscr{F}$ be a field of sets in a space X and let ν be a measure defined on $\mathscr{F}$. Suppose X can be written as the denumerable union of sets in $\mathscr{F}$ of finite measure. If $\mathscr{G}$ is the smallest Borel field containing $\mathscr{F}$, then ν can be extended in one and only one way to a measure defined on all of $\mathscr{G}$ which agrees with ν on sets of $\mathscr{F}$.

Before proving the theorem, we need some preliminary lemmas and definitions. The property in the statement of the theorem that X is the denumerable union of sets of finite measure is summarized by saying that ν is **sigma-finite**.

Let ν be a measure defined on a field of sets $\mathscr{F}$ in a space X, and suppose $X = \bigcup_{n=1}^{\infty} A_n$ with $A_n \in \mathscr{F}$ and $\nu(A_n) < \infty$. For each subset B of X, define $\mu(B) = \inf\{\sum \nu(B_n)\}$, where the infimum is taken over all denumerable coverings of B by sets $\{B_n\}$ of $\mathscr{F}$.

Lemma 1-20: The set function μ is non-negative. If A and B are subsets of X such that $A \subset B$, then $\mu(A) \leq \mu(B)$. If C is a set in $\mathscr{F}$, then $\mu(C) = \nu(C)$.

PROOF: We see that μ is non-negative because μ is the limit of non-negative quantities. If $A \subset B$, then $\mu(A) \leq \mu(B)$ because every covering of B is a covering of A. Let C be in $\mathscr{F}$. Then $\{C\}$ is a covering of C and $\mu(C) \leq \nu(C)$. And for any covering $\{C_n\}$,

$$\nu(C) \leq \sum_n \nu(C_n)$$

by Proposition 1-18. Therefore,

$$\nu(C) \leq \inf \sum_n \nu(C_n) = \mu(C).$$

Lemma 1-21: If $\{A_n\}$ is an arbitrary sequence of subsets of X and if $A = \bigcup_n A_n$, then $\mu(A) \leq \sum_n \mu(A_n)$.

PROOF: Let $\epsilon > 0$ be given. Let $\{B_k^{(n)}\}$ with $k = 1, 2, 3, \ldots$ be a denumerable covering of A_n such that $B_k^{(n)}$ is in $\mathscr{F}$ and $\sum_k \nu(B_k^{(n)}) \leq \mu(A_n) + \epsilon/2^n$. This choice is possible by the definition of μ. Then since all the B's form a covering of A, we have

$$\begin{aligned} \mu(A) &\leq \sum_n \sum_k \nu(B_k^{(n)}) \\ &\leq \sum_n \mu(A_n) + \epsilon \end{aligned}$$

and the assertion follows.

We define a set theoretic operation $\oplus$ for subsets of X by

$$A \oplus B = (A \cap \tilde{B}) \cup (B \cap \tilde{A}).$$

The set $A \oplus B$ is called the **symmetric difference** of A and B. A point is in $A \oplus B$ if it is in A or B but not both. We leave the details of the proof of the next lemma to the reader.

Lemma 1-22: The subsets of a space X form a ring under the operations $\oplus$ and $\cap$ with additive identity $\varnothing$ and multiplicative identity X. Every set is its own additive inverse.

Define a distance d between subsets of X by $d(A, B) = \mu(A \oplus B)$. We note that d has the properties

$$d(A, A) = \mu(\varnothing) = 0$$

and

$$d(A, B) = \mu(A \oplus B) = d(B, A).$$

Since

$$A \cup B = (A \oplus B) \cup B,$$

we have

$$A \subset (A \oplus B) \cup B$$

and by Lemmas 1-20 and 1-21

$$\mu(A) \leq d(A, B) + \mu(B).$$

Replacing A by $A \oplus B$ and B by $C \oplus B$, we obtain the triangle inequality

$$d(A, B) \leq d(A, C) + d(C, B).$$

Lemma 1-23: For any subsets A_1, A_2, B_1, B_2, A, and B, of X,

$$d((A_1 \cup A_2), (B_1 \cup B_2)) \leq d(A_1, B_1) + d(A_2, B_2)$$

$$d((A_1 \cap A_2), (B_1 \cap B_2)) \leq d(A_1, B_1) + d(A_2, B_2)$$

$$d(B, A) = d(\tilde{B}, \tilde{A}).$$

PROOF: We prove only the first and third assertions. First we observe that $(A_1 \cup A_2) \oplus (B_1 \cup B_2) \subset (A_1 \oplus B_1) \cup (A_2 \oplus B_2)$. For suppose $x \in (A_1 \cup A_2) \oplus (B_1 \cup B_2)$. We may assume without loss of generality that $x \in A_1 \cup A_2$ but $x \notin B_1 \cup B_2$. If $x \in A_1$, then $x \notin B_1$ so that $x \in A_1 \oplus B_1$. Similarly if $x \in A_2$, then $x \in A_2 \oplus B_2$ and the containment is established. The first assertion of the lemma now follows by applying Lemmas 1-20 and 1-21. For the third part, we have

$$B \oplus A = (A \cap \tilde{B}) \cup (\tilde{A} \cap B)$$

and

$$\tilde{B} \oplus \tilde{A} = (\tilde{A} \cap B) \cup (A \cap \tilde{B})$$

so that

$$B \oplus A = \tilde{B} \oplus \tilde{A}.$$

Definition 1-24: Convergence of sets in measure is defined by saying that $A_n \to A$ if $\lim_{n \to \infty} d(A_n, A) = 0$. Let $\mathscr{F}^*$ be the collection of all subsets A of X for which there exists a sequence $\{A_n\}$ of sets in $\mathscr{F}$ having the property that $A_n \to A$. Let $\mathscr{G}^*$ be the family of denumerable unions of sets in $\mathscr{F}^*$.

Lemma 1-25: If $\{A_n\}$ and $\{B_n\}$ are sequences of sets in $\mathscr{F}$ such that $A_n \to A$ and $B_n \to B$, then $A_n \cup B_n \to A \cup B$, $A_n \cap B_n \to A \cap B$, and $\tilde{A}_n \to \tilde{A}$. Therefore, $\mathscr{F}^*$ is a field of sets. For any $C_n \to C$, $\lim_n \mu(C_n) = \mu(C)$.

PROOF: Since by Lemma 1-23

$$d((A_n \cup B_n), (A \cup B)) \leq d(A_n, A) + d(B_n, B)$$

$$d((A_n \cap B_n), (A \cap B)) \leq d(A_n, A) + d(B_n, B)$$

$$d(A_n, A) = d(\tilde{A}_n, \tilde{A}),$$

we have $A_n \cup B_n \to A \cup B$, $A_n \cap B_n \to A \cap B$, and $\tilde{A}_n \to \tilde{A}$. The limit of $\mu(C_n)$ is established by the inequalities

$$\mu(C_n) \leq d(C_n, C) + \mu(C)$$

and

$$\mu(C) \leq d(C_n, C) + \mu(C_n).$$

Lemma 1-26: μ is additive on $\mathscr{F}^*$.

PROOF: Let A and B be disjoint sets in $\mathscr{F}^*$ and pick $\{A_n\}$ and $\{B_n\}$ in $\mathscr{F}$ such that $A_n \to A$ and $B_n \to B$. Then since ν is additive on $\mathscr{F}$ and since μ agrees with ν on sets of $\mathscr{F}$, we have

$$\mu(A_n \cup B_n) + \mu(A_n \cap B_n) = \mu(A_n) + \mu(B_n).$$

By Lemma 1-25,

$$\mu(A \cup B) + \mu(A \cap B) = \mu(A) + \mu(B)$$

or

$$\mu(A \cup B) = \mu(A) + \mu(B).$$

Lemma 1-27: If $A = \bigcup_n A_n$ with A in $\mathscr{G}^*$ and $\{A_n\}$ a sequence of disjoint sets in $\mathscr{F}^*$, then $\mu(A) = \sum_n \mu(A_n)$.

PROOF: Since $A \supset (A_1 \cup A_2 \cup A_3 \cup \cdots \cup A_k)$, we have, by Lemma 1-20, $\mu(A) \geq \mu(A_1 \cup A_2 \cup \cdots \cup A_k)$, and by Lemma 1-26 the right side equals $\sum_{n=1}^{k} \mu(A_n)$ for each k. Hence

$$\mu(A) \geq \sum_n \mu(A_n),$$

and equality holds by Lemma 1-21.

Lemma 1-28: If A is in $\mathscr{G}^*$ and if $\mu(A)$ is finite, then A is in $\mathscr{F}^*$.

REMARK: If A is in $\mathscr{F}^*$, then $\mu(A)$ is not necessarily finite.

PROOF OF LEMMA: Write $A = \bigcup_n A_n$ with $\{A_n\}$ in $\mathscr{F}^*$, A_n disjoint sets, and set $B_n = \bigcup_{k=1}^{n} A_k$. Then

$$\begin{aligned} d(A, B_n) &= \mu([A \cap \tilde{B}_n] \cup [\tilde{A} \cap B_n]) \\ &= \mu\left(\bigcup_{k=n+1}^{\infty} A_k\right), \end{aligned}$$

which by Lemma 1-27

$$= \sum_{k=n+1}^{\infty} \mu(A_k).$$

Since the last expression on the right is the tail of a convergent series, we have $B_n \to A$. Since $B_n \in \mathscr{F}^*$, we can find L_n in $\mathscr{F}$ such that $d(L_n, B_n) \leq 1/n$. Then $d(L_n, A) \leq 1/n + d(B_n, A)$, and hence $L_n \to A$. Thus A is in $\mathscr{F}^*$.

REMARK: If $A \in \mathscr{G}^*$ with $\mu(A)$ finite, then A is in $\mathscr{F}^*$; hence for every $\epsilon > 0$ there is a B in $\mathscr{F}$ such that $\mu(A \oplus B) \leq \epsilon$. Conversely, if there exists such a B_ϵ for any ϵ, then $B_{1/n} \to A$ so that $A \in \mathscr{F}^*$ and, *a fortiori*, $A \in \mathscr{G}^*$. These observations give a characterization of the sets A in $\mathscr{G}^*$ for which $\mu(A)$ is finite.

Lemma 1-29: μ is completely additive on $\mathscr{G}^*$, and $\mathscr{G}^*$ is a Borel field.

PROOF: Suppose

$$A = \bigcup_n A_n$$

is the union of disjoint sets $\{A_n\}$ in $\mathscr{G}^*$. Then $\mu(A) \geq \mu(A_n)$ for every n by Lemma 1-20, so that we may assume $\mu(A_n) < \infty$ for every n. The complete additivity of μ now follows from Lemmas 1-28 and 1-27. For the proof that $\mathscr{G}^*$ is a Borel field, we see clearly that $\mathscr{G}^*$ is closed under denumerable unions. It remains to be proved that $\mathscr{G}^*$ is closed under complementation. Since ν is sigma-finite, let

$$X = \bigcup_n A_n$$

with A_n in $\mathscr{F}$ and $\mu(A_n) = \nu(A_n) < \infty$. Let B in $\mathscr{G}^*$ be given and suppose $B = \bigcup_k B_k$ with B_k in $\mathscr{F}^*$. Since

$$A_n \cap B = \bigcup_k (A_n \cap B_k)$$

and since $A_n \cap B_k$ is in $\mathscr{F}^*$, $A_n \cap B$ is in $\mathscr{G}^*$. But by Lemma 1-20,

$$\mu(A_n \cap B) \leq \mu(A_n)$$

and we have assumed that $\mu(A_n) < \infty$. Thus by Lemma 1-28, $A_n \cap B \in \mathscr{F}^*$, and since $\mathscr{F}^*$ is a field,

$$A_n \cap (X - (A_n \cap B)) = A_n \cap \tilde{B}$$

is in $\mathscr{F}^*$. Therefore

$$\tilde{B} = X \cap \tilde{B} = \bigcup_n (A_n \cap \tilde{B})$$

is in $\mathscr{G}^*$, and the proof is complete.

We are now in a position to prove the Extension Theorem.

PROOF OF THEOREM 1-19: Existence of the extension of ν to a measure μ defined on $\mathscr{G}^*$ is proved by Lemmas 1-20 and 1-29. Since, by

Lemma 1-29, $\mathscr{G}^*$ is a Borel field containing $\mathscr{F}$, $\mathscr{G}^*$ contains $\mathscr{G}$. The extended measure restricted to sets of $\mathscr{G}$ has the desired properties. For uniqueness, suppose μ' is another measure on $\mathscr{G}$ that agrees with ν on $\mathscr{F}$. Since, by sigma-finiteness, X is the union of sets A_n in $\mathscr{F}$ of finite ν-measure, we may assume that X is a disjoint union of sets of finite measure by letting $B_n = A_n - \bigcup_{k<n} A_k$. Let C be any set in $\mathscr{G}$; we want to show $\mu'(C) = \mu(C)$. By definition,

$$\mu(C) = \inf\left\{\sum_n \nu(C_n)\right\}$$

with the infimum taken over covers $\{C_n\}$, where C_n is in $\mathscr{F}$. For any fixed cover $\{C_n\}$, we have

$$\mu'(C) \leq \mu'\left(\bigcup_n C_n\right) \leq \sum_n \mu'(C_n) = \sum_n \nu(C_n).$$

Therefore

$$\mu'(C) \leq \inf\left\{\sum_n \nu(C_n)\right\} = \mu(C).$$

Writing

$$\mu'(C) = \sum_n \mu'(C \cap B_n),$$

we see that it is sufficient to show that

$$\mu'(C \cap B_n) \geq \mu(C \cap B_n).$$

But

$$\mu'(C \cap B_n) + \mu'(\tilde{C} \cap B_n) = \nu(B_n) = \mu(C \cap B_n) + \mu(\tilde{C} \cap B_n).$$

Now we know that μ' is dominated by μ:

$$\mu'(\tilde{C} \cap B_n) \leq \mu(\tilde{C} \cap B_n) \leq \mu(B_n) < \infty.$$

If

$$\mu'(C \cap B_n) < \mu(C \cap B_n),$$

we obtain the contradiction

$$\mu'(C \cap B_n) + \mu'(\tilde{C} \cap B_n) < \mu(C \cap B_n) + \mu(\tilde{C} \cap B_n).$$

3. Measurable functions and Lebesgue integration

Let $\mathscr{F}$ be a Borel field of sets in a set X. The **measurable sets** of X are the sets of $\mathscr{F}$.

Definition 1-30: Let f be a function with domain X and with range the extended real number system. The function f is said to be a **measurable function** if for each real number c the set $\{x \mid f(x) < c\}$ is measurable.

The content of the next proposition is that the property $f(x) < c$ may be replaced by any of the conditions $f(x) \leq c$, $f(x) > c$, or $f(x) \geq c$. Therefore, if f is a measurable function, then the set

$$\{x \mid c \leq f(x) \leq d\}$$

is measurable; either or both of the signs $\leq$ may be replaced by $<$, and the set is still measurable.

Proposition 1-31: The following four conditions are equivalent:

(1) $\{x \mid f(x) < c\}$ is measurable for every c.
(2) $\{x \mid f(x) \leq c\}$ is measurable for every c.
(3) $\{x \mid f(x) > c\}$ is measurable for every c.
(4) $\{x \mid f(x) \geq c\}$ is measurable for every c.

PROOF: From

$$\{x \mid f(x) \leq c\} = \bigcap_{n=1}^{\infty} \left\{x \mid f(x) < c + \frac{1}{n}\right\},$$

$$\{x \mid f(x) > c\} = X - \{x \mid f(x) \leq c\},$$

$$\{x \mid f(x) \geq c\} = \bigcap_{n=1}^{\infty} \left\{x \mid f(x) > c - \frac{1}{n}\right\},$$

and

$$\{x \mid f(x) < c\} = X - \{x \mid f(x) \geq c\},$$

we see that (1) implies (2), that (2) implies (3), that (3) implies (4), and that (4) implies (1).

Proposition 1-32: Every constant function is measurable.

PROOF: If $f(x) = a$ identically, then $\{x \in X \mid f(x) < c\}$ is either $\varnothing$ or X.

In analogy with our procedure for matrices in Section 1, we define f^+ and f^- by

$$f^+(x) = \max\{f(x), 0\}$$

$$f^-(x) = -\min\{f(x), 0\}.$$

Proposition 1-33: If f is measurable, then so are f^+, f^-, and $|f|$.

PROOF:

$$\{x \mid f^+ \geq c\} = \{x \mid f \geq c\} \cup \{x \mid 0 \geq c\}$$

$$\{x \mid f^- \geq c\} = \{x \mid f \leq -c\} \cup \{x \mid 0 \geq c\}.$$

The set $\{x \mid 0 \geq c\}$ is either $\varnothing$ or X and is therefore measurable. For $|f|$ we have

$$\{x \mid |f| \leq c\} = \{x \mid -c \leq f(x) \leq c\}.$$

Proposition 1-34: Let f and g be measurable functions whose values are finite at all points. Then $f + g$ and $f \cdot g$ are measurable.

PROOF: We prove only the assertion about $f + g$. Order the rational numbers and call the nth one r_n. Then

$$\{x \mid (f + g)(x) > c\} = \bigcup_{n=1}^{\infty} (\{x \mid f(x) > c + r_n\} \cap \{x \mid g(x) > -r_n\}),$$

so that $f + g$ is measurable.

Corollary 1-35: If f is measurable, then so is cf for every constant c.

Proposition 1-36: Let $\{f_n\}$ be a sequence of measurable functions. Then the functions

(1) $\sup_n f_n(x)$
(2) $\inf_n f_n(x)$
(3) $\limsup_n f_n(x)$
(4) $\liminf_n f_n(x)$

are all measurable.

PROOF: The assertions follow from the observations that

$$\{x \mid \sup f_n(x) > c\} = \bigcup_{n=1}^{\infty} \{x \mid f_n(x) > c\},$$

$$\{x \mid \inf f_n(x) < c\} = \bigcup_{n=1}^{\infty} \{x \mid f_n(x) < c\},$$

$$\limsup_n f_n(x) = \inf_n \sup_{m>n} f_m(x)$$

and

$$\liminf_n f_n(x) = \sup_n \inf_{m>n} f_m(x).$$

The supremum of finitely many functions is their pointwise maximum. Therefore the maximum and minimum of finitely many measurable functions are both measurable.

Corollary 1-37: If $\{f_n\}$ is a sequence of measurable functions and if $f = \lim_n f_n$ exists at all points, then f is measurable.

We shall give three examples of measurable functions.

(1) Let $\mathscr{F}$ be the family of sets on the real line which are either finite unions of open and closed intervals or complements of such sets. Then $\mathscr{F}$ is a field of sets. Let $\mathscr{G}$ be the smallest Borel field containing $\mathscr{F}$. All continuous real-valued functions are measurable with respect to the Borel field $\mathscr{G}$.

(2) Let X be a space for which $\mathscr{B}$ is the family of all subsets of X. Then every function f defined on X is measurable.

(3) Let X be the union of a sequence of disjoint sets $\{A_n\}$, and let $\mathscr{B}$ be the family of all sets which are unions of sets in the sequence. Then a function f is measurable if and only if its restriction to the domain A_n is a constant function for each n. In particular, if $\mathscr{B} = \{X, \varnothing\}$, then the measurable functions are the constant functions.

Let A be any subset of X. The **characteristic function** of A, denoted $\chi_A(x)$, is defined by

$$\chi_A(x) = \begin{cases} 1 & \text{if } x \in A \\ 0 & \text{otherwise.} \end{cases}$$

A function that takes on only a finite number of values is called a **simple function.** It may be represented, uniquely, in the form

$$(*) \qquad s = \sum_{n=1}^{N} c_n \chi_{A_n}$$

where the c_n are the distinct values the function takes on and the sets A_n are disjoint. The simple function is measurable if and only if all of the sets $A_1, A_2, \ldots, A_N$ are measurable.

Proposition 1-38: For any non-negative function f defined on X, there exists a sequence of non-negative simple functions $\{s_n\}$ with the property that for each $x \in X$, $\{s_n(x)\}$ is a monotonically increasing sequence converging to $f(x)$. If f is measurable, the $\{s_n\}$ may be taken to be measurable.

PROOF: For every n and for $1 \leq j \leq n2^n$ set

$$A_{nj} = \left\{x \,\middle|\, \frac{j-1}{2^n} \leq f(x) < \frac{j}{2^n}\right\}$$

and

$$B_n = \{x \mid f(x) \geq n\}.$$

Then

$$s_n = \sum_{j=1}^{n2^n} \frac{j-1}{2^n} \chi_{A_{nj}} + n\chi_{B_n}$$

increases monotonically with n to f. If f is measurable, then so are $\{A_{nj}\}$ and B_n; thus s_n is measurable.

If μ is a measure defined on a Borel field $\mathscr{B}$ of subsets of X, we denote the measure space by the ordered triple $(X, \mathscr{B}, \mu)$. In $(X, \mathscr{B}, \mu)$ let E be a set of the family $\mathscr{B}$, and suppose s is a non-negative, measurable simple function, represented as in (*) above. Since s is measurable, A_n is measurable and $\mu(A_n \cap E)$ is defined for every n. Set

$$I_E(s) = \sum_{n=1}^{N} c_n\mu(A_n \cap E).$$

For any non-negative measurable function f, define the **Lebesgue integral** of f on the set E with respect to the measure μ by

$$\int_E f d\mu = \sup I_E(s),$$

where the supremum is taken over all simple functions s satisfying $0 \leq s \leq f$. We note that the value of the integral is non-negative and possibly infinite. It can be verified that if s is a non-negative measurable simple function, then

$$\int_E s d\mu = I_E(s).$$

If f is an arbitrary measurable function, then by Proposition 1-33, $\int_E f^+ d\mu$ and $\int_E f^- d\mu$ are both defined. If the integrals of f^+ and f^- are not both infinite, we define the **integral** of f by $\int_E f d\mu = \int_E f^+ d\mu - \int_E f^- d\mu$. The function f is said to be **integrable** on the set E if $\int_E f^+ d\mu$ and $\int_E f^- d\mu$ are both finite.

Following our examples of measure spaces and measurable functions, we give three examples of integration. A fourth example will arise in Chapter 2.

(1) Let $\mathscr{B} = \{\varnothing, X\}$ and suppose $\mu(X) = 1$. Only the constant functions are measurable and

$$\int_\varnothing c d\mu = 0; \qquad \int_X c d\mu = c.$$

(2) Let X be the real line, let $\mathscr{G}$ be the Borel field of sets constructed in the first example of measurable functions, and let μ be Lebesgue measure. Continuous functions are measurable, and it can be shown that the value of the Lebesgue integral of a continuous function on a closed interval agrees with the value of the Riemann integral. More generally one finds that every Riemann integrable function is Lebesgue integrable, but not conversely.

(3) Let X be the denumerable set of points described in Example 4 of measure spaces, and let π be a non-negative row vector defined on X. Then π defines a measure on X. If f is an arbitrary column vector defined on X, then f is a function on the points of X. Furthermore, f is measurable since all subsets of X are measurable sets. The reader should verify that the integral of f over the whole space X with respect to the measure π is the matrix product πf and that the condition for the integral to be defined is precisely the condition for the matrix product to be well defined. Because of this application of Lebesgue integration, we often speak of column vectors as **functions** and non-negative row vectors as **measures**. We shall return to this example in Section 5 of this chapter. The proof of the next proposition is left to the reader.

Proposition 1-39: The Lebesgue integral satisfies these seven properties:

(1) If c is a constant function,

$$\int_E c\,d\mu = c\mu(E).$$

(2) If f and g are measurable functions whose integrals are defined on E and if $f(x) \leq g(x)$ for all $x \in E$, then

$$\int_E f\,d\mu \leq \int_E g\,d\mu.$$

(3) If f is integrable on E and if c is a real number, then cf is integrable and $\int_E cf\,d\mu = c\int_E f\,d\mu$.

(4) If f is measurable and $\mu(E) = 0$, then $\int_E f\,d\mu = 0$.

(5) If E' and E are measurable sets with $E' \subset E$ and if f is a function for which $\int_E f\,d\mu$ is defined, then $\int_{E'} f\,d\mu$ is defined. In particular,

$$\int_{E'} f^+\,d\mu \leq \int_E f^+\,d\mu$$

and

$$\int_{E'} f^-\,d\mu \leq \int_E f^-\,d\mu.$$

(6) If $|f(x)| \leq c$ for all $x \in E$, if $\mu(E) < \infty$, and if f is measurable, then f is integrable on E.

(7) If f is measurable on E and if $|f| \leq g$ for a function g integrable on E, then f is integrable on E.

Corollary 1-40: If f is a non-negative measurable function with $\int_E f\,d\mu = 0$, then $f = 0$ a.e. on E.

PROOF: The subset of E where $f(x) \geq 1/n$ must have measure zero since otherwise f would have positive integral by (1) and (2) of Proposition 1-39. The set where $f \neq 0$ on E is the countable union on n of these sets.

4. Integration theorems

We shall make frequent use of four important facts about the Lebesgue integral. We develop these results as the four theorems of this section.

Theorem 1-41: Let f be a fixed measurable function and suppose that $\int_X f d\mu$ is defined. Then the set function $\rho(E) = \int_E f d\mu$ is completely additive.

PROOF: If we can prove the theorem for non-negative functions, we can write $f = f^+ - f^-$ and apply our result separately to f^+ and f^-. We therefore assume that f is non-negative. We must show that if $E = \bigcup_{n=1}^{\infty} E_n$ disjointly, then $\rho(E) = \sum_{n=1}^{\infty} \rho(E_n)$. If f is a characteristic function χ_A, then $\rho(E) = \int_E \chi_A d\mu = \mu(A \cap E)$ and the complete additivity of ρ is a consequence of the complete additivity of μ. If f is a simple function, the complete additivity of ρ is a consequence of the result for characteristic functions and of the fact that the limit of a sum is the sum of the limits. Thus, for general f we have for every simple function s satisfying $0 \leq s \leq f$,

$$\int_E s d\mu = \sum_{n=1}^{\infty} \int_{E_n} s d\mu \leq \sum_{n=1}^{\infty} \rho(E_n)$$

by property (2) of Proposition 1-39. Hence

$$\rho(E) = \sup_s \int_E s d\mu \leq \sum_{n=1}^{\infty} \rho(E_n).$$

We now prove the inequality in the other direction. By property (5) of Proposition 1-39, $\rho(E) \geq \rho(E_n)$ for every n since $f = f^+$. Thus if $\rho(E_n) = +\infty$ for any n, the desired result is proved. We therefore assume $\rho(E_n) < \infty$ for every n. Let $\epsilon > 0$ be given and choose a measurable simple function s satisfying $0 \leq s \leq f$,

$$\int_{E_1} s d\mu \geq \int_{E_1} f d\mu - \frac{\epsilon}{2},$$

and

$$\int_{E_2} s d\mu \geq \int_{E_2} f d\mu - \frac{\epsilon}{2}.$$

This choice is possible by the definition of the integral as a limit. Then

$$\rho(E_1 \cup E_2) = \int_{E_1 \cup E_2} f d\mu \geq \int_{E_1 \cup E_2} s d\mu = \int_{E_1} s d\mu + \int_{E_2} s d\mu$$

$$\geq \int_{E_1} f d\mu + \int_{E_2} f d\mu - \epsilon$$

$$= \rho(E_1) + \rho(E_2) - \epsilon.$$

Hence

$$\rho(E_1 \cup E_2) \geq \rho(E_1) + \rho(E_2).$$

By induction, we obtain

$$\rho(E_1 \cup E_2 \cup \cdots \cup E_n) \geq \rho(E_1) + \cdots + \rho(E_n)$$

and

$$\rho(E) \geq \rho(E_1) + \cdots + \rho(E_n) \quad \text{for every } n.$$

Hence

$$\rho(E) \geq \sum_{n=1}^{\infty} \rho(E_n).$$

The proofs of two corollaries of Theorem 1-41 are left as exercises. These results use only the additivity of the integral and not the complete additivity.

Corollary 1-42: If f is measurable, if $\int_E f d\mu$ is defined, and if

$$\mu(E \oplus F) = 0,$$

then $\int_F f d\mu$ is defined and $\int_E f d\mu = \int_F f d\mu$.

Corollary 1-43: If $\int_E f d\mu$ is defined, then

$$\left| \int_E f d\mu \right| \leq \int_E |f| d\mu.$$

If f is integrable on E, then so is $|f|$.

Let f and g be two functions whose integrals on E are defined. Suppose the set $A = \{x \in E \mid f(x) \neq g(x)\}$ is of measure zero; that is, suppose f and g are equal almost everywhere on E. Writing $E = A \cup (E \cap \tilde{A})$, we find by applying Corollary 1-42 twice,

$$\int_E f d\mu = \int_{E \cap \tilde{A}} f d\mu = \int_{E \cap \tilde{A}} g d\mu = \int_E g d\mu.$$

Functions which differ on a set of measure zero thus have equal integrals. Therefore, when we are thinking of a function f defined on X in terms of integration, it is sufficient that f be defined at almost all points of X. And, if we agree to augment Borel fields of sets by adjoining subsets of sets of measure zero, we see that if f and g differ on a set of measure zero and if f is measurable, then g is measurable. With the convention of augmenting the field, we obtain from Corollary 1-37, for example, the result that if $\{f_n\}$ is a sequence of measurable functions such that

$$f(x) = \lim_{n \to \infty} f_n(x)$$

almost everywhere, then f is measurable. Similarly the theorems to follow would be valid with convergence almost everywhere if the underlying Borel field were augmented; the necessary modifications in the proofs are easy.

We now state and prove the Monotone Convergence Theorem, which is due to B. Levi.

Theorem 1-44: Let E be a measurable set, and suppose $\{f_n\}$ is a sequence of measurable functions such that

$$0 \leq f_1 \leq f_2 \leq \cdots$$

and

$$f(x) = \lim_{n \to \infty} f_n(x).$$

Then

$$\int_E f d\mu = \lim_{n \to \infty} \int_E f_n d\mu.$$

PROOF: Since the $\{f_n\}$ increase with n, so do the $\left\{\int_E f_n d\mu\right\}$. Therefore

$$k = \lim_{n \to \infty} \int_E f_n d\mu$$

exists. Since f is non-negative and is the limit of measurable functions, we know that $\int_E f d\mu$ exists, and since $f_n \leq f$, we have

$$\int_E f_n d\mu \leq \int_E f d\mu$$

for every n. Therefore, $k \leq \int_E f d\mu$. Let c be a real number satisfying $0 < c < 1$, and let s be a measurable simple function such that $0 \leq s \leq f$. Set

$$E_n = \{x \in E \mid f_n(x) \geq cs(x)\}$$

so that $E_1 \subset E_2 \subset E_3 \subset \cdots$. Then $E = \bigcup_{n=1}^{\infty} E_n$. For any n we have

$$k \geq \int_E f_n d\mu \geq \int_{E_n} f_n d\mu \geq c \int_{E_n} s d\mu.$$

Since the integral is a completely additive set function (Theorem 1-41), we have by Proposition 1-16

$$\lim_{n \to \infty} c \int_{E_n} s d\mu = c \int_E s d\mu.$$

Thus, as $n \to \infty$,

$$k \geq c \int_E s d\mu.$$

Letting $c \to 1$, we have $k \geq \int_E s d\mu$, and taking the supremum over all s, we find $k \geq \int_E f d\mu$.

Proposition 1-45: Suppose $h = f + g$ with f and g integrable on E. Then h is integrable on E and

$$\int_E h d\mu = \int_E f d\mu + \int_E g d\mu.$$

PROOF: We first prove the assertion for f and g non-negative. For simple functions the assertion is trivial. If f and g are not both simple, apply Proposition 1-38 to find monotone sequences of non-negative measurable simple functions $\{t_n\}$ and $\{u_n\}$ converging to f and g. Then $\{s_n = t_n + u_n\}$ converges to h, and since

$$\int_E s_n d\mu = \int_E t_n d\mu + \int_E u_n d\mu,$$

the result follows from Theorem 1-44. Next, if $f \geq 0$, $g \leq 0$, and $h = f + g \geq 0$, we have $f = h + (-g)$ with $h \geq 0$ and $(-g) \geq 0$, so that

$$\int_E f d\mu = \int_E h d\mu + \int_E (-g) d\mu$$

or

$$\int_E h d\mu = \int_E f d\mu + \int_E g d\mu.$$

Since the right side is finite and since $h \geq 0$, h is integrable. For an arbitrary $h \geq 0$, decompose E into the disjoint union of three sets, one where $f \geq 0$ and $g \geq 0$, one where $f \geq 0$ and $g < 0$, and one where

$f < 0$ and $g \geq 0$. Theorem 1-41 then gives the desired result. Finally, for general h, write $h = h^+ - h^-$ and consider h^+ and h^- separately.

Corollary 1-46: Let E be a measurable set, and suppose $\{f_n\}$ is a sequence of non-negative measurable functions with

$$f(x) = \sum_{n=1}^{\infty} f_n(x).$$

Then

$$\int_E f d\mu = \sum_{n=1}^{\infty} \int_E f_n d\mu.$$

PROOF: The result follows from Proposition 1-45 and Theorem 1-44.

Proposition 1-47: If f is a non-negative integrable function, then for every $\epsilon > 0$ there is a $\delta > 0$ such that, if $\mu(E) < \delta$, then

$$\int_E f d\mu < \epsilon.$$

PROOF: Set $f_n = \min(f, n)$. By Theorem 1-44,

$$\lim_n \int_X f_n d\mu = \int_X f d\mu.$$

Since f is integrable, we may find an N such that

$$\int_X (f - f_N) d\mu < \frac{\epsilon}{2}$$

by Proposition 1-45. Let $\delta = \epsilon/(2N)$. If $\mu(E) < \delta$, then

$$\begin{aligned}
\int_E f d\mu &= \int_E (f - f_N) d\mu + \int_E f_N d\mu && \text{by Proposition 1-45} \\
&\leq \int_X (f - f_N) d\mu + N\mu(E) && \text{by Proposition 1-39} \\
&< \epsilon.
\end{aligned}$$

Our third theorem for this section is known as Fatou's Theorem.

Theorem 1-48: Let E be a measurable set, and let $\{f_n\}$ be a sequence of non-negative measurable functions. Then

$$\int_E \liminf_n f_n d\mu \leq \liminf_n \int_E f_n d\mu.$$

In particular, if $f(x) = \lim_n f_n(x)$,

$$\int_E f d\mu \leq \liminf_n \int_E f_n d\mu.$$

PROOF: Set $g_n(x) = \inf_{k \geq n} f_k(x)$. Then $0 \leq g_1(x) \leq g_2(x) \leq \cdots$, and g_n is measurable on E. We have $g_n(x) \leq f_n(x)$ and

$$\lim_n g_n(x) = \liminf_n f_n(x).$$

By Theorem 1-44,

$$\int_E \liminf_n f_n d\mu = \int_E \lim_n g_n d\mu = \lim_n \int_E g_n d\mu.$$

The result now follows from the inequality $\int_E g_n d\mu \leq \int_E f_n d\mu$.

The fourth basic integration theorem is the Lebesgue Dominated Convergence Theorem.

Theorem 1-49: Let E be a measurable set, and suppose $\{f_n\}$ is a sequence of measurable functions such that for some *integrable* g, $|f_n| \leq g$ for all n. If $f(x) = \lim_n f_n(x)$, then $\lim_n \int_E f_n d\mu$ exists and

$$\int_E f d\mu = \lim_n \int_E f_n d\mu.$$

PROOF: By property (7) of Proposition 1-39, f is integrable and so is f_n for every n. Apply Fatou's Theorem first to $f_n + g \geq 0$ to obtain $\int_E (f + g) d\mu \leq \liminf \int_E (f_n + g) d\mu$ or

$$\int_E f d\mu \leq \liminf_n \int_E f_n d\mu.$$

Apply the theorem once more to $g - f_n \geq 0$ to obtain

$$\int_E (g - f) d\mu \leq \liminf_n \int_E (g - f_n) d\mu$$

or

$$-\int_E f d\mu \leq \liminf_n \int_E (-f_n) d\mu$$

or

$$\int_E f d\mu \geq \limsup_n \int_E f_n d\mu.$$

Therefore $\lim_n \int_E f_n d\mu$ exists and has the value asserted.

Corollary 1-50: Let E be a set of finite measure and suppose $\{f_n\}$ is a sequence of measurable functions such that $|f_n| \leq c$ for all n. If $f(x) = \lim_n f_n(x)$, then $\int_E f d\mu = \lim_n \int_E f_n d\mu$.

Much of the discussion of this section has dealt with the following problem: A sequence of integrable functions f_n converges a.e. to a function f; we want to be able to conclude that $\int f_n d\mu$ tends to $\int f d\mu$. First we should note that at almost all points $f_n^+ \to f^+$ and $f_n^- \to f^-$, and hence it is sufficient to check the convergence of the integral separately for these two sequences. Thus we may consider the case $f_n \geq 0$ alone. For non-negative functions Fatou's inequality is the only general result; one cannot conclude equality without a further hypothesis. Two sufficient conditions are given by monotone and dominated convergence.

But if we restrict our attention to a space of finite total measure, then we can give a necessary and sufficient condition.

Definition 1-51: A sequence $\{f_n\}$ of non-negative integrable functions is said to be **uniformly integrable** if for each $\epsilon > 0$ there is a number k such that

$$\int_{\{x|f_n(x)>k\}} f_n d\mu < \epsilon$$

holds for every n.

Equivalently we may require that the inequality holds for all sufficiently large n. For suppose it holds for $n > N$ and for the number c. Then since each f_n is integrable, there is a k_n depending on n (and, of course, ϵ) such that

$$\int_{\{x|f_n(x)>k_n\}} f_n d\mu < \epsilon$$

and we may choose $k = \sup \{k_1, k_2, \ldots, k_N, c\}$.

Proposition 1-52: If $\{f_n\}$ is a sequence of non-negative integrable functions tending to f and if $\mu(E) < \infty$, then $\int_E f_n d\mu \to \int_E f d\mu$ if and only if the $\{f_n\}$ are uniformly integrable.

REMARK: The sequence $\{f_n\}$ need converge to f only almost everywhere for the proposition to be valid, provided f is assumed measurable. This measurability condition is always satisfied if the underlying Borel field is augmented.

PROOF: We shall use the notation $f^{(k)}$ for the function truncated at k; that is, $f^{(k)}(x) = \inf(f(x), k)$. Let

$$A_n = \int_E f d\mu - \int_E f_n d\mu$$

$$B^k = \int_E f d\mu - \int_E f^{(k)} d\mu$$

$$C_n^{\ k} = \int_E f^{(k)} d\mu - \int_E f_n^{\ (k)} d\mu$$

$$D_n^{\ k} = \int_E f_n d\mu - \int_E f_n^{\ (k)} d\mu = \int_{\{x \mid f_n > k\}} (f_n - k) d\mu.$$

We have

$$A_n = B^k + C_n^{\ k} - D_n^{\ k}.$$

Clearly $f^{(k)}$ increases monotonically to f, so that B^k tends to 0 by monotone convergence. Since $f_n \to f$, $f_n^{\ (k)} \to f^{(k)}$. But $f_n^{\ (k)}$ is bounded by k. Thus, on the totally finite measure space E, $\lim_n C_n^{\ k} = 0$ by Corollary 1-50. Hence by choosing a large k and then a sufficiently large n (depending on k), we can make the first two terms on the right side as small as desired. If the functions are uniformly integrable, then we can find a large k (perhaps larger than the one already chosen) for which $D_n^{\ k}$ will be small for all n. Hence, for all sufficiently large n, the left side is small. Thus the integrals converge.

Conversely, suppose that $A_n \to 0$. Then there is a k such that for all sufficiently large n, $D_n^{\ k} < \epsilon/2$. Thus for all n sufficiently large, we have

$$\begin{aligned} \epsilon/2 &> \int_{\{x \mid f_n > k\}} (f_n - k) d\mu \\ &\geq \int_{\{x \mid f_n > 2k\}} (f_n - k) d\mu \\ &\geq \frac{1}{2} \int_{\{x \mid f_n > 2k\}} f_n d\mu, \end{aligned}$$

since $f_n - k > \frac{1}{2} f_n$ on the set in the last two integrals. Taking $2k$ as the number in the equivalent definition, we see that we have uniform integrability.

5. Limit theorems for matrices

We have already said that if π is a row vector and if $\{f^{(k)}\}$ is a sequence of column vectors converging to f, then it is not necessarily true that

$\pi f^{(k)}$ converges to πf. Our object in this section is to obtain sufficient criteria to justify saying that $\pi f = \lim_{k\to\infty} \pi f^{(k)}$.

In the examples of Lebesgue integration, we noted that non-negative row vectors are measures and column vectors are functions. Functions are integrated by forming the matrix product of the measure and the function. Thus, the theorems of Section 4 immediately give us the following four results. In each of them it should be borne in mind that:

(1) There is a corresponding result if row vectors are thought of as functions and column vectors are thought of as measures.
(2) These results imply corresponding results about matrices which are not just row or column vectors. (Recall the discussion in Section 1 about proving matrix equalities entry-by-entry.)

Proposition 1-53: Let $\pi \geq 0$ be a row vector and suppose $\{f^{(k)}\}$ is a sequence of column vectors converging entry-by-entry to f and satisfying

$$0 \leq f^{(1)} \leq f^{(2)} \leq \ldots.$$

Then $\pi f = \lim_k \pi f^{(k)}$.

PROOF: This result is a restatement of the Monotone Convergence Theorem as it applies to matrices.

Corollary 1-54: Let $\pi \geq 0$ be a row vector and suppose $\{f^{(k)}\}$ is a sequence of non-negative column vectors such that

$$f = \sum_{k=1}^{\infty} f^{(k)}.$$

Then

$$\pi f = \sum_{k=1}^{\infty} \pi f^{(k)}.$$

PROOF: This corollary is immediate from Corollary 1-46.

Proposition 1-55: Let $\pi \geq 0$ be a row vector and suppose $\{f^{(k)}\}$ is a sequence of non-negative column vectors. Then

$$\pi(\liminf_k f^{(k)}) \leq \liminf_k (\pi f^{(k)}).$$

If $f = \lim_k f^{(k)}$ exists, then $\pi f \leq \liminf_k (\pi f^{(k)})$.

PROOF: This is Fatou's Theorem.

Proposition 1-56: Let $\pi \geq 0$ be a row vector such that $\pi\mathbf{1}$ is finite. If $\{f^{(k)}\}$ is a sequence of column vectors such that $|f^{(k)}| \leq c\mathbf{1}$ and $f = \lim_k f^{(k)}$ exists, then

$$\pi f = \lim_k \pi f^{(k)}.$$

PROOF: The result follows from Corollary 1-50.

A harder problem arises with a sequence of non-negative row vectors $\pi^{(k)}$ converging to a row vector π. It is not sufficient for $\pi^{(k)}\mathbf{1} \leq M$ and $|f| \leq c\mathbf{1}$ in order for $\pi f = \lim_k \pi^{(k)} f$. For set

$$\begin{aligned} \pi^{(1)} &= (1 \quad 0 \quad 0 \quad 0 \quad \ldots) \\ \pi^{(2)} &= (0 \quad 1 \quad 0 \quad 0 \quad \ldots) \\ \pi^{(3)} &= (0 \quad 0 \quad 1 \quad 0 \quad \ldots) \\ &\vdots \end{aligned}$$

and take $f = \mathbf{1}$. Then $\pi = 0$ so that $\pi f = 0$, while $\lim_k \pi^{(k)} f = 1$. We give two sufficient conditions for

$$\pi f = \lim_k \pi^{(k)} f;$$

our integration theorems do not provide us with quick proofs, however. The second proposition is closely related to the Silverman–Toeplitz Theorem on summability methods.

Proposition 1-57: If $\{\pi^{(k)}\}$ is a sequence of non-negative row vectors converging to π, if f is a column vector such that $0 \leq f \leq c\mathbf{1}$ for some c, and if $\pi\mathbf{1} = \lim_k \pi^{(k)}\mathbf{1}$ with $\pi\mathbf{1}$ finite, then

$$\pi f = \lim_k \pi^{(k)} f.$$

PROOF: Take f as a measure and $\{\pi^{(k)}\}$ as a sequence of non-negative functions and apply Fatou's Theorem. We have

$$\pi f \leq \liminf \pi^{(k)} f.$$

With $c\mathbf{1} - f$ as a measure and $\{\pi^{(k)}\}$ as a sequence of functions, Fatou's Theorem gives

$$\pi(c\mathbf{1} - f) \leq \liminf \pi^{(k)}(c\mathbf{1} - f).$$

Since $\pi\mathbf{1}$ is finite and $\lim \pi^{(k)}\mathbf{1} = \pi\mathbf{1}$, we find

$$-\pi f \leq \liminf(-\pi^{(k)} f)$$

or

$$\pi f \geq \limsup \pi^{(k)} f.$$

Proposition 1-58: Let $\{\pi^{(k)}\}$ be a sequence of row vectors converging to π and satisfying $|\pi^{(k)}|1 \leq M$. Suppose f is a column vector with the property that for any $\delta > 0$ only finitely many entries of f have absolute value greater than δ. Then

$$\pi f = \lim_k \pi^{(k)} f.$$

PROOF: The entries of f are clearly bounded, say by c. Numbering the entries, we have for every N

$$|\pi f - \pi^{(k)} f| \leq \sum_{j=1}^{N} |\pi_j - \pi_j^{(k)}|\,|f_j| + \sum_{j>N} (|\pi_j| + |\pi_j^{(k)}|)|f_j|.$$

Let $\epsilon > 0$ be given. Choose N sufficiently large that $|f_j| < \epsilon/4M$ for $j > N$. Choose k sufficiently large that $|\pi_j - \pi_j^{(k)}| < \epsilon/2cN$ for $j \leq N$. Then $|\pi f - \pi^{(k)} f| < \epsilon$, and the result is established.

As we noted in Section 1, results about general denumerable matrices can be reduced to results about row and column vectors. In particular, the propositions of the present section apply equally well to sequences of the forms $\{A^{(k)} f\}$ and $\{\pi A^{(k)}\}$.

6. Some general theorems from analysis

In this section we collect a variety of results from analysis which we shall need in later chapters. We prove only some of them. At first reading the reader may find it wise to skip this section, returning to it later as the material is required.

a. Stirling's formula. Stirling's formula gives an asymptotic expression for $m!$ The approximation is

$$m! \sim \frac{m^m}{e^m} \sqrt{2\pi m},$$

where the symbol $\sim$ indicates that the ratio of the two quantities tends to one as m increases. For a proof, see Courant and Hilbert [1953], pp. 522–524. Stirling's formula immediately gives an approximation for the **binomial coefficient** $\binom{rn}{n}$ for large n. The coefficient $\binom{a}{b}$ is defined as $\dfrac{a!}{b!(a-b)!}$.

Lemma 1-59: For $r > 1$,

$$\binom{rn}{n} \sim c \frac{1}{\sqrt{n}} \left(\frac{r^r}{(r-1)^{r-1}}\right)^n,$$

where c is a constant depending on r but not on n.

The multinomial coefficient $\binom{a}{b, c, \ldots, d}$ is defined to be $\dfrac{a!}{b!c! \cdots d!}$.

b. Difference equations. An nth order linear difference equation with constant coefficients is an expression of the form

$$y_{k+n} + c_{n-1}y_{k+n-1} + \cdots + c_1 y_{k+1} + c_0 y_k = r_k,$$

where y_k and r_k are functions defined on the integers and where the $c_{n-1}, \ldots, c_0$ are complex numbers. The equation is **homogeneous** if $r_k = 0$ and **nonhomogeneous** otherwise. For a nonhomogeneous solution, we refer to any single function y_k satisfying the equation as a **particular solution**, and we call the set of functions satisfying the same equation with $r_k = 0$ the **homogeneous solution**. The totality of solutions to any difference equation is known as the **general solution**.

Proposition 1-60: Every solution of the difference equation

$$y_{k+n} + c_{n-1}y_{k+n-1} + \cdots + c_0 y_k = 0$$

is a linear combination of n fixed functions, obtained as follows: If a is a root of multiplicity q of the **characteristic equation**

$$x^n + c_{n-1}x^{n-1} + \cdots + c_1 x + c_0 = 0,$$

then q of the functions are

$$a^k, ka^k, k^2a^k, \ldots, k^{q-1}a^k.$$

Conversely, each of these functions is a solution of the difference equation. Furthermore, each solution of the equation

$$y_{k+n} + c_{n-1}y_{k+n-1} + \cdots + c_0 y_k = r_k$$

is the sum of a fixed particular solution and some solution of

$$y_{k+n} + c_{n-1}y_{k+n-1} + \cdots + c_0 y_k = 0.$$

Conversely, every such sum is a solution of the nonhomogeneous equation.

For a proof of this proposition, see Hildebrand [1956], pp. 202–203.

c. Cesaro summability and Abel summability. Let $\{a_n\}$ be a sequence of real numbers. Define b_n to be the arithmetic mean of the first n terms of the sequence $\{a_n\}$. The sequence $\{b_n\}$ is called the sequence of **Cesaro averages** of the sequence $\{a_n\}$. The sequence $\{a_n\}$ is said to be **Cesaro summable** if its sequence, $\{b_n\}$, of Cesaro averages has a limit. If $A^{(n)}$ is a sequence of matrices, the sequence of Cesaro averages $B^{(n)}$

is defined entry-by-entry: $B_{ij}^{(n)}$ is the Cesaro average of $A_{ij}^{(1)}, A_{ij}^{(2)}, \ldots, A_{ij}^{(n)}$. The basic fact about Cesaro summability is the following proposition.

Proposition 1-61: If a sequence $\{a_n\}$ converges to a limit L, then the sequence of Cesaro averages $\{b_n\}$ converges and its limit is L.

PROOF: Let f be the column vector whose nth entry is $a_n - L$, and let $\pi^{(n)}$ be the row vector defined by

$$\pi_j^{(n)} = \begin{cases} 1/n & \text{for } 1 \leq j \leq n \\ 0 & \text{for } n > j. \end{cases}$$

Then

$$b_n = \pi^{(n)}f + L.$$

Now $|\pi^{(n)}|1 = 1$, $\lim_n \pi^{(n)} = 0$, and f is a column vector with entries tending to 0. Hence by Proposition 1-58, $\lim_n \pi^{(n)}f = 0$. Therefore $\lim_n b_n = L$.

The converse of Proposition 1-61 is false. The sequence $\{a_n\}$ defined by $a_{2n} = 0$, $a_{2n+1} = 1$ does not converge, but it is Cesaro summable.

Let $\{c_n\}$ be a sequence of real numbers. (In most applications the partial sums $c_0 + \cdots + c_n$ are assumed bounded.) If the limit

$$\lim_{t \uparrow 1} \sum_{n=0}^{\infty} c_n t^n$$

exists, the limit is called the **Abel sum** of the series $\sum c_n$ and the series is said to be **Abel summable**. Abel's Theorem is the following result.

Proposition 1-62: If the series $\sum c_n$ converges to a finite limit L, then it is Abel summable and its Abel sum is L.

PROOF: Since the partial sums converge to a finite limit, the c_n are bounded and $\sum c_n t^n$ converges absolutely for $t < 1$. Let $\{t_k\}$ be any sequence of positive reals less than one and increasing to one. Let f be the column vector whose nth entry is $(c_0 + \cdots + c_n) - L$, and let $\pi^{(k)}$ be the row vector defined by

$$\pi_j^{(k)} = (1 - t_k)t_k^{\,j}.$$

Then

$$\pi^{(k)}f = \sum c_n t_k^{\,n} - L.$$

Now $|\pi^{(k)}|1 = 1$, $\lim_k \pi^{(k)} = 0$, and f is a column vector with entries tending to 0. Hence by Proposition 1-58, $\lim \pi^{(k)}f = 0$. That is,

$$\lim_{k \to \infty} \sum c_n t_k^{\,n} = L.$$

This equality for every such sequence $\{t_k\}$ implies that

$$\lim_{t \uparrow 1} \sum c_n t^n = L.$$

d. Convergent subsequences of matrices. A bounded sequence of real numbers has a subsequence which converges to a finite limit. We shall obtain a generalization of this result to matrices.

Proposition 1-63: Let $\{A^{(n)}\}$ be a sequence of matrices with the property that for some pair of real numbers c and d, $cE \leq A^{(n)} \leq dE$ for all n. Then there exists a subsequence of matrices $\{A^{(n_\nu)}\}$ which converges in every entry.

PROOF: Since there are only denumerably many entries in each matrix, they can be numbered by a subset of the positive integers. Select a subsequence $\{A_1^{(n)}\}$ which converges in the first entry. Let $A_2^{(2)}, A_2^{(3)}, A_2^{(4)}, \ldots$ be a subsequence of $A_1^{(2)}, A_1^{(3)}, \ldots$ which converges in the second entry. In general, let $A_m^{(m)}, A_m^{(m+1)}, \ldots$ be a subsequence of $A_{m-1}^{(m)}, A_{m-1}^{(m+1)}, \ldots$ which converges in the mth entry. Finally set

$$A^{(n_\nu)} = A_\nu^{(\nu)}.$$

Then $\{A^{(n_\nu)}\}$ converges in every entry.

Corollary 1-64: Let $\{A^{(n)}\}$ be a sequence of matrices with the property that $cE \leq A^{(n)} \leq dE$ for all n. Then $\lim_n A^{(n)} = A$ exists if and only if $\lim_\nu A^{(n_\nu)} = A$ for every convergent subsequence $\{A^{(n_\nu)}\}$.

PROOF: The necessity of the condition is trivial. For the sufficiency suppose $\lim A_{ij}^{(n)}$ does not exist. Then $\liminf A_{ij}^{(n)} < \limsup A_{ij}^{(n)}$. Pick a subsequence of $\{L^{(n)}\}$ that converges in the i-jth entry to $\limsup A_{ij}^{(n)}$, and do the same for $\liminf A_{ij}^{(n)}$. Apply Proposition 1-63 to extract subsequences convergent in all entries from these sequences, and the result follows at once.

e. Sets of positive integers closed under addition. The greatest common divisor of a non-empty set of positive integers is the largest integer which divides all of them. If the set consists of $\{n_1, n_2, \ldots\}$, its greatest common divisor is denoted $(n_1, n_2, \ldots)$.

Lemma 1-65: If T is a set of positive integers with greatest common divisor d, then there exists a finite subset of T for which d is the greatest common divisor.

PROOF: Let $n_1 = k_1 d$ be an element of T. If $k_1 = 1$, $\{n_1\}$ is the required set. If not, choose n_2 such that $n_1 \nmid n_2$. Then $(n_1, n_2) = k_2 d$ with $k_2 < k_1$. If $k_2 = 1$, $\{n_1, n_2\}$ is the required set. Otherwise, find n_3 such that $k_2 d \nmid n_3$, and set $(n_1, n_2, n_3) = k_3 d$ with $k_3 < k_2$. Continuing in this way, we obtain a decreasing sequence of integers $k_1, k_2, \ldots$ bounded below by 1. It must terminate, and then we have constructed the finite set.

Lemma 1-66: Let T be a non-empty set of positive integers which is closed under addition and which has the greatest common divisor d. Then all sufficiently large multiples of d are in the set T.

PROOF: If $d \neq 1$, divide all the elements in T by d and reduce the problem to the case $d = 1$. By Lemma 1-65 there is a finite subset $\{n_1, \ldots, n_s\}$ of T with greatest common divisor 1. Then there exist integers $c_1, \ldots, c_s$ with the property

$$c_1 n_1 + \cdots + c_s n_s = 1.$$

If we collect the positive terms and the negative terms and note that T is closed under addition, we find that T contains non-negative integers m and n with $m - n = 1$. Suppose $q \geq n(n - 1)$. Then $q = an + b$ with $a \geq n - 1$ and $0 \leq b \leq n - 1$. Thus

$$q = (a - b)n + bm$$

and q is in T.

f. Renewal theorem. The Renewal Theorem, one of the important results in the elementary theory of probability, can be stated purely in terms of analysis.

Theorem 1-67: Let $\{f_n\}$ be a sequence of non-negative real numbers such that $\sum f_n = 1$ and $f_0 = 0$, and suppose the greatest common divisor of those indices k for which $f_k > 0$ is 1. Set $\mu = \sum n f_n$, $u_0 = 1$, and $u_n = \sum_{k=0}^{n-1} u_k f_{n-k}$. If μ is infinite, then $\lim_n u_n = 0$, and if μ is finite, then $\lim_n u_n = 1/\mu$.

For a proof, see Feller [1957], pp. 306–307.

g. Central Limit Theorem. Identically distributed independent random variables are defined in Sections 3-2 and 6-4. The mean of a random variable is its integral over its domain, and its variance is the integral of its square minus the square of its integral, a quantity which is always non-negative.

Theorem 1-68: Let $\{y_n\}$ be a sequence of identically distributed independent random variables with common mean μ and with common finite variance $\sigma^2 > 0$. Set $s_m = y_1 + \cdots + y_m$. Then for all real α and β with $\alpha < \beta$,

$$\lim_{n \to \infty} \Pr\left[\alpha < \frac{s_m - m\mu}{\sigma\sqrt{m}} < \beta\right] = \Phi(\beta) - \Phi(\alpha),$$

where

$$\Phi(x) = \frac{1}{\sqrt{2\pi}} \int_{-\infty}^{x} e^{-u^2/2} du.$$

For a proof, see Doob [1953], p. 140.

CHAPTER 2

STOCHASTIC PROCESSES

1. Sequence spaces

We shall introduce the concept of a stochastic process in this chapter and develop the basic tools needed to treat the processes. Before the formal development, we shall indicate the intuitive ideas underlying the formal definitions.

We imagine that a sequence of experiments is performed. The outcomes may be arbitrary elements of a specified set, such as the set {"yes," "no," "no opinion"}, the set {heads, tails}, the set {fair, cloudy, rainy, snowy}, or a set of numbers. The experiments may be quite general in nature, but we impose some natural restrictions:

(1) The set of possible outcomes is denumerable. (This restriction is natural for the present book. It would be removed in a more general treatment of stochastic processes.)

(2) The probability of an outcome for the nth experiment is completely determined by a knowledge of the outcomes of earlier experiments. Here "probability" is used heuristically, to motivate the later precise definition.

(3) The experimenter is, at each stage, aware of the outcomes of earlier experiments.

We shall first consider a sequence of n experiments, where n is specified at the beginning. Later we shall consider a denumerably infinite sequence of experiments. In each case we assume that the experiments do not stop earlier. However, this is an unimportant restriction; a process that terminates may be represented in our framework by allowing the outcome "stopped." The following are examples of such sequences of experiments:

(1) A sociologist wishes to find out whether people feel that television is turning us into a nation of illiterates. He asks a carefully selected sample of subjects and receives the answer "yes," "no," or "no opinion" in each case.

(2) A gambler flips a coin repeatedly, recording "heads" or "tails."

(3) A meteorologist records the daily weather for 23 years, classifying each day as "fair," "cloudy," "rainy," or "snowy."

(4) A physicist tries to determine the speed of light by making a series of measurements. (Since each measurement is recorded only to a certain number of decimal places, the possible outcomes are rational numbers, and hence the set is denumerable.)

(5) A physicist makes a count of the number of radioactive particles given off by an ounce of uranium. A measurement is made every second, and the outcome of the nth measurement is the total number of particles given off until then.

The exact way in which probabilities are determined from an experiment is a deep problem in the philosophy of science, and it will not concern us here. We will assume that the nature of the experiment yields us certain probabilities, namely the probability that the nth experiment results in an outcome c, given that the previous experiments resulted in outcomes $c_0, c_1, \ldots, c_{n-1}$. We then design a probability space in which one can compute the probability of a wide variety of statements concerning the experiments and in which the specified (conditional) probabilities turn out as given.

The elements of our probability space Ω are sequences of possible outcomes for the experiments (either sequences of length n, for a finite series of experiments, or infinite sequences). The elements of the Borel field $\mathscr{B}$ of measurable sets will be the truth sets of statements to which probabilities are to be assigned. (The **truth set** of a statement about the experiments is the set of all those sequences in Ω for which the statement is true.) A measure μ is constructed, and the probability of a statement is the measure of its truth set. In particular, $\mu(\Omega) = 1$ in a probability space, and hence the probability of a logically true statement is 1.

Let us first consider the case where n experiments are performed. The possible outcomes are conveniently represented by a tree, with each path through the tree representing a sequence of possible outcomes. In the diagram n equals 3, Ω has 8 elements, and $\mathscr{B}$ consists of all subsets of Ω.

The numbers on the branches, known as **branch weights**, represent the conditional probabilities mentioned above. For example, p_2 is the probability of heads, given that the first toss came up heads, while $1 - p_3$ is the probability that tails is the outcome that follows two heads. The weight assigned to the path HHT is taken to be the product of the branch weights, $p_1p_2(1 - p_3)$. The measure $\mu(A)$ of a set of paths A is the sum of the weights of the paths in A. In the usual

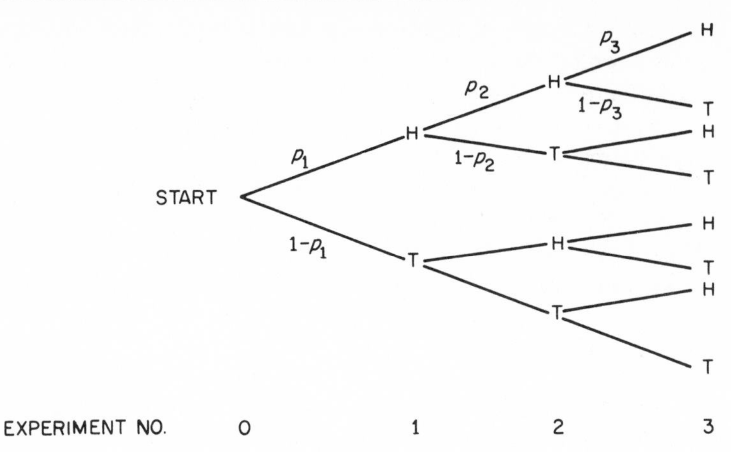

setup for coin tossing, each p is $\frac{1}{2}$, and the weights of the paths are each $\frac{1}{8}$.

The branch weights may be arbitrary non-negative numbers, but the sum of the weights of branches starting from a given branch-point must be one.

After we define conditional probabilities, it will be easy to verify that the numbers written on the branches do indeed turn out to be the desired conditional probabilities (see Kemeny, Mirkil, *et al.* [1959], Chapter 3).

Let us suppose that we have constructed a tree Ω_n for a series of n experiments. We consider k additional experiments, obtaining a tree Ω_{n+k}. We wish our probabilities to be consistent in the sense that a statement p about Ω_n has the same probability when computed in either measure space. Our method of computing measures has this consistency property. This assertion is easily verified for $k = 1$, and the result follows by induction.

In constructing an infinite tree Ω for a sequence of experiments, our measure is required to have the property that a statement about the first n outcomes has the same probability as if computed with respect to the finite tree Ω_n. (Of course, the same probability may be computed with respect to Ω_{n+k}, but the result is the same by consistency.) This convention assigns probabilities to many simple statements. We can then show that the probability of a much larger class of statements is uniquely determined. We will now consider the problem abstractly.

Let S be a denumerable set; S is called the **state space**. Let Ω be the set of all infinite sequences of elements of S. A typical element ω of Ω is represented as

$$\omega = (c_0, c_1, c_2, \ldots),$$

where $c_0, c_1, c_2, \ldots$ are elements of S. The points ω of Ω are called

paths, the whole space Ω is called a **sequence space**, and the value c_n on a path ω is called the nth **outcome** on ω. The function $x_n(\omega)$, defined from Ω to S by

$$x_n(c_0, c_1, c_2, \ldots, c_n, \ldots) = c_n$$

is called the nth **outcome function**, and the nth outcome is said to occur at **time** n.

Let $\mathscr{F}_n$ be the family of all unions of sets in Ω of the form

$$\{\omega \mid x_0(\omega) \in S_0 \wedge x_1(\omega) \in S_1 \wedge \cdots \wedge x_n(\omega) \in S_n\},$$

where $S_0, S_1, \ldots, S_n$ are subsets of the state space S. (Notice that the sets of $\mathscr{F}_n$ arise from the class of all subsets of the tree Ω_n described above.) It is clear that for each n, $\mathscr{F}_n$ is a Borel field. Let $\mathscr{F}$ be the family of sets defined by

$$\mathscr{F} = \bigcup_{n=0}^{\infty} \mathscr{F}_n.$$

Each set in $\mathscr{F}$ is a set of paths for which a finite number of outcomes are restricted to lie in certain subsets of S. All other outcomes are unrestricted. The reader should verify that $\mathscr{F}$ is a field of sets. In Section 3 we shall see that $\mathscr{F}$ is not a Borel field; in the meantime, we let $\mathscr{G}$ be the smallest Borel field containing $\mathscr{F}$ (Proposition 1-14). After we have defined a measure on $\mathscr{G}$, the Borel field $\mathscr{B}$ which we are seeking will be the augmented field obtained from $\mathscr{G}$ by adding subsets of sets of measure zero.

The sets of $\mathscr{F}$ are known as **cylinder sets**. If a cylinder set C is a set in $\mathscr{F}_n$, we note that C may be written as the denumerable union

$$C = \bigcup_i B_i^{(n)}$$

of (disjoint) **basic cylinder sets**

$$B_i^{(n)} = \{\omega \mid x_0(\omega) = c_0 \wedge x_1(\omega) = c_1 \wedge \cdots \wedge x_n(\omega) = c_n\}.$$

A basic cylinder set in $\mathscr{F}_n$ is the set of all possible continuations of a single path in Ω_n. We let $\nu(B_i^{(n)})$ equal the product of the branch weights on this path in Ω_n.

Recalling that the probability measures we assigned to Ω_n were defined consistently, we can show that ν is uniquely defined on the sets of $\mathscr{F}$. It has the properties that $\nu(\Omega) = 1$ and that the restriction of ν to $\mathscr{F}_n$ is a measure for every n.

We will next show that ν can be extended to a measure μ on the smallest Borel field containing $\mathscr{F}$. This result will be a consequence of Theorem 2-4. First we prove a series of lemmas. In each case $\mathscr{F}_n$, $\mathscr{F}$, and ν are as defined above.

Lemma 2-1: Let ν be a set function defined on $\mathscr{F} = \bigcup \mathscr{F}_n$ in such a way that the restriction of ν to $\mathscr{F}_n$ is a measure for every n. Then ν is non-negative and additive.

PROOF: Non-negativity is trivial. For additivity, let A and B be disjoint sets in $\mathscr{F}$. Then $A \in \mathscr{F}_m$ and $B \in \mathscr{F}_n$, say. Since the $\mathscr{F}_j$ are nested, A and B are both in $\mathscr{F}_r$, where $r = \sup(m, n)$. Since ν is a measure when restricted to $\mathscr{F}_r$,

$$\nu(A \cup B) = \nu(A) + \nu(B).$$

We shall in fact establish that ν is completely additive, a result due to Kolmogorov.

Lemma 2-2: Suppose $C_0 \supset C_1 \supset C_2 \supset \cdots$ is a sequence of sets in $\mathscr{F}$ such that $C_n \in \mathscr{F}_n$ and $\lim_n \nu(C_n) > 0$. Then for every m there is a basic cylinder set $B_i^{(m)}$ in $\mathscr{F}_m$ such that

(1) $\lim\limits_n \nu(C_n \cap B_i^{(m)}) > 0$

(2) $B_i^{(m)} \subset C_m$.

PROOF: By complete additivity of ν on $\mathscr{F}_r$, where $r = \sup(m, n)$, we have $\nu(C_n) = \sum_j \nu(C_n \cap B_j^{(m)})$ with $\nu(C_n \cap B_j^{(m)})$ monotonically decreasing in n. Then

$$0 < \lim_n \nu(C_n) = \lim_n \sum_j \nu(C_n \cap B_j^{(m)}) = \sum_j \lim_n \nu(C_n \cap B_j^{(m)}).$$

The interchange of limit and sum is justified by dominated convergence as follows: The functions of j, namely $\nu(C_n \cap B_j^{(m)})$, satisfy

$$\nu(C_n \cap B_j^{(m)}) \leq \nu(C_0 \cap B_j^{(m)}),$$

and we know that $\sum_j \nu(C_0 \cap B_j^{(m)}) = \nu(C_0)$ is finite since $\nu(\Omega) = 1$.

Thus, since a denumerable sum cannot be positive unless one of the terms is positive, we have for some $j = i$

$$\lim_n \nu(C_n \cap B_i^{(m)}) > 0.$$

Hence (1) is satisfied. But the terms in the sequence $\nu(C_n \cap B_i^{(m)})$ monotonically decrease to a positive limit, so that

$$\nu(C_m \cap B_i^{(m)}) > 0.$$

Now $C_m \in \mathscr{F}_m$ and is thus the union of basic cylinder sets. Since $C_m \cap B_i^{(m)}$ cannot be empty, we must have $B_i^{(m)} \subset C_m$.

Lemma 2-3: Suppose $A_0 \supset A_1 \dots$ is a sequence of sets in $\mathscr{F}$ such that $A_n \in \mathscr{F}_n$, and $\lim_n \nu(A_n) = L > 0$. Then there exists a sequence $\{B^{(n)}\}$ of basic cylinder sets such that

(1′) $B^{(n)}$ is a basic cylinder set of $\mathscr{F}_n$.
(2′) $\lim_m \nu(A_m \cap B^{(n)}) > 0$.
(3′) For $n > 0$, $B^{(n)} \subset B^{(n-1)}$.
(4′) $B^{(n)} \subset A_n$.

Remark: Property (3′) indicates that we are actually constructing a single path by adjoining branches one at a time.

Proof of Lemma: The construction is by induction. For $n = 0$ apply Lemma 2-2 to the sequence $\{A_n\}$ and the integer $m = 0$. The $B_i^{(0)}$ that the lemma gives is $B^{(0)}$. Property (2′) follows from (1), (3′) is vacuous, and (4′) follows from (2). Suppose that we have constructed $B^{(0)}, \dots, B^{(n)}$; we want $B^{(n+1)}$. Let

$$C_k = \begin{cases} A_k & \text{for } k < n \\ A_k \cap B^{(n)} & \text{for } k \geq n. \end{cases}$$

The sequence of sets $\{C_k\}$ is decreasing and $C_k \in \mathscr{F}_k$; we have

$$\lim_k \nu(C_k) > 0$$

by (2′) of the inductive hypothesis. Applying Lemma 2-2 to $\{C_k\}$ and the number $m = n + 1$, we obtain a basic cylinder set $B_i^{(n+1)}$, which we take as $B^{(n+1)}$. By (2),

$$B^{(n+1)} \subset C_{n+1} = A_{n+1} \cap B^{(n)}.$$

Hence (1′), (3′), and (4′) hold. By (1),

$$\begin{aligned} 0 < \lim_k \nu(C_k \cap B^{(n+1)}) &= \lim_k \nu(A_k \cap B^{(n)} \cap B^{(n+1)}) \\ &= \lim_k \nu(A_k \cap B^{(n+1)}) \qquad \text{by (3′)}. \end{aligned}$$

Hence (2′) holds.

Theorem 2-4: Let Ω be a sequence space, let $\mathscr{F}_n$ be the Borel field of sets generated by all statements

$$x_0 = c_0 \wedge \cdots \wedge x_n = c_n,$$

and let $\mathscr{F} = \bigcup_n \mathscr{F}_n$. Suppose for every n there is a probability measure ν_n defined on $\mathscr{F}_n$ with the property that the restriction of ν_{n+1} to $\mathscr{F}_n$ is ν_n. Let ν be the set function on $\mathscr{F}$ whose restriction to $\mathscr{F}_n$ is ν_n for all n. Then ν is a measure.

PROOF: By Lemma 2-1 and the contrapositive of the second half of Corollary 1-17, it is sufficient to show that if $A_0 \supset A_1 \supset A_2 \supset \cdots$ is a decreasing sequence of sets in $\mathscr{F}$ with $\lim_n \nu(A_n) = L > 0$, then $\bigcap_n A_n \neq \varnothing$. Since every set in $\mathscr{F}_n$ is a set in $\mathscr{F}_{n+1}$, we may assume that $A_n \in \mathscr{F}_n$ by repeating the same set several times in the sequence and by adding, if necessary, the set Ω a finite number of times at the beginning of the sequence. The hypotheses of Lemma 2-3 are satisfied. We then have $\bigcap_n B^{(n)} = \{\omega\}$, where ω is a single path of Ω. For every n, $\omega \in B^{(n)} \subset A_n$; hence $\omega \in \bigcap_n A_n$ and $\bigcap_n A_n \neq \varnothing$.

Applying Theorem 1-19 we extend the measure ν defined on $\mathscr{F}$ uniquely to a measure μ defined on the Borel field $\mathscr{G}$. Augmenting the field $\mathscr{G}$, we obtain the Borel field $\mathscr{B}$ with which we shall work. The extended measure μ is called **tree measure** and satisfies $\mu(\Omega) = 1$. This completes our construction of the sequence space $(\Omega, \mathscr{B}, \mu)$. From now on when we refer to a sequence space we shall mean $(\Omega, \mathscr{B}, \mu)$ constructed in the indicated manner.

2. Denumerable stochastic processes

We turn now to the definition of a denumerable stochastic process. After the definition we shall show that every sequence space defines in a natural way a stochastic process and that every denumerable stochastic process can, in a way to be described shortly, be considered as a sequence space.

Let S be a denumerable set, which will be called the **state space**, and let $(\Omega, \mathscr{B}, \mu)$ be a probability space. Points of Ω will be denoted ω.

Definition 2-5: Let $\{f_n\}$ be a sequence of functions with domain Ω and range in S, and let $\{\mathscr{B}_n\}$ be a sequence of Borel fields. The pair $(f_n, \mathscr{B}_n)$ is called a **denumerable stochastic process** on Ω if these two conditions are satisfied:

(1) $\mathscr{B}_0 \subset \mathscr{B}_1 \subset \mathscr{B}_2 \subset \cdots$; and for every n, $\mathscr{B}_n \subset \mathscr{B}$.
(2) For every fixed n and for each $s \in S$, the set $\{\omega \mid f_n(\omega) = s\}$ is a set in $\mathscr{B}_n$.

The second condition in the definition is a measurability requirement on f_n. If we were to think of the family $\mathscr{S}$ of all subsets of S as a Borel field, our condition would be equivalent to the demand that the inverse image under f_n of any set in $\mathscr{S}$ be a set in $\mathscr{B}_n$.

First we shall show that every sequence space defines a stochastic process in a natural way. Let $(\Omega, \mathscr{B}, \mu)$ be a sequence space. We take the outcome functions x_n as the sequence of functions and $\{\mathscr{F}_n\}$ as the sequence of Borel fields. It is clear that

$$\{\omega \mid x_n(\omega) = c\}$$

is a set of $\mathscr{F}_n$ and that $\mathscr{F}_0 \subset \mathscr{F}_1 \subset \cdots \subset \mathscr{F} \subset \mathscr{G} \subset \mathscr{B}$. The pair $(x_n, \mathscr{F}_n)$ therefore is a stochastic process and is referred to simply as $\{x_n\}$. We have thus shown that the outcomes of a sequence of experiments with a denumerable range form a denumerable stochastic process.

When we begin to discuss Markov chains, we shall wish to confine ourselves to stochastic processes defined on a sequence space. Therefore, our second task is to show that the restriction to sequence space is actually no restriction at all; this will indicate that our treatment of denumerable Markov chains is completely general.

Let $(f_n, \mathscr{B}_n)$ be a denumerable stochastic process on Ω' with state space S. Let μ' be a measure on Ω'. We shall construct a sequence space $(\Omega, \mathscr{B}, \mu)$ in such a way that the behavior of the process $(f_n, \mathscr{B}_n)$ on Ω' may be studied completely by studying the stochastic process $\{x_n\}$ on Ω.

Define Ω to be the space of all sequences of elements of S, and let $\mathscr{G}$ be the Borel field obtained by the construction in Section 2-1. We shall establish a correspondence between paths of Ω and subsets of Ω', and we define a measure μ on $\mathscr{G}$.

The correspondence we choose is

$$\omega \leftrightarrow \{\omega' \mid f_0(\omega') = x_0(\omega) \wedge \cdots \wedge f_n(\omega') = x_n(\omega) \wedge \cdots\}.$$

To assign a measure to Ω, we must assign a measure to cylinder sets, such as

$$A = \{\omega \mid x_0(\omega) \in S_0 \wedge \cdots \wedge x_n(\omega) \in S_n\}.$$

Noticing that the set A', defined by

$$A' = \{\omega' \mid f_0(\omega') \in S_0 \wedge \cdots \wedge f_n(\omega') \in S_n\},$$

is a set in $\mathscr{B}_n$ and is therefore μ'-measurable, we define

$$\mu(A) = \mu'(A').$$

The measure μ can then be extended to a measure defined on all of $\mathscr{B}$, and the construction of the space $(\Omega, \mathscr{B}, \mu)$ is complete. We thus see that an arbitrary stochastic process defined on Ω' may be considered as a process on a suitable sequence space Ω in which the f_n are outcome functions. The probability of any statement concerning the f_n can be computed in the sequence space.

3. Borel fields in stochastic processes

Probabilities are numbers assigned to statements about stochastic processes. We may now formally define the **probability** of a statement to be the measure of the statement's truth set. In symbols $\Pr[p] = \mu(P)$, where $P = \{\omega \mid p\}$. If the set P is not a set in the Borel field on which μ is defined, then $\Pr[p]$ is undefined. Statements for which $\Pr[p]$ is defined are called **measurable statements.**

In a stochastic process we see that a Borel field of sets represents a state of knowledge. The more sets there are in the Borel field, the more statements there are that we know how to assign probabilities to. Let us analyze briefly what this fact implies about Definition 2-5.

In a denumerable stochastic process we are given an increasing sequence of Borel fields $\mathscr{B}_n$ such that $\mathscr{B}_n \subset \mathscr{B}$ for every n. The field $\mathscr{B}_n$ represents the state of knowledge of the process up to time n. The fact that the Borel fields are increasing means that our knowledge of the process never decreases as time goes on, and the fact that all $\mathscr{B}_n$ are contained in $\mathscr{B}$ means that our total knowledge of the process necessarily includes knowledge of what happens in a finite number of steps.

Similarly, condition (2) in the definition is an abstract formulation of the requirement that in a stochastic process the present does not depend upon the future. We conclude, therefore, that our definition satisfies the conditions imposed at the beginning of Section 1.

We shall apply this insight about the role of Borel fields to a specific example to show that the field $\mathscr{F}$ in Section 1 is not the same as the Borel field $\mathscr{G}$. Let Ω be the sequence space constructed when S is taken as a two-element set. Measures $2^{-(n+1)}$ are assigned to each set B of paths of the form

$$B = \{\omega \mid x_0 = c_0 \wedge \cdots \wedge x_n = c_n\}.$$

This measure is eventually extended to a measure μ on the Borel field $\mathscr{B}$, and we obtain $\mu(\Omega) = 1$. The state space for this example is often taken as $S = \{\text{heads, tails}\}$, and the model for the stochastic process is the tossing of a balanced coin.

For the coin-tossing process a well-known example of a statement whose truth set is not in the field $\mathscr{F}$ but is in the Borel field $\mathscr{G}$ is involved in the Strong Law of Large Numbers (which we shall consider in more detail in Chapter 3). Let k and n be integers and let r_n be the fraction of the first n outcomes which are heads. Let p be the statement about r_n that

$$\frac{1}{2} - \frac{1}{k} < r_n < \frac{1}{2} + \frac{1}{k}.$$

Consider the statement q about $k > 0$ that

$$(\forall k)(\exists N)(\forall n)(n > N \rightarrow p).$$

We write the statement in this form to demonstrate that its truth set is in $\mathscr{G}$. In words, the statement q asserts that for any $k > 0$ there exists an N such that if $n > N$, then $|r_n - \frac{1}{2}| < 1/k$. That is, q says that $\lim_{n\to\infty} r_n = \frac{1}{2}$. The truth set of the statement q is in $\mathscr{G}$ but not $\mathscr{F}$, because the notion of limit cannot be expressed in terms of finitely many of the r_n. The Strong Law of Large Numbers asserts that

$$\Pr[q] = 1.$$

4. Statements of probability zero or one

In a probability space $(\Omega, \mathscr{B}, \mu)$ a statement with truth set Ω is logically true, whereas a statement with truth set $\varnothing$ is logically false. However, the Strong Law of Large Numbers asserts not that a certain set is Ω but that it has measure one. A statement whose truth set has measure one is said to be **almost always, almost everywhere, almost surely** true, or **true a.e.** Correspondingly, a statement with truth set of measure zero is almost surely false, and the negation $\sim p$ of a statement p which is almost surely false is almost surely true.

Two useful propositions are related to the subject of almost sure statements.

Proposition 2-6: Let $\{p_n\}$ be a denumerable set of statements and let q be the statement that p_n holds for all n. If $\Pr[p_n] = 1$ for all n, then $\Pr[q] = 1$.

PROOF: Let $\{P_n\}$ be the truth sets of the statements $\{p_n\}$. Applying Proposition 1-18, we have

$$1 - \Pr[q] = \mu\Big(X - \bigcap_n P_n\Big) = \mu\Big(\bigcup_n \tilde{P}_n\Big) \le \sum_n \mu(\tilde{P}_n) = 0$$

so that

$$\Pr[q] = 1.$$

The second result is one of the Borel–Cantelli Lemmas.

Proposition 2-7: If $\{p_n\}$ is a sequence of statements for which $\sum \Pr[p_n]$ is finite, then with probability one only finitely many of the p_n are true.

PROOF: Let q_n be the statement that all of the statements $p_n, p_{n+1}, \ldots$ are false. Let $\epsilon > 0$ be given. Choose N large enough so that $\sum_{n=N}^{\infty} \Pr[p_n] < \epsilon$. Then

$$\begin{aligned} 1 - \Pr[q_N] &= \Pr[\text{at least one of } p_N, p_{N+1}, \ldots \text{ occurs}] \\ &= \Pr[p_N \vee p_{N+1} \vee \cdots]. \end{aligned}$$

By Proposition 1-18 the right-side is

$$\leq \sum_{n=N}^{\infty} \Pr[p_n] < \epsilon.$$

Hence,

$$\begin{aligned} \Pr[\text{finitely many } p_n \text{ are true}] &= \Pr\left[\bigvee_{n=1}^{\infty} q_n\right] \\ &\geq \Pr[q_N] \\ &> 1 - \epsilon. \end{aligned}$$

Since this inequality holds for every $\epsilon > 0$, the probability must be 1.

5. Conditional probabilities

Let $(\Omega, \mathscr{B}, \mu)$ be a probability space. If p and q are statements such that $\Pr[q] \neq 0$, the **conditional probability of** p **given** q, written $\Pr[p \mid q]$, is defined by

$$\Pr[p \mid q] = \Pr[p \wedge q]/\Pr[q].$$

If $\Pr[q] = 0$, we shall normally agree that $\Pr[p \mid q] = 0$. (Alternatively, we might leave $\Pr[p \mid q]$ undefined if $\Pr[q] = 0$. Such a convention would be adopted in a more general context.)

The case $\Pr[q] = 0$ is not very interesting. Suppose $\Pr[q] \neq 0$, and let $Q \subset \Omega$ be the truth set of q. If Q is taken as a space of points, then $\mathscr{B}' = \{B' \mid B' = Q \cap B, B \in \mathscr{B}\}$ is a Borel field of sets in Q. For any set B' in $\mathscr{B}'$ we define a set function ν by

$$\nu(B') = \frac{\mu(B')}{\mu(Q)}.$$

The reader should verify that ν is a measure on $\mathscr{B}'$. Furthermore, $\nu(Q) = \mu(Q)/\mu(Q) = 1$. Therefore, $(Q, \mathscr{B}', \nu)$ is a probability space. We may thus speak of the probabilities of statements relative to Q, and we see that their values coincide with conditional probabilities given q and relative to Ω. That is, conditional probabilities possess the same properties as ordinary probabilities.

We shall apply this notion of conditional probability to the sequence space considered in the preceding sections. In this space the sets of $\mathscr{B}_n$ are denumerable disjoint unions of sets of the form

$$B_n = \{\omega \mid x_0 \in S_0 \wedge \cdots \wedge x_n \in S_n\}.$$

Since the state space S is denumerable, each of the sets $S_0, S_1, \ldots, S_n$ is denumerable and B_n is the denumerable union of disjoint sets of the form $\{\omega \mid x_0 = c_0 \wedge \cdots \wedge x_n = c_n\}$. By definition of conditional probability,

$$\begin{aligned}\Pr[x_0 = c_0 \wedge \cdots \wedge x_{n-1} = c_{n-1} \wedge x_n = c_n] \\ = \Pr[x_n = c_n \mid x_0 = c_0 \wedge \cdots \wedge x_{n-1} = c_{n-1}] \\ \times \Pr[x_0 = c_0 \wedge \cdots \wedge x_{n-1} = c_{n-1}].\end{aligned}$$

By induction, we find

$$\begin{aligned}\Pr[x_0 = c_0 \wedge \cdots \wedge x_n = c_n] = \Pr[x_0 = c_0]\cdot\Pr[x_1 = c_1 \mid x_0 = c_0] \\ \times \Pr[x_2 = c_2 \mid x_0 = c_0 \wedge x_1 = c_1] \\ \times \Pr[x_3 = c_3 \mid x_0 = c_0 \wedge x_1 = c_1 \wedge x_2 = c_2] \\ \times \cdots \times \Pr[x_n = c_n \mid x_0 = c_0 \wedge \cdots \wedge x_{n-1} = c_{n-1}].\end{aligned}$$

We have established the following result.

Proposition 2-8: The measure on a sequence space is completely determined by

(1) the starting probabilities, $\Pr[x_0 = c_0]$, and
(2) the transition probabilities,

$$\Pr[x_n = c_n \mid x_0 = c_0 \wedge \cdots \wedge x_{n-1} = c_{n-1}].$$

Conditional probabilities, as we anticipated, have their place in tree diagrams. To each branch we may assign a conditional probability. We see now the abstract formulation of the fact that the probabilities of statements like $x_0 = c_0 \wedge x_1 = c_1 \wedge \cdots \wedge x_n = c_n$ are computed simply by multiplying together the appropriate branch weights.

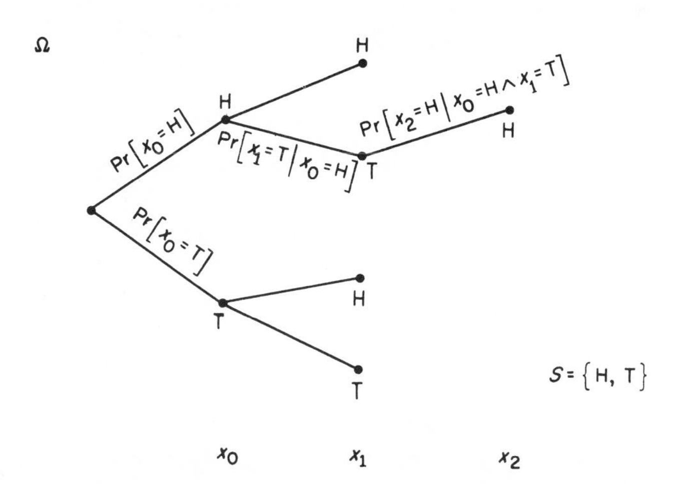

Two statements p and q are said to be **probabilistically independent** if $\Pr[p \wedge q] = \Pr[p]\Pr[q]$. A stochastic process defined on sequence space is called an **independent process** if the statements

$$x_n = c_n$$

and

$$x_0 = c_0 \wedge \cdots \wedge x_{n-1} = c_{n-1}$$

are independent for all $n \geq 1$ and for all $c_0, c_1, \ldots, c_n$. Coin tossing is an example. We shall see that an independent process is a special kind of Markov process. If, in addition, for each c the probability that $x_n = c$ does not depend on n, then the process is called an **independent trials process.**

6. Random variables and means

Let $(\Omega, \mathscr{B}, \mu)$ be a probability space. A measurable function $\mathbf{f}$ with domain Ω and range in the extended real-valued number system is called a **random variable**. We may apply all the properties of the measurable functions in Section 3 of Chapter 1.

If $\mathbf{f}(\omega)$ is an extended-real-valued function defined on the space Ω of a stochastic process and if Ω has the property that for some n, $\mathbf{f}(\omega)$ is measurable with respect to the Borel field $\mathscr{B}_n$, then $\mathbf{f}$ is a random variable because $\mathscr{B}_n \subset \mathscr{B}$. Such a function is said to be $\mathscr{B}_n$-measurable. For the special case in which Ω is sequence space, a function $\mathbf{f}$ is $\mathscr{B}_n$-measurable for some n if its values depend only on a bounded number of outcomes.

In terms of sequence space, we give two examples of random variables.

(1) Define a function $\mathbf{u}_j^{(n)}$ by

$$\mathbf{u}_j^{(n)} = \begin{cases} 1 & \text{if } x_n = j \\ 0 & \text{otherwise.} \end{cases}$$

Since the value of $\mathbf{u}_j^{(n)}$ depends only on outcome n, the function $\mathbf{u}_j^{(n)}$ is $\mathscr{B}_n$-measurable and is therefore a random variable. Letting $\mathbf{n}_j = \sum_{n=0}^{\infty} \mathbf{u}_j^{(n)}$ and noting that the limit of a sequence of random variables is a random variable, we see that $\mathbf{n}_j$ is a random variable. The function $\mathbf{n}_j(\omega)$ counts the number of times that the outcome j occurs on the path ω; it will appear again after we introduce Markov chains.

(2) Let $\mathbf{t}_j$ be the time to reach j. That is, define $\mathbf{t}_j(\omega)$ to be the first time n along path ω such that $x_n(\omega) = j$. If j is never reached, set $\mathbf{t}_j = +\infty$. Then $\mathbf{t}_j$ is a random variable because it is the limit on n of the function $\mathbf{t}_j^{(n)} = \inf(\mathbf{t}_j, n)$, which is $\mathscr{B}_n$-measurable.

Definition 2-9: A **random time t** is a random variable satisfying these two properties:

(1) Its range is in the non-negative integers with $\{+\infty\}$ adjoined.
(2) For each integer n the set $\{\omega \mid \mathbf{t}(\omega) = n\}$ is a set of $\mathscr{B}_n$.

The random variable $\mathbf{t}_j$ defined in the second example above is an example of a random time.

The **mean** of a random variable $\mathbf{f}$, denoted $\mathrm{M}[\mathbf{f}]$, is defined by $\mathrm{M}[\mathbf{f}] = \int_\Omega \mathbf{f}d\mu$, where $\mathrm{M}[\mathbf{f}]$ exists if and only if $\int_\Omega \mathbf{f}d\mu$ exists and where μ is the measure associated with the probability space. Since means are Lebesgue integrals, they satisfy all the properties of Lebesgue integrals. In particular, if $\{\mathbf{f}_n\}$ is a sequence of non-negative random variables, then by Corollary 1-46,

$$\mathrm{M}\left[\sum_{n=0}^{\infty} \mathbf{f}_n\right] = \sum_{n=0}^{\infty} \mathrm{M}[\mathbf{f}_n].$$

In addition, if $\mathbf{g}_n$ is a sequence of random variables with the properties that $|\mathbf{g}_n| \leq c$ for every n and that $\mathbf{g}_n$ converges to $\mathbf{g}$, then

$$\mathrm{M}[\mathbf{g}] = \lim_{n \to \infty} \mathrm{M}[\mathbf{g}_n]$$

by Corollary 1-50. An important application of these facts is the following result.

Proposition 2-10: In a sequence space,

$$\mathrm{M}[\mathbf{n}_j] = \sum_{n=0}^{\infty} \Pr[x_n = j].$$

PROOF:

$$\begin{aligned} \mathrm{M}[\mathbf{n}_j] &= \mathrm{M}\left[\sum_n \mathbf{u}_j^{(n)}\right] \\ &= \sum_n \mathrm{M}[\mathbf{u}_j^{(n)}] \\ &= \sum_n \int_\Omega \mathbf{u}_j^{(n)} d\mu \\ &= \sum_n \int_{\{\omega \mid x_n(\omega) = j\}} 1 d\mu \\ &= \sum_n \Pr[x_n = j]. \end{aligned}$$

7. Means conditional on statements

Suppose $(\Omega, \mathscr{B}, \mu)$ is a probability space on which a denumerable stochastic process is defined. We say that a denumerable family of subsets $\mathscr{R} = \{R_1, R_2, \ldots\}$ is a **partition** of Ω if the sets R_j are disjoint, exhaustive, and measurable. Each such subset R_j is called a **cell** of the partition; we allow the possibility that a cell may have measure zero. The reader should notice that the sets $\{\varnothing, R_1, R_2, \ldots\}$ together with all possible denumerable unions form a Borel field which we shall call $\mathscr{R}^*$. Since the sets in $\mathscr{R}$ are measurable, we see that $\mathscr{R}^* \subset \mathscr{B}$.

Two examples of partitions are typical.

(1) Let the process be coin tossing, and define a partition by

$$R_1 = \{\omega \mid x_0(\omega) \text{ is a head}\}, \qquad R_2 = \{\omega \mid x_0(\omega) \text{ is a tail}\}.$$

More generally, let $\mathscr{R}$ consist of disjoint exhaustive measurable sets which are in $\mathscr{B}_n$ for some fixed n.

(2) Suppose $\mathbf{f}$ is a random variable whose range is denumerable. If the range is $\{a_j\}$, define $R_j = \{\omega \mid \mathbf{f}(\omega) = a_j\}$. Then $\{R_j\}$ is a partition.

A denumerable set of statements $\{q_i\}$ about a stochastic process is said to be a **set of alternatives** if the truth sets Q_i of the statements form a partition. Since the integral is a completely additive set function, we obtain, for every random variable $\mathbf{f}$ whose mean exists, the relation

$$\mathrm{M}[\mathbf{f}] = \sum_i \int_{Q_i} \mathbf{f}\,d\mu.$$

Let p be a statement with measurable truth set P. If $\mathbf{f}$ is a random variable and if $\Pr[p] \neq 0$, then the **conditional mean** of $\mathbf{f}$ given p, written $\mathrm{M}[\mathbf{f} \mid p]$, is defined by

$$\mathrm{M}[\mathbf{f} \mid p] = \frac{1}{\Pr[p]} \int_P \mathbf{f}\,d\mu.$$

If $\Pr[p] = 0$, then $\mathrm{M}[\mathbf{f} \mid p]$ is defined to be zero. (In a more general setting, $\mathrm{M}[\mathbf{f} \mid p]$ is not defined when $\Pr[p] = 0$.)

Proposition 2-11: If $\mathrm{M}[\mathbf{f}]$ exists and $\{q_i\}$ is a set of alternatives with truth sets $\{Q_i\}$, then

$$\mathrm{M}[\mathbf{f}] = \sum_i \Pr[q_i] \cdot \mathrm{M}[\mathbf{f} \mid q_i].$$

PROOF: By definition of conditional mean, we have

$$\int_{Q_i} \mathbf{f}\,d\mu = \Pr[q_i] \cdot \mathrm{M}[\mathbf{f} \mid q_i].$$

Summing both sides of the equation on i gives the result immediately.

Corollary 2-12: Let p be a statement with measurable truth set P, and let $\{q_i\}$ be a set of alternatives with truth sets Q_i. Then

$$\Pr[p] = \sum_i \Pr[q_i] \cdot \Pr[p \mid q_i] = \sum_i \Pr[p \wedge q_i].$$

PROOF: Let $\mathbf{f}$ be the characteristic function of the set P and apply Proposition 2-11.

8. Problems

1. For coin tossing, show that
   (a) the probability of getting only finitely many "heads" is 0;
   (b) there will be infinitely many "heads" a.e.
2. Consider the experiment of selecting "1" with probability $\frac{1}{2}$, "2" with probability $\frac{1}{3}$, or "3" with probability $\frac{1}{6}$. If this experiment is repeated infinitely often, show that the probability of selecting "1" only finitely often is 0.
3. In Problem 2, let $\mathscr{B}_n = \mathscr{F}_n$. Which of the following f_n form a denumerable stochastic process $(f_n, \mathscr{B}_n)$?
   (a) f_n = the nth outcome.
   (b) f_n = time at which nth "3" is selected.
   (c) $f_n = s_n$ = sum of the first n numbers selected.
   (d) $f_n = \dfrac{s_n - a}{\sqrt{cn}}$, where a and c are constants.
4. Show that the following converse of the Borel–Cantelli Lemma is false: If $\sum \Pr[p_n] = +\infty$, then the probability that only finitely many p_n are true is less than 1.
5. Start with 4 jacks and 4 queens from a bridge deck. Find the probability of drawing two cards, both of which are jacks. Then compute the conditional probability of the same on each of the following conditions:
   (a) One card drawn is a jack.
   (b) One card drawn is a red jack.
   (c) One card drawn is the jack of hearts.
   (d) The first card drawn is the jack of hearts.
6. For coin tossing, define three random variables which are not measurable with respect to any finite tree Ω_n.
7. Let ${}^i\mathbf{n}_j(\omega)$ be the number of times on path ω that a j occurs before the first i occurs (if $i = j$, take ${}^i\mathbf{n}_j(\omega) = 0$). Prove that ${}^i\mathbf{n}_j$ is a random variable, and develop an infinite series representation for $\mathbf{M}[\mathbf{t}_i]$ in terms of it.
8. In coin tossing, let $\mathbf{t}$ be the first time that "heads" comes up.
   (a) Is $\mathbf{t}$ a random time?
   (b) Find $\mathbf{M}[\mathbf{t}]$.
   (c) Find $\mathbf{M}[\mathbf{t} \mid \text{first outcome is "tails"}]$.

9. In a randomly selected two-child family, let $\mathbf{f}$ = the number of boys. Find
   (a) $M[\mathbf{f}]$.
   (b) $M[\mathbf{f} \mid \text{first child is a boy}]$.
   (c) $M[\mathbf{f} \mid \text{there is at least one boy}]$.

10. For coin tossing, let $\mathbf{f}$ = the number of "heads" in the first three tosses. Let q_i be the statement that there were exactly i "heads" in the first two tosses ($i = 0, 1, 2$). Find $M[\mathbf{f}]$, $M[\mathbf{f} \mid q_0]$, $M[\mathbf{f} \mid q_1]$, and $M[\mathbf{f} \mid q_2]$. Verify that the first of these is a linear combination of the last three with appropriate probabilities as weights.

11. Let $\{q_i\}$ be a set of alternatives and let $\mathbf{f}$ be a non-negative random variable. Prove that if $M[\mathbf{f} \mid q_1] > M[\mathbf{f}]$, then there is an alternative q_j such that $M[\mathbf{f} \mid q_j] < M[\mathbf{f}]$.

12. Let X be the unit interval. Let $\mathscr{F}$ consist of all finite unions of finite intervals (with or without endpoints). We classify a point as an interval of length 0. The measure to be constructed in this problem is called **Lebesgue measure**, and the Borel field is the class of all **Borel sets**.
    (a) Show that $\mathscr{F}$ is a field.
    (b) Show that every set A in $\mathscr{F}$ can be written as a finite *disjoint* union of intervals $A_1, A_2, \ldots, A_n$.
    (c) If A is decomposed as in part (b), we define

$$\nu(A) = \sum_{k=1}^{n} l(A_k),$$

    where l denotes length. Show that ν is consistently defined and that ν is a non-negative additive set function.
    (d) Show that if A is a finite union of intervals and if $\epsilon > 0$ is given, then there is a finite union K of closed intervals and a finite union G of open intervals such that $K \subset A$, $A \subset G$, and

$$\nu(K) + \epsilon \geq \nu(A) \geq \nu(G) - \epsilon.$$

    [*Note:* An open interval here means a set which is the intersection of X with an open interval of the real line.]
    (e) Let A and A_n, $n = 1, 2, \ldots$, be finite unions of intervals with the A_n disjoint and with $\cup A_n = A$. Use parts (c) and (d) and the Heine–Borel Theorem to prove that, for any $\epsilon > 0$,

$$\nu(A) \leq \sum \nu(A_n) + \epsilon.$$

    (f) Deduce from part (e) that ν is completely additive on $\mathscr{F}$.
    (g) Apply Theorem 1-19 and describe the resulting measure space.
    (h) Using complete additivity, prove that a denumerable set of points has measure 0.
    (i) Why does the proof of (h) not show that every set has measure 0?
    (j) Show directly from the definition

$$\mu(B) = \inf \left[\sum \nu(A_n)\right]$$

    that every denumerable set of points has measure 0.

13. Let $(\Omega, \mathscr{B}, \mu)$ be the unit interval with Lebesgue measure. If $x \in \Omega$, let $f(x) = x^2$. Let p be the statement "$x \leq \frac{1}{2}$."
    (a) Show that f is a random variable.
    (b) Find $\mathrm{M}[f]$.
    (c) Find $\mathrm{M}[f \mid p]$ and $\mathrm{M}[f \mid \sim p]$.
    (d) Relate your answer in (b) to the solution of (c).

CHAPTER 3

MARTINGALES

1. Means conditional on partitions and functions

In this chapter we consider a natural abstraction of the idea of a fair game in gambling. We shall give several applications of the basic result, the Martingale Convergence Theorem.

We begin by defining what we mean by the conditional mean of **f** given a partition $\mathscr{R}$ of the domain. Let $(\Omega, \mathscr{B}, \mu)$ be a measure space. We shall normally assume $\mu(\Omega) \leq 1$, but such an assumption is not necessary as long as μ is finite.

Definition 3-1: Let $\omega \in \Omega$ and let r_j be the statement that ω is in a cell R_j of $\mathscr{R}$. If **f** is a random variable, then the **conditional mean of f given $\mathscr{R}$**, written $\mathrm{M}[\mathbf{f} \mid \mathscr{R}]$, is defined to be a function of ω whose value at every point in the celi R_j is the constant $\mathrm{M}[\mathbf{f} \mid r_j]$.

Next, we observe that if $\mathrm{M}[\mathbf{f}]$ exists and is finite, then $\mathrm{M}[\mathbf{f} \mid \mathscr{R}]$ exists and is finite. For, on cells of measure zero, the conditional mean is defined to be zero. If $\mu(R_j) > 0$, then

$$\mathrm{M}[|\mathbf{f}| \mid \mathscr{R}] = \mathrm{M}[|\mathbf{f}| \mid R_j] = \frac{1}{\mu(R_j)} \int_{R_j} |\mathbf{f}| d\mu \leq \frac{1}{\mu(R_j)} \mathrm{M}[|\mathbf{f}|] < \infty.$$

The next proposition provides an equivalent definition of the mean of **f** given $\mathscr{R}$.

Proposition 3-2: $\mathrm{M}[\mathbf{f} \mid \mathscr{R}]$ is characterized almost everywhere by these two properties:

(1) $\mathrm{M}[\mathbf{f} \mid \mathscr{R}]$ is constant on each cell of $\mathscr{R}$.

(2) $\int_{R_j} \mathbf{f} d\mu = \int_{R_j} \mathrm{M}[\mathbf{f} \mid \mathscr{R}] d\mu$.

PROOF: We must show that $\mathrm{M}[\mathbf{f} \mid \mathscr{R}]$ has these two properties and that any random variable satisfying these properties is equal to the conditional mean of **f** given $\mathscr{R}$ a.e. Set $\mathbf{g} = \mathrm{M}[\mathbf{f} \mid \mathscr{R}]$, and suppose first that **g** is known to be a conditional mean. Then **g** satisfies (1) by definition. If $\mu(R_j) = 0$, both sides of (2), are 0. If not, equality again follows from the definition of $\mathrm{M}[\mathbf{f} \mid \mathscr{R}]$.

Conversely, if a function **g** satisfies (1) and (2), then it agrees with $\mathrm{M}[\mathbf{f} \mid \mathscr{R}]$ on all paths in cells of positive measure; that is, it agrees with $\mathrm{M}[\mathbf{f} \mid \mathscr{R}]$ a.e.

We give two examples of conditional means.

(1) Let the stochastic process be coin tossing and let

$$R_1 = \{\omega \mid x_0(\omega) \text{ is a head}\}$$
$$R_2 = \{\omega \mid x_0(\omega) \text{ is a tail}\}.$$

Let **f** be the total number of heads on the zeroth and first tosses. Then

$$\mathrm{M}[\mathbf{f} \mid \mathscr{R}] = \begin{cases} \frac{3}{2} & \text{on } R_1 \\ \frac{1}{2} & \text{on } R_2 \end{cases}$$

and

$$\mathrm{M}[\mathbf{f}] = (\tfrac{1}{2})(\tfrac{3}{2}) + (\tfrac{1}{2})(\tfrac{1}{2}) = 1.$$

(2) For any stochastic process and for any denumerable-valued random variable **f**, let $\mathscr{R}$ be the trivial partition $\{\Omega\}$. Then

$$\mathrm{M}[\mathbf{f} \mid \mathscr{R}] = \mathrm{M}[\mathbf{f}]$$

on every path ω.

A partition $\mathscr{R}$ is said to be **contained** in $\mathscr{S}$, written $\mathscr{R} \subset \mathscr{S}$, if every cell of $\mathscr{R}$ is a union of cells of $\mathscr{S}$. If $\mathscr{R} \subset \mathscr{S}$, then $\mathscr{R}$ is the "coarser" subdivision, and $\mathscr{S}$ is called a **refinement** of $\mathscr{R}$.

Proposition 3-3: Conditional means satisfy these properties:

(1) $\mathrm{M}[\mathrm{M}[\mathbf{f} \mid \mathscr{S}] \mid \mathscr{R}] = \mathrm{M}[\mathbf{f} \mid \mathscr{R}]$ if $\mathscr{R} \subset \mathscr{S}$.
(2) $\mathrm{M}[\mathrm{M}[\mathbf{f} \mid \mathscr{S}]] = \mathrm{M}[\mathbf{f}]$ for any $\mathscr{S}$.
(3) If **g** is constant on each cell of $\mathscr{R}$, then $\mathrm{M}[\mathbf{g} \mid \mathscr{R}] = \mathbf{g}$ a.e., and if $\mathrm{M}[\mathbf{fg}]$ exists and is finite, then $\mathrm{M}[\mathbf{fg} \mid \mathscr{R}] = \mathbf{g}\mathrm{M}[\mathbf{f} \mid \mathscr{R}]$.
(4) If **f** and **g** assume only finite range values and if $\mathrm{M}[\mathbf{f} \mid \mathscr{R}]$ and $\mathrm{M}[\mathbf{g} \mid \mathscr{R}]$ both exist and are finite, then $\mathrm{M}[\mathbf{f} + \mathbf{g} \mid \mathscr{R}]$ exists, is finite, and is given by

$$\mathrm{M}[\mathbf{f} + \mathbf{g} \mid \mathscr{R}] = \mathrm{M}[\mathbf{f} \mid \mathscr{R}] + \mathrm{M}[\mathbf{g} \mid \mathscr{R}].$$

PROOF: For (1) it is sufficient to show that

$$\int_{R_j} \mathrm{M}[\mathrm{M}[\mathbf{f} \mid \mathscr{S}] \mid \mathscr{R}]\,d\mu = \int_{R_j} \mathrm{M}[\mathbf{f} \mid \mathscr{R}]\,d\mu$$

since both functions are constant on cells of $\mathscr{R}$. Applying property (2) of Proposition 3-2 three times, we have

$$\begin{aligned}\int_{R_j} \mathrm{M}[\mathrm{M}[\mathbf{f} \mid \mathscr{S}] \mid \mathscr{R}]d\mu &= \int_{R_j} \mathrm{M}[\mathbf{f} \mid \mathscr{S}]d\mu \\ &= \sum_i \int_{S_i \subset R_j} \mathrm{M}[\mathbf{f} \mid \mathscr{S}]d\mu \\ &= \sum_i \int_{S_i \subset R_j} \mathbf{f}d\mu \\ &= \int_{R_j} \mathbf{f}d\mu \\ &= \int_{R_j} \mathrm{M}[\mathbf{f} \mid \mathscr{R}]d\mu.\end{aligned}$$

The proofs of (3) and (4) use the same technique and are left to the reader. Property (2) follows from (1) with $\mathscr{R}$ taken as the trivial partition $\{\Omega\}$.

Definition 3-4: The **cross partition** $\mathscr{R} \otimes \mathscr{S}$ of two partitions $\mathscr{R}$ and $\mathscr{S}$ is the family of sets defined by

$$\mathscr{R} \otimes \mathscr{S} = \{R_i \cap S_j \mid R_i \in \mathscr{R}, S_j \in \mathscr{S}, R_i \cap S_j \neq \varnothing\}.$$

For example,

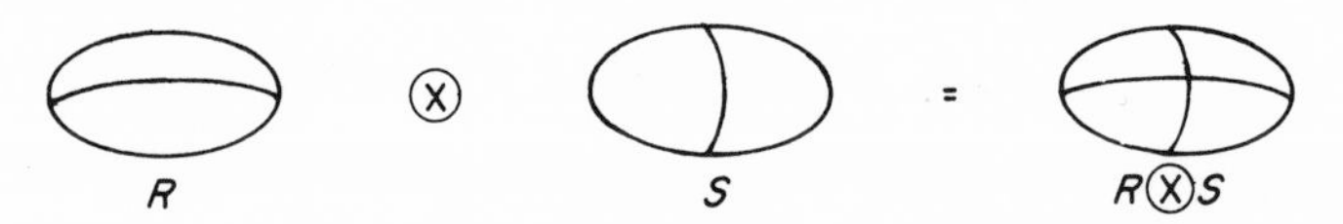

Since the intersection of measurable sets is measurable and since the sets $\{R_i \cap S_j\}$ are disjoint and exhaustive, a cross partition is a partition.

In Example 2 of Section 2-7 we noted that every denumerable-valued random variable determines a natural partition of Ω. We call the partition associated with the denumerable-valued random variable $\mathbf{g}$, $\mathscr{R}^g$. In terms of the natural partition induced by $\mathbf{g}$, we define the **conditional mean of f given g**, by

$$\mathrm{M}[\mathbf{f} \mid \mathbf{g}] = \mathrm{M}[\mathbf{f} \mid \mathscr{R}^g].$$

Then $\mathrm{M}[\mathbf{f} \mid \mathbf{g}]$ is constant on sets where $\mathbf{g}$ is constant, and on the set where $\mathbf{g} = c$ for some constant c, $\mathrm{M}[\mathbf{f} \mid \mathbf{g}]$ has the value $\mathrm{M}[\mathbf{f} \mid \mathbf{g} = c]$.

Observing that the operation of forming the cross partition of two partitions is both associative and commutative, we define more generally

$$\mathrm{M}[\mathbf{f}_{n+1} \mid \mathbf{f}_0 \wedge \cdots \wedge \mathbf{f}_n] = \mathrm{M}[\mathbf{f}_{n+1} \mid \mathscr{R}^{f_0} \otimes \cdots \otimes \mathscr{R}^{f_n}].$$

If p is a statement with truth set P, we define

$$\Pr[p \mid \mathbf{f}_0 \wedge \cdots \wedge \mathbf{f}_n] = \mathrm{M}[\chi_P \mid \mathbf{f}_0 \wedge \cdots \wedge \mathbf{f}_n],$$

where χ_P is the characteristic function of the set P.

A sequence of random variables $\{\mathbf{f}_n\}$ is said to be **independent** if, for every n and for every A,

$$\Pr[\mathbf{f}_{n+1} \in A \mid \mathbf{f}_0 \wedge \cdots \wedge \mathbf{f}_n] = \Pr[\mathbf{f}_{n+1} \in A]$$

almost everywhere. The reader should show that if $\{\mathbf{f}_n\}$ is a sequence of independent random variables, then

$$\mathrm{M}[\mathbf{f}_{n+1} \mid \mathbf{f}_0 \wedge \cdots \wedge \mathbf{f}_n] = \mathrm{M}[\mathbf{f}_{n+1}].$$

In the special case in which, for each A, $\Pr[\mathbf{f}_n \in A]$ does not depend on n, the random variables are said to be **identically distributed.**

2. Properties of martingales

With the background of Section 3-1, we proceed to define martingales and to give several examples of them. We still work with the probability space $(\Omega, \mathscr{B}, \mu)$ and a denumerable set of states S. We assume, however, that the set S is a subset of the extended real number system.

Definition 3-5: Let $\{f_n\}$ be a sequence of denumerable-valued random variables, and suppose $\mathscr{R}_0 \subset \mathscr{R}_1 \subset \cdots$ is an increasing sequence of partitions of Ω. The pair $(\mathbf{f}_n, \mathscr{R}_n)$ is called a **martingale** if three conditions are satisfied:

(1) $\mathrm{M}[|\mathbf{f}_n|]$ is finite for each n.
(2) $\mathbf{f}_n$ is constant on cells of $\mathscr{R}_n$.
(3) $\mathbf{f}_n = \mathrm{M}[\mathbf{f}_{n+1} \mid \mathscr{R}_n]$.

If (1) and (2) are satisfied and (3) is replaced by $\mathbf{f}_n \geq \mathrm{M}[\mathbf{f}_{n+1} \mid \mathscr{R}_n]$, then $(\mathbf{f}_n, \mathscr{R}_n)$ is a **supermartingale**. If (1) and (2) are satisfied and (3) is replaced by $\mathbf{f}_n \leq \mathrm{M}[\mathbf{f}_{n+1} \mid \mathscr{R}_n]$, then $(\mathbf{f}_n, \mathscr{R}_n)$ is a **submartingale**.

For a martingale, condition (3) in Definition 3-5 implies condition (2), but for a supermartingale or a submartingale it does not.

When we defined partitions in Section 2-7, we noted that every partition $\mathscr{R}_n$ determines a Borel field $\mathscr{R}_n{}^*$ and that the Borel field satisfied $\mathscr{R}_n{}^* \subset \mathscr{B}$. Since $\mathscr{R}_n \subset \mathscr{R}_{n+1}$ implies $\mathscr{R}_n{}^* \subset \mathscr{R}_{n+1}{}^*$, we see that a martingale is a stochastic process. The reader should notice that the condition that $\mathbf{f}_n$ is constant on cells of $\mathscr{R}_n$ is equivalent to the condition that $\mathbf{f}_n$ is measurable with respect to $\mathscr{R}_n{}^*$.

Throughout the discussion of martingales, it is convenient to keep in mind the idea of a fair game, which we shall introduce as our first example below. We shall see that the fair game is a special case of the following situation. Let $\mathbf{f}_0, \mathbf{f}_1, \ldots$ be a sequence of denumerable-valued random variables defined on a probability space such that $\mathrm{M}[|\mathbf{f}_j|]$ is finite. Setting $\mathscr{R}_n = \mathscr{R}^{f_0} \otimes \cdots \otimes \mathscr{R}^{f_n}$, we see that the sequence $\{\mathscr{R}_n\}$ is clearly increasing. Therefore, only condition (3) of Definition 3-5 need be satisfied for $(\mathbf{f}_n, \mathscr{R}_n)$ to be a martingale. In particular, we see that such a sequence of random variables forms a martingale if and only if

$$\mathrm{M}[\mathbf{f}_{n+1} \mid \mathbf{f}_0 \wedge \mathbf{f}_1 \wedge \cdots \wedge \mathbf{f}_n] = \mathbf{f}_n.$$

When the partitions $\mathscr{R}_n$ are obtained from the $\mathscr{R}^{f_n}$ in the way we have just described, we agree to refer to the pair $(\mathbf{f}_n, \mathscr{R}_n)$ simply as $\{\mathbf{f}_n\}$.

We shall give three examples of martingales at this time. More examples will appear after we introduce Markov chains.

(1) Let $\{\mathbf{y}_n\}$ be a sequence of independent random variables with denumerable range and let $\mathbf{s}_n = \mathbf{y}_0 + \cdots + \mathbf{y}_n$ represent the nth partial sum. Then $\{\mathbf{s}_n\}$ with its partition obtained from the $\mathscr{R}^{s_n}$ in the natural way is a martingale if and only if $\mathrm{M}[\mathbf{y}_k] = 0$ for every k. We have

$$\begin{aligned}
\mathrm{M}[\mathbf{s}_{n+1} \mid \mathbf{s}_0 \wedge \cdots \wedge \mathbf{s}_n] &= \mathrm{M}[\mathbf{y}_{n+1} + \mathbf{s}_n \mid \mathbf{s}_0 \wedge \cdots \wedge \mathbf{s}_n] \\
&= \mathrm{M}[\mathbf{y}_{n+1} \mid \mathbf{s}_0 \wedge \cdots \wedge \mathbf{s}_n] \\
&\quad + \mathrm{M}[\mathbf{s}_n \mid \mathbf{s}_0 \wedge \cdots \wedge \mathbf{s}_n] \\
&= \mathrm{M}[\mathbf{y}_{n+1} \mid \mathbf{s}_0 \wedge \cdots \wedge \mathbf{s}_n] + \mathbf{s}_n \\
&= \mathrm{M}[\mathbf{y}_{n+1} \mid \mathbf{y}_0 \wedge \cdots \wedge \mathbf{y}_n] + \mathbf{s}_n \\
&= \mathrm{M}[\mathbf{y}_{n+1}] + \mathbf{s}_n \quad \text{by independence} \\
&= \mathbf{s}_n \quad \text{if and only if } \mathrm{M}[\mathbf{y}_{n+1}] = 0.
\end{aligned}$$

A special case in which the $\mathbf{y}_k$ have identical distributions is the sequence of plays of a game of chance. A fair game is one in which the expected fortune $\mathrm{M}[\mathbf{s}_{n+1}]$ at any time $n + 1$ is equal to the actual fortune $\mathbf{s}_n$ at time n. Matching pennies is a fair game, whereas roulette is unfair. A game like roulette that is favorable to the house is a supermartingale. From the calculation above, we see that a game is fair if and only if the mean amount won in each round is zero.

(2) A particle moves on a line stopping at points whose coordinates are integers. At each step it moves n units to the right with probability $\{p_n\}$, where $n = \ldots, -2, -1, 0, 1, 2, \ldots$, only finitely many of the p_n are different from zero, $p_n \neq 0$ for some negative value and some positive value of n, and $\sum p_n = 1$. The particle's position after j steps is x_j. Set $f(s) = \sum_k p_k s^k$, and consider the positive roots of the equation $f(s) = 1$. Since $f''(s) > 0$ for $s > 0$, there are at most two roots of the equation, and since $f(1) = 1$, either one or two roots exist.

As $s \to 0$ or $\infty, f(s) \to \infty$. Hence either 1 is a minimum point or there are two positive roots. If 1 is a minimum point, then $f'(1) = 0$, and hence $\{x_j\}$ is a martingale, and if r is a root other than 1, then $\{r^{x_j}\}$ is a martingale. The details of verifying these assertions are left to the reader. We shall need to use these results later.

(3) Let Ω be the closed unit interval $[0, 1]$ on the real line, let the measure of an interval be its length, and let $\mathscr{R}_n$ be the partition $\{[0, 2^{-n}], (2^{-n}, 2\cdot 2^{-n}], \ldots, ((2^n - 1)2^{-n}, 1]\}$. Let f be a monotone increasing function on $[0, 1]$ and let $\mathbf{f}_n$ be a function which is constant on the interval $(c2^{-n}, (c + 1)2^{-n}]$ and whose value at any point in the interval is

$$2^n(f((c + 1)2^{-n}) - f(c2^{-n})).$$

Thus $\mathbf{f}_n$ is an approximation to the derivative f', if it exists. The reader should verify that $(\mathbf{f}_n, \mathscr{R}_n)$ is a martingale.

3. A first martingale systems theorem

In the first example in the preceding section, we saw that martingales bear some relation to gambling. A fair game is a martingale, a game favorable to the house is a supermartingale, and a game favorable to the player is a submartingale. A gambling system is a device to take advantage of the nature of a game of chance in order to increase the player's expected fortune. Systems theorems are theorems which prove that certain classes of gambling systems do not work. For our first systems theorem, which we shall need in the proof in the next section, we require a lemma.

Lemma 3-6: Let $\{\mathscr{R}_n\}$ be an increasing sequence of partitions and let $\{\mathbf{f}_n\}$ be a sequence of random variables. Suppose R_k is a set in the Borel field $\mathscr{R}_k{}^*$.

If $(\mathbf{f}_n, \mathscr{R}_n)$ is a submartingale, then $\int_{R_k} \mathbf{f}_{k+1} d\mu \geq \int_{R_k} \mathbf{f}_k d\mu$.
If $(\mathbf{f}_n, \mathscr{R}_n)$ is a martingale, then $\int_{R_k} \mathbf{f}_{k+1} d\mu = \int_{R_k} \mathbf{f}_k d\mu$.
If $(\mathbf{f}_n, \mathscr{R}_n)$ is a supermartingale, then $\int_{R_k} \mathbf{f}_{k+1} d\mu \leq \int_{R_k} \mathbf{f}_k d\mu$.

PROOF: We shall prove only the first assertion. By Proposition 3-2, $\int_{R_k} \mathbf{f}_{k+1} d\mu = \int_{R_k} \mathrm{M}[\mathbf{f}_{k+1} \mid \mathscr{R}_k] d\mu$, and since $\{\mathbf{f}_n, \mathscr{R}_n\}$ is a submartingale, we know that

$$\mathrm{M}[\mathbf{f}_{k+1} \mid \mathscr{R}_k] \geq \mathbf{f}_k.$$

The result follows immediately from integrating this inequality.

Proposition 3-7: Let $(\mathbf{f}_n, \mathscr{R}_n)$ be a submartingale, and suppose that $\{\boldsymbol{\epsilon}_n\}$ is a sequence of random variables such that $\boldsymbol{\epsilon}_n(\omega) = 1$ or 0 for every

ω and such that the set $\{\omega \mid \epsilon_n(\omega) = 1\}$ is a set in $\mathscr{R}_n^*$, the Borel field generated by $\mathscr{R}_n$. Define

$$\hat{\mathbf{f}}_n = \mathbf{f}_0 + \epsilon_0(\mathbf{f}_1 - \mathbf{f}_0) + \epsilon_1(\mathbf{f}_2 - \mathbf{f}_1) + \cdots + \epsilon_{n-1}(\mathbf{f}_n - \mathbf{f}_{n-1}).$$

Then $(\hat{\mathbf{f}}_n, \mathscr{R}_n)$ is a submartingale and $\mathrm{M}[\hat{\mathbf{f}}_n] \leq \mathrm{M}[\mathbf{f}_n]$.

REMARK: Analogous results hold for martingales and for supermartingales, but we need only what is proved here.

PROOF OF PROPOSITION: We first show that $\mathrm{M}[\hat{\mathbf{f}}_{n+1} \mid \mathscr{R}_n] \geq \hat{\mathbf{f}}_n$. We have

$$\begin{aligned} \mathrm{M}[\hat{\mathbf{f}}_{n+1} \mid \mathscr{R}_n] &= \mathrm{M}[\hat{\mathbf{f}}_n + \epsilon_n(\mathbf{f}_{n+1} - \mathbf{f}_n) \mid \mathscr{R}_n] \\ &= \mathrm{M}[\hat{\mathbf{f}}_n \mid \mathscr{R}_n] + \mathrm{M}[\epsilon_n(\mathbf{f}_{n+1} - \mathbf{f}_n) \mid \mathscr{R}_n] \\ &= \hat{\mathbf{f}}_n + \mathrm{M}[\epsilon_n(\mathbf{f}_{n+1} - \mathbf{f}_n) \mid \mathscr{R}_n]. \end{aligned}$$

Since $\{\omega \mid \epsilon_n(\omega) = 1\}$ is the union of cells of $\mathscr{R}_n$, ϵ_n is constant on cells of $\mathscr{R}_n$. Thus by Proposition 3-3, the above expression

$$\begin{aligned} &= \hat{\mathbf{f}}_n + \epsilon_n \mathrm{M}[(\mathbf{f}_{n+1} - \mathbf{f}_n) \mid \mathscr{R}_n] \\ &= \hat{\mathbf{f}}_n + \epsilon_n(\mathrm{M}[\mathbf{f}_{n+1} \mid \mathscr{R}_n] - \mathbf{f}_n), \end{aligned}$$

which is

$$\geq \hat{\mathbf{f}}_n$$

because $(\mathbf{f}_n, \mathscr{R}_n)$ is a submartingale and ϵ_n is non-negative. It remains to be shown that $\mathrm{M}[\mathbf{f}_n - \hat{\mathbf{f}}_n] \geq 0$; we prove the result by induction on n. For $n = 0$, $\mathbf{f}_0 = \hat{\mathbf{f}}_0$ and $\mathrm{M}[\mathbf{f}_0 - \hat{\mathbf{f}}_0] = 0$. Suppose we have proved that $\int_\Omega (\mathbf{f}_k - \hat{\mathbf{f}}_k) d\mu \geq 0$. Then we have $\hat{\mathbf{f}}_{k+1} = \hat{\mathbf{f}}_k + \epsilon_k(\mathbf{f}_{k+1} - \mathbf{f}_k)$, and when we subtract both sides from $\mathbf{f}_{k+1}$, we get

$$\begin{aligned} \mathbf{f}_{k+1} - \hat{\mathbf{f}}_{k+1} &= \mathbf{f}_{k+1} - \hat{\mathbf{f}}_k - \epsilon_k(\mathbf{f}_{k+1} - \mathbf{f}_k) \\ &= (1 - \epsilon_k)(\mathbf{f}_{k+1} - \mathbf{f}_k) + (\mathbf{f}_k - \hat{\mathbf{f}}_k). \end{aligned}$$

Thus

$$\begin{aligned} \int_\Omega (\mathbf{f}_{k+1} - \hat{\mathbf{f}}_{k+1}) d\mu &\geq \int_\Omega (1 - \epsilon_k)(\mathbf{f}_{k+1} - \mathbf{f}_k) d\mu \quad \text{by hypothesis} \\ &= \int_{\{\epsilon_k(\omega) = 0\}} (\mathbf{f}_{k+1} - \mathbf{f}_k) d\mu \\ &\geq 0 \quad \text{by Lemma 3-6.} \end{aligned}$$

To see the connection of Proposition 3-7 to gambling systems, we again consider Example 1 of Section 3-2. We take the partial sums $\mathbf{s}_n$ as the $\mathbf{f}_n$, and we observe that the differences $\mathbf{s}_k - \mathbf{s}_{k-1}$ in the definition of $\hat{\mathbf{s}}_n$ are simply the random variables $\mathbf{y}_k$. The $\hat{\mathbf{s}}_n$ become modified fortunes, fortunes changed by not playing in every round of the game. When $\epsilon_k = 1$, the player participates in the $k + 1$st game; and when $\epsilon_k = 0$, he does not. The whole set of ϵ's, therefore, represents a gambling system; the condition that the set of paths for which $\epsilon_n(\omega) = 1$ be the union of cells of $\mathscr{R}_n$ is the condition that the system not depend on any knowledge of the future. For the special case in which the process is a submartingale, the expected fortune after time $n + 1$ is greater than or equal to the fortune at time n and the game is favorable for the player. The content of Proposition 3-7 is that the player's expected fortune after time $n + 1$ would not have been increased by a system which caused him not to play in certain rounds.

4. Martingale convergence and a second systems theorem

In this section we shall prove two theorems which will be of great use in our treatment of Markov chains. The two results will indicate the value of recognizing martingales when they appear in our later work.

Definition 3-8: Let $\{\mathbf{f}_n\}$ be a sequence of random variables defined on points ω, and let r and s be numbers with $r < s$. An **upcrossing** of $[r, s]$ is said to occur between $n - k$ and n for the point ω if these conditions are satisfied:

(1) $\mathbf{f}_{n-k}(\omega) \leq r$
(2) $r < \mathbf{f}_{n-k+m}(\omega) < s$ for $0 < m < k$
(3) $\mathbf{f}_n(\omega) \geq s$.

The reader should notice that no two upcrossings overlap.

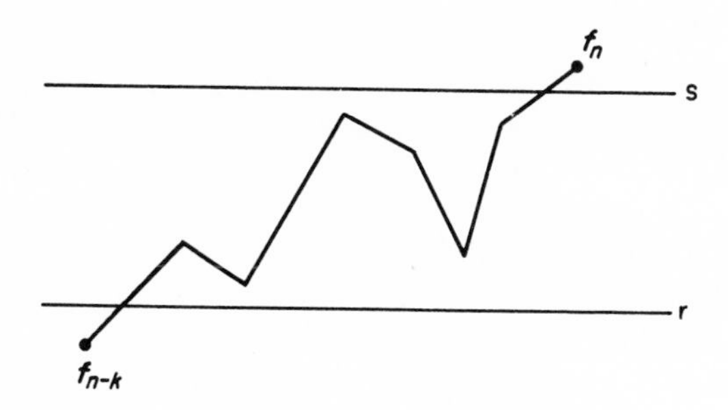

After three preliminary results, we proceed with a proof of the Martingale Convergence Theorem. Proposition 3-11 is known as the Upcrossing Lemma.

Lemma 3-9: If $(\mathbf{f}_n, \mathscr{R}_n)$ and $(\mathbf{g}_n, \mathscr{R}_n)$ are submartingales, then so is $(\sup (\mathbf{f}_n, \mathbf{g}_n), \mathscr{R}_n)$.

PROOF:

$$\mathrm{M}[|\sup (\mathbf{f}_n, \mathbf{g}_n)|] \leq \mathrm{M}[\sup (|\mathbf{f}_n|, |\mathbf{g}_n|)]$$

$$= \int_{|\mathbf{f}_n| \geq |\mathbf{g}_n|} |\mathbf{f}_n| d\mu + \int_{|\mathbf{f}_n| < |\mathbf{g}_n|} |\mathbf{g}_n| d\mu < \infty.$$

The function $\sup (\mathbf{f}_n, \mathbf{g}_n)$ is clearly constant on cells of $\mathscr{R}_n$ if $\mathbf{f}_n$ and $\mathbf{g}_n$ each are. Furthermore,

$$\mathrm{M}[\sup (\mathbf{f}_{n+1}, \mathbf{g}_{n+1}) \mid \mathscr{R}_n] \geq \mathrm{M}[\mathbf{f}_{n+1} \mid \mathscr{R}_n] \geq \mathbf{f}_n$$

and

$$\mathrm{M}[\sup (\mathbf{f}_{n+1}, \mathbf{g}_{n+1}) \mid \mathscr{R}_n] \geq \mathrm{M}[\mathbf{g}_{n+1} \mid \mathscr{R}_n] \geq \mathbf{g}_n$$

so that

$$\mathrm{M}[\sup (\mathbf{f}_{n+1}, \mathbf{g}_{n+1}) \mid \mathscr{R}_n] \geq \sup (\mathbf{f}_n, \mathbf{g}_n).$$

Lemma 3-10: If $(\mathbf{f}_n, \mathscr{R}_n)$ is a martingale, $\mathrm{M}[\mathbf{f}_n] = \mathrm{M}[\mathbf{f}_{n-1}]$. If $(\mathbf{f}_n, \mathscr{R}_n)$ is a supermartingale, $\mathrm{M}[\mathbf{f}_n] \leq \mathrm{M}[\mathbf{f}_{n-1}]$. If $(\mathbf{f}_n, \mathscr{R}_n)$ is a submartingale, $\mathrm{M}[\mathbf{f}_n] \geq \mathrm{M}[\mathbf{f}_{n-1}]$.

PROOF: The result is immediate from Lemma 3-6 when R_k is taken as Ω.

Proposition 3-11: Let $(\mathbf{f}_j, \mathscr{R}_j)$ be a submartingale, and let $\boldsymbol{\beta}(\omega)$ be the number of upcrossings of $[r, s]$ between times 0 and n. Then

$$\mathrm{M}[\boldsymbol{\beta}] \leq \frac{\mathrm{M}[(\mathbf{f}_n - r)^+]}{s - r} \leq \frac{\mathrm{M}[|\mathbf{f}_n|] + |r|}{s - r}.$$

PROOF: Consider first the special case $\mathbf{f}_k \geq 0$ for $0 \leq k \leq n$ and $r = 0$. Let $\hat{\mathbf{f}}_k$ be defined as in Proposition 3-7 with the $\boldsymbol{\epsilon}$'s to be specified. For a given path ω, define, by induction on m, $\boldsymbol{\epsilon}_m(\omega) = 1$ whenever $\mathbf{f}_m(\omega) = 0$, and let $\boldsymbol{\epsilon}_{m+k}(\omega) = 1$ as long as $\mathbf{f}_{m+k}(\omega) < s$. As soon as $\mathbf{f}_m(\omega) \geq s$, require that $\boldsymbol{\epsilon}_m(\omega) = 0$ until m is large enough so that $\mathbf{f}_m(\omega) = 0$ again. Then $\hat{\mathbf{f}}_m$ measures the increase in the sequence $\mathbf{f}_0(\omega), \ldots, \mathbf{f}_n(\omega)$ during upcrossings (plus a possible "partial upcrossing" at the end) and is greater than or equal to the minimum increase in each upcrossing multiplied by the number of upcrossings. That is,

$$s \cdot \boldsymbol{\beta}(\omega) \leq \hat{\mathbf{f}}_n(\omega).$$

Hence

$$s \cdot \mathrm{M}[\boldsymbol{\beta}] \leq \mathrm{M}[\hat{\mathbf{f}}_n],$$

which by Proposition 3-7 is

$$\leq \mathrm{M}[\mathbf{f}_n].$$

Therefore,

$$\mathrm{M}[\boldsymbol{\beta}] \leq (1/s)\mathrm{M}[\mathbf{f}_n]$$

and the special case is proved. For a general sequence $\{\mathbf{f}_n\}$ and general r, consider the function $(\mathbf{f}_k - r)^+$, which is the supremum of the zero function and the function $\mathbf{f}_k - r$. It is readily verified that constant functions are martingales and that the difference of a submartingale and a martingale is a submartingale. Thus $(\mathbf{f}_k - r, \mathscr{R}_k)$ is a submartingale and by Lemma 3-9 $((\mathbf{f}_k - r)^+, \mathscr{R}_k)$ is a submartingale. Applying the special case proved above to $(\mathbf{f}_k - r)^+$ and upcrossings of $[0, s - r]$, we find

$$\begin{aligned}(s - r)\mathrm{M}[\boldsymbol{\beta}] &\leq \mathrm{M}[(\mathbf{f}_n - r)^+] \\ &\leq \mathrm{M}[|\mathbf{f}_n - r|] \\ &\leq \mathrm{M}[|\mathbf{f}_n| + |r|] \\ &= \mathrm{M}[|\mathbf{f}_n|] + |r|\end{aligned}$$

and the proof of the Upcrossing Lemma is complete.

Theorem 3-12: If $(\mathbf{f}_n, \mathscr{R}_n)$ is a submartingale with the property that $\mathrm{M}[|\mathbf{f}_n|] \leq K < \infty$ for all n, then

$$\lim_{n\to\infty} \mathbf{f}_n(\omega)$$

exists *and is finite* for almost all points ω.

Proof: Failure of almost-everywhere convergence means that there exists a set of points ω of positive measure for which the sequence diverges. At least one of two things must happen. Either $\mathbf{f}_n(\omega)$ for each fixed ω in a set of positive measure oscillates infinitely often above and below rationals $r(\omega)$ and $s(\omega)$ with $r(\omega) < s(\omega)$, or else $|\mathbf{f}_n(\omega)|$ diverges to $+\infty$ on a set of positive measure. We consider the cases separately.

(1) Suppose $|\mathbf{f}_n(\omega)|$ diverges to $+\infty$ on a set E of positive measure m. Then by Fatou's Theorem

$$\begin{aligned}\int_E \liminf_n |\mathbf{f}_n(\omega)|\,d\mu &\leq \liminf_n \int_E |\mathbf{f}_n(\omega)|\,d\mu \\ &\leq \liminf_n \int_\Omega |\mathbf{f}_n(\omega)|\,d\mu \\ &= \liminf_n \mathrm{M}[|\mathbf{f}_n|] \\ &\leq K.\end{aligned}$$

But lim inf $|\mathbf{f}_n(\omega)| = +\infty$ on E, and E has positive measure m. Thus the left side of the inequality is infinite, and we have arrived at a contradiction.

(2) Suppose $\mathbf{f}_n(\omega)$ oscillates infinitely often above and below rationals $r(\omega)$ and $s(\omega)$ on a set of positive measure m. Order the set of all pairs of rationals (which is a denumerable set) and call the kth pair q_k. Consider the denumerable family of sets A_k defined by $A_k = \{\omega \mid \mathbf{f}_n(\omega)$ oscillates infinitely often above and below the rationals of the pair $q_k\}$. It is possible for more than one set to have the same point in it, but, on the other hand, every point ω for which $\mathbf{f}_n(\omega)$ oscillates infinitely often is in some A_k. Therefore,

$$\sum \mu(A_k) \geq \mu(\bigcup A_k) = m > 0$$

and there must exist a t for which $\mu(A_t) > 0$. That is, for every ω in a set A_t of positive measure, $\mathbf{f}_n(\omega)$ oscillates infinitely often above and below fixed rationals r and s with $r < s$. Let $\boldsymbol{\beta}_n(\omega)$ be the number of upcrossings of $[r, s]$ by $\mathbf{f}_0(\omega), \ldots, \mathbf{f}_n(\omega)$. By Proposition 3-11,

$$\begin{aligned} \mathrm{M}[\boldsymbol{\beta}_n] &\leq \frac{\mathrm{M}[|\mathbf{f}_n|] + |r|}{s - r} \\ &\leq \frac{K + |r|}{s - r} \\ &= c \quad \text{for every } n. \end{aligned}$$

Furthermore, the $\boldsymbol{\beta}_n$ are non-negative and increasing with n to a function $\boldsymbol{\beta}$, so that $\mathrm{M}[\boldsymbol{\beta}] = \lim \mathrm{M}[\boldsymbol{\beta}_n] \leq c$ by the Monotone Convergence Theorem. But $\mathrm{M}[\boldsymbol{\beta}] = +\infty$ since there are infinitely many upcrossings on a set of positive measure. This contradiction establishes the Martingale Convergence Theorem.

Corollary 3-13: Every non-negative supermartingale converges to a finite-valued function almost everywhere. In particular, every non-negative martingale converges almost everywhere.

PROOF: If $(\mathbf{f}_n, \mathscr{R}_n)$ is a non-negative supermartingale, then $(-\mathbf{f}_n, \mathscr{R}_n)$ is a submartingale. Since $\mathbf{f}_n \geq 0$, $|-\mathbf{f}_n| = \mathbf{f}_n$ and hence $\mathrm{M}[|-\mathbf{f}_n|] \leq \mathrm{M}[\mathbf{f}_0]$ by Lemma 3-10. Therefore, $\{-\mathbf{f}_n\}$ converges almost everywhere by Theorem 3-12, and so does $\{\mathbf{f}_n\}$.

We postpone a discussion of applications of Theorem 3-12 and Corollary 3-13 to the next section. We shall find that the corollary is used more frequently than the theorem itself.

A random time which is finite almost everywhere is called a **random stopping time** or simply a **stopping time**. If $\mathbf{t}$ is a stopping time, then the set $\bigcap_{n=1}^{\infty} \{\omega \mid \mathbf{t}(\omega) \geq n\}$ has measure zero. If $\{\mathbf{f}_n\}$ is a sequence of random variables, we define a function $\mathbf{f}_t$ almost everywhere by

$$\mathbf{f}_t(\omega) = \mathbf{f}_n(\omega) \quad \text{if} \quad \mathbf{t}(\omega) = n.$$

Since

$$\{\omega | \mathbf{f}_t(\omega) < c\} = \bigcup_{n=1}^{\infty} (\{\omega \mid \mathbf{t}(\omega) = n\} \cap \{\omega \mid \mathbf{f}_n(\omega) < c\}),$$

$\mathbf{f}_t$ is a random variable.

Lemma 3-14: If $(\mathbf{f}_n, \mathscr{R}_n)$ is a martingale and if $\mathbf{t}$ is a stopping time for which $\int_\Omega \mathbf{f}_t d\mu$ exists, then for any n

$$\int_\Omega \mathbf{f}_t d\mu = \int_{\{\mathbf{t} \leq n\}} \mathbf{f}_n d\mu + \int_{\{\mathbf{t} > n\}} \mathbf{f}_t d\mu.$$

PROOF: We have

$$\int_\Omega \mathbf{f}_t d\mu = \sum_{k=0}^{n} \int_{\{\mathbf{t} = k\}} \mathbf{f}_t d\mu + \int_{\{\mathbf{t} > n\}} \mathbf{f}_t d\mu$$

$$= \sum_{k=0}^{n} \int_{\{\mathbf{t} = k\}} \mathbf{f}_k d\mu + \int_{\{\mathbf{t} > n\}} \mathbf{f}_t d\mu,$$

which by Lemma 3-6

$$= \sum_{k=0}^{n} \int_{\{\mathbf{t} = k\}} \mathbf{f}_n d\mu + \int_{\{\mathbf{t} > n\}} \mathbf{f}_t d\mu$$

$$= \int_{\{\mathbf{t} \leq n\}} \mathbf{f}_n d\mu + \int_{\{\mathbf{t} > n\}} \mathbf{f}_t d\mu.$$

Theorem 3-15: If $(\mathbf{f}_n, \mathscr{R}_n)$ is a martingale and if $\mathbf{t}$ is a stopping time, then $M[\mathbf{f}_t] = M[\mathbf{f}_0]$ if

(1) $M[|\mathbf{f}_t|] < \infty$, and

(2) $\lim_n \int_{\{\mathbf{t} > n\}} \mathbf{f}_n d\mu = 0.$

REMARK: Analogous results hold for submartingales and for supermartingales. Inequalities replace the equality in the conclusion.

PROOF OF THEOREM: By (1), $\int_\Omega \mathbf{f}_t d\mu$ exists, so that Lemma 3-14 applies. Thus, for any n,

$$\int_\Omega \mathbf{f}_t d\mu = \int_{\{\mathbf{t} \leq n\}} \mathbf{f}_n d\mu + \int_{\{\mathbf{t} > n\}} \mathbf{f}_t d\mu$$
$$= \int_\Omega \mathbf{f}_n d\mu - \int_{\{\mathbf{t} > n\}} \mathbf{f}_n d\mu + \int_{\{\mathbf{t} > n\}} \mathbf{f}_t d\mu,$$

which by Lemma 3-10

$$= \int_\Omega \mathbf{f}_0 d\mu - \int_{\{\mathbf{t} > n\}} \mathbf{f}_n d\mu + \int_{\{\mathbf{t} > n\}} \mathbf{f}_t d\mu.$$

Using condition (1) together with Corollary 1-17 and the complete additivity of the integral as a set function, we see that

$$\lim_n \int_{\{\mathbf{t} > n\}} \mathbf{f}_t d\mu = 0.$$

Since $\int_{\{\mathbf{t} > n\}} \mathbf{f}_n d\mu \to 0$ by hypothesis, we have $\int_\Omega \mathbf{f}_t d\mu = \int_\Omega \mathbf{f}_0 d\mu$.

Corollary 3-16: Suppose $(\mathbf{f}_n, \mathscr{R}_n)$ is a martingale defined on a space Ω of finite total measure and $\mathbf{t}$ is a stopping time. If $|\mathbf{f}_n| \leq K$ for all n, then $\mathrm{M}[\mathbf{f}_t] = \mathrm{M}[\mathbf{f}_0]$.

PROOF: We must show the two conditions of Theorem 3-15 are satisfied. For (1) we have $|\mathbf{f}_t| \leq K$ by definition, and hence $\mathbf{f}_t$ is integrable. For (2) we have

$$\left| \int_{\{\mathbf{t} > n\}} \mathbf{f}_n d\mu \right| \leq \int_{\{\mathbf{t} > n\}} |\mathbf{f}_n| d\mu$$
$$\leq \int_{\{\mathbf{t} > n\}} K d\mu$$
$$= K\mu(\{\omega \mid \mathbf{t}(\omega) > n\})$$
$$\to 0.$$

In terms of gambling systems, the result of Lemma 3-10, namely that for martingales $\mathrm{M}[\mathbf{f}_n] = \mathrm{M}[\mathbf{f}_0]$, states that the expected fortune at any fixed stopping time is equal to the initial fortune if the game is fair. That is, the fairness of a game is not altered by deciding in advance to stop playing at some fixed time. But what about a system where the player stops according to how the game is going? The system he adopts is represented by the random time $\mathbf{t}$, and Theorem 3-15 and

Corollary 3-16 give sufficient criteria for the game still to be fair. Corollary 3-16 by itself is a general result; it covers the situation, for example, where the game stops if either the player or the house goes bankrupt. If the game does stop under such circumstances, the corollary states that the fairness of the game is not altered by any gambling system whatsoever. Similar remarks apply to supermartingales. If the amount of money that a player has is limited, no system that he adopts for stopping according to how the game is going will make an unfair game favorable.

The following proposition is useful in proving that certain random times are stopping times.

Proposition 3-17: Let $(\mathbf{f}_n, \mathscr{R}_n)$ be a martingale, let $\mathbf{t}$ be a random time, and let $\hat{\mathbf{f}}_n$ be the stopped process with

$$\hat{\mathbf{f}}_n(\omega) = \mathbf{f}_{\min(n,\mathbf{t}(\omega))}(\omega).$$

Then $(\hat{\mathbf{f}}_n, \mathscr{R}_n)$ is a martingale.

PROOF: We first note that $\mathrm{M}[|\hat{\mathbf{f}}_n|] < \infty$ since $|\hat{\mathbf{f}}_n| \leq \sum_{j=0}^n |\mathbf{f}_j|$ and each $\mathbf{f}_j$ is integrable. Next, let R be a cell in $\mathscr{R}_n$ with $\mu(R) \neq 0$. In R we have

$$\begin{aligned}
\mathrm{M}[\hat{\mathbf{f}}_{n+1} \mid \mathscr{R}_n] &= \frac{1}{\mu(R)} \int_R \hat{\mathbf{f}}_{n+1} d\mu \\
&= \frac{1}{\mu(R)} \left[\int_{R \cap \{\mathbf{t} \leq n\}} \hat{\mathbf{f}}_{n+1} d\mu + \int_{R \cap \{\mathbf{t} > n\}} \hat{\mathbf{f}}_{n+1} d\mu \right] \\
&= \frac{1}{\mu(R)} \left[\int_{R \cap \{\mathbf{t} \leq n\}} \hat{\mathbf{f}}_n d\mu + \int_{R \cap \{\mathbf{t} > n\}} \mathbf{f}_{n+1} d\mu \right]
\end{aligned}$$

by definition of $\hat{\mathbf{f}}_n$. Since $(\mathbf{f}_n, \mathscr{R}_n)$ is a martingale and $\{\mathbf{t} > n\}$ is in $\mathscr{R}_n$, the above expression by Lemma 3-6 is

$$\begin{aligned}
&= \frac{1}{\mu(R)} \left[\int_{R \cap \{\mathbf{t} \leq n\}} \hat{\mathbf{f}}_n d\mu + \int_{R \cap \{\mathbf{t} > n\}} \mathbf{f}_n d\mu \right] \\
&= \frac{1}{\mu(R)} \left[\int_{R \cap \{\mathbf{t} \leq n\}} \hat{\mathbf{f}}_n d\mu + \int_{R \cap \{\mathbf{t} > n\}} \hat{\mathbf{f}}_n d\mu \right] \\
&= \frac{1}{\mu(R)} \int_R \hat{\mathbf{f}}_n d\mu.
\end{aligned}$$

The last expression equals $\hat{\mathbf{f}}_n$ since $\hat{\mathbf{f}}_n$ is constant on R.

The application of Proposition 3-17 is this: Let $(\mathbf{f}_n, \mathscr{R}_n)$ be an integer-valued martingale, and let S be a set of integers. Let the martingale almost surely have the property that it can be constant from some time on only for values in S (and possibly for no values). We stop $\mathbf{f}_n$ the first time it takes on a value in S. That is, we let $\mathbf{t}$ be the first time that $\mathbf{f}_n$ is in S, and we introduce the stopped process $\hat{\mathbf{f}}_n$. If the values of $\hat{\mathbf{f}}_n$ are bounded from below or from above, then the "stopped process," is almost sure to stop. The proof proceeds as follows.

First, assume that $\hat{\mathbf{f}}_n \geq 0$. Then $(\hat{\mathbf{f}}_n, \mathscr{R}_n)$ is a non-negative martingale, by Proposition 3-17, which must converge a.e. to a finite value depending on ω. Since by hypothesis these values must be in S for a.e. ω, the process $\{\hat{\mathbf{f}}_n\}$ almost surely stops. Next, if $\{\hat{\mathbf{f}}_n\}$ is bounded below, apply the result for non-negative martingales to $\hat{\mathbf{f}}_n$ plus a suitable constant. Finally, if $\hat{\mathbf{f}}_n \leq c$, apply the result for $\{\hat{\mathbf{f}}_n\}$ bounded below to $\{-\hat{\mathbf{f}}_n\}$.

These results are used in the next section in Examples 1 and 4. In Example 1 a fair game is stopped when it leaves a certain finite set, whereas in Example 4 it is stopped when a positive value is reached for the first time. By the above argument these random times are stopping times.

Proposition 3-18: Let $\mathscr{R}_0 \subset \mathscr{R}_1 \subset \cdots$ be an increasing sequence of partitions and let $\mathscr{R}^*$ be the smallest augmented Borel field containing the field $\bigcup \mathscr{R}_n^*$. Let $\mathbf{f}$ be a random variable measurable with respect to $\mathscr{R}^*$ and having finite mean, and set $\mathbf{g}_n = \mathrm{M}[\mathbf{f} \mid \mathscr{R}_n]$. Then $(\mathbf{g}_n, \mathscr{R}_n)$ is a martingale, and

$$\lim_{n\to\infty} \mathbf{g}_n = \mathbf{f}$$

almost everywhere.

PROOF: We may assume that $\mathbf{f} \geq 0$ since the general case follows by considering $\mathbf{f}^+$ and $\mathbf{f}^-$ separately. Then $\mathbf{g}_n \geq 0$ and $\mathrm{M}[|\mathbf{g}_n|] = \mathrm{M}[\mathbf{f}] < \infty$ by conclusion (2) of Proposition 3-3. Since, in addition,

$$\begin{aligned}\mathrm{M}[\mathbf{g}_{n+1} \mid \mathscr{R}_n] &= \mathrm{M}[\mathrm{M}[\mathbf{f} \mid \mathscr{R}_{n+1}] \mid \mathscr{R}_n]\\ &= \mathrm{M}[\mathbf{f} \mid \mathscr{R}_n] \quad \text{by (1) of Proposition 3-3}\\ &= \mathbf{g}_n,\end{aligned}$$

we see that $(\mathbf{g}_n, \mathscr{R}_n)$ is a non-negative martingale. By Corollary 3-13, $\mathbf{g} = \lim \mathbf{g}_n$ exists a.e. We shall prove that $\mathbf{g} = \mathbf{f}$ a.e.

First we prove that the $\mathbf{g}_n$ are uniformly integrable. Let $\epsilon > 0$ be given. Choose, by Proposition 1-47, $\delta > 0$ small enough so that $\mu(E) < \delta$ implies $\int_E \mathbf{f}\, d\mu < \epsilon$. Now

$$N\mu(\{\mathbf{g}_n \geq N\}) \leq \mathrm{M}[\mathbf{g}_n] = \mathrm{M}[\mathbf{f}]$$

or

$$\mu(\{\mathbf{g}_n \geq N\}) \leq \frac{\mathrm{M}[\mathbf{f}]}{N}.$$

Choose N large enough so that the right side is less than δ. Then for all n we have

$$\int_{\{\mathbf{g}_n \geq N\}} \mathbf{f} d\mu < \epsilon.$$

Since

$$\int_{\{\mathbf{g}_n \geq N\}} \mathbf{g}_n d\mu = \int_{\{\mathbf{g}_n \geq N\}} \mathbf{f} d\mu$$

by Lemma 3-6, we conclude the $\mathbf{g}_n$ are uniformly integrable.

Let E be any subset of $\mathscr{R}_n{}^*$. For $m \geq n$,

$$\int_E \mathbf{g}_m d\mu = \int_E \mathbf{f} d\mu.$$

By uniform integrability and Proposition 1-52,

$$\lim_m \int_E \mathbf{g}_m d\mu = \int_E \mathbf{g} d\mu.$$

Therefore

$$\int_E \mathbf{f} d\mu = \int_E \mathbf{g} d\mu$$

for all E in $\bigcup \mathscr{R}_n{}^*$. The two sides of this last equation, considered as set functions, are equal on $\bigcup \mathscr{R}_n{}^*$. By the uniqueness half of Theorem 1-19 they must be equal on $\mathscr{R}^*$. That is, $\mathbf{f}$ and $\mathbf{g}$ are measurable with respect to $\mathscr{R}^*$ and satisfy

$$\int_E \mathbf{f} d\mu = \int_E \mathbf{g} d\mu.$$

Taking E successively to be the set where $\mathbf{f} \geq \mathbf{g}$ and the set where $\mathbf{f} \leq \mathbf{g}$ and applying Corollary 1-40, we find that $\mathbf{f} = \mathbf{g}$ a.e.

5. Examples of convergent martingales

Four examples will serve at present to illustrate Theorems 3-12 and 3-15. Each of the first three refers to the correspondingly numbered example in Section 2.

EXAMPLE 1: The sequence $\{\mathbf{s}_n\}$ of nth partial sums of the independent random variables $\mathbf{y}_n$ forms a martingale if $\mathrm{M}[\mathbf{y}_n] = 0$. Suppose the $\mathbf{y}_n$ have identical distributions and mean zero, suppose they assume only the values 0, 1, and -1, and suppose that the process is stopped whenever $\mathbf{s}_n(\omega) = M$ or $\mathbf{s}_n(\omega) = -N$. (Mean zero implies that the outcomes $+1$ and -1 are equally likely.) The player of the fair game is ruined if $s_n(\omega)$ ever equals $-N$ and he breaks the bank if $s_n(\omega) = M$. Set

$$p = \text{Pr[player breaks the bank]}$$

and

$$q = \text{Pr[player is eventually ruined]}.$$

By the remarks following Proposition 3-17, $p + q = 1$. In this situation, Corollary 3-16 applies and

$$\begin{aligned} 0 = \mathrm{M}[s_0] = \mathrm{M}[s_t] &= p \cdot M + q(-N) \\ &= pM + (1 - p)(-N) \end{aligned}$$

or

$$p = \frac{N}{M + N}$$

and

$$q = \frac{M}{M + N}.$$

Example 2: With the particle moving on the line, suppose there is an $r > 0$ which is a root of the equation $f(s) = 1$. We shall assume that $r < 1$. Then $\{r^{x_n}\}$ is a non-negative martingale, and by Corollary 3-13, $\{r^{x_n}\}$ converges to a limiting function a.e. Since x_n is integer-valued, this convergence means either

(1) for almost all ω, there is an N such that if $n \geq N$, then $x_n = x_N$, or
(2) $\lim x_n = +\infty$.

Now $x_N = x_{N+1} = x_{N+2} = \cdots = x_{N+k}$ means that the particle fails to move for k consecutive steps. Since such a thing happens with probability ${p_0}^k < 1$, the probability that $x_n = x_N$ for all $n \geq N$ is zero. Thus case 1 is eliminated, and we have established that $\lim x_n = +\infty$ a.e. That is, for any k and for almost all ω, $x_n(\omega) = k$ for only finitely many n.

Example 3: When the f_n's are the difference quotients of a monotone function f, the pair $(f_n, \mathscr{R}_n)$ forms a non-negative martingale. By Corollary 3-13, f_n converges to a limiting function at almost all points. The limiting function will turn out to be the derivative of f. However, our argument considered only nested partitions, and hence it provides only part of the proof of the existence of f' a.e.

Next we consider an example where Theorem 3-15 is *not* applicable.

Example 4: Suppose that a player plays the fair game of Example 1 and that he stops the first time that he is ahead. The process is stopped when $s_n(\omega) = 1$. We have already seen that this is a stopping time. Then $s_0 = 0$, and $s_t = 1$ a.e. Hence $M[s_0] = 0 \neq 1 = M[s_t]$. Why is the theorem not applicable? Condition (1) is clearly satisfied. However,

$$0 = \int s_n d\mu = \int_{\{t \leq n\}} s_n d\mu + \int_{\{t > n\}} s_n d\mu.$$

The first term equals the probability that 1 has been reached and tends to 1. Hence the second term tends to -1, not to 0. Thus condition (2) is violated.

In practice this gambling strategy cannot be implemented, since the gambler would need infinite capital to be able to absorb arbitrarily large losses.

6. Law of Large Numbers

The Strong Law of Large Numbers, which may be derived from the Martingale Convergence Theorem, is formulated as follows.

Theorem 3-19: Let $\{y_n\}$ be a sequence of independent identically distributed random variables with finite mean $a = M[y_k]$. If $s_n = y_1 + \cdots + y_n$ and $s^* = s_n/n$, then

$$\Pr[\lim_n s_n^* = a] = 1.$$

We shall prove the theorem for the special case where the random variables have finite range; say, $\Pr[y_k = j] = p_j$ for a finite number of j's. For more generally applicable proofs the reader is referred to the bibliography. (See Feller [1957], pp. 244–245, for an elementary proof in the case y_n is denumerable-valued; and see Doob [1953], pp. 334–342, for a general proof using the Martingale Convergence Theorem.)

We introduce a useful tool, the generating function

$$\varphi(t) = \sum_j p_j \cdot t^j.$$

It is a well-behaved function satisfying $\varphi(1) = 1$ and $\varphi'(1) = a$. Let

$$f(k, n) = \frac{t^k}{\varphi(t)^n}$$

for some $t > 0$ to be specified. We shall show that $\{f(s_n, n)\}$ is a martingale. The conditional mean of $f(s_{n+1}, n+1)$ given $s_n = k$ is

$$\frac{\sum_j p_j \cdot t^{k+j}}{\varphi(t)^{n+1}} = f(k, n) \cdot \frac{\sum_j p_j \cdot t^j}{\varphi(t)} = f(k, n) = f(s_n, n).$$

Since $M[f(\mathbf{s}_0, 0)] = 1 < \infty$, $\{f(\mathbf{s}_n, n)\}$ is a martingale, and it is clearly non-negative. Thus, by the Martingale Convergence Theorem, $f(\mathbf{s}_n, n)$ converges to a finite limit a.e., where

$$f(\mathbf{s}_n, n) = t^{\mathbf{s}_n}/\varphi(t)^n = [t^{\mathbf{s}_n *}/\varphi(t)]^n.$$

Fix $\epsilon > 0$, let $b = a + \epsilon$, and form the function $g(t) = t^b/\varphi(t)$. Since $g(1) = 1$ and $g'(1) = b - a > 0$, we have $g(t_0) > 1$ for some sufficiently small t_0 greater than 1. If $\mathbf{s}_n^*(\omega) \geq b$, we have

$$[t_0^{\mathbf{s}_n*(\omega)}/\varphi(t_0)] \geq g(t_0) > 1,$$

and hence if $\mathbf{s}_n^*(\omega) \geq b$ for infinitely many n, then $f(\mathbf{s}_n(\omega), n)$ has a subsequence tending to $+\infty$. By the convergence of f, we conclude that

$$\limsup \mathbf{s}_n^*(\omega) \leq b = a + \epsilon$$

a.e. Similarly, by choosing a suitable $t < 1$ we would find that

$$\liminf \mathbf{s}_n^*(\omega) \geq a - \epsilon$$

a.e. Since ϵ is arbitrary, $\mathbf{s}_n^*$ converges to a with probability one.

7. Problems

1. Show that if $\{\mathbf{f}_n\}$ are denumerable-valued independent random variables, then
$$M[\mathbf{f}_{n+1} \mid \mathbf{f}_0 \wedge \cdots \wedge \mathbf{f}_n] = M[\mathbf{f}_{n+1}].$$
Show also that
$$\Pr[\mathbf{f}_0 \in A_0 \wedge \mathbf{f}_1 \in A_1 \wedge \cdots \wedge \mathbf{f}_n \in A_n] = \Pr[\mathbf{f}_0 \in A_0] \cdot \cdots \cdot \Pr[\mathbf{f}_n \in A_n].$$
2. Let $\{\mathbf{f}_n\}$ be a sequence of denumerable-valued independent random variables and let $\{g_n\}$ be a sequence of Borel-measurable functions defined on the real line. Show that $\{g_n(\mathbf{f}_n)\}$ is a sequence of independent random variables. If the $\mathbf{f}_n$ are identically distributed and all the g_n's are equal to g, show that the $g(\mathbf{f}_n)$'s are identically distributed.
3. Verify that Examples 2 and 3 of Section 2 are martingales.
4. Prove that if $(\mathbf{f}_n, \mathscr{R}_n)$ and $(\mathbf{g}_n, \mathscr{R}_n)$ are martingales, then so is $(\mathbf{f}_n + \mathbf{g}_n, \mathscr{R}_n)$. Does the same hold for $(\mathbf{f}_n\mathbf{g}_n, \mathscr{R}_n)$?
5. Prove that if $(\mathbf{f}_n, \mathscr{R}_n)$ and $(\mathbf{g}_n, \mathscr{R}_n)$ are non-negative supermartingales, then so is $(\min(\mathbf{f}_n, \mathbf{g}_n), \mathscr{R}_n)$.
6. Prove that if $(\mathbf{f}_n, \mathscr{R}_n)$ is a martingale, then $(|\mathbf{f}_n|, \mathscr{R}_n)$ is a submartingale. If the $\mathbf{f}_n$ form a martingale on their cross partitions, do the $|\mathbf{f}_n|$ form a submartingale on their own cross partitions?
7. Let $(\mathbf{f}_n, \mathscr{R}_n)$ be a submartingale. Prove that, for any $c > 0$,
$$c \Pr\left[\max_{i \leq n} \mathbf{f}_i \geq c\right] \leq M[|\mathbf{f}_n|].$$
[*Hint:* Take as stopping time the first time c is surpassed, or n, whichever comes first.]

8. Prove that every submartingale can be written as the sum of a martingale and an increasing submartingale. [*Hint:* If $(\mathbf{x}_n, \mathscr{R}_n)$ is given, put $\mathbf{f}_0 = 0$, $\mathbf{f}_n = \mathrm{M}[\mathbf{x}_n \mid \mathscr{R}_{n-1}] - \mathbf{x}_{n-1}$, $\mathbf{z}_n = \mathbf{f}_0 + \cdots + \mathbf{f}_n$, and $\mathbf{y}_n = \mathbf{x}_n - \mathbf{z}_n$.]
9. Consider the following stochastic process: A white and a black ball are placed in an urn. One ball is drawn, and this ball is replaced by two of the same color. Let $\mathbf{f}_n$ be the fraction of white balls after n experiments.
   (a) Prove that $\{\mathbf{f}_n\}$ is a martingale.
   (b) Prove that it converges.
   (c) Prove that the limiting distribution has mean $\frac{1}{2}$.
   (d) Prove that the probability of ever reaching a fraction $\frac{3}{4}$ of white balls is at most $\frac{2}{3}$. [*Hint:* Use Problem 7.]
10. We consider an experiment with each outcome one of two possible outcomes H or T. We have two different hypotheses A and B as to how the underlying measure for a stochastic process should be assigned. For a given sequence HHT...H we denote by $p_n(\text{HHT}\ldots\text{H})$ the assignment under hypothesis A and by $r_n(\text{HHT}\ldots\text{H})$ the assignment under hypothesis B. Let f_n be defined by

$$f_n(\text{HHT}\ldots\text{H}) = \frac{p_n(\text{HHT}\ldots\text{H})}{r_n(\text{HHT}\ldots\text{H})}.$$

   (a) Show that if the measure is defined by hypothesis B, then $\{f_n\}$ is a martingale and hence converges a.e.
   (b) Specialize to the case of tossing a biased coin. Let hypotheses A and B be that the probability of heads is, respectively, p and r. Show that if the measure is defined by hypothesis B, then $\{f_n\}$ converges to 0 a.e. if $p \neq r$.

Problems 11 to 13 concern a type of stochastic process employed by psychologists in learning theory. The state space consists of the rational points on the unit interval, and we are given two rational parameters, $0 < b \leq a < 1$. From a point x the process moves to $bx + (1 - a)$ with probability x, or to bx with probability $1 - x$. It is started at an interior point x_0.

11. Show that if $b = a$, then $\{x_n\}$ is a martingale.
12. Prove that the process converges either to 0 or to 1, and compute the probability of going to 1 as a function of the starting position x_0.
13. Show that if $b < a$, then $\{x_n\}$ is a supermartingale and the process converges to 0 a.e.

Problems 14 to 18 concern the notion of conditional mean given a Borel field and show how it generalizes the notion of conditional mean given a partition. It will be necessary to know the Radon–Nikodym Theorem to solve Problem 14. Let $(\Omega, \mathscr{B}, \mu)$ be a probability space in which every subset of a set of measure zero is measurable.

14. If $\mathbf{f}$ is a random variable such that $\mathrm{M}[|\mathbf{f}|] < \infty$ and if $\mathscr{G}$ is a Borel field contained in $\mathscr{B}$, show that there exists a function $\mathrm{M}[\mathbf{f} \mid \mathscr{G}]$ defined on Ω satisfying
   (i) $\mathrm{M}[\mathbf{f} \mid \mathscr{G}]$ is measurable with respect to $\mathscr{G}'$, the augmented field obtained from $\mathscr{G}$.
   (ii) If G is a set in $\mathscr{G}$, then $\int_G \mathbf{f}\, d\mu = \int_G \mathrm{M}[\mathbf{f} \mid \mathscr{G}]\, d\mu$.

Show that $\mathrm{M}[\mathbf{f} \mid \mathscr{G}]$ is unique in this sense: Any two functions satisfying (i) and (ii) differ only on a set of μ-measure zero. We can therefore define any determination of $\mathrm{M}[\mathbf{f} \mid \mathscr{G}]$ to be the conditional mean of $\mathbf{f}$ given $\mathscr{G}$.

15. Show that if $\mathscr{G}$ is the Borel field generated by a partition $\mathscr{R}$, then $\mathrm{M}[\mathbf{f} \mid \mathscr{G}] = \mathrm{M}[\mathbf{f} \mid \mathscr{R}]$ a.e. Show that if $\mathscr{G} = \mathscr{B}$, then $\mathrm{M}[\mathbf{f} \mid \mathscr{G}] = \mathbf{f}$ a.e.
16. State and prove a result for these conditional means in analogy with Proposition 3-3. [*Hint:* In (3), the condition that $\mathbf{g}$ be constant on cells of $\mathscr{R}$ should be replaced with the condition that $\mathbf{g}$ be measurable with respect to $\mathscr{G}$.]
17. Generalize the definition of martingale in Definition 3-5, using these conditional means. Verify that the statements and proofs of Lemma 3-6, Proposition 3-7, Lemmas 3-9 and 3-10, Proposition 3-11, Theorem 3-12, and Corollary 3-13 apply with only minor changes to this generalized notion of martingale. Perform the same verification for Lemma 3-14, Theorem 3-15, Corollary 3-16, and Proposition 3-17.
18. Prove the following generalization of Proposition 3-18: Let $\mathscr{G}_0 \subset \mathscr{G}_1 \subset \cdots$ be an increasing sequence of Borel fields in $\mathscr{B}$, let $\mathscr{G}$ be the least Borel field containing $\bigcup \mathscr{G}_n$, and let $\mathbf{f}$ be a random variable with $\mathrm{M}[|\mathbf{f}|] < \infty$. Then $(\mathrm{M}[\mathbf{f} \mid \mathscr{G}_n], \mathscr{G}_n)$ is a martingale, and

$$\lim_{n \to \infty} \mathrm{M}[\mathbf{f} \mid \mathscr{G}_n] = \mathrm{M}[\mathbf{f} \mid \mathscr{G}]$$

a.e.

CHAPTER 4

PROPERTIES OF MARKOV CHAINS

1. Markov chains

During all of our discussion of Markov chains, we shall wish to confine ourselves to stochastic processes defined on a sequence space. We have shown that an arbitrary stochastic process may be considered as a process on a suitable Ω in which the outcome functions f_n are coordinate functions. We see, therefore, that in a sense no generality is lost by discussing Markov chains in terms of sequence space.

Definition 4-1: Let $(\Omega, \mathscr{B}, \mu)$ be a sequence space with a denumerable stochastic process $\{x_n\}$ defined from Ω to a denumerable state space S of more than one element. The process is called a **denumerable Markov process** if

$$\Pr[x_{n+1} = c_{n+1} \mid x_0 = c_0 \wedge \cdots \wedge x_{n-1} = c_{n-1} \wedge x_n = c_n] = \Pr[x_{n+1} = c_{n+1} \mid x_n = c_n]$$

for any n and for any $c_0, \ldots, c_{n+1}$ such that

$$\Pr[x_0 = c_0 \wedge \cdots \wedge x_n = c_n] > 0.$$

The condition that defines a Markov process is known as the **Markov property**. If a denumerable Markov process has the property that for any m and n and for any c_n, c_{n+1} such that $\Pr[x_m = c_n] > 0$ and $\Pr[x_n = c_n] > 0$,

$$\Pr[x_{n+1} = c_{n+1} \mid x_n = c_n] = \Pr[x_{m+1} = c_{n+1} \mid x_m = c_n]$$

holds, then the process is called a **denumerable Markov chain.** The condition that defines a Markov chain is called the **Markov chain property.**

All Markov chains that we shall discuss will be denumerable. From Proposition 2-8 we immediately have the following result.

Proposition 4-2: The measure on the space for a Markov chain is completely determined by

(1) the starting probabilities, $\Pr[x_0 = i]$, and
(2) the one-step transition probabilities, the common value of $\Pr[x_{n+1} = j \mid x_n = i]$ for all n such that $\Pr[x_n = i] > 0$.

If S is the set of states for a Markov chain, we customarily denote representative elements of S by $i, j, k, \ldots$, and 0. For any Markov chain we define on the set S a row vector π and a square matrix P by

$$\pi_i = \Pr[x_0 = i],$$

$$P_{ij} = \Pr[x_{n+1} = j \mid x_n = i], \quad \text{where} \quad \Pr[x_n = i] > 0.$$

The vector π is the **starting distribution**, and the matrix P is the **transition matrix** for the chain. They satisfy the properties $\pi \geq 0$, $\pi\mathbf{1} = 1$, and $P \geq 0$. If $\Pr[x_n = i] = 0$ for all n, then the ith row of P is not covered by the above definition, and we shall agree to take $P_{ij} = 0$ for all j in this case.

The definition of P implies that, for each i, $(P\mathbf{1})_i = 1$ or 0. It will be convenient, however, to think of Markov chains from a point of view which allows P to be any matrix with $P \geq 0$ and $P\mathbf{1} \leq \mathbf{1}$. To do so, we shall admit the possibility that some of the paths in the sequence space are of finite length. Intuitively a path of finite length is one along which the Markov chain can "disappear"; the process disappears from state i with probability $1 - (P\mathbf{1})_i$. Mathematically paths of finite length can be introduced as follows: Suppose a Markov chain with state space S has a distinguished state 0 for which $\pi_0 = 0$ and $P_{00} = 1$. We shall sometimes identify entry to state 0 with the act of disappearing in a process with state space $S - \{0\}$ which also will be called a Markov chain. The transition matrix for the new Markov chain is the same as the original one except that the 0th row and column are omitted. Any path in the original process which has 0 as an outcome is now thought of as a path of finite length which terminates before the first occurrence of 0. The original process can be recovered from the new process by re-introducing state 0 to the state space and by requiring that the transition probabilities to state 0 in the original process be the same as the probabilities of disappearing in the new process. With these conventions a Markov chain determines a vector π and a matrix P with $\pi \geq 0$, $\pi\mathbf{1} = 1$, $P \geq 0$, and $P\mathbf{1} \leq \mathbf{1}$.

Conversely, if π is a row vector defined on S for which $\pi \geq 0$ and $\pi\mathbf{1} = 1$ and if P is a square matrix defined on S for which $P \geq 0$ and $P\mathbf{1} \leq \mathbf{1}$, then π and P define a unique Markov chain with state space S by Theorem 2-4. If $(P\mathbf{1})_i < 1$, then the process has positive probability equal to $1 - (P\mathbf{1})_j$ of disappearing each time it is in state j. Whenever convenient, the act of disappearing may be thought of as entry to an ideal state adjoined to S.

Any state i of a Markov chain P for which $P_{ii} = 1$ is said to be an **absorbing state.** If outcome i occurs at some time, the process is said to enter the absorbing state and to become **absorbed.** It is easily seen that once the process has been absorbed, it is impossible for it to leave the absorbing state.

If P is a Markov chain with starting distribution π and if q is a statement about the process, we denote the probability of q by $\Pr_\pi[q]$. If

$$\pi_k = \begin{cases} 1 & \text{when } k = i \\ 0 & \text{otherwise,} \end{cases}$$

we may alternatively write $\Pr_i[q]$. Similarly if $\mathbf{f}$ is a random variable, we write $\mathrm{M}_\pi[\mathbf{f}]$ or $\mathrm{M}_i[\mathbf{f}]$, depending on the starting distribution. With this notational convention, we are free to discuss a whole class of Markov chains at once. The class contains all chains whose transition matrices are some fixed matrix P, and two chains of the class differ only in their starting distributions. Most of our treatment of Markov chains will be on this more natural level, where a matrix P, but no distribution π, is specified.

We conclude this section with a simple but useful proposition. Its proof is left to the reader.

Proposition 4-3: If P is a Markov chain, then for $n \geq 0$,

$$\Pr_i[x_n = j] = (P^n)_{ij}$$

and

$$\Pr_\pi[x_n = j] = (\pi P^n)_j.$$

We shall use the notation $P_{ij}^{(n)}$ for $(P^n)_{ij}$, the n-step probability from i to j.

2. Examples of Markov chains

We give ten examples of Markov chains; we shall refer to all of them from time to time.

EXAMPLE 1: Weather in the Land of Oz.

The Land of Oz is blessed by many things, but good weather is not one of them. They never have two nice days in a row. If they have a nice day, they are just as likely to have snow as rain the next day. If they have snow (or rain), they have an even chance of having the same the next day. If there is a change from snow or rain, only half of the time is this a change to a nice day.

The weather is conveniently represented as a Markov chain with the three states $S = \{\text{Rain, Nice, Snow}\}$. The transition matrix becomes

$$P = \begin{matrix} & \text{R} & \text{N} & \text{S} \\ \text{R} & \frac{1}{2} & \frac{1}{4} & \frac{1}{4} \\ \text{N} & \frac{1}{2} & 0 & \frac{1}{2} \\ \text{S} & \frac{1}{4} & \frac{1}{4} & \frac{1}{2} \end{matrix}.$$

EXAMPLE 2: Chain with a set of states E made absorbing.

Let P be an arbitrary Markov chain with S the set of states. Let a subset E of S be specified. We modify the original process by requiring that if the process is ever in a state j of E, it does not leave that state. The new process is also a Markov chain; its transition matrix P' differs from the P-matrix in that $P'_{jj} = 1$ and $P'_{ji} = 0$ for every $j \in E$ and for every $i \neq j$. The new process is called the chain with E made absorbing.

EXAMPLE 3: Finite drunkard's walk.

A drunkard walks randomly on a street between his house and a lake, starting at a bar in the middle. He has some idea of which way is home. The steps along the way are labeled by the integers from 0 to n; the bar, some integer i between 0 and n, is the starting state, and the drunkard moves one step toward home (state n) with probability p and one step toward the lake (state 0) with probability $q = 1 - p$. States 0 and n are absorbing. We assume that $p \neq 0$ and $p \neq 1$. The transition matrix is

$$P = \begin{matrix} & 0 & 1 & 2 & 3 & & n-1 & n \\ 0 & 1 & 0 & 0 & 0 & \cdots & 0 & 0 \\ 1 & q & 0 & p & 0 & \cdots & 0 & 0 \\ 2 & 0 & q & 0 & p & \cdots & 0 & 0 \\ & & \cdots & & & & & \cdots \\ n-1 & 0 & 0 & 0 & 0 & \cdots & 0 & p \\ n & 0 & 0 & 0 & 0 & \cdots & 0 & 1 \end{matrix}.$$

The reader should verify that if $p = \frac{1}{2}$, then $\{x_n\}$ is a martingale and that if $p \neq \frac{1}{2}$, then $\{(q/p)^{x_n}\}$ is a martingale.

EXAMPLE 4: Infinite drunkard's walk.

For this process, which is an extension of the one in Example 3, the states are the non-negative integers, and state 0 is absorbing. For each $i > 0$, we have

$$P_{i,i+1} = p, \quad P_{i,i-1} = q; \quad \text{and} \quad p + q = 1.$$

We assume $p \neq 0$ and $q \neq 0$. The transition matrix is

$$P = \begin{matrix} & \begin{matrix} 0 & 1 & 2 & 3 & \end{matrix} \\ \begin{matrix} 0 \\ 1 \\ 2 \\ \\ \end{matrix} & \begin{pmatrix} 1 & 0 & 0 & 0 & \cdots \\ q & 0 & p & 0 & \cdots \\ 0 & q & 0 & p & \cdots \\ \vdots & & & & \ddots \end{pmatrix} \end{matrix}.$$

If $p = \frac{1}{2}$, then $\{x_n\}$ is a martingale, and if $p \neq \frac{1}{2}$, then $\{(q/p)^{x_n}\}$ is a martingale.

EXAMPLE 5: Basic example.

A sequence of tasks is to be performed in a certain order, each with its own probability of success. Success means that the process goes to the next state; failure means that the process must start over at state 0. Thus the states are the non-negative integers, and with each positive integer i we associate two probabilities p_i and q_i such that $p_i + q_i = 1$. The value p_i is the transition probability from state $i - 1$ to state i, and q_i is the transition probability from state $i - 1$ to 0. Thus p_i is the probability of succeeding in the ith task. We assume that $p_i < 1$ for infinitely many i, and we normally assume that $p_i > 0$ for every i. The transition matrix is

$$P = \begin{matrix} & \begin{matrix} 0 & 1 & 2 & 3 & \end{matrix} \\ \begin{matrix} 0 \\ 1 \\ 2 \\ \\ \end{matrix} & \begin{pmatrix} q_1 & p_1 & 0 & 0 & \cdots \\ q_2 & 0 & p_2 & 0 & \cdots \\ q_3 & 0 & 0 & p_3 & \cdots \\ \vdots & & & & \ddots \end{pmatrix} \end{matrix}.$$

In connection with this example, we define a row vector β by

$$\beta_0 = 1$$

$$\beta_i = \prod_{k=1}^{i} p_k.$$

Then β_i is the probability of i successes in a row after the process starts at 0. The reader should verify that a necessary and sufficient condition for $\beta = \beta P$ is that $\lim_{i \to \infty} \beta_i = 0$. This Markov chain will be referred to hereafter as "the basic example."

EXAMPLE 6: Sums of independent random variables.

The states of a Markov chain P are the elements of an index set I on which an operation of addition is defined in such a way that I becomes an abelian group. A probability distribution $\{p_i\}$ defined on I satisfies $p_i \geq 0$ and $\sum p_i = 1$. The Markov chain P is defined by $p_{ij} = p_{j-i}$.

The name of this Markov chain is derived from thinking of performing independent experiments which have probability p_i of outcome i. The states of the chain are the partial sums of these results, and the sum changes from i to j with probability $P_{i,i+k} = p_k$ if $j = i + k$.

For the case in which the index set I is the set of integers with the usual concept of addition defined on them, martingales arise as in Example 1 of Section 3-2. We shall apply these ideas in Chapter 5.

EXAMPLE 7: Two classes of random walks.

We shall be concerned especially with two kinds of random walks. The symmetric random walk in n-dimensions is defined to be a sums of independent random variables process on the lattice of integer points in n-dimensional Euclidean space. The transition probability from one lattice point to another is $(2n)^{-1}$ if the two points are a Euclidean distance of one unit apart; the transition probability is zero otherwise. Thus, from each point the process moves to one of $2n$ neighboring points with probability $(2n)^{-1}$.

A second kind of random walk with which we shall be concerned is a sums of independent random variables process on the integers with $P_{i,i+1} = p$ and $P_{i,i-1} = q$ for every i. We shall call this process the p-q random walk. If $p = q = \frac{1}{2}$, then $\{x_n\}$ is a martingale, and if $p \neq \frac{1}{2}$, then $\{(q/p)^{x_n}\}$ is a martingale.

EXAMPLE 8: General random walks on the line.

The state space for a random walk on the line is the set of integers, and for each integer i, three probabilities p_i, q_i, and r_i with $p_i + q_i + r_i = 1$ are specified. A Markov chain is defined by

$$P_{i,i+1} = p_i$$
$$P_{i,i-1} = q_i$$
$$P_{i,i} = r_i.$$

The drunkard's walk and the p-q random walk are both special cases.

An important case of random walks on the line which we have not discussed yet is the reflecting random walk. For this chain the process is started at a state which is a non-negative integer, and the assumption is made that $q_0 = 0$. The process never reaches the negative integers, and the state space may just as well be taken as $\{0, 1, 2, \ldots\}$.

EXAMPLE 9: Branching process.

The state space is the set of non-negative integers, and a fixed probability distribution $p = \{p_0, p_1, p_2, \ldots\}$ is specified. Suppose the mean $\sum kp_k$ of p is m. Let $\{\mathbf{y}_n\}$ be a sequence of non-negative integer-valued independent random variables with common distribution p, and set $\mathbf{s}_n = \mathbf{y}_1 + \cdots + \mathbf{y}_n$. Let $p_j^{(n)} = \Pr[\mathbf{s}_n = j]$. Then the branching process is defined to be a Markov chain with transition probabilities $P_{ij} = p_j^{(i)}$.

The usual model is the following. A species of bacteria has the distribution p representative of the number of offspring one such bacterium has before it dies. The value of $\mathbf{y}_k$ represents the number of offspring the kth bacterium has while it is alive, and $\mathbf{s}_n$ represents the total number of bacteria produced by n bacteria in one generation of the colony. The rth position, x_r, of the stochastic process is the number of bacteria in the rth offspring generation.

As we have noted, the branching process is a Markov chain. Let $\{x_n\}$ be the outcome functions for the chain started in state 1 (that is, with one bacterium in the colony initially), and suppose the mean m is finite. Then $\{x_n/m^n\}$ with its natural partition forms a martingale. The reader should verify that $\mathrm{M}[|x_n/m^n|]$ is finite; we shall show that $\mathrm{M}[x_{n+1}/m^{n+1} \mid x_1/m \wedge \cdots \wedge x_n/m^n] = x_n/m^n$. First we note that

$$\mathrm{M}[\mathbf{s}_n] = \sum \mathrm{M}[\mathbf{y}_i] = nm,$$

so that if we know that the process is in state r, then the mean state that it is in after the next step is rm. Then

$$\begin{aligned}
&\mathrm{M}[x_{n+1}/m^{n+1} \mid x_1/m \wedge \cdots \wedge x_n/m^n] \\
&\qquad = \mathrm{M}[x_{n+1}/m^{n+1} \mid x_n/m^n] \quad \text{by the Markov property} \\
&\qquad = \mathrm{M}[x_{n+1}/m^{n+1} \mid x_n] \\
&\qquad = (1/m^{n+1})\mathrm{M}[x_{n+1} \mid x_n] \\
&\qquad = (1/m^{n+1})x_n m \quad \text{by the remarks above} \\
&\qquad = x_n/m^n.
\end{aligned}$$

EXAMPLE 10: Tree process.

Let $\{x_n\}$ be a denumerable stochastic process defined on sequence space, and let S be the set of states. Define a set T to be the set of all finite sequences of elements in S. Define a new stochastic process as follows: If t and u are elements of T for which

$$t = (c_0, c_1, c_2, \ldots, c_n)$$

and

$$u = (c_0, c_1, c_2, \ldots, c_n, c_{n+1}),$$

define

$$\Pr[y_{n+1} = u \mid y_n = t] = \Pr[x_{n+1} = c_{n+1} \mid x_0 = c_0 \wedge \cdots \wedge x_n = c_n].$$

The process $\{y_n\}$ defined from the same space to the set T is a Markov chain; the entire history of the original process up to time n is contained in the knowledge of the value of the nth outcome function y_n for the new process.

An example of a tree process is obtained by considering an individual's voting history in successive years. Letting D and R represent the political parties, we see that his possible histories can be conveniently represented as a tree:

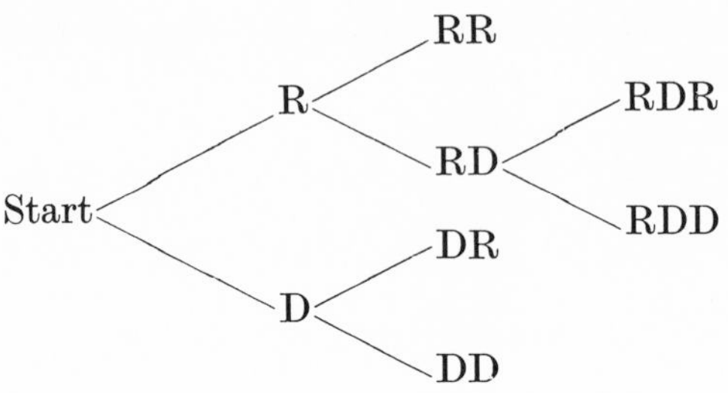

The chain is in each state—D, R, DD, DRR, etc.—at most once.

3. Applications of martingale ideas

Let P be the transition matrix for a Markov chain. A column vector f is said to be a P-**regular function**, or simply a **regular function**, if $f = Pf$. The function is **superregular** if $f \geq Pf$; it is **subregular** if $f \leq Pf$.

The reader should convince himself that the regularity of a function is a condition of the following form: At each point of the domain, the value of the function is equal to the average value of the function at neighboring points. By neighboring points we mean those states that it is possible for the process to reach in one step, and by average value we mean the average obtained by weighting the function values at neighboring points by the transition probabilities to those states. A function f is said to be **regular at a point** j if $f_j = (Pf)_j$.

Regular measures may be defined analogously with regular functions. A non-negative row vector π is a **regular measure** if $\pi = \pi P$; it is **superregular** if $\pi \geq \pi P$ and **subregular** if $\pi \leq \pi P$.

Let h be a P-regular function and let $h(x_n)$ denote h_j if $x_n = j$. Suppose $\mathrm{M}[|h(x_n)|]$ is finite. We shall show that $(h(x_n), \mathscr{R}_n)$ is a martingale, where $\mathscr{R}_n$ is the cross partition $\mathscr{R}^{x_0} \otimes \mathscr{R}^{x_1} \otimes \cdots \otimes \mathscr{R}^{x_n}$ determined by the outcome functions $x_0, x_1, \ldots, x_n$ for the Markov chain. It is sufficient to show that

$$\mathrm{M}[h(x_{n+1}) \mid x_0 \wedge \cdots \wedge x_n] = h(x_n).$$

On the cell of the cross-partition where $x_0 = i, \ldots, x_n = j$ and where $\Pr[x_0 = i \wedge \cdots \wedge x_n = j] > 0$,

$$\begin{aligned}
M[h(x_{n+1}) \mid x_0 \wedge \cdots \wedge x_n] &= M[h(x_{n+1}) \mid x_0 = i \wedge \cdots \wedge x_n = j] \\
&= \sum_k \Pr[x_{n+1} = k \mid x_0 = i \wedge \cdots \wedge x_n = j] \cdot h_k \\
&= \sum_k P_{jk} h_k \quad \text{by the Markov chain property} \\
&= h_j \quad \text{since } h \text{ is regular} \\
&= h(x_n).
\end{aligned}$$

Thus $(h(x_n), \mathscr{R}_n)$ is a martingale. Similarly, superregular functions are associated naturally with supermartingales and subregular functions correspond to submartingales. The proofs differ from the above proof only by insertion of the appropriate inequality sign in the next to the last step.

Most of our applications of martingale ideas we shall leave to the next few chapters. We shall, however, settle some things about branching processes at this time. Let $\{x_n\}$ be the outcome functions for a branching process started in state 1, and suppose the mean $m = \sum kp_k$ is finite. As we noted in Section 2, $\{x_n/m^n\}$ forms a non-negative martingale, which by Corollary 3-13 converges almost everywhere to a finite limiting function g. One can show that g is not a constant function; that is, the value of the limit of $\{x_n(\omega)/m^n\}$ very much depends upon the early history of the path. The exact distribution of g, however, is an unsolved problem.

On the other hand, information about whether the process dies out (by being absorbed at state 0) is not hard to obtain. Let $\varphi(s) = \sum_j p_j s^j$ and suppose $r = \varphi(r)$, $r \geq 0$, and $r \neq 1$. Then $\{r^{x_n}\}$ is a non-negative martingale. First suppose $r > 1$. Since $\{r^{x_n}\}$ is a non-negative martingale, it converges to a finite limit almost everywhere, and since $r > 1$, x_n itself converges with probability one. Since x_n is integer-valued, x_n is constant on almost all paths from some point on. It is left as an exercise to show that the constant must be zero and that the process therefore dies out with probability one. Next suppose $r < 1$. Then $\{r^{x_n}\}$ is bounded and converges almost everywhere to a limiting function r^{x_∞}, which must be 0 or 1 (that is, $x_\infty = \infty$ or 0) almost everywhere. By dominated convergence we have

$$\begin{aligned}
r = M[r^{x_0}] = M[r^{x_\infty}] = 1 \cdot &\Pr[\text{process dies out}] \\
&+ 0 \cdot \Pr[\text{process does not die out}],
\end{aligned}$$

so that r is the probability that the process dies out in the long run.

Finally, suppose $r = 1$ is the only non-negative root of $s = \varphi(s)$. Then $m = 1$ and $\{x_n\}$ is a non-negative martingale. Once again we must have $x_n \to 0$ with probability one, and the process is almost certain to die out.

The reader should notice that the case $r = 1$ has the property that $\mathrm{M}[x_n] = 1$ for all n, whereas $\mathrm{M}[\lim x_n] = 0$. The process in this case is an example of a fair game whose final expected fortune is strictly less than the starting fortune.

4. Strong Markov property

The **strong Markov property** is a rigorous formulation of the following assertion about a Markov chain: If the present is known, then the probability of any statement depending on the future is independent of what additional information about the past is known. In this section we shall state and prove this result; our procedure will be first to prove a conceptually simpler special case and then to obtain the general theorem as an easy consequence. In the special case the time of the present will be a fixed time n, whereas in the general case the time of the present will be allowed to depend on the past history of the process. That is, the time of the present will be a random time. Knowledge of the present, then, means knowledge of the outcome at the time of the present.

If $\omega = (c_0, c_1, c_2, \ldots, c_{n-1}, c_n, c_{n+1}, \ldots)$ is a point in a sequence space, we agree to call the path

$$(c_n, c_{n+1}, \ldots)$$

by the name ω_n.

Lemma 4-4: Let $\{p_k\}$ be a sequence of statements whose truth sets are disjoint in pairs, let $\bigvee p_k$ be their disjunction, and suppose $\Pr[\bigvee p_k] > 0$. If p is a statement for which $\Pr[p \mid p_k] = c$ whenever $\Pr[p_k] > 0$, then $\Pr[p \mid \bigvee p_k] = c$.

PROOF: For each k,

$$c \cdot \Pr[p_k] = \Pr[p \wedge p_k].$$

Thus

$$c \sum_k \Pr[p_k] = \sum_k \Pr[p \wedge p_k].$$

Since the p_k are disjoint statements, it follows from complete additivity that $\sum \Pr[p_k] = \Pr[\bigvee p_k]$ and that $\sum \Pr[p \wedge p_k] = \Pr[p \wedge (\bigvee p_k)]$. Thus $c = \Pr[p \wedge (\bigvee p_k)]/\Pr[\bigvee p_k]$ and the lemma follows.

Throughout the remainder of this section let $(\Omega, \mathscr{B}, \mu)$ be a fixed sequence space and $\{x_n\}$ a fixed denumerable Markov chain defined on Ω. The field of cylinder sets will be denoted by $\mathscr{F}$, as in Section 2-1, and the smallest Borel field containing $\mathscr{F}$ will be called $\mathscr{G}$.

Definition 4-5: The **tail-field** $\mathscr{T}_n$ is the smallest augmented Borel field containing all truth sets of statements $x_n = c_n \wedge \cdots \wedge x_m = c_m$, $m \geq n$. [Thus $\mathscr{T}_0 = \mathscr{B}$ and $\mathscr{T}_n \subset \mathscr{T}_{n-1}$.]

A statement relative to the field $\mathscr{F}_n$ defined in Section 2-1 is one whose truth set depends only on outcomes $x_0, \ldots, x_n$, whereas a statement relative to $\mathscr{T}_n$ is one whose truth set does not depend on outcomes $x_0, \ldots, x_{n-1}$. Specifically, a set R in $\mathscr{B}$ is in $\mathscr{T}_n$ if and only if, whenever $\omega \in R$ and ω' is such that $\omega'_n = \omega_n$, then $\omega' \in R$.

We note that the class of sets $\mathscr{T}_n \cap \mathscr{F}$, being the intersection of fields, is again a field. Moreover, $\mathscr{T}_n$ is the smallest augmented Borel field containing $\mathscr{T}_n \cap \mathscr{F}$, so that the uniqueness statement of Theorem 1-19 applies: A probability measure on $\mathscr{T}_n$ is completely determined by its values on $\mathscr{T}_n \cap \mathscr{F}$.

Lemma 4-6: Let $\{x_n\}$ be a Markov chain with starting distribution π, let q be a statement relative to $\mathscr{F}_{n-1}$, let r be a statement relative to $\mathscr{T}_{n+1} \cap \mathscr{F}$, and suppose $\Pr_\pi[q \wedge x_n = i] > 0$. Then

$$\Pr_\pi[r \mid q \wedge x_n = i] = \Pr_\pi[r \mid x_n = i] = \Pr_i[r'],$$

where r' is so chosen that $\omega \in R$ if and only if $\omega_n \in R'$.

REMARK: Such an r' exists (and is unique), since r is a statement relative to $\mathscr{T}_{n+1} \cap \mathscr{F}$.

PROOF:

Case 1: r is of the form $x_{n+1} = j$. Write q as a disjunction $q = \bigvee q_m$, where

$$q_m: \quad x_0 = c_0^{(m)} \wedge \cdots \wedge x_{n-1} = c_{n-1}^{(m)}.$$

For each m such that $\Pr_\pi[q_m \wedge x_n = i] > 0$,

$$\Pr_\pi[r \mid q_m \wedge x_n = i] = \Pr_\pi[x_{n+1} = j \mid q_m \wedge x_n = i] = P_{ij}$$

by Definition 4-1. Hence, since $\Pr_\pi[q \wedge x_n = i] > 0$,

$$\Pr_\pi[r \mid q \wedge x_n = i] = P_{ij} = \Pr_\pi[r \mid x_n = i]$$

by Lemma 4-4. Taking r' as $x_1 = j$, we have

$$P_{ij} = \Pr_i[x_1 = j] = \Pr_i[r'].$$

Case 2: r is of the form $x_{n+1} = c_{n+1} \wedge \cdots \wedge x_m = c_m$, $m > n$. We have

$$
\begin{aligned}
\Pr_\pi[r \mid q \wedge x_n = i] \\
&= \Pr_\pi[x_{n+1} = c_{n+1} \mid q \wedge x_n = i] \\
&\times \Pr_\pi[x_{n+2} = c_{n+2} \mid q \wedge x_n = i \wedge x_{n+1} = c_{n+1}] \\
&\times \cdots \times \Pr_\pi[x_m = c_m \mid q \wedge x_n = i \wedge \cdots \wedge x_{m-1} = c_{m-1}].
\end{aligned}
$$

The general factor on the right is

$$\Pr_\pi[x_{n+k+1} = c_{n+k+1} \mid q \wedge x_n = i \wedge \cdots \wedge x_{n+k} = c_{n+k}].$$

First, suppose that none of these factors is zero. Then we may apply Case 1 with $n + k$ in place of n and $q \wedge x_n = i \wedge \cdots \wedge x_{n+k-1} = c_{n+k-1}$ in place of q. The q's drop out of the conditions, and the product of the new conditional probabilities is $\Pr_\pi[r \mid x_n = i]$.

Next, suppose that at least one of the factors is zero; let the first such factor from the left be

$$\Pr_\pi[x_{n+k+1} = c_{n+k+1} \mid q \wedge x_n = i \wedge \cdots \wedge x_{n+k} = c_{n+k}].$$

We must show that $\Pr_\pi[r \mid x_n = i] = 0$. If $k = 0$, then by Case 1

$$0 = \Pr_\pi[x_{n+1} = c_{n+1} \mid q \wedge x_n = i] = \Pr_\pi[x_{n+1} = c_{n+1} \mid x_n = i],$$

and hence $\Pr_\pi[r \mid x_n = i] = 0$. If $k > 0$, then

$$\Pr_\pi[q \wedge x_n = i \wedge \cdots \wedge x_{n+k-1} = c_{n+k-1}] > 0,$$

and Case 1 gives

$$
\begin{aligned}
0 &= \Pr_\pi[x_{n+k+1} = c_{n+k+1} \mid q \wedge x_n = i \wedge \cdots \wedge x_{n+k} = c_{n+k}] \\
&= \Pr_\pi[x_{n+k+1} = c_{n+k+1} \mid x_n = i \wedge \cdots \wedge x_{n+k} = c_{n+k}].
\end{aligned}
$$

Hence $\Pr_\pi[r \mid x_n = i] = 0$.

Finally r' is the statement $x_1 = c_{n+1} \wedge \cdots \wedge x_{m-n} = c_m$, and, since $\Pr_\pi[x_n = i] \geq \Pr_\pi[q \wedge x_n = i] > 0$, we have

$$\Pr_\pi[r \mid x_n = i] = P_{i,c_{n+1}} \cdot P_{c_{n+1},c_{n+2}} \cdot \cdots \cdot P_{c_{m-1},c_m} = \Pr_i[r'].$$

Case 3: r is arbitrary in $\mathscr{T}_{n+1} \cap \mathscr{F}$. A general statement r reduces to the denumerable union of the type statements in Case 2, and the result follows from the complete additivity of the probability measure.

The lemma to follow is the strong Markov property for the case in which the time of the present is a fixed time n.

Lemma 4-7: Let $\{x_n\}$ be a Markov chain with starting distribution π, let q be a statement relative to $\mathscr{F}_n$, let r be a statement relative to $\mathscr{T}_n$, and suppose $\Pr_\pi[q \wedge x_n = i] > 0$. Then

$$\Pr_\pi[r \mid q \wedge x_n = i] = \Pr_\pi[r \mid x_n = i] = \Pr_i[r'],$$

where $\omega \in R$ if and only if $\omega_n \in R'$.

Proof: Write

$$q = \bigvee_m (x_0 = c_0^{(m)} \wedge \cdots \wedge x_{n-1} = c_{n-1}^{(m)} \wedge x_n = c_n^{(m)}).$$

If we set $q^* = \bigvee_m (x_0 = c_0^{(m)} \wedge \cdots \wedge x_{n-1} = c_{n-1}^{(m)})$, where the disjunction is taken over just those m such that $c_n^{(m)} = i$, then

$$(q^* \wedge x_n = i) \equiv (q \wedge x_n = i)$$

and q^* is a statement relative to $\mathscr{F}_{n-1}$. In the special case where r is relative to $\mathscr{T}_n \cap \mathscr{F}$, we may write

$$r = \bigvee_m (x_n = c_n^{(m)} \wedge \cdots \wedge x_N = c_N^{(m)}) \qquad (N \text{ fixed})$$

and

$$r^* = \bigvee_m (x_{n+1} = c_{n-1} \wedge \cdots \wedge x_N = c_N^{(m)})$$

with the second disjunction taken over only those m such that $c_n^{(m)} = i$. Then

$$\Pr_\pi[r \mid q \wedge x_n = i] = \Pr_\pi[r^* \mid q^* \wedge x_n = i],$$

$$\Pr_\pi[r \mid x_n = i] = \Pr_\pi[r^* \mid x_n = i],$$

and

$$\Pr_i[r'] = \Pr_i[r^{*\prime}].$$

By Lemma 4-6,

$$\Pr_\pi[r^* \mid q^* \wedge x_n = i] = \Pr_\pi[r^* \mid x_n = i] = \Pr_i[r^{*\prime}].$$

Hence

$$\Pr_\pi[r \mid q \wedge x_n = i] = \Pr_\pi[r \mid x_n = i] = \Pr_i[r'].$$

We have thus established the lemma for every r measurable with respect to $\mathscr{T}_n \cap \mathscr{F}$. But $\mathscr{T}_n$ is the smallest augmented Borel field containing $\mathscr{T}_n \cap \mathscr{F}$, and by Theorem 1-19 any two measures on $\mathscr{T}_n$ which agree on $\mathscr{T}_n \cap \mathscr{F}$ must agree on all of $\mathscr{T}_n$. Thus

$$\Pr_\pi[r \mid q \wedge x_n = i], \quad \Pr_\pi[r \mid x_n = i], \quad \text{and} \quad \Pr_i[r'],$$

which define such measures as r varies, are equal for every r measurable with respect to $\mathscr{T}_n$.

Turning to the general case of the strong Markov property, let $\mathbf{t}$ be a random time. We define ω_t pointwise to be ω_n at all points where

$\mathbf{t}(\omega) = n$. We do not define ω_t if $\mathbf{t}(\omega) = \infty$. Similarly the outcome function x_t is defined to be $x_n(\omega)$ if $\mathbf{t}(\omega) = n$, and it is not defined for $\mathbf{t}(\omega) = \infty$.

Definition 4-8: The field $\mathscr{F}_t$ is the Borel field of all sets A such that, for each n, $A \cap \{\omega \mid \mathbf{t}(\omega) = n\}$ is in $\mathscr{F}_n$. The tail-field $\mathscr{T}_t$ is the smallest augmented Borel field containing all truth sets of statements

$$x_t = c_t \wedge \cdots \wedge x_{t+k} = c_{t+k}, \qquad k \geq 0.$$

A statement q relative to $\mathscr{F}_t$ is one such that, for each n, the statement $q \wedge \mathbf{t} = n$ depends only on outcomes $x_0, \ldots, x_n$. A statement r relative to $\mathscr{T}_t$ is one whose truth set does not depend on outcomes before time $\mathbf{t}$. Specifically, a set R in $\mathscr{B}$ is in $\mathscr{T}_t$ if and only if whenever $\omega \in R$ and ω' is such that $\omega'_t = \omega_t$, then $\omega' \in R$.

We state the strong Markov property as the next theorem.

Theorem 4-9: Let $\{x_n\}$ be a Markov chain with starting distribution π, let $\mathbf{t}$ be a random time, let q be a statement relative to $\mathscr{F}_t$, let r be a statement relative to $\mathscr{T}_t$, and suppose $\Pr_\pi[q \wedge x_t = i] > 0$. Then

$$\Pr_\pi[r \mid q \wedge (x_t = i)] = \Pr_\pi[r \mid x_t = i] = \Pr_i[r'],$$

where $\omega \in R$ if and only if $\omega_t \in R'$.

PROOF: We shall prove the theorem for any statement r measurable with respect to $\mathscr{T}_t \cap \mathscr{F}$. The theorem for general r will then follow, as in the proof of Lemma 4-7, from the uniqueness half of Theorem 1-19. Since $x_t \neq i$ when $\mathbf{t} = \infty$, we have

$$(q \wedge x_t = i) \equiv \bigvee_{n=0}^{\infty} (q \wedge x_n = i \wedge \mathbf{t} = n).$$

We are going to apply Lemma 4-4 with p the statement r, with p_n the statement $q \wedge x_n = i \wedge \mathbf{t} = n$, and with c the constant $\Pr_i[r']$. To do so, we must show that

$$\Pr_\pi[r \mid q \wedge x_n = i \wedge \mathbf{t} = n] = \Pr_i[r']$$

whenever $\Pr_\pi[q \wedge x_n = i \wedge \mathbf{t} = n] > 0$, and we will have proved that

$$\Pr_\pi[r \mid q \wedge x_t = i] = \Pr_i[r'].$$

The fact that $\Pr_\pi[r \mid x_t = i]$ equals both of these quantities will follow by taking q to be a tautology.

Thus we first note that $q \wedge \mathbf{t} = n$ is measurable with respect to $\mathscr{F}_n$. In addition there exists a statement $\hat{r}$ measurable with respect to $\mathscr{T}_n$ such that

$$(r \wedge \mathbf{t} = n) \equiv (\hat{r} \wedge \mathbf{t} = n);$$

this is so because r is the denumerable union of statements

$$x_t = c_t \wedge \cdots \wedge x_{t+N} = c_{t+N},$$

and we may take $\hat{r}$ to be the same union over the statements

$$x_n = c_t \wedge \cdots \wedge x_{n+N} = c_{t+N}.$$

In this notation the statement r' is the union of the statements

$$x_0 = c_t \wedge \cdots \wedge x_N = c_{t+N},$$

and we have that ω is in the truth set of $\hat{r}$ if and only if ω_n is in the truth set of r'. Hence

$$\begin{aligned}
\Pr_\pi[r \mid q \wedge x_n = i \wedge \mathbf{t} = n] &= \Pr_\pi[r \wedge \mathbf{t} = n \mid q \wedge x_n = i \wedge \mathbf{t} = n] \\
&= \Pr_\pi[\hat{r} \wedge \mathbf{t} = n \mid q \wedge x_n = i \wedge \mathbf{t} = n] \\
&= \Pr_\pi[\hat{r} \mid (q \wedge \mathbf{t} = n) \wedge x_n = i] \\
&= \Pr_\pi[\hat{r} \mid x_n = i] \\
&= \Pr_i[r'],
\end{aligned}$$

the last two equalities following from Lemma 4-7.

An equivalent way of stating the first equality of the conclusion of the preceding theorem is

$$\Pr_\pi[q \wedge r \mid x_t = i] = \Pr_\pi[q \mid x_t = i] \Pr_\pi[r \mid x_t = i].$$

This is the form in which the theorem asserts that if the present is known, then the past and future are independent.

5. Systems theorems for Markov chains

As immediate consequences of the strong Markov property, we can prove two systems theorems for Markov chains. The first states that if p is a statement depending on outcomes only beyond some random time $\mathbf{t}$, then one may compute $\Pr_\pi[p]$ as if the chain were started with the initial distribution $\Pr_\pi[x_t = j]$.

Theorem 4-10: Let $\{x_n\}$ be a Markov chain, and let p be a measurable statement with truth set P satisfying

(1) $\Pr_\pi[p \wedge (\mathbf{t} = \infty)] = 0$, and
(2) there exists a statement p' with truth set P' such that if $\mathbf{t}(\omega) < +\infty$, then $\omega \in P$ if and only if $\omega_t \in P'$.

Then

$$\Pr_\pi[p] = \sum_{k \in S} \Pr_\pi[x_t = k] \Pr_k[p'].$$

PROOF: By (1) we have

$$\begin{aligned}
\Pr_\pi[p] &= \sum_k \Pr_\pi[\omega \in P \wedge x_t = k] \\
&= \sum_k \Pr_\pi[\omega_t \in P' \wedge x_t = k] \\
&= \sum_k \Pr_\pi[x_t = k] \Pr_\pi[\omega_t \in P' \mid x_t = k] \\
&= \sum_k \Pr_\pi[x_t = k] \Pr_k[p'] \quad \text{by Theorem 4-9.}
\end{aligned}$$

Theorem 4-10, which is a result about probabilities of statements, can also be thought of as a result about means of characteristic functions. Then Theorem 4-11 to follow becomes a straightforward generalization to arbitrary functions.

Theorem 4-11: Let $\{x_n\}$ be a Markov chain, and let $\mathbf{f}$ be a random variable satisfying

(1) $\Pr_\pi[\mathbf{f} \neq 0 \wedge \mathbf{t} = \infty] = 0$, and
(2) there exists a random variable $\mathbf{f}'$ such that if $\mathbf{t}(\omega) < \infty$, then $\mathbf{f}(\omega) = \mathbf{f}'(\omega_t)$.

Then

$$\mathrm{M}_\pi[\mathbf{f}] = \sum_k \Pr_\pi[x_t = k]\, \mathrm{M}_k[\mathbf{f}'].$$

PROOF: If $\mathbf{f}$ assumes negative values, we may prove the result for $\mathbf{f}^+$ and $\mathbf{f}^-$ separately. We therefore assume $\mathbf{f} \geq 0$. Let $p_j^{(m)}$ be the statement $j/2^m \leq \mathbf{f} < (j+1)/2^m$, for $1 \leq j < m \cdot 2^m$, let $p_0^{(m)}$ be the statement $0 < \mathbf{f} < 1/2^m$, and let $p_{m2^m}^{(m)}$ be the statement $m \leq \mathbf{f}$. Define statements $p_j^{(m)'}$ similarly for $\mathbf{f}'$. Then $p_j^{(m)}$ and $p_j^{(m)'}$ satisfy the hypotheses of Theorem 4-10, so that

$$\Pr_\pi[p_j^{(m)}] = \sum_k \Pr_\pi[x_t = k] \Pr_k[p_j^{(m)'}].$$

Hence

$$\begin{aligned}
\mathrm{M}_\pi[\mathbf{f}] &= \lim_m \sum_{j=0}^{m2^m} \frac{j}{2^m} \Pr_\pi[p_j^{(m)}] \\
&= \lim_m \sum_k \Pr_\pi[x_t = k] \sum_{j=0}^{m2^m} \frac{j}{2^m} \Pr_k[p_j^{(m)'}] \\
&= \sum_k \Pr_\pi[x_t = k] \lim_m \sum_{j=0}^{m2^m} \frac{j}{2^m} \Pr_k[p_j^{(m)'}] \quad \text{by monotone convergence} \\
&= \sum_k \Pr_\pi[x_t = k]\, \mathrm{M}_k[\mathbf{f}'].
\end{aligned}$$

6. Applications of systems theorems

The theorems of Section 5 will play an important role in our study of denumerable Markov chains. At this time we shall not illustrate the full power of the theorems but shall be content instead to use them in developing some of the machinery needed for the classification of states in Section 7.

We begin by introducing some notation. Define

$$\delta_{ij} = \begin{cases} 1 & \text{if } i = j \\ 0 & \text{otherwise.} \end{cases}$$

Let h_j be the statement about a Markov chain that state j is eventually reached. We have already defined the random variables $\mathbf{n}_j$ and $\mathbf{t}_j$ for general stochastic processes (see Section 2-6); $\mathbf{n}_j$ is the number of times in state j, and $\mathbf{t}_j$ is the time to reach state j. Let $f_j^{(k)}$ be the statement that $\mathbf{t}_j(\omega) = k$.

Confining ourselves to Markov chains, we associate the quantities $\bar{h}_j$, $\bar{\mathbf{n}}_j$, $\bar{\mathbf{t}}_j$, and $\bar{f}_j^{(k)}$ with h_j, $\mathbf{n}_j$, $\mathbf{t}_j$, and $f_j^{(k)}$. They are defined as follows:

$$\begin{aligned} &\bar{h}_j\text{:} && h_j \text{ is true for } \omega_1 \\ &\bar{\mathbf{n}}_j(\omega) && = \mathbf{n}_j(\omega_1) \\ &\bar{\mathbf{t}}_j(\omega) && = \mathbf{t}_j(\omega_1) + 1 \\ &\bar{f}_j^{(k)}\text{:} && \bar{\mathbf{t}}_j(\omega) = k. \end{aligned}$$

In terms of these quantities, we define a collection of matrices. We note that, in general, an expression of the form $\{\mathrm{M}_i[\mathbf{g}_j]\}$ stands for a matrix.

$$\begin{aligned} H_{ij} &= \Pr_i[h_j] \\ N_{ij} &= \mathrm{M}_i[\mathbf{n}_j] \\ F_{ij}^{(k)} &= \Pr_i[f_j^{(k)}] \\ \bar{H}_{ij} &= \Pr_i[\bar{h}_j] \\ \bar{N}_{ij} &= \mathrm{M}_i[\bar{\mathbf{n}}_j] \\ \bar{F}_{ij}^{(k)} &= \Pr_i[\bar{f}_j^{(k)}]. \end{aligned}$$

It is trivial to verify that $H_{ii} = 1$, that $F^{(0)} = I$, that $F^{(1)} = P - P_{dg}$, and that $N = I + \bar{N}$.

Proposition 4-12: If P is a Markov chain, then

$$N = \sum_{k=0}^{\infty} P^k.$$

Proof: The result follows immediately from Propositions 2-10 and 4-3.

Proposition 4-13: If P is a Markov chain, then $\bar{N} = PN$ and $N = I + PN$.

PROOF: The second identity follows from the first by adding I to both sides. To obtain the first one, we apply Theorem 4-11 with $\mathbf{f} = \bar{\mathbf{n}}_j$ and the random time identically one. Then $\mathbf{f}' = \mathbf{n}_j$ since $\bar{\mathbf{n}}_j(\omega) = \mathbf{n}_j(\omega_1)$ by definition, and thus

$$\begin{aligned} \mathrm{M}_i[\bar{\mathbf{n}}_j] &= \sum_k \Pr_i[x_1 = k]\, \mathrm{M}_k[\mathbf{n}_j] \\ &= \sum_k P_{ik}\, \mathrm{M}_k[\mathbf{n}_j] \end{aligned}$$

or

$$\bar{N} = PN.$$

Proposition 4-14: If P is a Markov chain, then

$$H = \sum_{k=0}^{\infty} F^{(k)}, \quad \bar{H} = \sum_{k=1}^{\infty} \bar{F}^{(k)}, \quad \text{and} \quad \bar{H} = PH.$$

PROOF: The first two assertions follow from the complete additivity of μ; we have $h_j \equiv \bigvee f_j^{(k)}$ and $\bar{h}_j \equiv \bigvee \bar{f}_j^{(k)}$ disjointly. For the third assertion we apply Theorem 4-10 with $p = \bar{h}_j$ and the random time identically one. Then p' is the statement h_j and

$$\begin{aligned} \bar{H}_{ij} = \Pr_i[p] &= \sum_k \Pr_i[x_1 = k] \Pr_k[p'] \\ &= \sum_k P_{ik} H_{kj}. \end{aligned}$$

Proposition 4-15: If P is a Markov chain, then

$$\begin{aligned} N_{ij} &= H_{ij} N_{jj} \\ \bar{N}_{ij} &= \bar{H}_{ij} N_{jj} \\ N_{ii} &= 1 + \bar{H}_{ii} N_{ii}. \end{aligned}$$

PROOF: The third assertion follows from the second and the identity $N = I + \bar{N}$ with $j = i$. For the first assertion we apply Theorem 4-11 with $\mathbf{f} = \mathbf{n}_j$ and the random time equal to $\mathbf{t}_j$. It is clear that $\mathbf{n}_j(\omega) = \mathbf{n}_j(\omega_{t_j})$ if $\mathbf{t}_j(\omega) < \infty$. Therefore $\mathbf{f}' = \mathbf{n}_j$, and

$$\begin{aligned} N_{ij} = \mathrm{M}_i[\mathbf{n}_j] &= \sum_k \Pr_i[x_{t_j} = k]\, \mathrm{M}_k[\mathbf{n}_j] \\ &= \Pr_i[x_{t_j} = j]\, \mathrm{M}_j[\mathbf{n}_j] \\ &= \Pr_i[\mathbf{t}_j < \infty]\, \mathrm{M}_j[\mathbf{n}_j] \\ &= H_{ij} N_{jj}. \end{aligned}$$

Similarly, for the second assertion we apply Theorem 4-11 with $\mathbf{f} = \bar{\mathbf{n}}_j$, $\mathbf{f}' = \mathbf{n}_j$, and the random time equal to $\bar{\mathbf{t}}_j$. By the same kind of argument, we find

$$\mathrm{M}_i[\bar{\mathbf{n}}_j] = \bar{H}_{ij}\,\mathrm{M}_j[\mathbf{n}_j].$$

Proposition 4-16: Let p be the statement that a Markov chain reaches state j and then state k with $j \neq k$. Then $\mathrm{Pr}_i[p] = H_{ij}H_{jk}$.

PROOF: In the notation of Theorem 4-10, if $\mathbf{t}_j$ is taken as the random time, then p' is the statement h_k. The theorem applies and

$$\begin{aligned}\mathrm{Pr}_i[p] &= \sum_m \mathrm{Pr}_i[x_{t_j} = m]\,\mathrm{Pr}_m[p'] \\ &= \mathrm{Pr}_i[x_{t_j} = j]\,\mathrm{Pr}_j[h_k] \\ &= H_{ij}H_{jk}.\end{aligned}$$

In our discussion of Markov chains, we shall make frequent use of the following notational devices. Let k and j be states of a Markov chain. By ${}^k\mathbf{n}_j(\omega)$ is meant the number of times on the path ω that the process is in state j before (and not including) the first time that the process is in k. We define ${}^k\bar{\mathbf{n}}_j$ as the number of visits to j before the process reaches k *after* time 0. Notice that ${}^j\mathbf{n}_j(\omega) = 0$, but ${}^j\bar{\mathbf{n}}_j(\omega)$ is 1 if ω starts with j. For fixed k we introduce the corresponding matrices ${}^kN$ and ${}^k\bar{N}$ by

$$\begin{aligned}{}^kN_{ij} &= \mathrm{M}_i[{}^k\mathbf{n}_j(\omega)], \\ {}^k\bar{N}_{ij} &= \mathrm{M}_i[{}^k\bar{\mathbf{n}}_j(\omega)].\end{aligned}$$

They are related as follows:

$${}^k\bar{N}_{ij} = \delta_{ij} + \mathrm{M}_i[{}^k\mathbf{n}_j(\omega_1)].$$

We further define ${}^kH_{ij}$ to be the probability of hitting j before k, having started in i; ${}^k\bar{H}_{ij}$ is the probability of hitting j before hitting k after time 0, having started in i.

We will later want a more general notation than ${}^k\mathbf{n}_j(\omega)$. By ${}^E\mathbf{n}_j(\omega)$ we shall mean the number of times on the path ω that the process is in j before it is in any state of the set E. It is sometimes convenient, in this connection, to think in terms of the modified chain in which the states of E have been made absorbing. Again we have matrices ${}^EN$, and we also introduce the matrices ${}^EH$ and ${}^E\bar{H}$ analogously.

If E is a subset of the set of states S for which neither E nor $\tilde{E}$ is empty, we shall decompose the P matrix into

$$P = \begin{array}{c} \\ E \\ \tilde{E} \end{array}\!\!\begin{array}{c} \begin{array}{cc} E & \tilde{E} \end{array} \\ \begin{pmatrix} T & U \\ R & Q \end{pmatrix} \end{array}$$

according to the method discussed at the end of Section 1-1. If A is an arbitrary matrix indexed by the set S, we write A_E for the restriction of A to a matrix indexed only by E. As an example, we note that $P_E = T$.

7. Classification of states

We introduce a partial ordering on the states S of a Markov chain P. Two states i and j are said to be R-related, written $R(i, j)$ if $H_{ij} > 0$, that is, if it is possible to reach j from i. If $R(i, j)$ and $R(j, i)$, we say that i and j **communicate** and write $i \sim j$. To see that R is a partial ordering, we note that

(1) $H_{ii} = 1 > 0$ so that $R(i, i)$.
(2) If $R(i, j)$ and $R(j, k)$, then $R(i, k)$ because $H_{ik} \geq H_{ij}H_{jk} > 0$ by Proposition 4-16.

The reader should verify that $\sim$ is an equivalence relation.

The relation $\sim$ therefore partitions the states of S into equivalence classes within the ordering, and movement from state to state is within a class or upward through the ordering. We do not assert the existence of maximal classes; we shall see an example later where no maximal classes are present. (The reader should then be able to exhibit an example of a chain having no minimal classes.)

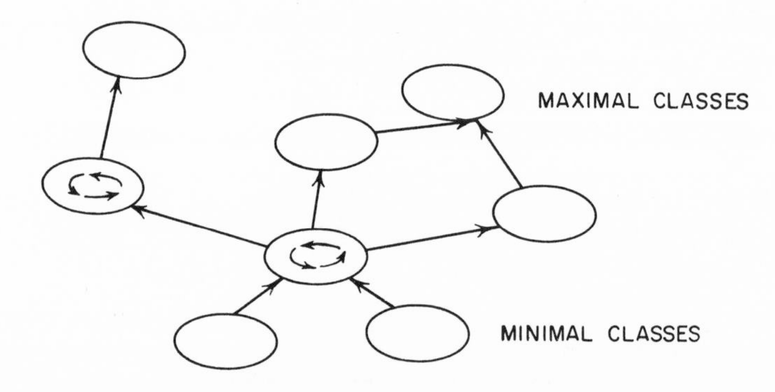

FLOW IN A MARKOV CHAIN

Proposition 4-17: States i and j are R-related if and only if there exists an $n \geq 0$ for which $(P^n)_{ij} > 0$.

PROOF: Suppose $n \geq 0$ is the smallest exponent for which $(P^n)_{ij} > 0$. Then $F_{ij}^{(n)} = (P^n)_{ij} > 0$, and since $H_{ij} = \sum_n F_{ij}^{(n)}$, we have $R(i, j)$. Conversely, if $R(i, j)$, then $H_{ij} > 0$ and it must be true that $F_{ij}^{(n)} > 0$ for some n. Thus $(P^n)_{ij} > 0$.

Definition 4-18: A state i is said to be **recurrent** if $\bar{H}_{ii} = 1$; it is said to be **transient** if $\bar{H}_{ii} < 1$.

The lemma to follow contains some identities connecting H and $\bar{H}$ which will be used in the next few propositions. The reader should study these examples of the use of the strong Markov property in order to develop his intuition.

Lemma 4-19: The following statements hold:

(1) The probability starting in i of returning to state i at least k times is $(\bar{H}_{ii})^k$. (Use the convention $0^0 = 1$.)

(2) The probability starting in i of returning at least k times to i before hitting j is $({}^j\bar{H}_{ii})^k$, provided $i \neq j$.

(3) The probability starting in i of returning to i *via* j is ${}^i\bar{H}_{ij}H_{ji}$, provided $i \neq j$.

(4) The probability starting in i of reaching j for the first time after n returns to i is $({}^j\bar{H}_{ii})^n \, {}^i\bar{H}_{ij}$, provided $i \neq j$.

(5) The probability starting in i of being in state j at least n times is $H_{ij}(\bar{H}_{jj})^{n-1}$.

PROOF: The proofs are all by Theorem 4-10.

(1) Use induction on k. For $k = 0$ the result is trivial; assume that it holds for $k - 1$. Let p be the statement that the process returns to i at least k times, and let $\mathbf{t} = \bar{\mathbf{t}}_i$. Then p' is the statement that the process returns to i at least $k - 1$ times.

$$\begin{aligned}
\Pr_i[p] &= \sum_j \Pr_i[x_{\bar{t}_i} = j] \Pr_j[p'] \\
&= \Pr_i[x_{\bar{t}_i} = i] \Pr_i[p'] \\
&= \Pr_i[\bar{\mathbf{t}}_i < \infty] \Pr_i[p'] \\
&= \bar{H}_{ii}(\bar{H}_{ii})^{k-1} \quad \text{by inductive hypothesis.}
\end{aligned}$$

(2) If $i \neq j$, the result is the same as (1) for the chain in which the single state j has been made absorbing (see Example 2, Section 4-2).

(3) Let p be the statement that the process returns to i *via* j, and let $\mathbf{t}$ be the time that j is reached if j is reached before a return to i, or $+\infty$ if j is not reached before i. Then p' is the statement that i is reached, and

$$\begin{aligned}
\Pr_i[p] &= \sum_k \Pr_i[x_t = k] \Pr_k[p'] \\
&= \Pr_i[x_t = j] H_{ji} \\
&= \Pr_i[\mathbf{t} < \infty] H_{ji} \\
&= {}^i\bar{H}_{ij}H_{ji}.
\end{aligned}$$

(4) The argument is the same as in (3). Use the systems theorem with $\mathbf{t}$ equal to the time of the nth return to i if the return occurs before j is reached, or $+\infty$ otherwise.

(5) The proof is by induction on n and is the same as in (1) except that the random time becomes $\mathbf{t} = \bar{\mathbf{t}}_j$.

Proposition 4-20: State i is transient if and only if $N_{ii} < +\infty$. Then $N_{ii} = 1/(1 - \bar{H}_{ii})$.

PROOF:

$$N_{ii} = \sum_{k=1}^{\infty} k \Pr_i[\mathbf{n}_i = k],$$

which upon rearrangement of terms becomes

$$= \sum_{k=1}^{\infty} \sum_{m=k}^{\infty} \Pr_i[\mathbf{n}_i = m],$$

which by complete additivity is

$$= \sum_{k=1}^{\infty} \Pr_i[\mathbf{n}_i \geq k]$$

$$= \sum_{k=1}^{\infty} (\bar{H}_{ii})^{k-1} \quad \text{by conclusion (1) of Lemma 4-19.}$$

The right side is finite if and only if $\bar{H}_{ii} < 1$.

Corollary 4-21: If j is a transient state, then $N_{ij} < \infty$ for all states i in the chain, and $N_{ij} = H_{ij}N_{jj}$.

PROOF: From Proposition 4-15 we have

$$N_{ij} = H_{ij}N_{jj} \leq N_{jj}.$$

The result now follows from Proposition 4-20.

We are now in a position to put together the ideas of recurrence and transience with the partial ordering R and the equivalence relation $\sim$. We need two lemmas before we can prove our fundamental result—that all states in an equivalence class are of the same type, recurrent or transient.

Lemma 4-22: If $i \neq j$ and $R(i, j)$, then ${}^j\bar{H}_{ii} < 1$ and ${}^jN_{ii} < +\infty$.

PROOF: Suppose ${}^j\bar{H}_{ii} = 1$. By conclusion (2) of Lemma 4-19 the probability of returning n times before hitting j is $({}^j\bar{H}_{ii})^n = 1$. Hence, by Proposition 2-6, there is probability one for returning infinitely often before hitting j, in contradiction to the relation $R(i, j)$. For the second half of the lemma, we have, as in Proposition 4-20,

$$^jN_{ii} = \sum_{n=0}^{\infty} ({}^j\bar{H}_{ii})^n < +\infty.$$

Lemma 4-23: If i is recurrent and $R(i, j)$, then $H_{ji} = 1$ and $H_{ij} = 1$.

PROOF: The result is obvious if $i = j$. If $i \neq j$, consider returning to i with and without first reaching j. By conclusion (3) of Lemma 4-19,

$$1 = \bar{H}_{ii} = {}^i\bar{H}_{ij}H_{ji} + {}^j\bar{H}_{ii}$$

Since ${}^i\bar{H}_{ij} + {}^j\bar{H}_{ii} \leq 1$, this equation is a contradiction unless ${}^j\bar{H}_{ii} = 1$ or $H_{ji} = 1$. The first alternative is ruled out by Lemma 4-22. Thus $H_{ji} = 1$ and ${}^i\bar{H}_{ij} = 1 - {}^j\bar{H}_{ii}$. Next, since i is recurrent, one may compute H_{ij} by summing the probabilities of reaching j for the first time after n returns to i, where $n = 0, 1, 2, \ldots$. By conclusion (4) of Lemma 4-19,

$$H_{ij} = \sum_{n=0}^{\infty} ({}^j\bar{H}_{ii})^n \, {}^i\bar{H}_{ij} = (1 - {}^j\bar{H}_{ii}) \sum_{n=0}^{\infty} ({}^j\bar{H}_{ii})^n = 1.$$

The last equality holds, since ${}^j\bar{H}_{ii} < 1$ by Lemma 4-22.

Proposition 4-24: All states in an equivalence class are of the same type, recurrent or transient.

PROOF: It is sufficient to show that if one state in an equivalence class is recurrent, so are all others. Let i be a recurrent state, and suppose $j \sim i$, $j \neq i$. Then $\bar{H}_{jj} \geq H_{ji}H_{ij}$, since the probability of returning is at least as great as the probability of returning via i. (We have used an argument familiar from Proposition 4-16 to compute the latter.) Hence $\bar{H}_{jj} = 1$ by Lemma 4-23.

Corollary 4-25: If i is recurrent and $i \sim j$, then $H_{ij} = H_{ji} = 1$.

PROOF: The corollary follows from Lemma 4-23.

Because of Proposition 4-24 we are free to speak of transient and recurrent classes of states. We shall mention a few simple results about classes of states. By a **closed class** we mean one that it is impossible to leave. A process cannot disappear when it is in a closed class.

Proposition 4-26: Recurrent classes are closed and maximal with respect to the partial ordering R.

PROOF: It is sufficient to prove that a recurrent class S' is closed, since closed classes are clearly maximal. Suppose the class can be left, say from a state $j \in S'$. If k is a state outside S' for which $P_{jk} > 0$, then it is not true that $R(k, j)$ because j and k do not communicate. Thus $\bar{H}_{jj} \leq 1 - P_{jk} < 1$, and j is not recurrent.

Proposition 4-27: If a Markov chain is started in a recurrent class S', then the chain is in every state of S' infinitely often with probability one. In particular, if i and j are in S', then $N_{ij} = +\infty$.

PROOF: Suppose the chain is started in state i. Then, by conclusion (5) of Lemma 4-19, the probability of being in state j at least n times is $H_{ij}(\bar{H}_{jj})^{n-1} = 1$. By Proposition 2-6 the chain is in state j infinitely often with probability one. Again by Proposition 2-6 it is in *every* state infinitely often with probability one.

Proposition 4-28: A Markov chain is in a finite subset of transient states only finitely often, with probability one.

PROOF: If the chain were in a finite set S' infinitely often with positive probability, it would be in one state j of S' infinitely often with positive probability. Such an occurrence would imply that N_{ij} is infinite for some i, in contradiction to Corollary 4-21 if j is transient.

We single out two kinds of Markov chains for special attention. We note that every absorbing state forms a one-element recurrent class, and conversely.

Definition 4-29: A Markov chain is said to be a **recurrent chain** if its states comprise a single equivalence class and if that class is recurrent. A chain is called a **transient chain** if all of its recurrent states are absorbing.

If P is an arbitrary Markov chain with r recurrent classes, then all properties of P can be deduced from the properties of one transient and r recurrent chains. This assertion follows from the observations:

(1) If the process P starts in a recurrent state j, movement from state to state is confined to the single equivalence class to which j belongs. The properties of the chain started in j are the properties of a chain while it is in one recurrent class; they are thus the properties of a recurrent chain.

(2) If the process P starts in a transient state, its behavior while in transient states is the same as the behavior of the transient chain P' obtained from P by making all recurrent states absorbing. If P enters a recurrent state, then P' becomes absorbed. And after P has entered a recurrent state, its properties are those of a recurrent chain. Thus the properties of P may be studied by considering the one transient chain P' and the r separate recurrent chains.

Because of these observations, we shall restrict our discussion in subsequent chapters to Markov chains which are either transient or recurrent.

The reader should notice that every chain whose states form only one equivalence class is either a transient chain or a recurrent chain. Shortly we shall examine the basic example, in which all pairs of states communicate, to determine when it is transient and when it is recurrent.

First we discuss some properties of maximal classes for a moment. Not every chain has maximal classes; a tree process, for example, consists of infinitely many transient classes of one state each. None of the classes is maximal. Even if a chain does have a maximal class, that class does not have to be closed. The process may have a positive probability of disappearing from some state in the maximal class.

Nor is it true that all closed classes are recurrent. An additional condition is needed.

Proposition 4-30: All closed equivalence classes consisting of finitely many states are recurrent.

PROOF: Let the states be the first n positive integers, and suppose the class is transient. Then N_{ij} is finite for every i and j in the class. Therefore

$$c = M_i\left[\sum_{j=1}^{n} \mathbf{n}_j\right] = \sum_{j=1}^{n} N_{ij}$$

is finite. But c is the mean total number of steps taken in the class, and c is infinite because the class is closed. This contradiction establishes the proposition.

To see that infinite closed equivalence classes need not be recurrent, we consider the basic example, whose states form a single equivalence class. Let $\bar{H}_{00}^{(n)}$ be the probability that the chain, started in state 0, returns to 0 at some time up to and including time n. Then

$$\bar{H}_{00} = \lim_n \bar{H}_{00}^{(n)}.$$

But

$$1 - \bar{H}_{00}^{(n)} = p_1 p_2 \cdots p_n = \beta_n,$$

since a single step other than away from zero returns the process to zero at once. Now in order for the process to be recurrent it is necessary and sufficient that $\bar{H}_{00} = 1$ or that

$$\lim_n \beta_n = \lim_n (1 - \bar{H}_{00}^{(n)}) = 0.$$

The reader should be able to construct examples where $\lim_n \beta_n = 0$ and where $\lim_n \beta_n \neq 0$. Thus the basic example may be either transient or recurrent.

8. Problems

1. Find an expression analogous to that in Proposition 4-3 for $\Pr_i[x_n = j]$ in a Markov process.
2. Let p_m be the Poisson distribution with mean m on the non-negative integers. A game is played as follows: A random integer n_1 is selected with probabilities determined by p_1. A second random integer n_2 is selected with probabilities determined by p_{n_1}. The ith random integer is selected with probabilities determined by $p_{n_{i-1}}$. Prove that with probability one the integer 0 is eventually selected.
3. Show that if $h \geq 0$ is a column vector for which $P^n h$ converges, then the limit function is non-negative superregular.
4. Let j be an absorbing state. Prove that the probability starting at i of ever reaching j is a regular function.
5. Show that an independent trials process is a Markov chain in which P_{ij} is independent of i. Let 0 be any fixed state and let $\mathbf{t}$ be any stopping time. Show that $\Pr_\pi[x_{\mathbf{t}+1} = j] = P_{0j}$, and give an example to show that $\Pr_\pi[x_{\mathbf{t}} = j]$ does not have to equal P_{0j}.
6. If the symmetric random walk in 3 dimensions is started at the origin, the probability of being at the origin after n steps is 0 if n is odd and is of the order of magnitude of $n^{-3/2}$ for n even. Prove that the probability of returning to the origin is less than 1.
7. Consider the following random walk in the plane. If the process is not on an axis, it is equally likely to move to any of the four neighboring states. If it is at the origin, it stays at the origin. Otherwise, on the x-axis it takes a step away from the origin, whereas on the y-axis it takes a step toward the origin. Give a complete classification of the states.
8. Let j be a transient state in a closed class. Prove that there must be a state i in the class such that $H_{ij} < 1$.
9. Prove that every tree-process is a transient chain and that each equivalence class of states is a unit-set.
10. Prove or disprove: In a chain with a minimal class and with no closed class, there is no non-zero non-negative regular measure.

Problems 11 to 14 refer to a reflecting random walk, that is, a random walk on the non-negative integers with $q_0 = 0$.

11. Prove that the only regular functions are constants.

12. Let

$$\beta_i = p_0 p_1 \cdots p_{i-1}, \qquad \gamma_i = q_1 q_2 \cdots q_i, \qquad \alpha_i = c\beta_i/\gamma_i.$$

Show that, for any choice of the constant c, α is a regular measure.

13. Show that all regular measures are of the form given in the previous problem.

14. Show that $\alpha_j P_{ji}/\alpha_i = P_{ij}$ for all i and j.

Problems 15 to 18 refer to a branching process and use the notation of Sections 2 and 3 of the text.

15. Show that the roots of the equation $\varphi(r) = r$ satisfy the following conditions:
(a) There is an $r < 1$ if $m > 1$.
(b) There is an $r > 1$ if $m < 1$.
(c) $r = 1$ is the only root if $m = 1$.

16. Show that $\{r^{x_n}\}$ is a martingale if and only if $\varphi(r) = r$.

17. Show that $\{x_n\}$ is a martingale if and only if $m = 1$.

18. What condition on m will assure that the branching process has positive probability of survival (of not dying out)?

Problems 19 to 24 concern space-time processes and martingales. If P is a Markov chain with state space S, we define the **space-time process** to be a Markov chain whose states are pairs (i, n), where i is in S and n is a non-negative integer, and which moves from (i, n) to $(j, n + 1)$ with probability P_{ij}.

19. Prove that any space-time process is transient. What can be said about classification of states?

20. Prove that if $f(i, n)$ is a finite-valued non-negative regular function for the space-time process, then $f(x_n, n)$ is a martingale for the process P started at a given state 0.

21. Specialize to the case of sums of independent random variables on the integers with $p_k = 0$ for $k < 0$. Define $\varphi(t) = \sum_k p_k t^k$ for all $t \geq 0$ for which the right side is finite. Show that $\varphi(t)$ is defined at least for $0 \leq t \leq 1$. Fix a t for which $\varphi(t)$ is defined and put

$$f(i, n) = \frac{t^i}{[\varphi(t)]^n}.$$

Prove that $f(i, n)$ is regular for the space-time process.

22. In Problem 21 show that $f(x_n, n)$ converges a.e. if the process is started at 0.

23. Specialize further to the case where $p_0 = p_1 = \frac{1}{2}$, and define, for $0 \leq t \leq 1$,

$$g(i, n) = 2^n t^i (1 - t)^{n-i}.$$

Show by change of variable in Problem 22 that $g(x_n, n)$ converges almost everywhere in the process started at 0.

24. Using only the result of Problem 23, prove that if p is any number between 0 and 1, not equal to $\frac{1}{2}$, then the probability that $x_n = [np]$ for infinitely many n is 0. Here $[np]$ is the nearest integer to np.

CHAPTER 5

TRANSIENT CHAINS

1. Properties of transient chains

Recall that a transient Markov chain is a Markov chain all of whose recurrent states are absorbing. Its transition matrix satisfies $P\mathbf{1} \leq \mathbf{1}$. For any transient state j in the chain, we have seen that $\bar{H}_{jj} < 1$ and $N_{ij} < +\infty$ for every i. If E is any set of states, we can put the transition matrix in the canonical form

$$P = \begin{array}{c} \\ E \\ \tilde{E} \end{array} \begin{array}{c} \begin{array}{cc} E & \tilde{E} \end{array} \\ \begin{pmatrix} T & U \\ R & Q \end{pmatrix} \end{array}.$$

In the special case in which P is a transient chain and E is the set of absorbing states, we find that $T = I$ and $U = 0$. (If there are no absorbing states, we agree to write $P = Q$. We shall assume that not all states are absorbing, however.) Thus, for a transient chain,

$$P = \begin{pmatrix} I & 0 \\ R & Q \end{pmatrix}.$$

The matrices R and Q for a transient chain will always be associated with this standard decomposition. We observe that Q itself is the transition matrix for a transient chain and that this chain has only transient states. Some authors actually define a transient chain to be one with all states transient. However, in the study of these chains, it is often convenient to add absorbing states to ensure $P\mathbf{1} = \mathbf{1}$. And as we saw in Chapter 4, the decomposition of general Markov chains into transient and recurrent chains depends on allowing absorbing states in transient chains. For these reasons we have adopted the slightly more general definition of transient chain which permits absorbing states.

Let P be the transition matrix of a transient chain, and consider the quantity N_{ij}, the mean number of times in state j when the process is started in state i. If j is an absorbing state, then this quantity is infinite if j can be reached from i with positive probability and 0 otherwise. If i is absorbing, it is 0 unless $i = j$, and then it is infinite. Hence N_{ij} is of interest only when i and j are transient. Thus we shall agree to restrict N to these entries: The matrix so restricted is called the **fundamental matrix** for the chain. We shall show that the restricted matrix is the matrix $\{N_{ij}\}$ for the chain determined by Q. In what follows, N always denotes the restricted matrix associated with P.

Lemma 5-1: If P is a transient chain and if $\tilde{E}$ is the set of transient states, then $(P^k)_{\tilde{E}} = Q^k$.

PROOF: We readily verify by induction that

$$P^k = \begin{array}{c} \\ E \\ \tilde{E} \end{array} \begin{array}{c} \begin{array}{cc} \qquad\qquad E \qquad\qquad & \tilde{E} \end{array} \\ \begin{pmatrix} I & 0 \\ (I + Q + \cdots + Q^{k-1})R & Q^k \end{pmatrix} \end{array},$$

and the result follows at once.

Proposition 5-2:

$$N = \sum_{k=0}^{\infty} Q^k.$$

PROOF: For transient states i and j, we have in the P-process

$$N_{ij} = \sum_k (P^k)_{ij} = \sum_k (Q^k)_{ij},$$

by Proposition 4-12 and Lemma 5-1.

Proposition 5-3: N is finite-valued, and $\lim_k Q^k = 0$.

PROOF: N_{ij} in the P-process is finite when j is transient; hence N is finite-valued. Therefore $\lim_k (Q^k) = 0$ by Proposition 5-2.

We recall that $\bar{N}_{ij} = \mathrm{M}_i[\bar{\mathbf{n}}_j]$, and $N = I + \bar{N}$. Hence $\bar{N} = \sum_{k=1}^{\infty} Q^k$.

Proposition 5-4: If P is a transient chain, then

$$\begin{aligned} \bar{N} &= QN \\ N &= I + QN \\ N_{ij} &= H_{ij}N_{jj} \\ \bar{N}_{ij} &= \bar{H}_{ij}N_{jj} \\ N_{ii} &= 1 + \bar{H}_{ii}N_{ii} \\ N_{ii} &= 1/(1 - \bar{H}_{ii}). \end{aligned}$$

PROOF: The first two assertions are restatements of Proposition 4-13 for the case where Q is our transition matrix. The last four results are a restatement of Proposition 4-15.

Note that the conclusions of Proposition 5-4 show how to compute N from H and $\bar{H}$. Our next result establishes a method of finding N without using the H-matrix: For finite matrices the knowledge that N is a (two-sided) inverse of $I - Q$ is sufficient to determine N uniquely, but for infinite matrices it is not. For if r is a Q-regular column vector and β is a Q-regular row vector, then $N + r\beta$ is a second two-sided inverse of $I - Q$. We shall see that such regular vectors r and β often exist.

In Section 2 we shall obtain a refinement of Proposition 5-5 by proving that N is the unique *minimum* non-negative inverse of $I - Q$ on each side.

Proposition 5-5: $N(I - Q) = (I - Q)N = I$ and $QN = NQ \leq N$. In particular, every row of N is a Q-superregular measure, and every column of N is a Q-superregular function.

PROOF: The second and third assertions follow from the first, and $QN = N - I$ by Proposition 5-4. Also $NQ = N - I$ by Proposition 5-2 and monotone convergence. Since N has finite entries, the first assertion follows.

If P is a transient chain with a non-empty set E of absorbing states, we define the **absorption matrix** B to have index sets $\tilde{E}$ and E and to have entries

$$B_{ij} = \Pr_i[\text{process is absorbed at } j].$$

The B-matrix is not square; it has the same index sets as the R-matrix.

Proposition 5-6: If P is a transient chain with a non-empty set of absorbing states, then $B = NR$.

PROOF: Let i be transient and let j be absorbing. By Theorem 4-10 with the random time equal to the constant n and with the statement p taken as the assertion that the process is absorbed at j on the n + 1st step, we have

$$\begin{aligned}\Pr_i[p] &= \sum_k (P^n)_{ik} R_{kj} \\ &= \sum_k (Q^n)_{ik} R_{kj}.\end{aligned}$$

Summing on n, we find

$$B = \sum_{n=0}^{\infty} (Q^n R),$$

which by monotone convergence

$$= \left(\sum_{n=0}^{\infty} Q^n\right) R$$
$$= NR.$$

As a result, we see from the proof of Lemma 5-1 that if P is a transient chain, then

$$\lim_k P^k = \begin{pmatrix} I & 0 \\ B & 0 \end{pmatrix}.$$

Let P be an arbitrary Markov chain, let E be a subset of the set of states, and let s_E be the statement that the process is in states of E infinitely often. Define s^E by $s_i^E = \Pr_i[s_E]$.

Proposition 5-7: For any subset E of states in a Markov chain P, s^E is a P-regular function.

PROOF: Letting p be the statement s_E and taking the random time to be identically one, we see that p' in Theorem 4-10 is also s_E and that

$$\Pr_i[s_E] = \sum_k P_{ik} \Pr_k[s_E]$$

or

$$s^E = Ps^E.$$

For any Markov chain P we define a **hitting vector** h^E and an **escape vector** e^E by

$$h_i^E = \Pr_i[\text{process eventually reaches } E]$$

and

$$e_i^E = \Pr_i[\text{process goes on first step from } E \text{ to } \tilde{E} \text{ and then never returns to } E].$$

We notice that if $i \in E$, then $h_i^E = 1$, and that if $j \in \tilde{E}$ then $e_j^E = 0$.

The **absorption matrix** B^E for the set E is defined to be a square matrix with index set the set of all states and with entries defined by

$$B_{ij}^E = \Pr_i[\text{process at some time enters } E \text{ and first entry is at state } j].$$

We see that the B^E-matrix is computed by finding the entries of the B-matrix for the process with states of E made absorbing. Specifically, if E is the set of all absorbing states, then

$$B^E = \begin{array}{c} \\ E \\ \tilde{E} \end{array} \begin{array}{c} \begin{array}{cc} E & \tilde{E} \end{array} \\ \begin{pmatrix} I & 0 \\ B & 0 \end{pmatrix} \end{array}.$$

The matrices s^E, h^E, e^E, and B^E are interrelated as in the following proposition, whose proof is left to the reader.

Proposition 5-8: Let P be an arbitrary Markov chain. Then

(1) $h^E = B^E \mathbf{1}$.
(2) $h^E = e^E + Ph^E$ and hence $e^E = (I - P)h^E$.
(3) $s^E = \mathbf{1}$ if and only if $h^E = \mathbf{1}$ and $P\mathbf{1} = \mathbf{1}$.
(4) If $E \subset F$, then $h^E \leq h^F$ and $s^E \leq s^F$.
(5) $s^E = B^E s^E$.
(6) If $E \subset F$, then $B^F B^E = B^E$.
(7) $s^E = \lim P^n h^E$.

2. Superregular functions

Superregular measures and functions were defined in Section 4-3; a vector is P-superregular if $h \geq Ph$. Let P be a transient chain, and let Q be the restriction of P to transient states. As we have seen before, Q is a transition matrix. Our object in this section is to obtain a standard decomposition of non-negative Q-superregular functions and to use it in a consideration of the solutions to the equation $(I - Q)x = f$. Our results will hold equally well for Q-superregular measures, but we shall not supply the proofs. A way of transforming rigorously theorems and proofs about functions into theorems and proofs about measures will emerge later when we discuss duality. Generalizations of the present results will arise in the study of potential theory.

The transformation later of theorems about functions into theorems about measures by duality will require the existence of a positive finite-valued Q-superregular measure. Any row of N will suffice if all pairs of transient states communicate, but if not, we proceed as follows: Number the states, beginning with 1, and take

$$\beta = \sum_i 2^{-i} N^{(i)},$$

where $N^{(i)}$ is the ith row of N. It is clear that β is superregular because it is the sum of non-negative superregular measures; β is positive because $\beta_j = \sum_i 2^{-i} N_{ij} \geq 2^{-j} N_{jj} \geq 2^{-j} > 0$. Finally β is finite-valued because

$$\beta_j = \sum_i 2^{-i} N_{ij} = \sum_i 2^{-i} H_{ij} N_{jj} \leq \sum_i 2^{-i} N_{jj} \leq N_{jj} < \infty.$$

The lemma and theorem to follow hold for arbitrary Markov chains. In the transient case they will most often be applied to the chain Q. The theorem has an analog in classical potential theory, but we postpone a discussion of this point until the end of Section 8-1 after we have introduced Markov chain potentials.

Lemma 5-9: Let P be any Markov chain and let $N = \sum P^n$. If Nf is well defined and finite-valued, then $(I - P)(Nf) = f$.

PROOF: Write $f = f^+ - f^-$. Then Nf^+ and Nf^- are both finite-valued by hypothesis. Since $PN + I = N$, we have $PN \leq N$ and hence $PNf^+ \leq Nf^+$ and $PNf^- \leq Nf^-$. Therefore, by Corollary 1-5,

$$\begin{aligned} (I - P)(Nf) = Nf - P(Nf) &= Nf - (PN)f \\ &= Nf - (N - I)f \\ &= f. \end{aligned}$$

Theorem 5-10: Let P be any Markov chain and let $N = \sum P^n$. Any non-negative P-superregular finite-valued function h has a unique representation $h = Nf + r$, where r is regular. In the representation f and r are both non-negative, and $f = (I - P)h$.

PROOF: Since h is P-superregular,

$$h \geq Ph \geq P^2 h \geq \cdots \geq 0.$$

Thus $P^n h$ converges to a non-negative function r. By the Dominated Convergence Theorem,

$$Pr = P(\lim P^n h) = \lim P^{n+1} h = r.$$

Hence r is regular. Also

$$h = P^{n+1} h + (I + P + \cdots + P^n)(h - Ph).$$

Since $h - Ph \geq 0$, we may apply monotone convergence in passing to the limit on n; we obtain

$$h = r + N(h - Ph).$$

Set $f = h - Ph$, and existence follows.

For uniqueness, suppose that $h = r' + Nf'$ with r' regular. Then Nf and Nf' are finite-valued since h is. Multiplying the equation

$$r + Nf = r' + Nf'$$

through by $I - P$ and applying Lemma 5-9, we obtain

$$f = f'.$$

Hence also $r = r'$.

We return now to the special case of transient chains, where $N = \sum Q^n$. A solution g to an equation is the **minimum non-negative solution** if whenever h is a non-negative solution, we have $h \geq g \geq 0$.

Proposition 5-11: If $f \geq 0$ and if Nf is finite, then Nf is the minimum non-negative solution of $(I - Q)x = f$.

PROOF: By Lemma 5-9, Nf is a solution. Let x be any non-negative solution. Then x is finite-valued and superregular. By Theorem 5-10, $x = Nf + r$ where $r \geq 0$. Hence $x \geq Nf$.

It follows that N is the minimum non-negative right inverse of $(I - Q)$. To prove that the jth column of N is minimum, define f by $f_i = \delta_{ij}$ and then apply Proposition 5-11. After the analog of Proposition 5-11 for measures has been established, we find similarly that N is the minimum non-negative left inverse of $(I - Q)$.

3. Absorbing chains

A class of Markov chains of special interest is the class of absorbing chains. We shall use the material developed in the two preceding sections to establish the basic facts about absorbing chains.

Definition 5-12: A Markov chain P is said to be **absorbing** if, for every starting state, the probability of ending in an absorbing state is one.

If P is a Markov chain containing a recurrent nonabsorbing state i, then the process cannot be absorbed if it is started in state i. That is, all absorbing chains are transient chains. It is not true, however, that all transient chains are absorbing. The property $P\mathbf{1} = \mathbf{1}$ is a necessary condition. But even it is not sufficient, since the basic example can be transient but is never absorbing.

The proposition to follow is the special case of the identity $B^E\mathbf{1} = h^E$ in which E is the set of all absorbing states.

Proposition 5-13: If P is a transient chain, then P is absorbing if and only if $B1 = 1$.

The next two propositions give two ways in which absorbing chains arise.

Proposition 5-14: If P is a finite transient chain such that $P1 = 1$, then P is absorbing.

PROOF:

$$\begin{aligned} B1 &= (NR)1 = N(R1) \\ &= N[(I - Q)1] \quad \text{since } (R1 + Q1)_i = (P1)_i = 1 \\ &= [N(I - Q)]1 \quad \text{by Corollary 1-6} \\ &= 1. \end{aligned}$$

Let $\mathbf{a}(\omega)$ be the time on the path ω of a chain P that absorption takes place. If the process is not absorbed along ω, define $\mathbf{a}(\omega) = +\infty$. Since $\mathbf{a}(\omega) = \sum_j \mathbf{n}_j(\omega)$, where the sum is taken over the transient states, we see that $\mathbf{a}$ is measurable and we conclude that $\mathbf{a}$ is a random time. Define the column vector a by $a_i = M_i[\mathbf{a}]$. The vector a is indexed by the transient states. It is clear that the chain P is absorbing if and only if $\mathbf{a}$ is finite a.e.

Proposition 5-15: If P is a recurrent chain and if ${}^EP$ is the Markov chain obtained by making a non-empty set E of states absorbing, then ${}^EP$ is absorbing.

PROOF: Let $j \in E$. Since $H_{ij} = 1$ for every i, $\mathbf{t}_j(\omega)$ is finite almost everywhere. But $\mathbf{a}(\omega) \le \mathbf{t}_j(\omega)$, and ${}^EP$ is thus absorbing.

The notation ${}^EP$ will be used in later sections to refer either to the chain P with the states made absorbing or to the chain P made so that it disappears instead of entering E. If

$$P = \begin{pmatrix} T & U \\ R & Q \end{pmatrix},$$

then these two chains are, respectively,

$$\begin{pmatrix} I & 0 \\ R & Q \end{pmatrix} \quad \text{and} \quad \begin{pmatrix} 0 & 0 \\ 0 & Q \end{pmatrix}.$$

It will be clear from the context which one is meant.

There is one more important way in which absorbing chains arise. Suppose that $P1 \neq 1$. If we add an absorbing state 0, the state "stopped," to the state space S and define $\bar{P}$ by

$$\begin{aligned}
\bar{P}_{ij} &= P_{ij} \quad \text{if } i \neq 0 \text{ and } j \neq 0 \\
P_{0j} &= \delta_{0j} \\
\bar{P}_{i0} &= 1 - \sum_{k \in S} P_{ik} \quad \text{if } i \neq 0,
\end{aligned}$$

then $\bar{P}$, called the **enlarged chain**, may be absorbing. If P is a finite transient chain, then $\bar{P}$ necessarily will be absorbing by Proposition 5-14.

With this set of propositions to indicate how absorbing chains arise, we conclude with an investigation of the properties of the vector a.

Proposition 5-16: If P is a transient chain, then $a = N1$.

PROOF:

$$\begin{aligned}
(N1)_i &= \sum_j N_{ij} \quad \text{summed over the transient states} \\
&= \sum_j M_i[\mathbf{n}_j] \\
&= \mathrm{M}_i\Big[\sum_j \mathbf{n}_j\Big] \quad \text{by monotone convergence.}
\end{aligned}$$

But $\mathbf{a} = \sum_j \mathbf{n}_j$, where the sum is taken over transient states j. Thus $(N1)_i = \mathrm{M}_i[\mathbf{a}] = a_i$.

Corollary 5-17: If P is a transient chain for which $P1 = 1$ and if a has only finite entries, then $x = a$ is the unique minimum non-negative solution of the equation $(I - Q)x = 1$.

PROOF: It is the unique minimum non-negative solution by Proposition 5-16 and Proposition 5-11.

4. Finite drunkard's walk

The finite drunkard's walk is a Markov chain defined on the integers $\{0, 1, \ldots, n\}$ with states 0 and n absorbing and with transition probabilities

$$P_{i,i+1} = p$$

and

$$P_{i,i-1} = q = 1 - p \quad \text{for } 0 < i < n.$$

If we set $r = q/p$, two cases arise. Either $r = 1$ and $\{x_n\}$ is a martingale or $r \neq 1$ and $\{r^{x_n}\}$ is a martingale.

We shall use the second martingale systems theorem (Theorem 3-15) to compute the entries of B, H, N, and a for the case $r = 1$.

To compute the entries of the B matrix when $r = 1$, we note that $B\mathbf{1} = \mathbf{1}$ by Proposition 5-14; therefore $B_{i0} = 1 - B_{in}$ for each transient state i. Since $\{x_n\}$ is a bounded martingale, Corollary 3-16 applies with the time taken as the time of absorption (which is a stopping time because P is absorbing if and only if $\mathbf{a}$ is finite a.e.). Then

$$i = 0B_{i0} + nB_{in}$$

so that

$$B_{in} = i/n$$

and

$$B_{i0} = 1 - i/n.$$

To find the entry H_{ij} of the H-matrix, we make state j absorbing and consider the resulting process. If $i \leq j$, the modified process is the drunkard's walk on the integers $\{0, \ldots, j\}$ with j absorbing. Hence $H_{ij} = i/j$. If $i \geq j$, the modified process is the drunkard's walk on $\{j, \ldots, n\}$. Renumbering the states, we can consider the process as starting at $i - j$ and taking place on the states $\{0, \ldots, n - j\}$. Thus,

$$H_{ij} = 1 - \frac{i - j}{n - j} = \frac{n - i}{n - j}.$$

To get $\bar{H}_{ii}$, where i is transient, we use the fact that $\bar{H} = PH$, so that

$$\begin{aligned} \bar{H}_{ii} &= pH_{i+1,i} + qH_{i-1,i} \\ &= 1 - \frac{n}{2i(n - i)} \qquad \text{since } p = q = \tfrac{1}{2}. \end{aligned}$$

The N-matrix is determined as a function of $\bar{H}$ and H by

$$N_{jj} = \frac{1}{1 - \bar{H}_{jj}}$$

and

$$N_{ij} = H_{ij}N_{jj}.$$

We find

$$N_{ij} = \begin{cases} \dfrac{2}{n}(i)(n - j) & \text{for } i \leq j \\ \\ \dfrac{2}{n}(j)(n - i) & \text{for } i \geq j. \end{cases}$$

Finally, we have

$$a_i = \mathrm{M}_i[\mathbf{a}] = (N\mathbf{1})_i = \sum_{j=1}^{n-1} N_{ij} = i(n - i).$$

The case $r \neq 1$ proceeds in the same way. By the systems theorem

$$r^i = r^0 B_{i0} + r^n B_{in}.$$

From this equation one easily deduces that

$$B_{i0} = \frac{r^i - r^n}{1 - r^n}.$$

After first computing H and N, we find

$$a_i = \mathrm{M}_i[\mathbf{a}] = \frac{1}{q - p}\left[i - n\frac{1 - r^i}{1 - r^n}\right].$$

When $r > 1$, this process is sometimes known as "gambler's ruin" because of the following interpretation. A gambler walks into a gambling house with i dollars in his pocket, and the house has $n - i$ dollars to bet against him. In a given game the gambler has probability p of winning. Since the house fixes the odds, we have $p < \frac{1}{2}$ and therefore $r > 1$. If the game is played repeatedly, x_k in the above Markov chain represents the gambler's cash after k games, and B_{i0} is the probability of his eventual ruin. Since $r > 1$,

$$B_{i0} = \frac{1 - r^{-(n-i)}}{1 - r^{-n}} > 1 - r^{-(n-i)}$$

is nearly 1 when $n - i$ (the house's capital) is large. Thus the gambler is nearly sure to be ruined, no matter how rich he is. However, a_i is approximately $i/(q - p)$, which is very large if i is substantial and p is near to $\frac{1}{2}$. Thus the gambler is likely to have a long run for his money.

5. Infinite drunkard's walk

Extending the finite drunkard's walk to a process P defined on all of the non-negative integers, we set

$$\begin{aligned} P_{0i} &= \delta_{0i} \\ P_{i,i+1} &= p & \text{for } 0 < i < \infty \\ P_{i,i-1} &= q = 1 - p & \text{for } 0 < i < \infty. \end{aligned}$$

Again we take $r = q/p$.

Our first problem is to establish the sense in which the infinite drunkard's walk P is the limiting case of the finite drunkard's walk.

Let ${}^nB$, ${}^nN$, and ${}^na$ denote, respectively, the absorption matrix, the fundamental matrix, and the mean time to absorption vector for the finite drunkard's walk on the integers $\{0, \ldots, n\}$. Define in connection with the infinite drunkard's walk the random variable ${}^k\mathbf{n}_j$ to be the number of times the process is in state j up to the first time it is in state k. Let p be the statement that the process is not absorbed at state 0, and let p_k be the statement that the process is not absorbed at state 0 at any time up to the time it reaches state k.

Proposition 5-18: In the infinite drunkard's walk

$$B_{i0} = \lim_n {}^nB_{i0},$$

$$N_{ij} = \lim_n {}^nN_{ij},$$

and

$$a_i = \lim_n {}^na_i.$$

PROOF: We have $\Pr[p] = 1 - B_{i0}$ and $\Pr[p_n] = 1 - {}^nB_{i0}$. Since the union of the truth sets of the p_n is the truth set of p, we have, by Proposition 1-16,

$$1 - B_{i0} = \lim_n (1 - {}^nB_{i0}).$$

For the N-matrix we note that

$${}^nN_{ij} = \mathrm{M}_i[{}^n\mathbf{n}_j]$$

and

$$\mathbf{n}_j = \lim_n {}^n\mathbf{n}_j \quad \text{monotonically.}$$

The result for N therefore follows from the Monotone Convergence Theorem. Since $a_i = \mathrm{M}_i[\mathbf{a}] = \sum_j N_{ij}$, the assertion about a_i is also a consequence of monotone convergence.

Taking the limits of some of the quantities computed for the finite drunkard's walk we find that

$$B_{i0} = \begin{cases} 1 & \text{if } r \geq 1 \\ r^i & \text{if } r \leq 1 \end{cases}$$

and

$$a_i = \begin{cases} \dfrac{i}{q - p} & \text{if } r > 1 \\ +\infty & \text{if } r \leq 1. \end{cases}$$

The value of B_{i0} shows that the chain is absorbing when $p \leq q$; that is, $\mathbf{a}$ is finite almost everywhere when $p \leq q$. However, $\mathrm{M}_i[\mathbf{a}]$ is finite only when $p < q$.

If we have calculated B_{i0} and have seen that **a** is a stopping time when $p \leq q$, we may compute $\mathrm{M}_i[\mathbf{a}]$ directly from martingales without any knowledge of the N-matrix. Let $p \leq q$, and for $0 < s \leq 1$ define

$$f(k, n) = \frac{s^k}{(ps + qs^{-1})^n},$$

where n represents time and k represents position. Now

$$f(k, n) = s^k(ps + qs^{-1})^{-n} \leq (ps + qs^{-1})^{-n},$$

which is maximized when $s = \sqrt{q/p}$. Thus

$$\begin{aligned} f(k, n) &\leq (p/q)^{n/2} \\ &\leq 1 \quad \text{since } p \leq q. \end{aligned}$$

Hence f is bounded. It is easily seen that $\{f(x_n, n)\}$ is a martingale: Since f is bounded, $\mathrm{M}[f]$ is finite; the reader may verify the regularity property by showing that

$$p \cdot f(x_n + 1, n + 1) + q \cdot f(x_n - 1, n + 1) = f(x_n, n).$$

Let **a** be the stopping time of Corollary 3-16. Taking i as the starting state, we have

$$\frac{s^i}{(ps + qs^{-1})^0} = \sum \Pr[\mathbf{a} = n] \frac{s^0}{(ps + qs^{-1})^n}.$$

Set $1/u = ps + qs^{-1}$. Then

$$s = \frac{1 - \sqrt{1 - 4pqu^2}}{2pu}$$

and

$$\sum \Pr[\mathbf{a} = n]u^n = \left(\frac{1 - \sqrt{1 - 4pqu^2}}{2pu}\right)^i.$$

Defining

$$\varphi(u) = \sum \Pr[\mathbf{a} = n]u^n,$$

we note that

$$\varphi'(u) = \sum n \Pr[\mathbf{a} = n]u^{n-1}$$

and that

$$\varphi'(1) = \sum n \Pr[\mathbf{a} = n] = \mathrm{M}_i[\mathbf{a}] = a_i.$$

Using the fact that $\sqrt{1 - 4pq} = q - p$ to calculate $\varphi'(1)$, we find that $a_i = i/(q - p)$, in agreement with the result obtained by the longer method.

The present method further allows us to find the probability distribution of $\Pr[\mathbf{a} = n]$ by expanding $[(1 - \sqrt{1 - 4pqu^2})/(2pu)]^i$ as a power series in u. We thus find that

$$\begin{aligned} \Pr[\mathbf{a} &= n] = 0 \quad \text{for } n < i \\ \Pr[\mathbf{a} &= i] = q^i \\ \Pr[\mathbf{a} &= i + 1] = 0 \\ \Pr[\mathbf{a} &= i + 2] = ipq^{i+1} \end{aligned}$$

and so on.

6. A zero-one law for sums of independent random variables

Historically, the first infinite Markov chain that was studied was the sums of independent random variables process. We gather some of the results in the next few sections, beginning with two propositions and a corollary of rather general applicability.

Proposition 5-19: If P is a Markov chain for which the only bounded P-regular functions are constant vectors, then, for each subset of states E, $\Pr_i[s_E] = 0$ or 1, independently of the starting state i.

PROOF: By Proposition 5-7, s^E is regular and it is clearly bounded; therefore $s^E = c\mathbf{1}$. On the other hand, by Proposition 5-8,

$$s^E = B^E s^E$$

so that

$$c\mathbf{1} = cB^E\mathbf{1} = ch^E.$$

Therefore, either $c = 0$ or $h^E = \mathbf{1}$. In the latter case, $s^E = \mathbf{1}$ by Proposition 5-8.

As in Example 6 of Section 4-6, we let $p_k = P_{i,i+k}$, which, by assumption, is independent of i.

Proposition 5-20: Let P be the transition matrix of a Markov chain obtained from sums of independent random variables. If, for each pair of states q and r, there is a state s such that q can be reached from s or s can be reached from q *and* such that r can be reached from s or s can be reached from r, then the only bounded regular functions are constant functions. In particular, the hypothesis is satisfied if all pairs of states communicate.

PROOF: Let f be non-constant regular and suppose $f_q \neq f_r$. We shall assume that s can be reached from both q and r; the proof in the other cases is completely analogous. Let $q, q + a_1, q + a_1 + a_2, \ldots,$

$q + a_1 + \cdots + a_m$ and $r, r + b_1, \ldots, r + b_1 + \cdots + b_n$ be respective paths of positive probability leading from q to s and from r to s. Then

$$f_{q+a_1+\cdots+a_m} = f_s = f_{r+b_1+\cdots+b_n},$$

so that at least one of the equalities in the two chains

$$f_q = f_{q+a_1} = f_{q+a_1+a_2} = \cdots = f_{q+a_1+\cdots+a_m}$$

and

$$f_r = f_{r+b_1} = f_{r+b_1+b_2} = \cdots = f_{r+b_1+\cdots+b_n}$$

must be false, since otherwise $f_q = f_r$. Without loss of generality, let $f_{q+\cdots+a_{k-1}} \neq f_{q+\cdots+a_k}$ and let $a = a_k$. Then $p_a > 0$. Let $g_i = f_{i+a} - f_i$. Then g is not identically 0. Further, g is regular because

$$\begin{aligned}\sum_j P_{ij} g_j &= \sum_j P_{ij}(f_{j+a} - f_j)\\ &= \sum_j P_{ij} f_{j+a} - \sum_j P_{ij} f_j\\ &= \sum_j P_{i+a,j+a} f_{j+a} - \sum_j P_{ij} f_j\\ &= f_{i+a} - f_i\\ &= g_i.\end{aligned}$$

Suppose that for all i, $|f_i| \leq c$. Then $|g_i|$ is bounded by $2c$. Since multiplying g by -1 affects neither its regularity nor its boundedness, we may assume $b = \sup_i g_i$ is positive and finite. For any i and any $m > 0$,

$$\left|\sum_{k=0}^{m-1} g_{i+ka}\right| = |f_{i+ma} - f_i| \leq 2c.$$

Choose N so that $N \cdot b/2 > 2c$. Let $p^{(n)} = P^{(n)}_{i,i+na} \geq (p_a)^n > 0$, let $p = \min_{n<N} p^{(n)}$, and let t be a state such that $g_t > b(1 - p/2)$. A choice for t exists since b is finite and since $p > 0$. Then for $n < N$,

$$\begin{aligned}b\left(1 - \frac{p^{(n)}}{2}\right) \leq b\left(1 - \frac{p}{2}\right) &< g_t\\ &= \sum_k P^{(n)}_{tk} g_k\\ &= P^{(n)}_{t,t+na} g_{t+na} + \sum_{k \neq t+na} P^{(n)}_{tk} g_k\\ &\leq p^{(n)} g_{t+na} + (1 - p^{(n)})b.\end{aligned}$$

Thus

$$g_{t+na} > b/2 \quad \text{for } n < N.$$

Hence

$$\sum_{k=0}^{N-1} g_{t+ka} > 2c,$$

a contradiction.

Corollary 5-21: If P is a sums of independent random variables Markov chain in which all pairs of states communicate, then for each subset of states E, $\Pr_i[s_E] = 0$ or 1, independently of the starting state i.

7. Sums of independent random variables on the line

Let P be a sums of independent random variables Markov chain indexed by the integers and defined by the probability distribution $\{p_k\}$. The set of integers k for which $p_k > 0$ we shall call the set of k-values associated with the chain. We shall assume that the greatest common divisor of the k-values is one. Thus, if both positive and negative k-values exist, we see (from Lemma 1-66, for example) that all pairs of states communicate.

The mean m for the process is defined by $m = \sum_k kp_k$ and is said to exist if and only if the positive and negative parts of the sum are not both infinite. In this section we shall establish the following result.

Proposition 5-22: If P is a Markov chain representing sums of independent random variables on the line, if there are finitely many k-values and if they have greatest common divisor one, if $\sum p_k = 1$, and if $m = \sum kp_k$, then in order for the chain to be recurrent it is necessary and sufficient that $m = 0$.

Before we come to the proof, two comments are in order. The first is that the proposition can be generalized to the case where there are infinitely many k-values as long as the mean m still exists and the k-values still have greatest common divisor one. The same condition $m = 0$ is necessary and sufficient for P to be recurrent. The second comment is that the necessity of the condition $m = 0$ is an immediate consequence of the Strong Law of Large Numbers (Theorem 3-19) and that the special added assumption we used in the proof of that theorem translates exactly into the condition that there are only finitely many k-values. Nevertheless, we give a different proof.

For the proof we may assume that both positive and negative k-values exist. Otherwise, the chain is obviously transient. Recalling our discussion in Example 2 of Section 3-2, we observe that if both positive and negative k-values exist, then there are either two distinct real roots or one double real root of the equation

$$f(s) = \sum_k p_k s^k = 1.$$

If there is a root r other than $s = 1$, then $\{r^{x_n}\}$ is a non-negative martingale. And if $s = 1$ is a double root, then $\{x_n\}$ is a martingale.

But $f'(1) = (\sum kp_k s^k)_{s=1} = \sum kp_k = m$, and $f(s) = 1$ has a double root at $s = 1$ if and only if $m = 0$. In the case $m \neq 0$, $\{r^{x_n}\}$ converges a.e. and must converge to the zero function. Thus, according as $r < 1$ or $r > 1$, we have $\lim x_n = +\infty$ or $\lim x_n = -\infty$. Hence, the process returns to each state only finitely often with probability one. But if the chain were recurrent, each state would be reached infinitely often with probability one. Therefore, if $m \neq 0$, the chain is transient.

In the case in which $m = 0$, let $-u$ be the smallest k-value and let v be the largest k-value. Let E be the set of states $\{-u, \ldots, -2, -1\}$ and let E' be the set of states $\{j, j+1, \ldots, j+v-1\}$ for some fixed j. Start the process in state i with $0 \leq i < j$, and let $\mathbf{t}$ be the time to reach the set $E \cup E'$. The chain stopped at time $\mathbf{t}$ is absorbing by Proposition 5-14, and $\mathbf{t}$ is therefore a stopping time. Since $\{x_n\}$ is a bounded martingale before time $\mathbf{t}$, Corollary 3-16 applies. Therefore, $\mathrm{M}[x_0] = \mathrm{M}[x_t]$, and for $0 \leq i < j$, we have

$$\begin{aligned} i = \mathrm{M}[x_0] = \mathrm{M}[x_t] &= \sum_{k \in E} B_{ik}^E k + \sum_{k \in E'} B_{ik}^{E'} k \\ &\geq -u \sum_{k \in E} B_{ik}^E + j \sum_{k \in E'} B_{ik}^{E'} \\ &= -uh_i^E + jh_i^{E'}, \end{aligned}$$

which, by Proposition 5-13,

$$= -uh_i^E + j(1 - h_i^E).$$

Then

$$(j + u)h_i^E \geq j - i$$

and

$$h_i^E \geq \frac{j - i}{j + u}.$$

Letting $j \to \infty$, we find that $h_i^E = 1$ for all $i \geq 0$.

Reversing the argument for $i \leq 0$ and $F = \{1, \ldots, v\}$, we find similarly that $h_i^F = 1$ for all $i \leq 0$. Thus, for any state i, $h_i^{E \cup F} = 1$. By Proposition 5-8, $s_i^{E \cup F} = 1$. Since $E \cup F$ is a finite set, Proposition 4-28 applies, and the chain is recurrent.

8. Examples of sums of independent random variables

Calculations with sums of independent random variables on the line normally involve either martingales or difference equations. We shall illustrate in this section each of these methods with an example.

Example 1: Let the defining distribution for a Markov chain P representing sums of independent random variables on the integers be $\{p_1 = q, p_2 = p\}$. The process is obviously transient since $\bar{H}_{jj} = 0$

and $N_{jj} = 1$ for all j. Since $N_{ij} = H_{ij}N_{jj} = H_{ij} = H_{0,j-i}$, we will have determined the H-matrix and the N-matrix completely by finding the value of H_{0k} for all k. We note first that $H_{0k} = 0$ if k is negative. For the case $k \geq 0$, $\{r^{x_n}\}$ is a nonconstant martingale if r is a nonzero root other than one of the equation $\sum p_k s^k = qs + ps^2 = 1$. Thus $\{(-1/p)^{x_n}\}$ is a martingale. Taking the stopping time $\mathbf{t}$ as the time when the process reaches or passes state k, we find from Corollary 3-16 that for the process started at state 0

$$(-1/p)^0 = H_{0k}(-1/p)^k + (1 - H_{0k})(-1/p)^{k+1}.$$

Therefore,

$$H_{0k} = \frac{1}{1+p}(1 - (-p)^{k+1}).$$

It is interesting to note that

$$\lim_{k\to\infty} H_{0k} = \frac{1}{1+p}.$$

This result can also be obtained from the Renewal Theorem of Section 1-6 if we observe that

$$m = \sum kp_k = q + 2p = 1 + p$$

so that

$$\frac{1}{1+p} = \frac{1}{m}.$$

EXAMPLE 2: Let $p_{-1} = \frac{2}{9}$, $p_1 = \frac{3}{9}$, and $p_2 = \frac{4}{9}$. Then

$$m = \sum kp_k = 1,$$

and the process is transient by Proposition 5-22. Let the transition matrix be called P.

If g is a P-regular vector at state i, then $g_i = (Pg)_i$ and

$$g_i = \tfrac{2}{9}g_{i-1} + \tfrac{3}{9}g_{i+1} + \tfrac{4}{9}g_{i+2}.$$

We shall need a characterization of such vectors in the calculation of the H-matrix. The difference equation we have just formed may be rewritten as

$$4g_{i+2} + 3g_{i+1} - 9g_i + 2g_{i-1} = 0.$$

Its characteristic equation (see Section 1-6b) is

$$4k^3 + 3k^2 - 9k + 2 = 0,$$

whose solutions are $k = 1, \frac{1}{4}$, and -2. Thus,

$$g_i = A + B(\tfrac{1}{4})^i + C(-2)^i.$$

In finding the entries of the H-matrix, we know that $H_{ij} = H_{i-j,0}$; therefore, it is sufficient to consider only the entries H_{i0}. We shall look at the cases $i \geq 0$ and $i \leq 0$ separately.

Suppose $i \geq 0$. Since the only negative k-value is -1, it is impossible for the process to start in state $i + 1$ and reach state 0 without first passing through state i. Thus, for $i \geq 0$,

$$\begin{aligned} H_{i+1,0} &= H_{i+1,i}H_{i,0} \\ &= H_{1,0}H_{i,0} \end{aligned}$$

with $H_{00} = 1$. The result is a first-order difference equation whose solution is $H_{i,0} = c(H_{1,0})^i$. Setting $i = 0$ shows that $c = 1$. Thus $H_{i,0}$ is exponential.

But $\bar{H} = PH$, so that $H_{i,0} = (PH)_{i,0}$ for all $i > 0$. Thus $H_{i,0}$ satisfies the difference equation

$$g_i = \tfrac{2}{9}g_{i-1} + \tfrac{3}{9}g_{i+1} + \tfrac{4}{9}g_{i+2}$$

for $i > 0$. It therefore satisfies

$$g_{i+1} = \tfrac{2}{9}g_i + \tfrac{3}{9}g_{i+2} + \tfrac{4}{9}g_{i+3}$$

for $i \geq 0$. Hence $H_{i,0} = A + B(\tfrac{1}{4})^i + C(-2)^i$ for all $i \geq 0$. Since $H_{i,0}$ is known to be exponential, two of the coefficients A, B, and C are zero and the other is one. The alternatives $C = 1$ and $A = 1$ are eliminated, respectively, by the facts that -2 is not a probability and that P drifts to the right a.e. Thus,

$$H_{i,0} = (\tfrac{1}{4})^i \quad \text{for } i \geq 0.$$

For $i \leq 0$ we again use the fact that $\bar{H} = PH$, and we find that $H_{i,0}$ is a solution of the equation

$$g_{i-2} = \tfrac{2}{9}g_{i-3} + \tfrac{3}{9}g_{i-1} + \tfrac{4}{9}g_i$$

for all $i \leq 1$. Therefore,

$$H_{i,0} = A + B(\tfrac{1}{4})^i + C(-2)^i$$

for all $i \leq 1$. Known values for $H_{i,0}$ when $i = 0$ and $i = 1$ give us two conditions on the three unknowns A, B, and C. The fact that $H_{i,0} \leq 1$ as $i \to -\infty$ tells us that $B = 0$. We have as a result

$$H_{i,0} = \begin{cases} (\tfrac{1}{4})^i & \text{for } i \geq 0 \\ \tfrac{3}{4} + \tfrac{1}{4}(-2)^i & \text{for } i \leq 1. \end{cases}$$

From a knowledge of the H-matrix, we can compute $\bar{H}$ by $\bar{H} = PH$. The entries of the N-matrix follow from

$$N_{jj} = 1/(1 - \bar{H}_{jj})$$

and

$$N_{ij} = H_{ij}N_{jj}.$$

9. Ladder process for sums of independent random variables

For a sums of independent random variables Markov chain defined on the integers, we define a sequence $\mathbf{s}_i(\omega)$ of **positive step times** inductively as follows: $\mathbf{s}_0(\omega)$ is the least n such that $x_0(\omega) > 0$, and $\mathbf{s}_i(\omega)$ is the least n such that $x_n(\omega) > x_{\mathbf{s}_{i-1}(\omega)}(\omega)$. If we construct a stochastic process by watching the old Markov chain only at the positive step times—that is, by calling the nth outcome in the new process the $\mathbf{s}_n$th outcome in the old process—then the strong Markov property as formulated in Theorem 4-9 implies that the new process is a Markov chain. We shall go through this implication in detail.

Proposition 5-23: If P is a sums of independent random variables Markov chain defined on the integers, then the stochastic process whose nth outcome is the $\mathbf{s}_n$th outcome in P is a Markov chain P^+. Moreover, $P^+_{ij} = P^+_{0,j-i}$.

PROOF: The times $\mathbf{s}_n$ are random times. Applying Theorem 4-9 to the time $\mathbf{s}_n$ and the statement $r \equiv (x_{\mathbf{s}_{n+1}} = c_{n+1})$, we find that, if $\Pr_\pi[x_{\mathbf{s}_0} = c_0 \wedge \cdots \wedge x_{\mathbf{s}_n} = c_n] > 0$, then

$$\Pr_\pi[x_{\mathbf{s}_{n+1}} = c_{n+1} \mid x_{\mathbf{s}_0} = c_0 \wedge \cdots \wedge x_{\mathbf{s}_n} = c_n] = \Pr_\pi[x_{\mathbf{s}_{n+1}} = c_{n+1} \mid x_{\mathbf{s}_n} = c_n].$$

And if $\Pr_\pi[x_{\mathbf{s}_n} = c_n] > 0$, then

$$\Pr_\pi[x_{\mathbf{s}_{n+1}} = c_{n+1} \mid x_{\mathbf{s}_n} = c_n] = \Pr_{c_n}[x_{\mathbf{s}_1} = c_{n+1}].$$

Thus the process is a Markov chain P^+. The fact that $P^+_{ij} = P^+_{0,j-i}$ follows from the fact that P represents sums of independent random variables.

The chain P^+ is called the **ladder process** for P. The ladder process moves from i to j if j is the first state greater than i that is reached in the original process. If the mean step m in P is positive, then the process reaches or passes any given positive state with probability one, so that the $\mathbf{s}_i$ are finite a.e. Hence $P^+\mathbf{1} = \mathbf{1}$, and the ladder process represents sums of independent random variables.

As an example, we shall compute explicitly the ladder process associated with Example 2 of the preceding section. For the given chain we have $p_{-1} = \frac{2}{9}$, $p_1 = \frac{3}{9}$, and $p_2 = \frac{4}{9}$. The ladder process has two k-values, namely 1 and 2, and thus has a distribution $\{p_1^+, p_2^+\}$. Since the positive step times are finite a.e., we have $\sum p_k^+ = 1$ or

$$p_1^+ + p_2^+ = 1.$$

To find the values of p_1^+ and p_2^+, we note that

$$\begin{aligned} p_2^+ &= {}^1H_{02} \\ &= P_{02} + P_{0,-1}p_1^+p_2^+ \\ &= p_2 + p_{-1}p_1^+p_2^+. \end{aligned}$$

Putting in the known values for p_{-1} and p_2, we find

$$p_1^+ = p_2^+ = \tfrac{1}{2}.$$

The ladder process for our Example 2 is therefore an instance of Example 1 in the same section.

10. The basic example

The basic example is a Markov chain with state space the non-negative integers and with transition probabilities

$$\begin{aligned} P_{i-1,i} &= p_i, \quad i > 0 \\ P_{i-1,0} &= q_i = 1 - p_i. \end{aligned}$$

We normally assume that none of the p_i's is 0. A row vector β is defined by

$$\begin{aligned} \beta_0 &= 1 \\ \beta_i &= p_i\beta_{i-1} \quad \text{for } i > 0; \end{aligned}$$

it is regular if and only if $\lim_{i\to\infty} \beta_i = 0$, and the process is recurrent or transient according as the limit is or is not 0.

In this section we shall compute the H and N matrices for the basic example when it is transient, and we shall show that a transient basic example has no non-zero regular row vector.

The process cannot leave the set $\{0, 1, \ldots, j\}$ without hitting j. Hence $H_{ij} = 1$ if $i \leq j$. If $i > j$, then j can be reached only via 0, so that

$$H_{ij} = H_{i0}H_{0j} = H_{i0}$$

by Proposition 4-16. Thus we need only find H_{i0}. The only way the process can fail to reach 0 is to continue moving to the right from i.

Let $\beta_\infty = \lim_{i\to\infty} \beta_i = \prod_{i=1}^{\infty} p_i$. Then

$$1 - H_{i0} = \prod_{j>i} p_i = \frac{\beta_\infty}{\beta_i},$$

and we find

$$H_{ij} = \begin{cases} 1 & \text{if } i \le j \\ 1 - \dfrac{\beta_\infty}{\beta_i} & \text{if } i > j. \end{cases}$$

Then

$$\begin{aligned} \bar{H}_{jj} &= p_{j+1}H_{j+1,j} + q_{j+1}H_{0j} \\ &= p_{j+1}\left(1 - \frac{\beta_\infty}{\beta_{j+1}}\right) + q_{j+1} \\ &= 1 - \frac{\beta_\infty}{\beta_j}. \end{aligned}$$

Suppose now that the process is transient—that is, that $\beta_\infty > 0$. Then

$$N_{jj} = \frac{1}{1 - \bar{H}_{jj}} = \frac{\beta_j}{\beta_\infty},$$

so that

$$N_{ij} = H_{ij}N_{jj} = \begin{cases} \dfrac{\beta_j}{\beta_\infty} & \text{if } i \le j \\ \dfrac{\beta_j}{\beta_\infty} - \dfrac{\beta_j}{\beta_i} & \text{if } i > j. \end{cases}$$

If $\beta_\infty > 0$, we know that β is not regular. Indeed, a transient basic example has no non-zero regular row vector. For if α is regular, then

$$\sum_i \alpha_i q_{i+1} = \alpha_0$$

and

$$\alpha_{j-1}p_j = \alpha_j \quad \text{for } j > 0.$$

From the second condition we find by induction that $\alpha_j = \alpha_0\beta_j$. Then the first condition yields

$$\alpha_0 = \alpha_0 \sum_i \beta_i q_{i+1} = \alpha_0 \sum_i (\beta_i - \beta_{i+1}) = \alpha_0(1 - \beta_\infty).$$

Thus $\alpha_0 = 0$ and $\alpha = 0$.

11. Problems

1. Consider the finite Markov chain with states

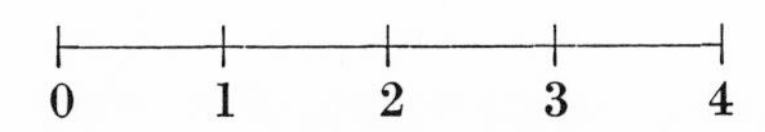

States 0 and 4 are absorbing. At each of the other states the process takes a step to the right with probability $\frac{2}{3}$, or a step to the left with probability $\frac{1}{3}$. Compute P, N, and B by means of Propositions 5-5 and 5-6.

2. If the states of a transient chain form a single closed set, show that each column of N is a non-constant positive superregular function. [*Note:* We shall see later that there are no such functions for recurrent chains; hence their existence is a necessary and sufficient condition for a closed set to be transient.]

3. Prove Proposition 5-8. Prove also that $h^E = Ne^E + s^E$. Interpret each result.

4. In the basic example, let $E = \{0, 1, 2\}$. Compute B^E, h^E, e^E, and s^E. Check formulas (1), (2), and (3) in Proposition 5-8.

5. Prove an analog of Theorem 5-10 for row vectors. Use it to show that if $\pi \geq \pi P \geq 0$ in a transient basic example, then there is a measure μ such that $\pi = \mu N$.

6. For a transient chain let

$$x_i = \mathrm{M}_i[\mathbf{n}_E],$$

where $\mathbf{n}_E$ is the number of times the chain is in the finite set of states E. Use a systems theorem to find an equation of the form

$$(I - P)x = y,$$

and prove that x is the minimum non-negative solution.

7. Find the probability in the p-q random walk started at 0 of reaching $+n$ before $-n$. [*Hint:* Use the results obtained for the finite drunkard's walk.] If $p > q$, what happens to this probability as n increases?

8. The one-dimensional symmetric random walk is a process to which Corollary 5-21 applies. If E is the set of primes, is s^E equal to $\mathbf{1}$ or is it equal to 0?

9. Let $x_0, x_1, x_2, \ldots$ be the outcome functions for the symmetric random walk on the integers started at 0. Show that there is no non-constant non-negative function $f(n)$ defined on the integers such that $f(x_0)$, $f(x_1), \ldots$ is a martingale.

10. Show by direct computation that the sums of independent random variables process on the integers with $p_2 = \frac{1}{3}$ and $p_{-1} = \frac{2}{3}$ is recurrent.

11. Find H and N for sums of independent random variables on the integers with $p_{-1} = p_2 = \frac{1}{2}$.

Problems 12 to 19 refer to sums of independent random variables on the integers with $p_{-1} = \frac{1}{3}$ and $p_1 = \frac{2}{3}$.

12. Find H and N.

13. Describe the long-range behavior of the chain.

14. Give two examples of infinite sets E to illustrate the two possibilities $s^E = \mathbf{1}$ and $s^E = 0$.

15. Find all non-negative regular functions.

16. Give a necessary and sufficient condition on a non-negative function f that Nf be finite-valued.

17. In the previous problem, let $g = Nf$. Choose an f satisfying your condition, and show that $(I - P)g = f$.

18. Let

$$h_i = \begin{cases} 2 + (\frac{1}{2})^i & \text{if } i < 0 \\ (3 + i)(\frac{1}{2})^i & \text{if } i \geq 0. \end{cases}$$

Show that h is superregular, and decompose h as in Theorem 5-10.

19. Use the function of Problem 18 together with the Martingale Convergence Theorem to prove that the process is to the left of 0 only finitely often a.e.

Problems 20 to 22 refer to the game of tennis. It will be necessary to know how one keeps score in tennis. A match is being played between A and B, and A has probability p of winning any one point.

20. Set up a single game as a transient chain with the two absorbing states "A wins" and "B wins." [Minimize the number of states, e.g., identify "30–30" with "deuce."] Compute the probability that A wins the game as a function of p.

21. Suppose that A has probability p' of winning a game. What is the probability that he wins a set? What of winning the match (if he is required to win three sets)?

22. What is the probability that A wins the match if $p = 0.6$? What if $p = 0.51$?

CHAPTER 6

RECURRENT CHAINS

1. Mean ergodic theorem for Markov chains

Recurrent chains are Markov chains such that the set of states is a single recurrent class. They have the properties that $P1 = 1$, $\bar{H} = E$, and $M_i[\mathbf{n}_j] = \infty$. The study of recurrent chains begins with a characterization of finite-valued non-negative superregular measures and functions; the reader should turn back to Sections 1-6c and 1-6d for the terms referred to in what follows.

We shall apply Proposition 1-63 and Corollary 1-64 to the sequence of matrices obtained as the Cesaro sums of the powers of a recurrent chain P. Define

$$L^{(n)} = \frac{1}{n} \sum_{k=0}^{n-1} P^k.$$

Then

$$0 \leq L^{(n)} \leq E \quad \text{for all } n.$$

Theorem 6-1: If P is the matrix of a recurrent chain, then the sequence of powers of P is Cesaro summable to a limiting matrix L with the properties $L \geq 0$ and $LP = L = PL = L^2$.

Proof: We shall show that every convergent subsequence converges to the same limit L. The proof proceeds in four steps.

(1) Since

$$L^{(n)} = \frac{1}{n}(I + P + \cdots + P^{n-1}),$$

we have

$$PL^{(n)} = \frac{1}{n}(P + P^2 + \cdots + P^n) = L^{(n)}P$$

$$= L^{(n)} + \frac{1}{n}(P^n - I).$$

Let $\{L^{(n_\nu)}\}$ be a convergent subsequence; such a sequence exists by Proposition 1-63. Set $L = \lim_\nu L^{(n_\nu)}$. Then $L \geq 0$. Since

$$\lim_n \frac{1}{n}(P^n - I) = 0,$$

we have $\lim_\nu PL^{(n_\nu)} = L = \lim_\nu L^{(n_\nu)}P$.

(2) By Proposition 1-56 (dominated convergence), we have

$$\lim_\nu PL^{(n_\nu)} = P \lim_\nu L^{(n_\nu)} = PL$$

and thus

$$PL = L.$$

By Proposition 1-55 (Fatou's Theorem), we may further conclude

$$(\lim_\nu L^{(n_\nu)})P \leq \lim_\nu (L^{(n_\nu)}P) = L$$

and

$$LP \leq L.$$

(3) Suppose LP is not equal to L. Then for some i and j, $(LP)_{ij} < L_{ij}$. Summing the inequalities $(LP)_{ik} \leq L_{ik}$ on k, we obtain

$$\sum_k (LP)_{ik} < \sum_k L_{ik}$$

since strict inequality holds in the jth entry. Thus $[(LP)\mathbf{1}]_i < (L\mathbf{1})_i$. Since L, P, and $\mathbf{1}$ are non-negative, associativity holds and $(LP)\mathbf{1} = L(P\mathbf{1}) = L\mathbf{1}$. Therefore, $[(LP)\mathbf{1}]_i = (L\mathbf{1})_i$, and we have a contradiction. Hence $LP = L = PL$. By induction, we readily see that $LP^n = L = P^nL$ for every n. Adding these results, we obtain finally

$$LL^{(n)} = L = L^{(n)}L.$$

(4) Let $\{L^{(n_\mu)}\}$ be a convergent subsequence with limit $\bar{L}$. It is sufficient to show that $L = \bar{L}$. From step (3) we have $L = LL^{(n_\mu)}$ for any μ, and by Fatou's Theorem $L\mathbf{1} \leq \mathbf{1}$ and $\bar{L}\mathbf{1} \leq \mathbf{1}$. Thus, by dominated convergence,

$$L = \lim_\mu (LL^{(n_\mu)}) = L\bar{L}.$$

Interchanging the roles of L and $\bar{L}$, we find $\bar{L} = \bar{L}L$. But by Fatou's Theorem $\bar{L}L \leq L$ and $L\bar{L} \leq \bar{L}$. Therefore,

$$\bar{L} = \bar{L}L \leq L \quad \text{and} \quad L = L\bar{L} \leq \bar{L}.$$

Hence $L = \bar{L}$ and $L = L^2$.

Definition 6-2: If P is a recurrent chain, P is said to be a **null chain** if $L = 0$. If $L \neq 0$, P is said to be an **ergodic chain** and the limit matrix L is called A.

Proposition 6-3: If P is a recurrent chain, every constant function is regular and the only non-negative (finite-valued) superregular functions are constants.

PROOF: Constant functions are trivially regular since $P1 = 1$. Let h be a finite-valued non-negative superregular function, and let the chain be started in any fixed state i. Since $\mathrm{M}[|h(x_0)|] = h_i < \infty$, $(h(x_n), \mathscr{R}_n)$ is a non-negative supermartingale (see Section 4-3). Thus $\lim_n h(x_n)$ exists and is finite with probability one by Corollary 3-13. If h is not a constant function, then $h_j \neq h_k$ for some j and k. Since the chain is in states j and k infinitely often a.e., $h(x_n) = h_j$ and $h(x_n) = h_k$ for infinitely many n with probability one. Thus $h(x_n)$ diverges a.e., a contradiction.

To prove the corresponding result for measures, we introduce the dual matrix $\hat{P}$, defined whenever a positive finite-valued P-superregular measure α exists. The entries of $\hat{P}$ are $\hat{P}_{ij} = \alpha_j P_{ji}/\alpha_i$. Although we shall investigate $\hat{P}$ more fully in the next section, we mention some of its properties here. Suppose P is recurrent. Since $\hat{P}_{ij} \geq 0$ and since

$$\sum_j \hat{P}_{ij} = \frac{1}{\alpha_i} \sum_j \alpha_j P_{ji} \leq \frac{1}{\alpha_i} \alpha_i = 1,$$

$\hat{P}$ is a transition matrix. Since all pairs of states communicate in P, they do in $\hat{P}$. Now, using induction on n, we note that if

$$(\hat{P}^{n-1})_{ij} = \frac{\alpha_j (P^{n-1})_{ji}}{\alpha_i} \quad \text{for all } i \text{ and } j,$$

then

$$(\hat{P}^n)_{ij} = \sum_k \hat{P}_{ik} (\hat{P}^{n-1})_{kj} = \frac{\alpha_j}{\alpha_i} \sum_k (P^{n-1})_{jk} P_{ki} = \frac{\alpha_j (P^n)_{ji}}{\alpha_i}.$$

Summing on n, we see that $\mathrm{M}_i[\hat{\mathbf{n}}_i] = +\infty$ because $\mathrm{M}_i[\mathbf{n}_i] = \infty$. Hence $\hat{P}$ is recurrent.

Proposition 6-4: If P is a recurrent chain, all (finite-valued) non-negative superregular measures are regular and are uniquely determined up to multiplication by a constant. A non-zero non-negative superregular measure is positive.

PROOF: We prove the second assertion first. Suppose $\alpha \geq \alpha P$. Then $\alpha \geq \alpha P^n$ for every n. If $\alpha_i = 0$ and $\alpha_j > 0$, find n such that $(P^n)_{ji} > 0$. We have

$$\alpha_i \geq \sum_m \alpha_m (P^n)_{mi} \geq \alpha_j (P^n)_{ji} > 0,$$

a contradiction. Hence $\alpha > 0$. For the first assertion of the proposition, let α and β be non-zero non-negative finite-valued superregular measures. Then α and β are positive. Use α to form the recurrent chain $\hat{P}$. Then $\hat{P}1 = 1$ since $\hat{P}$ is recurrent. Therefore,

$$1 = \sum_j \hat{P}_{ij} = \sum_j \frac{\alpha_j P_{ji}}{\alpha_i} = \frac{1}{\alpha_i} \sum_j \alpha_j P_{ji}.$$

Thus $\alpha_i = \sum_j \alpha_j P_{ji}$ and α is regular. If we can show that $\{\beta_j/\alpha_j\}$ is a superregular function for $\hat{P}$, we will have shown that $\beta = c\alpha$, and the proof will be complete. We have

$$\sum_j \hat{P}_{ij} \frac{\beta_j}{\alpha_j} = \sum_j \frac{\alpha_j}{\alpha_i} P_{ji} \frac{\beta_j}{\alpha_j} = \frac{1}{\alpha_i} \sum_j \beta_j P_{ji} \leq \frac{\beta_i}{\alpha_i}.$$

Proposition 6-5: If P is ergodic, then $A = 1\alpha$, $\alpha 1 = 1$, and α is regular.

PROOF: We have $PA = A$. Thus every column of A is regular and must be constant by Proposition 6-3. Hence $A = 1\alpha$. Since $AP = A$, every row of A is regular and α must be regular. It therefore remains to be shown that $\alpha 1 = 1$. Now $A^2 = A$ so that $(1\alpha)(1\alpha) = (1\alpha)$. By associativity $1(\alpha(1\alpha)) = 1\alpha$ so that $\alpha(1\alpha) = \alpha$. But $\alpha(1\alpha) = (\alpha 1)\alpha$ so that $(\alpha 1)\alpha = \alpha$. If $\alpha_j \neq 0$, then from $(\alpha 1)\alpha_j = \alpha_j$ we may conclude $\alpha 1 = 1$.

The existence of a positive regular measure for ergodic chains is thus an easy matter to prove. For null chains, however, the proof is harder since the limiting matrix $L = 0$ is no help. The technique we shall use is to watch the recurrent chain P only while it is in a subset E of the set of states.

Let E be a subset of states and let P^E be the stochastic process whose nth outcome is the outcome of P the nth time the process P is in the set E. We shall see in Lemma 6-6 that P^E is a Markov chain. From its interpretation it is clear that P^E is recurrent if P is recurrent. Moreover, if $E \subset F$, then $(P^F)^E = P^E$.

The index set for the matrix P^E is taken to be E. Writing P as

$$P = \begin{array}{c} \\ E \\ \tilde{E} \end{array} \!\!\begin{array}{c} \begin{array}{cc} E & \tilde{E} \end{array} \\ \begin{pmatrix} T & U \\ R & Q \end{pmatrix} \end{array},$$

we have the following relationship between P^E and P.

Lemma 6-6: For an arbitrary Markov chain P, P^E is a Markov chain and

$$P^E = T + UNR,$$

where $N = \sum_{k=0}^{\infty} Q^k$.

REMARK: The lemma holds even if N has infinite entries, provided we agree as usual that $0 \cdot \infty = 0$.

PROOF: Let y_n be the nth outcome in the P^E-process. If

$$\Pr_\pi[y_0 = c_0 \wedge \cdots \wedge y_{n-1} = c_{n-1}] > 0,$$

let $\mathbf{t}$ be the random time of outcome $n - 1$ and apply Theorem 4-9. Then

$$\begin{aligned}\Pr_\pi[y_n &= c_n \mid y_0 = c_0 \wedge \cdots \wedge y_{n-1} = c_{n-1}] \\ &= \Pr_\pi[y_n = c_n \mid y_0 = c_0 \wedge \cdots \wedge y_{n-1} = c_{n-1} \wedge x_{\mathbf{t}} = c_{n-1}] \\ &= \Pr_{c_{n-1}}[y_1 = c_n],\end{aligned}$$

and it follows that P^E is a Markov chain. Now let i and j be in E. Applying Theorem 4-10 with the random time identically one and with the statement that E is hit after time 0 first at state j, we have

$$\begin{aligned}P_{ij}^E &= \sum_k P_{ik} B_{kj}^E \\ &= \sum_{k \in E} P_{ik} B_{kj}^E + \sum_{k \notin E} P_{ik} B_{kj}^E \\ &= P_{ij} + \sum_{k \notin E} P_{ik} B_{kj}^E.\end{aligned}$$

The result then follows from Proposition 5-6.

Lemma 6-7: For an arbitrary Markov chain P, if E is a subset of states and β is a finite-valued non-negative P-superregular measure, then β_E is P^E-superregular.

PROOF: Since $\beta \geq \beta P$, multiplication of the submatrices of β by the submatrices of P gives the two relations

$$\beta_E \geq \beta_E T + \beta_{\tilde{E}} R$$

and

$$\beta_{\tilde{E}} \geq \beta_E U + \beta_{\tilde{E}} Q.$$

We may rewrite the second relation as

$$\beta_{\tilde{E}}(I - Q) \geq \beta_E U \geq 0.$$

The proof of Theorem 5-10 translates directly into a proof for row vectors. From it we find

$$\beta_{\tilde{E}} = \gamma N + \rho,$$

where $\gamma = \beta_{\bar{E}}(I - Q)$, $N = \sum Q^n$, and ρ is non-negative and Q-regular. Hence

$$\beta_{\bar{E}} \geq \gamma N \geq (\beta_E U)N.$$

Thus $\beta_E P^E = \beta_E T + \beta_E UNR \leq \beta_E T + \beta_{\bar{E}} R \leq \beta_E$.

Lemma 6-8: No finite null chains exist.

PROOF: We have $L^{(n)}\mathbf{1} = \mathbf{1}$ or $\sum_j L_{ij}^{(n)} = 1$ for every n. Since the limit of a finite sum is the sum of the limits,

$$(L\mathbf{1})_i = \sum_j L_{ij} = \sum_j \lim_n L_{ij}^{(n)} = \lim_n \sum_j L_{ij}^{(n)} = 1.$$

Theorem 6-9: Every recurrent chain P has a positive finite-valued regular measure α which is unique up to multiplication by a scalar. Furthermore, $\alpha\mathbf{1} < \infty$ if and only if P is ergodic.

PROOF: Order the states by the positive integers, let E be the first n of the states, and let F be the first $n + 1$. Then P^E and P^F are ergodic chains and have regular measures α^E and α^F. Also $(P^F)^E = P^E$. Thus α_E^F is P^E-regular by Lemma 6-7, and we may choose α^F such that $\alpha_E^F = \alpha^E$ by the uniqueness part of Proposition 6-4. The procedure of adding a single state to F may be continued by induction, and we set $\alpha = \lim_{E \to S} (\alpha^E \quad 0)$. Now for any of these sets E we have

$$\alpha_E T \leq \alpha_E T + \alpha_E UNR = \alpha_E P^E = \alpha_E$$

or

$$\alpha_E T \leq \alpha_E.$$

Thus,

$$\alpha \begin{pmatrix} T & 0 \\ 0 & 0 \end{pmatrix} = (\alpha_E T \quad 0) \leq (\alpha_E \quad 0) \leq \alpha.$$

As $E \to S$, the entries of $\begin{pmatrix} T & 0 \\ 0 & 0 \end{pmatrix}$ increase monotonically from zero to the entries of P. Hence, by monotone convergence, $\alpha P \leq \alpha$ and, by Proposition 6-4, α is regular. Clearly $\alpha > 0$, and we know that if P is ergodic then $\alpha\mathbf{1} < \infty$. Conversely, suppose $\alpha\mathbf{1} < \infty$. Then, by dominated convergence,

$$\begin{aligned} \alpha L &= \lim_n \alpha L^{(n)} \\ &= \lim_n \frac{1}{n} \alpha(I + \cdots + P^{n-1}) \\ &= \alpha \\ &\neq 0 \end{aligned}$$

and $L \neq 0$.

2. Duality

The proof of Proposition 6-4 is somewhat artificial without further explanation. What was used was a standard method of converting proofs about functions into proofs about measures. We had proved uniqueness for non-negative finite-valued superregular functions, and the idea was to take advantage of this fact in the result for measures.

The isomorphism that exists between row vectors and column vectors is known as **duality**. Not only does duality make rigorous the correspondence between row and column vectors, but also it provides easy proofs of some new results.

Definition 6-10: Let P be an arbitrary Markov chain transition matrix and suppose there exists a positive finite-valued P-superregular measure α. The **α-dual matrix** of P is a matrix $\hat{P}$ defined by

$$\hat{P}_{ij} = \frac{\alpha_j P_{ji}}{\alpha_i}.$$

Let D be a diagonal matrix with diagonal entries $1/\alpha_i$.

We note that $\hat{P} = DP^T D^{-1}$.

We cannot define duality in general, because we are not always assured of the existence of a positive superregular measure. However, there are only two important special cases, and we know that a superregular measure α exists for each of them:

(1) P is recurrent. Then there exists a unique α-dual of P. We call $\hat{P}$ the **dual** of P or the **reverse chain**. We shall investigate the properties of the reverse chain in some detail in Section 8.

(2) P has only transient states. Then, as we saw in Section 5-2, a positive superregular α exists. All duality statements are relative to such a vector α, but there is no assurance that α is unique.

Proposition 6-11: If P is a transition matrix, then so is $\hat{P}$. If all pairs of states in P communicate, then all pairs of states in $\hat{P}$ communicate.

PROOF: It is clear that $\hat{P} \geq 0$. For $\hat{P}\mathbf{1} \leq \mathbf{1}$ we have

$$\sum_j \hat{P}_{ij} = \sum_j \frac{\alpha_j P_{ji}}{\alpha_i} = \frac{1}{\alpha_i} \sum_j \alpha_j P_{ji} \leq \frac{1}{\alpha_i} \alpha_i = 1.$$

If i and j communicate in P by the routes

$$i, m_1, m_2, \ldots, m_r, j$$

and

$$j, n_1, n_2, \ldots, n_s, i,$$

then they communicate in $\hat{P}$ by

$$i, n_s, \ldots, n_1, j,$$
$$j, m_r, \ldots, m_1, i.$$

Proposition 6-12: If P is a transition matrix, then

$$\hat{P}^n = D(P^n)^T D^{-1} \quad \text{and} \quad \{\mathrm{M}_i[\hat{\mathbf{n}}_j]\} = D\{\mathrm{M}_i[\mathbf{n}_j]\}^T D^{-1}.$$

If P is either (1) recurrent or (2) transient with only transient states, then $\hat{P}$ is of the same type. If P is of the second type, then

$$\hat{N} = DN^T D^{-1}.$$

PROOF: The proof of the first assertion is by induction on n. The case $n = 1$ is Definition 6-10. Suppose that

$$\hat{P}^{k-1} = D(P^{k-1})^T D^{-1}.$$

Then

$$\hat{P}^k = \hat{P}\hat{P}^{k-1} = (DP^T D^{-1})(D(P^{k-1})^T D^{-1}) = D(P^k)^T D^{-1}.$$

Associativity holds because all the matrices are non-negative. Now

$$\{\mathrm{M}_i[\hat{\mathbf{n}}_j]\} = \sum_k \hat{P}^k = D \sum_k ((P^k)^T) D^{-1} = D\{\mathrm{M}_i[\mathbf{n}_j]\}^T D^{-1}.$$

In particular, if $\mathrm{M}_j[\mathbf{n}_j]$ is infinite, then so is $\mathrm{M}_j[\hat{\mathbf{n}}_j]$. Hence, by Proposition 6-11, if P is recurrent, so is $\hat{P}$.

Definition 6-13: Let P be a transition matrix, and let α be a positive finite-valued superregular measure. Let Y be any square matrix, let β be any row vector, and let f be any column vector all indexed by the set of states. Define D to be a diagonal matrix whose diagonal entries are $1/\alpha_i$. The **duals** of Y, β, and f are defined by

$$\begin{aligned} \text{dual } Y &= DY^T D^{-1} \\ \text{dual } \beta &= D\beta^T \\ \text{dual } f &= f^T D^{-1}. \end{aligned}$$

The **dual** of a number is that number.

We see that the dual of a row vector is a column vector and that the dual of a column vector is a row vector. The reader should note that $\hat{P}$ is identical with dual P and that part of the content of Proposition 6-12 is that $\mathrm{M}_i[\mathbf{n}_j]$ transforms to the $\hat{P}$ chain in the same way that P_{ij} does:

$$\{\mathrm{M}_i[\hat{\mathbf{n}}_j]\} = \text{dual } \{\mathrm{M}_i[\mathbf{n}_j]\}.$$

The fundamental properties of duals are listed in the next proposition.

Proposition 6-14: Let X and Y be square matrices, row vectors, or column vectors indexed by the set of states for a Markov chain P. Suppose $\hat{P}$ is the α-dual of P. Then

(1) dual dual $X = X$.
(2) dual $(X + Y) =$ dual X + dual Y.
(3) dual $(cX) = c$ dual X.
(4) dual $(XY) =$ dual Y dual X.
(5) dual $I = I$ and dual $0 = 0$.
(6) If $X \geq 0$, then dual $X \geq 0$; and if $X > 0$, then dual $X > 0$.
(7) If $X \geq Y$, then dual $X \geq$ dual Y; and if $X > Y$, then dual $X >$ dual Y.
(8) If f is a P-superregular (or subregular) column vector, then dual f is a $\hat{P}$-superregular (or subregular) row vector; and if β is a P-superregular (or subregular) row vector, then dual β is a $\hat{P}$-superregular (or subregular) column vector.
(9) dual $\mathbf{1} = \alpha$ and dual $\alpha = \mathbf{1}$. The measure α is $\hat{P}$-superregular.
(10) If $\lim_n X^{(n)} = X$, then $\lim_n$ dual $X^{(n)} =$ dual X.

PROOF: We shall prove only (1) and (4); the rest of the proof is left to the reader. For (1) we have

$$\begin{aligned} \text{dual dual } X &= \text{dual } (DX^TD^{-1}) \\ &= D(DX^TD^{-1})^TD^{-1} \\ &= DD^{-1}X^{TT}DD^{-1} \\ &= X. \end{aligned}$$

Associativity holds because D and D^{-1} are diagonal matrices.

For (4) we have

$$\begin{aligned} \text{dual } XY &= D(XY)^TD^{-1} \\ &= DY^TX^TD^{-1} \\ &= (DY^TD^{-1})(DX^TD^{-1}) \\ &= \text{dual } Y \text{ dual } X. \end{aligned}$$

We may summarize Proposition 6-14 by saying that the operation dual is its own inverse, it reverses products, and it preserves sums, equalities, inequalities, regularity, and limits. We know, for example, that the dual of a recurrent chain is recurrent, and since dual is one-one, a dual recurrent chain is the most general recurrent chain. Hence a proof "for all recurrent chains $\hat{P}$" is a proof for all recurrent chains.

The essential feature of duality lies in this last statement; we shall apply it to the proof of Proposition 6-4. We start with a recurrent chain P and two positive superregular measures α and β. Forming $\hat{P}$, the

α-dual of P, we observe that since $\hat{P}$ is the most general recurrent chain and since $\mathbf{1}$ is $\hat{P}$-regular, the dual of $\mathbf{1}$ must be P-regular. Thus α is regular. Now since β is P-superregular and non-negative, dual β is $\hat{P}$-superregular and non-negative. Hence dual β is a constant vector, and the proof that β is a constant multiple of α is complete.

To form the α-dual of the restriction of a matrix, we use the appropriate restrictions of D and D^{-1} which make the matrix products defined. For example, write

$$P = \begin{matrix} & E & \tilde{E} \\ E & T & U \\ \tilde{E} & R & Q \end{matrix}.$$

By definition,

$$\text{dual } T = D_E T^T D_E^{-1},$$

$$\text{dual } U = D_{\tilde{E}} U^T D_E^{-1},$$

$$\text{dual } R = D_E R^T D_{\tilde{E}}^{-1},$$

and

$$\text{dual } Q = D_{\tilde{E}} Q^T D_{\tilde{E}}^{-1}.$$

Note that

$$\begin{pmatrix} \hat{T} & \hat{U} \\ \hat{R} & \hat{Q} \end{pmatrix} = \hat{P} = DP^TD^{-1}$$

$$= \begin{pmatrix} D_E & 0 \\ 0 & D_{\tilde{E}} \end{pmatrix} \begin{pmatrix} T^T & R^T \\ U^T & Q^T \end{pmatrix} \begin{pmatrix} D_E^{-1} & 0 \\ 0 & D_{\tilde{E}}^{-1} \end{pmatrix}$$

$$= \begin{pmatrix} D_E T^T D_E^{-1} & D_E R^T D_{\tilde{E}}^{-1} \\ D_{\tilde{E}} U^T D_E^{-1} & D_{\tilde{E}} Q^T D_{\tilde{E}}^{-1} \end{pmatrix},$$

so that

$$\text{dual } T = \hat{T},$$

$$\text{dual } U = \hat{R},$$

$$\text{dual } R = \hat{U},$$

and

$$\text{dual } Q = \hat{Q}.$$

To make effective use of duality, it is convenient to know what interpretation, if any, the duals of the matrices associated with P have in terms of the $\hat{P}$-process. At this time we shall calculate the duals of ${}^EP$, P^E, and B^E.

If we let ${}^EP$ be the process P watched until it enters E, then ${}^EP$ has transition matrix

$$ {}^EP = \begin{pmatrix} 0 & 0 \\ 0 & Q \end{pmatrix} $$

and fundamental matrix

$$ {}^EN = \begin{pmatrix} 0 & 0 \\ 0 & \sum Q^n \end{pmatrix}. $$

Hence dual ${}^EP = {}^E\hat{P}$ and dual ${}^EN = {}^E\hat{N}$.

The duals of P^E and B^E are not so trivial to settle, and we shall state what they are as the next two propositions.

Proposition 6-15: The dual of P^E is $\hat{P}^E$.

PROOF: By Lemma 6-6,

$$ P^E = T + U\left(\sum Q^n\right)R. $$

Hence

$$ \begin{aligned} \text{dual } P^E &= \text{dual } T + (\text{dual } R)\left(\sum (\text{dual } Q)^n\right)(\text{dual } U) \\ &= \hat{T} + \hat{U}\left(\sum \hat{Q}^n\right)\hat{R} \\ &= \hat{P}^E. \end{aligned} $$

Proposition 6-16:

$$ (\text{dual } B^E)_{ij} = \begin{cases} {}^E\hat{\hat{N}}_{ij} & \text{if } i \in E \\ 0 & \text{if } i \notin E, \end{cases} $$

where ${}^E\hat{\hat{N}}_{ij}$ is the number of times that the α-dual process started at i is in j before returning to E.

PROOF: Let $N = \sum Q^n$. Then

$$ B^E = \begin{pmatrix} I & 0 \\ NR & 0 \end{pmatrix}, $$

so that

$$ \text{dual } B^E = \begin{pmatrix} I & \hat{U}\hat{N} \\ 0 & 0 \end{pmatrix}. $$

Thus if $i \notin E$, then $(\text{dual } B^E)_{ij} = 0$, and if i and j are in E, then

$$ (\text{dual } B^E)_{ij} = \delta_{ij} = {}^E\hat{\hat{N}}_{ij}. $$

If $i \in E$ and $j \notin E$, then the result that

$$ (\hat{U}\hat{N})_{ij} = {}^E\hat{\hat{N}}_{ij} $$

follows from Theorem 4-11 with the random time identically one.

If we define ${}^E\bar{N}$ to be a matrix indexed by E and S whose i-jth entry is the mean number of times starting in i that the process is in j before returning to E, then we may rewrite the result of Proposition 6-16 as

$$\text{dual } B^E = \begin{pmatrix} {}^E\hat{N} \\ 0 \end{pmatrix}.$$

The rest of this section contains applications of Propositions 6-15 and 6-16. We begin by deriving two identities relating B^E to other matrices, and we shall then dualize the first identity to obtain a result which will be used in Chapters 8 and 9. Finally we shall apply Proposition 6-16 in a different way to get a probabilistic interpretation for α_j/α_i.

Proposition 6-17: For any set E,

$$(I - P)B^E = \begin{pmatrix} I - P^E & 0 \\ 0 & 0 \end{pmatrix}.$$

If $N^{(n)} = I + P + \cdots + P^n$, then

$$B^E N^{(n)} = N^{(n)} + {}^E N(P^{n+1} - I).$$

PROOF: Set $N = \sum Q^n$. For the first identity we have

$$(I - P)B^E = \begin{pmatrix} I - T & -U \\ -R & I - Q \end{pmatrix}\begin{pmatrix} I & 0 \\ NR & 0 \end{pmatrix} = \begin{pmatrix} I - (T + UNR) & 0 \\ -R + (I - Q)NR & 0 \end{pmatrix}$$

$$= \begin{pmatrix} I - P^E & 0 \\ 0 & 0 \end{pmatrix} \quad \text{since } (I - Q)N = I.$$

For the second identity, we have

$${}^E N = \begin{pmatrix} 0 & 0 \\ 0 & N \end{pmatrix}$$

and hence

$${}^E NP = \begin{pmatrix} 0 & 0 \\ NR & N - I \end{pmatrix}$$

and

$${}^E N(P - I) = \begin{pmatrix} 0 & 0 \\ NR & -I \end{pmatrix} = B^E - I.$$

Therefore ${}^E N(P^{n+1} - I) = [{}^E N(P - I)]N^{(n)} = (B^E - I)N^{(n)}$.

Proposition 6-18: For any set E

$$\begin{pmatrix} {}^E\bar{N} \\ 0 \end{pmatrix}(I - P) = \begin{pmatrix} I - P^E & 0 \\ 0 & 0 \end{pmatrix}.$$

PROOF: Apply duality to the first identity of Proposition 6-17, using Propositions 6-15 and 6-16. Then

$$\begin{pmatrix} {}^E\hat{\bar{N}} \\ 0 \end{pmatrix}(I - \hat{P}) = \begin{pmatrix} I - \hat{P}^E & 0 \\ 0 & 0 \end{pmatrix}.$$

Since this identity holds for all reverse processes $\hat{P}$, it holds for all processes.

Using Proposition 6-16, we can obtain a simple interpretation for the ratio α_j/α_i. The case in which P is recurrent is of special importance because α is unique up to multiplication by a constant. But first we prove a more general result.

Proposition 6-19: Let α be a positive finite-valued superregular measure for P, and let $\hat{P}$ be the α-dual for P. Then for any set E,

$$\sum_{i \in E} \alpha_i \, {}^E\bar{N}_{ij} = \alpha_j \hat{h}_j^E.$$

PROOF: By Proposition 6-16,

$$(\text{dual } \hat{B}^E)_{ij} = \begin{cases} {}^E\bar{N}_{ij} & \text{for } i \in E \\ 0 & \text{for } i \notin E. \end{cases}$$

Therefore

$$\begin{aligned} \text{dual } \hat{h}^E = \text{dual } (\hat{B}^E 1) &= \alpha(\text{dual } \hat{B}^E) \\ &= \sum_{i \in E} \alpha_i \, {}^E\bar{N}_{ij}. \end{aligned}$$

Corollary 6-20: Let α be a positive finite-valued superregular measure for P, and let $\hat{P}$ be the α-dual of P. Then

$${}^i\bar{N}_{ij} = \frac{\alpha_j}{\alpha_i} \hat{H}_{ji}.$$

PROOF: Set $E = \{i\}$ in Proposition 6-19.

In particular, ${}^i\bar{N}_{ij} \leq \alpha_j/\alpha_i$ for any such α.

Corollary 6-21: Let α be a positive finite-valued regular measure for a recurrent chain P. Then

$$ {}^{i}\bar{N}_{ij} = \alpha_j/\alpha_i. $$

PROOF: Since P is recurrent, so is $\hat{P}$. Thus $\hat{\bar{H}}_{ji} = 1$, and we may apply Corollary 6-20.

Corollary 6-22: Let α be a positive finite-valued regular measure for a recurrent chain P. Then

$$ \sum_{i \in E} \alpha_i \, {}^{E}\bar{N}_{ij} = \alpha_j \qquad \text{for all } j. $$

PROOF: Apply Proposition 6-19. Then $\hat{h}_j^E = 1$.

Definition 6-23: Let P be a recurrent chain. Set

$$ M_{ij} = \mathrm{M}_i[\mathbf{t}_j] $$

and

$$ \bar{M}_{ij} = \mathrm{M}_i[\bar{\mathbf{t}}_j]. $$

The matrix M is called the **mean first passage time matrix**. Similarly M_{iE} is the mean time from i until E is reached, and $\bar{M}_{iE}$ is the mean time from i to return to E.

Proposition 6-24: If P is a recurrent chain with positive regular measure α, then

$$ \sum_{i \in E} \alpha_i \bar{M}_{iE} = \alpha \mathbf{1}. $$

PROOF: We have

$$ \begin{aligned} \sum_{i \in E} \alpha_i \bar{M}_{iE} &= \sum_{i \in E} \sum_j \alpha_i \, {}^{E}\bar{N}_{ij} = \sum_j \sum_{i \in E} \alpha_i \, {}^{E}\bar{N}_{ij} \\ &= \sum_j \alpha_j = \alpha \mathbf{1}, \end{aligned} $$

the next to last equality following from Corollary 6-22.

Proposition 6-25: If P is a recurrent chain with positive regular measure α, then

$$ \bar{M}_{ii} = \begin{cases} \dfrac{1}{\alpha_i} & \text{if } P \text{ is ergodic and } \alpha \mathbf{1} = 1 \\ +\infty & \text{if } P \text{ is null.} \end{cases} $$

PROOF: Set $E = \{i\}$ in Proposition 6-24.

3. Cyclicity

Let P be a recurrent Markov chain and let i be a fixed state of the chain. Define a set of positive integers T by

$$T = \{k \mid (P^k)_{ii} > 0, k > 0\}.$$

Let d be the greatest common divisor of the integers in T.

Lemma 6-26: T is non-empty and is closed under addition.

PROOF: T is clearly non-empty since P is recurrent. Suppose m and n are integers in T. Then $(P^m)_{ii} > 0$ and $(P^n)_{ii} > 0$, so that

$$\begin{aligned}(P^{m+n})_{ii} &= \sum_k (P^m)_{ik}(P^n)_{ki} \\ &\geq (P^m)_{ii}(P^n)_{ii} \\ &> 0.\end{aligned}$$

Hence $m + n$ is in T.

Noting the discussion in Section 1-6e, we arrive at the following result, using Lemma 1-66.

Lemma 6-27: T contains all sufficiently large multiples of its greatest common divisor d.

The integer d we shall call the **period of the chain for the state** i.

Proposition 6-28: The period of a recurrent chain for the state i is a constant independent of the state i.

PROOF: Let i and j be any two states in the chain. Since the chain is recurrent, i and j communicate. Let d be the period associated with state i and let $\bar{d}$ be the period associated with state j. Suppose the minimum possible time for the process to go from state i to state j is s, and suppose the minimum time for the process to go from j to i is t. By Lemma 6-27 let N be large enough so that the process can return to j in $n\bar{d}$ steps for all $n \geq N$. Then the process can go from i to j in s steps, return to j in $N\bar{d}$ steps, and go back to i in t steps. Hence $d \mid (s + N\bar{d} + t)$. Similarly, $d \mid (s + (N + 1)\bar{d} + t)$. Thus d divides the difference, or $d \mid \bar{d}$. Reversing the roles of i and j, we find that $\bar{d} \mid d$. Therefore, $d = \bar{d}$.

We may thus speak of the **period of a recurrent chain** without ambiguity. Every recurrent chain has a period, and that period is finite. If, for example, P is a recurrent chain in which $P_{ii} > 0$ for some i, then P has period one.

Definition 6-29: A recurrent chain is said to be **non-cyclic** if its period is one and **cyclic** if its period is greater than one.

Let P be a recurrent chain of period d. Define a relation R on the states of P by the following: We say that $i\ R\ j$ if and only if, starting at i, the process can reach j in md steps for some m. From the definition of the period d, it follows that $i\ R\ i$. The symmetry of R follows from the fact md plus the time to return from j to i must be a multiple of d. To see that R is transitive, we note that if j can be reached from i in md steps and if k can be reached from j in nd steps, then k can be reached from i in $(m + n)d$ steps.

Thus R partitions the states into cyclic subclasses. The reader may verify that there are d distinct subclasses and that the nth class contains all those states which it is possible to reach from the starting state only at times which are congruent to n modulo d. The process moves cyclically through the classes in the specified order. Furthermore, if the chain is watched after every dth step, the resulting process is again a Markov chain (by the strong Markov property), and its behavior will be noncyclic. The transition matrix for the new process is P^d, and its form is that of d separate recurrent chains:

$$P^d = \begin{pmatrix} \square & & \\ & \ddots & \\ & & \square \end{pmatrix}, \quad d \text{ blocks.}$$

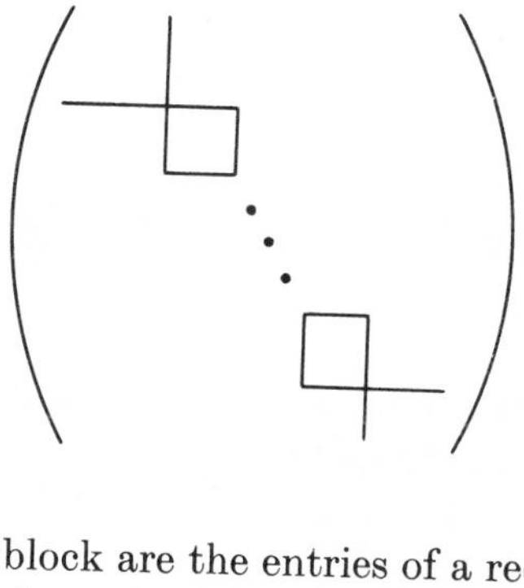

The entries in each block are the entries of a recurrent noncyclic chain, and the entries which are not in any block are all zeros.

The observation that P^d is really d separate recurrent noncyclic chains enables us to study representatively the properties of all recurrent chains by considering only noncyclic chains. Thus, it is to noncyclic chains that we now turn our attention. The main tool in their study will be chains representing sums of independent random variables.

4. Sums of independent random variables

We have already investigated some of the properties of sums of independent random variables Markov chains. Such processes are especially important because of how they arise from general recurrent chains (see Proposition 6-32), and it is for this reason that we now discuss their origin.

For concreteness we shall confine ourselves to sums of independent random variables chains defined on the integers. Before recalling the definition of independent random variables, we remark that if **y** is a real-valued function defined on a probability space and having a denumerable range, then a necessary and sufficient condition for **y** to be measurable (and hence to be a random variable) is that the inverse image under **y** of every one-point set be measurable. The condition is necessary because $\{\omega \mid \mathbf{y}(\omega) = c\} = \{\omega \mid c \leq \mathbf{y}(\omega) \leq c\}$ must be measurable, and it is sufficient because $\{\omega \mid \mathbf{y}(\omega) < c\}$ is a countable union of such sets. Therefore, if **y** is a denumerable-valued random variable and if E is an arbitrary set of real numbers, the set $\{\omega \mid \mathbf{y}(\omega) \in E\}$ is measurable.

Definition 6-30: The denumerable-valued random variables $\mathbf{y}_1, \mathbf{y}_2, \mathbf{y}_3, \ldots$ defined on Ω are **independent** if, for every finite collection of sets $E_1, E_2, \ldots, E_m$ of reals, it is true that

$$\Pr[\mathbf{y}_{n_k}(\omega) \in E_k \text{ for } k = 1, \ldots, m] = \prod_{k=1}^{m} \Pr[\mathbf{y}_{n_k}(\omega) \in E_k].$$

The random variables are **identically distributed** if, for any m and n and for any set E of reals, it is true that

$$\Pr[\mathbf{y}_m(\omega) \in E] = \Pr[\mathbf{y}_n(\omega) \in E].$$

An independent process $\{y_n\}$ was defined in Section 2-5 as one in which the statements $y_0 = c_0 \wedge \cdots \wedge y_{n-1} = c_{n-1}$ and $y_n = c_n$ are probabilistically independent for every $n > 0$ and for every choice of the c's. We see that an independent process is that special case of a Markov process in which $\Pr_\pi[y_{n+1} = j \mid y_n = i]$ is independent of i. Moreover, an independent process is a Markov chain if and only if it is an independent trials process.

Proposition 6-31: Let $\{\mathbf{y}_n\}$ be a stochastic process defined from a sequence space Ω to a denumerable set of real numbers S. The stochastic process is an independent process if and only if the $\{\mathbf{y}_n\}$ are independent random variables. It is an independent trials process if and only if the $\{\mathbf{y}_n\}$ are independent and identically distributed.

PROOF: We are to prove first that the $\{\mathbf{y}_n\}$ are independent if and only if the statements $\mathbf{y}_0 = c_0 \wedge \cdots \wedge \mathbf{y}_{n-1} = c_{n-1}$ and $\mathbf{y}_n = c_n$ are probabilistically independent for any $n \geq 1$ and for any choice of the c's. Independence of the $\mathbf{y}$'s means

$$\Pr[\mathbf{y}_0 = c_0 \wedge \cdots \wedge \mathbf{y}_n = c_n] = \prod_{k=1}^{n} \Pr[\mathbf{y}_k = c_k]$$

and

$$\Pr[\mathbf{y}_0 = c_0 \wedge \cdots \wedge \mathbf{y}_{n-1} = c_{n-1}] = \prod_{k=1}^{n-1} \Pr[\mathbf{y}_k = c_k]$$

for all n. This statement holds if and only if

$$\begin{aligned}\Pr[\mathbf{y}_0 &= c_0 \wedge \cdots \wedge \mathbf{y}_n = c_n] \\ &= \Pr[\mathbf{y}_0 = c_0 \wedge \cdots \wedge \mathbf{y}_{n-1} = c_{n-1}]\Pr[\mathbf{y}_n = c_n]\end{aligned}$$

for all n. Second, we are to prove that the $\{\mathbf{y}_n\}$ are also identically distributed if and only if

$$\Pr[\mathbf{y}_n = c_n] = \Pr[\mathbf{y}_m = c_m]$$

for any n and m. But this assertion is clear from Definition 6-30.

Let $\{\mathbf{y}_n\}$ for $n \geq 0$ be a sequence of independent random variables which are identically distributed for $n \geq 1$ and which have range in the union of the integers and $\{-\infty, +\infty\}$, and define inductively

$$x_0 = \mathbf{y}_0$$

and

$$x_{n+1} = \mathbf{y}_{n+1} + x_n \quad \text{for } n \geq 0.$$

If the $\mathbf{y}_n$ are finite-valued a.e., we claim that the random variables x_n are the outcome functions for a sums of independent random variables process on the integers with starting distribution $\pi_i = \Pr[\mathbf{y}_0 = i]$. Setting

$$p_k = \Pr[\mathbf{y}_n = k], \quad n > 0,$$

which is a constant not depending on any other function in the sequence (by independence) and not depending on n (by identical distributions), we see that $\sum p_k = 1$ since $\mathbf{y}_n$ is finite-valued a.e. Moreover, if $\Pr[x_0 = a \wedge \cdots \wedge x_{n-1} = i] > 0$ with $n > 0$, then

$$\begin{aligned}\Pr[x_n &= j \mid x_0 = a \wedge x_1 = b \wedge \cdots \wedge x_{n-1} = i] \\ &= \Pr[\mathbf{y}_n = j - i \mid \mathbf{y}_0 = a \wedge y_1 = b - a \wedge \cdots \wedge \mathbf{y}_{n-1} = i - h] \\ &= \Pr[\mathbf{y}_n = j - i] \\ &= p_{j-i}.\end{aligned}$$

Hence the process is a Markov chain representing sums of independent random variables.

Conversely, let P be the transition matrix for a sums of independent random variables Markov chain with state space the integers and with outcome functions x_n. Let $\mathbf{y}_n = x_n - x_{n-1}$ for $n > 0$. We shall show that x_0 and the $\mathbf{y}_n$ are independent and that the $\mathbf{y}_n$ are identically distributed; it is clear that x_0 and the $\mathbf{y}_n$ are finite-valued a.e. In fact, we have

$$\begin{aligned}\Pr_\pi[x_0 = c_0 \wedge (\mathbf{y}_k = c_k \text{ for } 1 \le k \le n)] &\\ &= \pi_{c_0} P_{c_0, c_0 + c_1} \cdot \cdots \cdot P_{c_0 + \cdots + c_{n-1}, c_0 + \cdots + c_n}\\ &= \pi_{c_0} p_{c_1} \cdots p_{c_n}\\ &= \Pr_\pi[x_0 = c_0] \cdot \prod_{k=1}^{n} \Pr_\pi[\mathbf{y}_k = c_k],\end{aligned}$$

and independence follows by taking countable disjoint unions of such statements; since

$$\Pr_\pi[\mathbf{y}_n = j] = p_j,$$

the $\mathbf{y}_n$ are identically distributed.

Sums of independent random variables appear in a natural way in the study of recurrent chains. The result to follow associates to every recurrent chain P a sums of independent random variables chain P^* with state space the integers.

Proposition 6-32: Let P be a recurrent chain with outcome functions x_n. For a fixed state s let $\mathbf{t}_n(\omega)$ be the $(n + 1)$st time on the path ω that state s is reached. Then the random times $\mathbf{t}_n$ for $n \ge 0$ are the outcome functions for a sums of independent random variables ladder process P^* with state space the integers.

PROOF: If $\Pr_\pi[\mathbf{t}_0 = c_0 \wedge \cdots \wedge \mathbf{t}_{n-1} = c_{n-1}] > 0$, then

$$\begin{aligned}\Pr_\pi[\mathbf{t}_n = c_n \mid \mathbf{t}_0 = c_0 \wedge \cdots \wedge \mathbf{t}_{n-1} = c_{n-1}] &\\ &= \Pr_\pi[x_{c_{n-1}+1} \ne s \wedge \cdots \wedge x_{c_n - 1} \ne s \wedge x_{c_n} = s \mid x_0 \ne s\\ &\qquad \wedge \cdots \wedge x_{c_0 - 1} \ne s \wedge x_{c_0} = s \wedge x_{c_0 + 1} \ne s\\ &\qquad \wedge \cdots \wedge x_{c_{n-1}} = s]\\ &= \Pr_s[x_1 \ne s \wedge \cdots \wedge x_{c_n - c_{n-1}} = s] \quad \text{by Theorem 4-9}\\ &= \Pr_s[\bar{\mathbf{t}}_s = c_n - c_{n-1}],\end{aligned}$$

where $\bar{\mathbf{t}}_s$ is the time to return to state s. Hence

$$\Pr_\pi[\mathbf{t}_n = j \mid \mathbf{t}_0 = a \wedge \cdots \wedge \mathbf{t}_{n-1} = i] = \bar{F}_{ss}^{(j-i)}.$$

Set $P^*_{ij} = \bar{F}^{(j-i)}_{ss}$. Since by recurrence of P

$$(P^*1)_k = \sum_j P^*_{kj} = \sum_j \bar{F}^{(j-k)}_{ss} = \bar{H}_{ss} = 1,$$

P^* is a sums of independent random variables Markov chain.

The following is a converse to the preceding result.

Proposition 6-33: Let P' be a sums of independent random variables ladder process on the integers. There exists a recurrent chain P and a state s such that the times to return to s are the outcome functions for P'.

PROOF: Let $p'_k = P'_{0k}$ for $k \geq 1$; then $\sum_{k=1}^{\infty} p'_k = 1$. We take P to be a basic example and s to be state 0; the values of p_i and q_i in the basic example are yet to be specified. Define recursively the q_i's by the relations

$$p'_1 = q_1 = \beta_0 - \beta_1$$

and

$$p'_n = p_1 \cdots p_{n-1} q_n = \beta_{n-1} q_n = \beta_{n-1} - \beta_n.$$

In P we have $\Pr[\mathbf{t}_1 - \mathbf{t}_0 = k] = p'_k$ as required; it remains to be proved that P is recurrent. We have

$$\sum_{k=1}^{n} p'_k = \sum_{k=1}^{n} (\beta_{k-1} - \beta_k) = \beta_0 - \beta_n = 1 - \beta_n,$$

and since $\sum_{k=1}^{\infty} p'_k = 1$, we must have $\lim_n \beta_n = 0$. Hence P is recurrent.

We close this section with two remarks about sums of independent random variables and their relation to recurrence. First, we have seen in Proposition 5-22 that a sums of independent random variables process on the integers with finitely many k-values is a recurrent chain if and only if the k-values have mean zero and their greatest common divisor is one. Second, we note that an infinite recurrent chain representing sums of independent random variables must be null, since $\alpha = 1^T$ is regular and $1^T 1 = \infty$.

5. Convergence theorem for noncyclic chains

By restricting our attention to noncyclic recurrent chains, we can prove a stronger result than the Mean Ergodic Theorem, namely that P^n itself converges with n to a limiting matrix. We shall give two proofs of this convergence theorem—the first a matrix proof using sums of independent random variables and the second the classical proof using the Renewal Theorem of Section 1-6f. We shall further show that, conversely, the truth of the convergence theorem just when P is the basic example implies the full validity of the Renewal Theorem.

We begin by proving two lemmas needed in both proofs of the convergence theorem; their effect is to formulate noncyclicity in a number-theoretic way.

Lemma 6-34: For any Markov chain and any states i and j,

$$P_{ij}^{(0)} = \delta_{ij}$$

$$\bar{F}_{ij}^{(0)} = 0,$$

and

$$P_{ij}^{(n)} = \sum_{k=1}^{n} \bar{F}_{ij}^{(k)} P_{jj}^{(n-k)} = \sum_{k=0}^{n-1} P_{jj}^{(k)} \bar{F}_{ij}^{(n-k)} \quad \text{for } n > 0.$$

PROOF: The first two statements are obvious; for the third we first note that if $\Pr_i[\mathfrak{t}_j = k] > 0$, then for $n \geq k$,

$$\begin{aligned} \Pr_i[x_n = j \mid \mathfrak{t}_j = k] &= \Pr_i[x_n = j \mid x_k = j \wedge x_{k-1} \neq j \wedge \cdots \wedge x_1 \neq j] \\ &= \Pr_i[x_n = j \mid x_k = j] \quad \text{by Lemma 4-6} \\ &= \Pr_j[x_{n-k} = j] \quad \text{by Lemma 4-6} \\ &= P_{jj}^{(n-k)}. \end{aligned}$$

Hence no matter what the value of $\Pr_i[\mathfrak{t}_j = k]$, it is true that, for $n \geq k$,

$$\Pr_i[\mathfrak{t}_j = k] \Pr_i[x_n = k \mid \mathfrak{t}_j = k] = \bar{F}_{ij}^{(k)} P_{jj}^{(n-k)}.$$

Using $x_n = j \wedge \mathfrak{t}_j = k$, $1 \leq k \leq n$, as a set of alternatives for $x_n = j$, we have

$$\begin{aligned} P_{ij}^{(n)} &= \sum_{k=1}^{n} \Pr_i[\mathfrak{t}_j = k] \cdot \Pr_i[x_n = j \mid \mathfrak{t}_j = k] \\ &= \sum_{k=1}^{n} \bar{F}_{ij}^{(k)} P_{jj}^{(n-k)} \\ &= \sum_{k=0}^{n-1} P_{jj}^{(k)} \bar{F}_{ij}^{(n-k)} \quad \text{by a change of variable.} \end{aligned}$$

Lemma 6-35: A recurrent chain is noncyclic if and only if the set $Z = \{k \mid \bar{F}_{ii}^{(k)} > 0\}$ has greatest common divisor one.

PROOF: If Z has greatest common divisor one, then the period for the state i is one. Conversely, suppose that the greatest common divisor is c. We shall show that c divides d, the period of the chain. Hence, if $c > 1$, the chain is cyclic. Let n be the smallest integer for which $P_{ii}^{(n)} > 0$ and $c \nmid n$, and write $n = qc + r$ with $0 < r < c$. By Lemma 6-34,

$$P_{ii}^{(n)} = \sum_{k=1}^{n} \bar{F}_{ii}^{(k)} P_{ii}^{(n-k)}$$

$$= \sum_{j=1}^{q} \bar{F}_{ii}^{(jc)} P_{ii}^{((q-j)c+r)}.$$

Then the right side is zero since every term $P_{ii}^{((q-j)c+r)}$ is zero, a contradiction. Thus $P_{ii}^{(n)} > 0$ only if $c \mid n$.

The next two lemmas lead to the convergence theorem; the first one is a consequence of Proposition 6-32 and the zero-one law for sums of independent random variables.

Lemma 6-36: Let P be a noncyclic recurrent chain, and for a fixed state s let E and F be any two sets of integers whose union is the set of all non-negative integers. Then either

$$\Pr_\pi[x_n = s \quad \text{for infinitely many } n \in E] = 1$$

or

$$\Pr_\pi[x_n = s \quad \text{for infinitely many } n \in F] = 1 \quad \text{(or both)},$$

and whichever alternative holds is independent of the starting distribution π.

PROOF: Form the process P^* of Proposition 6-32. We shall first show that for any two states i and j there is a state k which it is possible to reach from both i and j; for this purpose it is sufficient to show that from state 0 it is possible to reach all sufficiently large states, since P^* represents sums of independent random variables. Now the set of states which can be reached from 0 is non-empty and is closed under addition (since P^* represents sums of independent random variables); its greatest common divisor is one by Lemma 6-35. Hence by Lemma 1-66 all sufficiently large states can be reached.

By the zero-one law, which is Propositions 5-19 and 5-20,

$$\Pr_i[\mathbf{t}_n \in E \text{ infinitely often}] = \Pr_i[x_n = s \text{ for infinitely many } n \in E]$$

is zero or one and is independent of i. Thus

$$\Pr_\pi[x_n = s \text{ for infinitely many } n \in E]$$

is zero or one and is independent of π. If

$$\Pr_\pi[x_n = s \text{ for infinitely many } n \in E] = 0$$

and

$$\Pr_\pi[x_n = s \text{ for infinitely many } n \in F] = 0,$$

then

$$\Pr_\pi[x_n = s \text{ infinitely often}] = 0,$$

in contradiction to the recurrence of P.

The following lemma uses the notation $\|\beta\| = \sum_i |\beta_i|$.

Lemma 6-37: Let P be a noncyclic recurrent chain, and let β and γ be probability vectors. Then $\lim_{n\to\infty} \|(\beta - \gamma)P^n\| = 0$.

Proof: Let $E = \{n \mid (\beta P^n)_s \leq (\gamma P^n)_s\}$ and $F = \{n \mid (\beta P^n)_s \geq (\gamma P^n)_s\}$. By Lemma 6-36, either

$$\Pr_\beta[x_n = s \text{ for infinitely many } n \in E] = 1$$

or

$$\Pr_\gamma[x_n = s \text{ for infinitely many } n \in F] = 1,$$

and by symmetry we may assume that the former alternative holds.

Let $h^{(n)}$ be the statement that $x_m = s$ for some $m \in E$ with $m < n$, and let

$$\beta_k^{(n)} = \Pr_\beta[x_n = k \wedge \sim h^{(n)}].$$

Then

$$\|\beta^{(n)}\| = \beta^{(n)}\mathbf{1} = \Pr_\beta[\sim h^{(n)}] \to 0$$

by the above assumption. Also

$$\beta^{(0)} = \beta$$

and

$$\beta_k^{(n+1)} = \begin{cases} \sum_j \beta_j^{(n)} P_{jk} & \text{if } n \notin E \\ \sum_{j \neq s} \beta_j^{(n)} P_{jk} & \text{if } n \in E. \end{cases}$$

We may represent this last identity conveniently by using ϵ, a row vector such that $\epsilon_i = \delta_{is}$, and by defining

$$\delta^{(n)} = \begin{cases} \beta_s^{(n)}\epsilon & \text{if } n \in E \\ 0 & \text{otherwise.} \end{cases}$$

Then

$$\beta^{(n+1)} = (\beta^{(n)} - \delta^{(n)})P.$$

Next, we define

$$\gamma^{(n)} = \beta^{(n)} + (\gamma - \beta)P^n.$$

From this relation,

$$\gamma^{(n-1)}P = \beta^{(n-1)}P + (\gamma - \beta)P^n$$

and

$$\gamma^{(0)} = \gamma$$

and hence

$$\gamma^{(n+1)} = \gamma^{(n)}P + \beta^{(n+1)} - \beta^{(n)}P = (\gamma^{(n)} - \delta^{(n)})P.$$

We shall show by induction that $\gamma^{(n)} \geq \delta^{(n)}$. First, $\gamma^{(0)} = \gamma \geq 0$. If $0 \notin E$, then $\delta^{(0)} = 0 \leq \gamma^{(0)}$. If $0 \in E$, then $\gamma_s^{(0)} = \gamma_s \geq \beta_s = \beta_s^{(0)}$ by the definition of E, and $\beta_s^{(0)} = \delta_s^{(0)}$. Thus $\gamma^{(0)} \geq \delta^{(0)}$ in either case. Suppose $\gamma^{(n-1)} \geq \delta^{(n-1)}$. Then

$$\gamma^{(n)} = (\gamma^{(n-1)} - \delta^{(n-1)})P \geq 0.$$

If $n \notin E$, then $\delta^{(n)} = 0$ and $\gamma^{(n)} \geq \delta^{(n)}$. If $n \in E$, then $[(\gamma - \beta)P^n]_s \geq 0$ by the definition of E, and hence

$$\gamma_s^{(n)} \geq \beta_s^{(n)} = \delta_s^{(n)}$$

by the definition of $\gamma^{(n)}$. Thus $\gamma^{(n)} \geq \delta^{(n)}$ for every n.

In particular, we have $\gamma^{(n)} \geq 0$. Thus

$$\begin{aligned}
\|\gamma^{(n)}\| &= \gamma^{(n)}\mathbf{1} \\
&= \beta^{(n)}\mathbf{1} + [(\gamma - \beta)P^n]\mathbf{1} \\
&= \beta^{(n)}\mathbf{1} + (\gamma - \beta)(P^n\mathbf{1}) \\
&= \beta^{(n)}\mathbf{1} + (\gamma - \beta)\mathbf{1} \\
&= \beta^{(n)}\mathbf{1} \\
&= \|\beta^{(n)}\| \to 0.
\end{aligned}$$

Finally

$$\|(\beta - \gamma)P^n\| = \|\beta^{(n)} - \gamma^{(n)}\|$$

by the definition of $\gamma^{(n)}$, and the right side is

$$\leq \|\beta^{(n)}\| + \|\gamma^{(n)}\| \to 0.$$

Theorem 6-38: If P is a noncyclic recurrent chain, then $\lim_{n\to\infty} P^n$ exists. If P is ergodic, then $\lim P^n = A = \mathbf{1}\alpha$ and $\lim_n \|\pi P^n - \alpha\| = 0$ for every probability vector π. If P is null, then $\lim_n (\pi P^n) = 0$ for every probability vector π.

PROOF: Every recurrent chain is either ergodic or null; taking π to be a vector with 1 in the ith entry and zeros elsewhere, we see that the existence of $\lim P^n$ follows from the other assertions of the theorem.

Let P be ergodic and let $A = \mathbf{1}\alpha$ be the Cesaro limit of the powers of P. We have $\alpha P^n = \alpha$ for every n. Letting $\beta = \pi$ and $\gamma = \alpha$ in Lemma 6-37, we obtain the desired result.

Let P be null and suppose the assertion of the theorem is false. Then by Corollary 1-64, for some probability vector π, there is an increasing sequence $\{n_k\}$ of positive integers and there is a row vector $\rho \neq 0$ such that

$$\lim_k (\pi P^{n_k})_i = \rho_i \quad \text{for every state } i.$$

Certainly $\rho_i \geq 0$. Summing on i, we obtain

$$\rho\mathbf{1} = \sum_i \rho_i = \sum_i \lim_k (\pi P^{n_k})_i \leq \lim_k \sum_i (\pi P^{n_k})_i = 1,$$

the inequality following from Fatou's Theorem. Applying Lemma 6-37 with $\beta = \pi$ and $\gamma = \pi P$, we see that

$$\lim_k (\pi P^{n_k+1})_i = \rho_i \quad \text{for each } i.$$

By Fatou's Theorem,

$$\rho P = (\lim \pi P^{n_k})P \leq \lim \pi P^{n_k+1} = \rho.$$

Hence ρ is non-negative superregular and satisfies $\rho\mathbf{1} < \infty$; ρ must be regular by Proposition 6-4, and the fact that P is null then contradicts Theorem 6-9.

Corollary 6-39: If P is a null chain (not necessarily noncyclic), then $\lim P^n = 0$.

PROOF: Let P have period d. By Theorem 6-38

$$\lim_n P^{nd} = 0.$$

By dominated convergence, $\lim_n P^{nd+r} = 0$ for each $r = 0, 1, 2, \ldots, d-1$. Hence,

$$\lim_n P^n = 0.$$

The classical proof of Theorem 6-38 that follows proves only that $\lim P^n$ exists.

SECOND PROOF OF THEOREM 6-38: We first prove the theorem for the ith diagonal entry. Set $f_n = \bar{F}_{ii}^{(n)}$, $u_n = (P^n)_{ii}$, and

$$\mu = \sum_n nf_n = \sum_n n\bar{F}_{ii}^{(n)} = \sum_n n \Pr_i[\mathbf{t}_i = n] = \mathrm{M}_i[\mathbf{t}_i] = \bar{M}_{ii}.$$

Lemmas 6-34 and 6-35 establish all the hypotheses of Theorem 1-67 except for the fact that $\sum_n f_n = 1$:

$$\sum_n f_n = \sum_n \bar{F}_{ii}^{(n)} = \bar{H}_{ii} = 1.$$

Therefore $u_n \to 1/\mu$ or 0 according as μ is finite or infinite, and the value of the limit for the diagonal entries follows from Proposition 6-25. For the off-diagonal entries, let i and j be any two distinct states. Define a row vector β and a sequence of column vectors $\{g^{(n)}\}$ by

$$\beta_m = \bar{F}_{ij}^{(m)}$$

$$g_m^{(n)} = \begin{cases} (P^{n-m})_{jj} & \text{if } n \geq m \\ 0 & \text{if } n < m. \end{cases}$$

Then $\lim_n g_m^{(n)} = L_{jj}$ exists since we have proved the theorem for diagonal entries. Furthermore, by Lemma 6-34,

$$\begin{aligned} (P^n)_{ij} &= \sum_{m=1}^{n} \beta_m (P^{n-m})_{jj} \\ &= \sum_{m=1}^{\infty} \beta_m g_m^{(n)} \\ &= \beta g^{(n)}. \end{aligned}$$

Since $\beta 1 = 1$ and $g^{(n)} \leq 1$, the Dominated Convergence Theorem applies and

$$\begin{aligned} \lim_n (P^n)_{ij} &= \lim_n \beta g^{(n)} \\ &= \beta \lim_n g^{(n)} \\ &= \beta(L_{jj}1) \\ &= L_{jj}. \end{aligned}$$

As a converse to the second proof of Theorem 6-38, we shall show that the convergence of P^n for every noncyclic recurrent chain implies the truth of the Renewal Theorem. This result is of particular interest because all that is needed is convergence of the diagonal entries of P^n, when P is a noncyclic recurrent case of the basic example.

Proposition 6-40: If every noncyclic recurrent chain converges to a limiting matrix, then the Renewal Theorem holds.

PROOF: Let the sequence $\{f_n\}$ be given. Define $r_i = \sum_{k>i} f_k$; the r_i tend to 0 because $\sum_k f_k = 1$. As long as $r_i > 0$, define

$$p_{i+1} = \frac{r_{i+1}}{r_i} \quad \text{for } i = 0, 1, 2, \ldots.$$

If $r_i = 0$ for some i, then $p_i = 0$ and the p_k for $k > i$ are irrelevant. Set $q_i = 1 - p_i$ and let the p_i and the q_i represent the transition probabilities associated with the basic example. We have

$$\beta_i = p_1 p_2 \cdots p_i = \frac{r_1}{r_0}\frac{r_2}{r_1}\cdots\frac{r_i}{r_{i-1}} = r_i.$$

That is, $\beta_i = r_i$. Since $r_i \to 0$, $\beta_i \to 0$ and the chain is recurrent. Now

$$\begin{aligned} \bar{F}_{00}^{(n)} &= \beta_{n-1}(1 - p_n) \\ &= \beta_{n-1} - \beta_n \\ &= \sum_{k>n-1} f_k - \sum_{k>n} f_k \\ &= f_n. \end{aligned}$$

Thus $\mu = \sum_n n f_n = \bar{M}_{00}$. By Lemma 6-34 we see that $u_n = P_{00}^{(n)}$. The Markov chain is noncyclic by Lemma 6-35 because the greatest common divisor of the k's for which $f_k > 0$ is 1. Therefore $\lim u_n = \lim P_{00}^{(n)}$ exists. On the other hand, by Proposition 6-25 the Cesaro limit of $P_{00}^{(n)}$ is $1/\bar{M}_{00} = 1/\mu$ if $\bar{M}_{00} < \infty$ or 0 if $\bar{M}_{00} = +\infty$. Hence by Proposition 1-61

$$\lim u_n = \begin{cases} 1/\mu & \text{if } \mu < \infty \\ 0 & \text{if } \mu = +\infty. \end{cases}$$

6. Mean first passage time matrix

The matrices M and $\bar{M}$ have already been defined by

$$\begin{aligned} M_{ij} &= \mathrm{M}_i[\mathbf{t}_j] \\ \bar{M}_{ij} &= \mathrm{M}_i[\bar{\mathbf{t}}_j]. \end{aligned}$$

In Proposition 6-25 we saw that

$$\bar{M}_{ii} = \begin{cases} 1/\alpha_i & \text{if } P \text{ is ergodic and } \alpha 1 = 1. \\ \infty & \text{if } P \text{ is null.} \end{cases}$$

Proposition 6-41: In any recurrent chain, $\bar{M} = E + PM$.

PROOF: Apply Theorem 4-11 with the random time equal to one. Then

$$\begin{aligned} \mathrm{M}_i[\bar{\mathbf{t}}_j] &= \sum_k P_{ik}\mathrm{M}_k[\mathbf{t}_j + 1] \\ &= \sum_k P_{ik}(M_{kj} + 1) \\ &= \sum_k P_{ik}M_{kj} + \sum_k P_{ik} \\ &= (PM)_{ij} + 1. \end{aligned}$$

For an ergodic chain P we define D to be the diagonal matrix whose diagonal entries are $1/\alpha_i$, where $\alpha 1 = 1$. From Proposition 6-25 we see that

$$\bar{M} = D + M.$$

Proposition 6-42: If P is an ergodic chain, the mean first passage time matrix M satisfies these properties:

(1) $M_{dg} = 0$ and $M \geq 0$.
(2) M is finite-valued.
(3) $(I - P)M = E - D$.

PROOF: The first statement is obvious, and the third follows immediately from Proposition 6-41 and the identity $\bar{M} = D + M$ if we can show that M is finite-valued. The problem thus reduces to proving (2). We know that $M_{ii} = 0$; therefore let i and j be distinct states. We shall show that M_{ij} is finite. Let $\mathbf{t} = \min(\bar{\mathbf{t}}_i, \bar{\mathbf{t}}_j)$. Then

$$\begin{aligned} \frac{1}{\alpha_j} = \bar{M}_{jj} &= \mathrm{M}_j[\bar{\mathbf{t}}_j] \\ &\geq \Pr_j[x_{\mathbf{t}} = i]\mathrm{M}_j[\bar{\mathbf{t}}_j \mid x_{\mathbf{t}} = i] \\ &\geq \Pr_j[x_{\mathbf{t}} = i]M_{ij} \quad \text{by Theorem 4-11} \\ &= {}^j\bar{H}_{ji}M_{ij}. \end{aligned}$$

If we can show that ${}^j\bar{H}_{ji} > 0$, we will have

$$M_{ij} \leq \frac{\bar{M}_{jj}}{{}^j\bar{H}_{ji}} < \infty.$$

But $0 < \alpha_i/\alpha_j = {}^jN_{ji} = {}^j\bar{H}_{ji}\,{}^jN_{ii}$ by Corollary 6-21 and Proposition 4-15, so that ${}^j\bar{H}_{ji} > 0$.

The remarkable fact about the mean first passage time matrix M for ergodic chains is that the converse of Proposition 6-42—namely

Theorem 6-43—is also true. Thus once a candidate for M has been found, even by guessing, we need check only that it satisfies (1), (2), and (3).

Theorem 6-43: If P is an ergodic chain, the mean first passage time matrix M is characterized by these properties:

(1) $M_{dg} = 0$ and $M \geq 0$.
(2) M is finite valued.
(3) $(I - P)M = E - D$.

PROOF: Proposition 6-42 shows that M satisfies these properties. Let Y be any matrix for which (1), (2), and (3) hold. Let 0 be an arbitrary fixed state of the chain. It is sufficient to show that y, the zeroth column of Y, is equal to m, the zeroth column of M. Forming the chain ${}^0P$, in which state 0 has been made absorbing, and writing

$$ {}^0P = \begin{pmatrix} 1 & 0 \quad 0 & \cdots \\ \vdots & Q & \end{pmatrix} \quad \text{and} \quad y = \begin{pmatrix} y_0 \\ \bar{y} \end{pmatrix} \quad \text{and} \quad m = \begin{pmatrix} m_0 \\ \bar{m} \end{pmatrix}, $$

we see that $m = \{\mathrm{M}_i[\mathbf{a}]\}$ and by Corollary 5-17 that $\bar{m}$ is the minimum finite-valued non-negative solution of the equation $(I - Q)\bar{x} = \mathbf{1}$. We first show that $\bar{y}$ is another finite-valued non-negative solution. We know that $\bar{y}$ is finite-valued and non-negative by hypothesis. The identity $(I - P)Y = E - D$ yields in the zeroth column

$$ \begin{pmatrix} 1 - P_{00} & \cdots \\ \vdots & I - Q \end{pmatrix} \begin{pmatrix} y_0 \\ \bar{y} \end{pmatrix} = \begin{pmatrix} 1 - 1/\alpha_0 \\ \mathbf{1} \end{pmatrix}. $$

But $y_0 = 0$ so that $(I - Q)\bar{y} = \mathbf{1}$. We conclude that $\bar{y} \geq \bar{m}$. Since $y_0 = m_0 = 0$, we have $y \geq m$. Hence

$$ (I - P)(y - m) = (I - P)y - (I - P)m = 0 $$

and $y - m$ is a finite-valued non-negative P-regular function. Thus $y - m = c\mathbf{1}$ by Proposition 6-3. Looking at the zeroth entries, we see that $0 = y_0 - m_0 = c$. Therefore, $y = m$.

7. Examples of the mean first passage time matrix

In this section we shall compute the mean first passage time matrix associated with two infinite recurrent chains. The first example is a reflecting random walk, and the second is the basic example.

EXAMPLE 1: Reflecting random walk.

A random walk on the non-negative integers is defined by the transition probabilities

$$P_{00} = q, \quad \text{where } q \neq 0 \text{ and } q \neq 1,$$
$$P_{i,i+1} = p = 1 - q \quad \text{for } i \geq 0$$
$$P_{i,i-1} = q \quad \text{for } i > 0.$$

We note that the process P with 0 made absorbing is the infinite drunkard's walk P'. For the present chain we have

$$\bar{H}_{00} = pH_{10} + qH_{00} = pH_{10} + q.$$

But H_{10} is the absorption probability B_{10} for the infinite drunkard's walk. And $B_{10} = 1$ if $p \leq q$, and $B_{10} < 1$ if $p > q$. Therefore

$$\bar{H}_{00} \begin{cases} = 1 & \text{if } p \leq q \\ < 1 & \text{if } p > q, \end{cases}$$

and P is recurrent if and only if $p \leq q$.

A similar relation holds for $\bar{M}_{00}$; we have

$$\bar{M}_{00} = 1 + pM_{10} + qM_{00} = 1 + pM_{10}.$$

Since M_{10} is the mean time to absorption $M_1[\mathbf{a}]$ in the P' chain, we see that $\bar{M}_{00}$ is finite if and only if $p < q$. That is,

$$P \text{ is} \begin{cases} \text{transient} & \text{if } p > q \\ \text{null} & \text{if } p = q \\ \text{ergodic} & \text{if } p < q. \end{cases}$$

The chain is never cyclic, since $P_{00} > 0$.

We shall compute M for the ergodic case. Let $r = q/p > 1$. A P-regular measure α must satisfy $\alpha = \alpha P$, or

$$\alpha_0 = \alpha_0 q + \alpha_1 q$$
$$\alpha_i = \alpha_{i-1} p + \alpha_{i+1} q \quad \text{for } i > 0.$$

From the first equation we obtain $\alpha_1 = \alpha_0/r$, and from the second, which is a second-order difference equation, we obtain

$$\alpha_i = A + Br^{-i} \quad \text{for } i \geq 0.$$

From the two equations we find $\alpha_0 = A + B$ and $\alpha_0/r = \alpha_1 = A + B/r$. Therefore, $A/r = A$, and since $r > 1$, we must have $A = 0$. Choosing B so that $\alpha 1 = 1$, we arrive at the result

$$\alpha_i = (1 - 1/r)r^{-i}.$$

The entries M_{i0} of the mean first passage time matrix are easily found once the value of α_0 is known. Letting m be the zeroth column of the M matrix, we see from Proposition 6-42 that

$$(I - P)m = 1 - \{\delta_{i0}/\alpha_0\}$$

or that

$$m_0 - qm_0 - pm_1 = 1 - 1/\alpha_0$$

and

$$m_i - qm_{i-1} - pm_{i+1} = 1 \quad \text{for } i > 0.$$

Since $\alpha_0 = 1 - p/q$, we have $1 - 1/\alpha_0 = -p/(q - p)$. The fact that $m_0 = 0$ then reduces the first relation to

$$-pm_1 = -p/(q - p),$$

so that

$$m_1 = 1/(q - p).$$

The difference equation for m_i has as a general solution

$$m_i = A + B(q/p)^i + i/(q - p) \quad \text{for } i \geq 0.$$

The boundary conditions on m_0 and m_1 imply that $A = B = 0$ and that

$$M_{i0} = i/(q - p).$$

The computation of M_{0i} uses the same general methods. First, we note that if $i < j$, then the process must pass through i from 0 to get to j. Hence

$$M_{0i} + M_{ij} = M_{0j}$$

or

$$M_{ij} = M_{0j} - M_{0i}.$$

Now

$$M_{01} = p + q(1 + M_{01})$$

so that

$$M_{01} = 1/p.$$

For $0 < i < j$,

$$M_{ij} = p(1 + M_{i+1,j}) + q(1 + M_{i-1,j})$$

or

$$M_{0j} - M_{0i} = 1 + p(M_{0j} - M_{0,i+1}) + q(M_{0j} - M_{0,i-1}).$$

Thus for $i > 0$,

$$pM_{0,i+1} - M_{0i} + qM_{0,i-1} = 1,$$

and for $i \geq 0$,

$$pM_{0,i+2} - M_{0,i+1} + qM_{0i} = 1.$$

Solving this equation and using the relations $M_{00} = 0$ and $M_{01} = 1/p$, we find

$$M_{0i} = \frac{q}{(q-p)^2}(r^i - 1) - \frac{i}{q-p}.$$

Algebraic manipulation yields the alternate formula

$$M_{0i} = \frac{1}{p}\sum_{k=0}^{i-1} r^k(i-k).$$

Since $M_{i0} = M_{ij} + M_{j0}$ when $i > j$ and since $M_{0i} + M_{ij} = M_{0j}$ when $i < j$, we may summarize our results as follows:

$$M_{ij} = \begin{cases} \dfrac{i-j}{q-p} & \text{if } i \geq j \\[2ex] \dfrac{q}{(q-p)^2}(r^j - r^i) - \dfrac{j-i}{q-p} & \text{if } i \leq j. \end{cases}$$

EXAMPLE 2: Basic Example.

The vector β defined for the basic example has the property that $\beta P = \beta$ if and only if $\lim_{i\to\infty} \beta_i = 0$. Furthermore, P is recurrent if and only if $\lim_{i\to\infty} \beta_i = 0$. When P is recurrent, it is null if $\sum_i \beta_i$ is infinite and ergodic if $\sum_i \beta_i$ is finite. In the ergodic case the regular measure α for which $\alpha 1 = 1$ has entries

$$\alpha_i = \frac{\beta_i}{\sum_i \beta_i}.$$

Entries M_{ij} of the mean first passage time matrix for the basic example satisfy the equations

$$M_{0i} + M_{ij} = M_{0j} \qquad \text{for } i < j$$

and

$$M_{ij} = M_{i0} + M_{0j} \quad \text{for } i > j.$$

Since

$$M_{i0} + M_{0i} = \bar{M}_{ii} = \frac{\sum \beta_i}{\beta_i} \quad \text{for } i > 0,$$

it is sufficient to compute M_{0i} for the chain. Taking the statements {the process moves from 0 to $k \leq i - 1$ and then to zero, the process moves directly to i} as a set of alternatives, we find that

$$M_{0i} = \beta_i i + \sum_{k=0}^{i-1} \beta_k q_{k+1}(k + 1 + M_{0i}).$$

Solving this equation with the aid of the relation $\beta_k q_{k+1} = \beta_k - \beta_{k+1}$, we obtain

$$M_{0i} = \frac{1}{\beta_i} \sum_{k<i} \beta_k.$$

Therefore, for $i > 0$,

$$M_{i0} = \frac{1}{\beta_i} \sum_{k \geq i} \beta_k.$$

The general entries may be computed from

$$M_{ij} = M_{0j} - M_{0i} \quad \text{if } i < j$$
$$M_{ij} = M_{i0} + M_{0j} \quad \text{if } i > j.$$

8. Reverse Markov chains

Let $\{x_n\}$ be the outcome functions for a denumerable Markov process defined on a space Ω and with range in S. The outcome functions appear in a certain order and represent the forward passage of time. One may well wonder, however, if, when the functions are looked at in reverse order, the process is in any sense still Markovian. It is the purpose of this section to discuss this question; as a by-product of the discussion, we shall gain an interpretation for the dual of an ergodic Markov chain.

The sense in which a Markov process reversed in time is still a Markov process is the following.

Proposition 6-44: Let $\{x_n\}$ be a denumerable Markov process and let N be a fixed positive integer. Define $y_n = x_{N-n}$ for $0 \leq n \leq N$ and $y_n =$ "stop" for $n > N$. Then the functions y_n are the outcome functions for a denumerable Markov process with the same state space with "stop" adjoined.

PROOF: We shall show that the functions y_n satisfy the Markov property. Clearly, this needs to be checked only for $n \leq N$. If $\Pr[y_0 = c_0 \wedge \cdots \wedge y_{n-1} = c_{n-1}] > 0$, then

$$\begin{aligned}
\Pr[y_n = c_n \mid y_0 = c_0 \wedge \cdots \wedge y_{n-1} = c_{n-1}] \\
&= \Pr[x_{N-n} = c_n \mid x_N = c_0 \wedge \cdots \wedge x_{N-n+1} = c_{n-1}] \\
&= \frac{\Pr[x_{N-n} = c_n \wedge x_{N-n+1} = c_{n-1} \wedge \cdots \wedge x_N = c_0]}{\Pr[x_{N-n+1} = c_{n-1} \wedge \cdots \wedge x_N = c_0]}.
\end{aligned}$$

The numerator is

$$\begin{aligned}
\Pr[x_{N-n} = c_n \wedge x_{N-n+1} = c_{n-1}] \\
&\times \Pr[x_{N-n+2} = c_{n-2} \mid x_{N-n} = c_n \wedge x_{N-n+1} = c_{n-1}] \\
&\times \cdots \times \Pr[x_N = c_0 \mid x_{N-n} = c_n \wedge \cdots \wedge x_{N-1} = c_1],
\end{aligned}$$

which by the Markov property is

$$\Pr[x_{N-n} = c_n \wedge x_{N-n+1} = c_{n-1}] \cdot P_{c_{n-1}c_{n-2}} \cdot \cdots \cdot P_{c_1 c_0}.$$

The denominator similarly reduces to

$$\Pr[x_{N-n+1} = c_{n-1}] \cdot P_{c_{n-1}c_{n-2}} \cdot \cdots \cdot P_{c_1 c_0}.$$

Dividing, we obtain

$$\begin{aligned}\Pr[y_n = c_n \mid y_0 = c_0 \wedge \cdots \wedge y_{n-1} = c_{n-1}] &= \frac{\Pr[x_{N-n} = c_n \wedge x_{N-n+1} = c_{n-1}]}{\Pr[x_{N-n+1} = c_{n-1}]} \\ &= \Pr[x_{N-n} = c_n \mid x_{N-n+1} = c_{n-1}] \\ &= \Pr[y_n = c_n \mid y_{n-1} = c_{n-1}].\end{aligned}$$

It is not true in general that, if the original process is a Markov chain P, then the new process is also a Markov chain. Let P be started with distribution π. We then have, if $\Pr[y_{n-1} = i] > 0$,

$$\begin{aligned}\Pr[y_n = j \mid y_{n-1} = i] &= \Pr_\pi[x_{N-n} = j \mid x_{N-n+1} = i] \\ &= \frac{\Pr_\pi[x_{N-n} = j \wedge x_{N-n+1} = i]}{\Pr_\pi[x_{N-n+1} = i]} \\ &= \frac{(\pi P^{N-n})_j \cdot P_{ji}}{(\pi P^{N-n+1})_i}.\end{aligned}$$

The last quantity on the right need not be independent of n. Nevertheless, if P is ergodic, there is a case where we *can* state a positive result—a result which gives us an interpretation for the dual of P.

Proposition 6-45: Let $\{x_n\}$ be the outcome functions for an ergodic chain P, let N be a fixed positive integer, and let α be the unique P-regular probability measure. If P is started with distribution α, then the process $\{y_n = x_{N-n}, 0 \leq n \leq N\}$ is an initial segment of the Markov chain with transition matrix $\hat{P}$ and with starting distribution α.

PROOF: If $\Pr[y_{n-1} = i] > 0$, then $\sigma = \alpha P^N = \alpha$ and

$$\begin{aligned}\Pr_\alpha[y_n = j \mid y_{n-1} = i] &= \frac{(\alpha P^{N-n})_j \cdot P_{ji}}{(\alpha P^{N-n+1})_i} \\ &= \frac{\alpha_j P_{ji}}{\alpha_i} \\ &= \hat{P}_{ij}\end{aligned}$$

independently of n for $n \leq N$.

The motivation for calling $\hat{P}$ the reverse chain when P is recurrent now becomes clear.

9. Problems

1. Compute for the Land of Oz example P^2, P^4, and P^8. What is $A = \lim P^n$? Show that each row of A is a regular measure and that each column is a regular function.

2. Let

$$P = \begin{array}{c} \\ 1 \\ 2 \\ 3 \\ 4 \\ 5 \\ 6 \end{array} \begin{array}{c} \begin{array}{cccccc} 1 & 2 & 3 & 4 & 5 & 6 \end{array} \\ \begin{pmatrix} \frac{1}{2} & 0 & 0 & \frac{1}{4} & \frac{1}{4} & 0 \\ 0 & 1 & 0 & 0 & 0 & 0 \\ 0 & 0 & \frac{3}{5} & 0 & \frac{2}{5} & 0 \\ \frac{1}{2} & 0 & 0 & 0 & 0 & \frac{1}{2} \\ 0 & 0 & \frac{3}{5} & 0 & \frac{2}{5} & 0 \\ 0 & \frac{1}{4} & \frac{1}{4} & \frac{1}{4} & \frac{1}{4} & 0 \end{pmatrix} \end{array}.$$

The process is started in state 1. Find the probability of being in the various states in the long run.

3. In the basic example, let

$$p_i = \left(\frac{i}{i+1}\right)^2.$$

Is the chain transient, null, or ergodic?

4. Prove that $\alpha = \mathbf{1}^T$ is regular for any sums of independent random variables process. Give a careful statement as to the existence of transient, null, and ergodic examples.

5. Establish the following relationships between a chain with transition matrix P and one with matrix P^E:
   (a) If P is transient, then P^E is transient.
   (b) If P is recurrent, then P^E is recurrent and $\alpha^E = c\alpha_E$.
   (c) If P is ergodic, then P^E is ergodic, but the converse is not always true.

6. Prove that if a recurrent P has column-sums equal to 1, then $\hat{P} = P^T$.

7. Consider sums of independent random variables on the integers with $p_{-1} = \frac{1}{3}$ and $p_1 = \frac{2}{3}$. Choose two essentially different positive regular measures α, and show that each gives a correct expression for ${}^i\bar{N}_{ij}$ in Corollary 6-20.

8. Show that if $P_{ii} > 0$ for a single state in a recurrent chain, then the chain is noncyclic.

9. Show by an example that $\hat{M}_{ij} = \alpha_j M_{ji}/\alpha_i$ need not be true.

10. Show that in an ergodic chain αM may be either finite-valued or infinite-valued.

11. Determine whether the following chain is transient or recurrent:

$$P = \begin{pmatrix} \frac{1}{2} & 0 & \frac{1}{4} & \frac{1}{4} \\ 0 & 1 & 0 & 0 \\ \frac{1}{4} & \frac{1}{4} & \frac{1}{4} & \frac{1}{4} \\ 0 & 0 & 0 & 1 \end{pmatrix}.$$

If transient: Put into standard form, and find N, B, and a.
If recurrent: Is it cyclic? Find α, M, $\hat{P}$, $\hat{M}$.

12. Do the same for

$$P = \begin{pmatrix} 1-c & c \\ 1 & 0 \end{pmatrix}, \qquad 0 < c \leq 1.$$

13. Find α and M for an independent trials process by the methods of this chapter, and check your answers by a direct computation. [See Problem 5 in Chapter 4.]
14. (a) Complete the work of finding M for the basic example.
(b) Find the reverse of the basic example (when recurrent), and compute M for this chain.
15. A light bulb in a fixture lasts j time units with probability f_j. It is replaced with a similar bulb when it burns out. Assume that $\sum f_j = 1$, $f_0 = 0$, and $f_1 > 0$. Let x_n be the length of time that the bulb in use at time n has lasted (taken to be 0 if there is a replacement at time n). Show that $\{x_n\}$ is the set of outcome functions for a Markov chain and discuss the connection with the basic example. Show that the probability that a bulb is replaced at time n tends to a limit as $n \to \infty$.

Problems 16 to 26 refer to sums of independent random variables on the circle. Mark n ($n \geq 3$) points on a circle, labeled $i = 0, 1, \ldots, n-1$ clockwise. The process either moves one step clockwise with probability $\frac{2}{3}$, or it moves one step counterclockwise with probability $\frac{1}{3}$.

16. Prove that the chain is ergodic. Is it cyclic?
17. Find a positive regular measure α with $\alpha\mathbf{1} = 1$. Interpret it.
18. Construct the reverse chain.
19. Compute M by means of Theorem 6-43. [It is sufficient to find M_{i0}.]
20. Show that for large n,

$$M_{i0} \sim 3(n - i - n(\tfrac{1}{2})^i).$$

Compare this result with the absorption times of a drunkard's walk on $\{0, 1, \ldots, n\}$ with $p = \frac{2}{3}$.
21. Show that the approximation in the previous problem is excellent for $n = 50$.
22. Use the approximation of Problem 20 to show that the maximum value of M_{i0} occurs approximately at

$$i \doteq \frac{\log n}{\log 2}.$$

Check this conclusion for $n = 50$.
23. Let n be even, and let E be the set of even-numbered states. Compute P^E.
24. For $n = 3$, compute P^2, P^4, P^8, and P^9. What is the limit of P^n?
25. Repeat Problem 24 for $n = 4$.
26. Show that $\alpha M = c_n \mathbf{1}^T$, and find an asymptotic expression for c_n.

CHAPTER 7

INTRODUCTION TO POTENTIAL THEORY

1. Brownian motion

One of the fruitful achievements of probability theory in recent years has been the recognition that two seemingly unrelated theories in physics—one for Brownian motion and one for potentials—are mathematically equivalent. That is, the results of the two theories are in one-to-one correspondence and any proof of a result in one theory can be translated directly into a proof of the corresponding result in the other theory. In this chapter we shall sketch how this equivalence comes about, and we shall see that Brownian motion is a process which is like a Markov chain except that it does not have a denumerable state space and time does not proceed in discrete steps. The details of this equivalence can be found in Knapp [1965]. The important thing to notice will be that the definitions of potential-theoretic concepts in terms of Brownian motion do not depend on isolated specific properties of the process but depend only on the Markovian character of Brownian motion. In other words, there is reasonable hope of defining for an arbitrary Markov chain a potential theory in which analogs of the classical theorems hold.

We begin by describing Brownian motion. In 1826 the botanist Robert Brown observed that microscopic particles, when left alone in a liquid, are seen to move constantly in the fluid along erratic paths. Much later Albert Einstein investigated this movement of particles from a theoretical point of view. Einstein was able to derive statistical laws which estimate how a large number of particles spread over a period of time, and his predictions were verified.

In setting up a probabilistic model for this so-called Brownian motion, we simply replace Einstein's estimate of what happens to a large number of particles by a probability for what happens to one particle. We are then to require that

$$\Pr[\text{particle started at } u \text{ is in } E \text{ at time } t] = \int_E \frac{1}{(2\pi t)^{3/2}} e^{-|u-y|^2/2t} dy,$$

where E is a Borel set in three-dimensional (Euclidean) space R^3 and $|u - y|$ denotes the Euclidean distance from u to y. If we use the notation $\Pr_u[x_t \in E]$ for the left side and the notation $p^t(u, E)$ for the right side, we have

$$\Pr_u[x_t \in E] = p^t(u, E).$$

By Theorem 1-41, $p^t(u, F)$ is a measure depending on t and u and defined on the smallest Borel field containing the open sets of R^3. Therefore, we may write

$$\Pr_u[x_t \in E] = \int_E p^t(u, dy).$$

The physical theory also makes us require that if $t_1 < t_2 < \cdots < t_n$, then $\{x_{t_1}, x_{t_2} - x_{t_1}, \ldots, x_{t_n} - x_{t_{n-1}}\}$ should behave like a set of independent random variables with $x_{t+s} - x_t$ having the same distribution as x_s. That is, we require that

$$\begin{aligned}\Pr_u[x_{t_1} \in E_1 \wedge \cdots \wedge (x_{t_n} - x_{t_{n-1}}) \in E_n] &= \Pr_u[x_{t_1} \in E_1] \cdot \cdots \cdot \Pr_u[x_{t_n} - x_{t_{n-1}} \in E_n] \\ &= \Pr_u[x_{t_1} \in E_1] \cdot \cdots \cdot \Pr_u[x_{(t_n - t_{n-1})} \in E_n].\end{aligned}$$

This identity implies that we must have

$$\begin{aligned}\Pr_u[x_q \in E \wedge x_r \in F \wedge \cdots \wedge x_s \in G \wedge x_t \in H] \\ = \int_E p^q(u, dw) \int_F p^{r-q}(w, dx) \int \cdots \int_H p^{t-s}(y, dz).\end{aligned}$$

We note that with these various requirements we have given more than one definition for $\Pr_u[x_t \in E]$ and that we must check, for example, that

$$\Pr_u[x_s \in R^3 \wedge x_t \in E] = \Pr_u[x_t \in E]$$

and

$$\Pr_u[x_s \in F \wedge x_t \in R^3] = \Pr_u[x_s \in F].$$

Such identities can be verified by direct calculation. It should not be too surprising that such consistency conditions arise since they arose with denumerable stochastic processes earlier: In the proof of the Kolmogorov Extension Theorem in Chapter 2 we required that the measures on cylinder sets all be consistent.

Now for any denumerable Markov chain P we have

$$0 \le P_{ij} = \Pr_i[x_1 = j], \tag{1}$$

$$\Pr_i[x_1 \in S] = \sum_j P_{ij} \le 1, \tag{2}$$

$$\Pr_i[x_1 = j \wedge x_2 = k \wedge \cdots \wedge x_{n-1} = r \wedge x_n = s] = P_{ij}P_{jk} \cdots P_{rs}. \tag{3a}$$

The last equality (3a) implies and is implied by

$$\text{(3b)}\quad \Pr_i[x_1 \in E \wedge x_2 \in F \wedge \cdots \wedge x_{n-1} \in G \wedge x_n \in H] = \sum_{j \in E} P_{ij} \sum_{k \in F} P_{jk} \cdots \sum_{s \in H} P_{rs}.$$

It is easy to prove both the Markov property and the Markov chain property from (3a), and hence (1), (2), and (3b) give an equivalent definition of Markov chain. The analogous statements for Brownian motion are

$$\text{(1')}\qquad 0 \leq \int_E p^t(u, dy) = \Pr_u[x_t \in E],$$

$$\text{(2')}\qquad \Pr_u[x_t \in R^3] = \int_{R^3} p^t(u, dy) = 1,$$

and

$$\text{(3')}\quad \Pr_u[x_q \in E \wedge x_r \in F \wedge \cdots \wedge x_t \in H] = \int_E p^q(u, dw) \int_F p^{r-q}(w, dx) \int \cdots \int_H p^{t-s}(y, dz).$$

As expected, these statements imply that the position of a particle at time $t + s$ depends only on s and the particle's position at time t, and not on the value of t or what happened to the particle before time t. (This assertion can be formulated precisely in terms of means of functions given a Borel field, which are a technical generalization of conditional means of functions given a partition.)

As with denumerable Markov chains we need not require that a Brownian motion particle be started deterministically at a state u. If we start the particle according to probabilities assigned by a measure μ on R^3, then we have

$$\begin{aligned}\Pr_\mu[x_t \in E] &= \int_{R^3} \int_E \frac{1}{(2\pi t)^{3/2}} e^{-|u-y|^2/2t} dy\, d\mu(u) \\ &= \int_{R^3} \int_E p^t(u, dy)\, d\mu(u).\end{aligned}$$

A similar expression holds for the probability of being in a finite sequence of sets at specified times.

In Section 3 we shall need a formal definition of Brownian motion, and we use the formula for $\Pr_\mu[x_t \in E]$ to motivate it. We define a transformation P^t of the measures μ on R^3 with $\mu(R^3) = 1$ into themselves by

$$(\mu P^t)(E) = \Pr_\mu[x_t \in E] = \int_E \left[\int_{R^3} \frac{1}{(2\pi t)^{3/2}} e^{-|x-y|^2/2t} d\mu(x)\right] dy.$$

Later we shall replace the expression in brackets by $f(y, t)$ for simplicity of notation. That μP^t is a measure follows from Theorem 1-41, and that $(\mu P^t)(R^3) = 1$ follows from the identity

$$(\mu P^t)(R^3) = \int_{R^3} \left[\int_{R^3} \frac{1}{(2\pi t)^{3/2}} e^{-|x-y|^2/2t} dy \right] d\mu(x)$$

$$= \int_{R^3} \left[\frac{1}{(2\pi)^{3/2}} \int_{R^3} e^{-|u|^2/2} du \right] d\mu(x) = 1$$

after a change of variable.

Definition 7-1: The totality of theorems about the operators P^t, the measures μ on R^3 with $\mu(R^3) = 1$, and quantities definable in terms of them and properties of R^3 is called **Brownian motion theory**.

We immediately extend P^t by linearity to be defined on all finite measures on R^3 and all differences of two finite measures.

2. Potential theory

Classical potential theory begins as a study of Coulomb's law of attraction of electrical charges in physics. This law states that every two charges in the universe attract (or repel) each other with a force whose direction is the line connecting them and whose magnitude is proportional to the magnitude of each of them and inversely proportional to the square of the distance between them. That is,

$$F = \epsilon_0 \frac{Qq}{r^2},$$

where ϵ_0 is a constant depending on the units. As an aid in the study, one introduces the notion of potential: The potential at a point x due to a charge q is the work (or energy) required to bring a unit charge from infinity to the point x. It can be shown that this potential is independent of the path along which the charge is brought to the point x and that its value is

$$\frac{1}{2\pi} \frac{q}{|x - x_0|},$$

where x_0 is the position of the charge and where the constant $1/2\pi$ has been fixed after a certain choice of units.

More generally one defines a **charge distribution** to be the difference of any two finite measures defined on the Borel sets of R^3, that is, the smallest Borel field in R^3 containing all open sets. The potential at x due to the charge distribution is again the work required to bring a unit charge from infinity to the point x. Since force (and hence work) are additive, the potential due to a charge distribution consisting of charges $q_1, \ldots, q_n$ at points $x_1, \ldots, x_n$ is

$$\frac{1}{2\pi} \sum_{i=1}^{n} \frac{q_i}{|x - x_i|}.$$

Passing to the limit in an appropriate sense, we would expect the potential due to an arbitrary charge distribution μ to be

$$\frac{1}{2\pi} \int_{R^3} \frac{d\mu(y)}{|x - y|}.$$

After checking that such an expression is always well defined, we shall define a potential to be any function of this form.

Lemma 7-2: If μ is a charge distribution, then

$$g(x) = \frac{1}{2\pi} \int_{R^3} \frac{1}{|x - y|} \, d\mu(y)$$

is finite a.e. with respect to Lebesgue measure.

PROOF: It suffices to prove the lemma for the case where μ is a measure, since the general case follows by taking differences. Let K_n denote the closed ball about the origin of radius n, and form

$$\int_{K_n} g(x)dx = \frac{1}{2\pi} \int_{K_n} \int_{R^3} \frac{1}{|x - y|} \, d\mu(y)dx.$$

By Fubini's Theorem we may interchange the order of integration to get

$$\int_{K_n} g(x)dx = \frac{1}{2\pi} \int_{R^3} \left[\int_{K_n} \frac{1}{|x - y|} \, dx \right] d\mu(y).$$

The inside integral on the right is bounded by its value at the origin, which is some finite number c. Thus the right side does not exceed

$$\frac{1}{2\pi} \, c\mu(R^3) < \infty,$$

and g must be finite a.e. on K_n. Since the countable union of the sets K_n is R^3, we conclude that g is finite a.e.

Definition 7-3: The function

$$\frac{1}{2\pi} \int_{R^3} \frac{1}{|x - y|} \, d\mu(y)$$

for μ a charge distribution will be called the **potential** of μ.

The class of theorems relating charges and potentials and quantities definable in terms of them and properties of R^3 is called **classical potential theory**. The operator transforming a charge into its potential is called the **potential operator.** The kernel of the potential operator is called the **Green's function**.

As we have defined it, potential theory contains the subject known in physics as electrostatics. Our definition includes the notions of distance, charge, and potential, and all the quantities commonly arising in electrostatics are definable in terms of these three notions. As an illustration, Table 7-1 shows how some of the quantities arising in electrostatics are related dimensionally to distance, charge, and potential. The table uses the notation

$$\begin{aligned} \text{distance} &= x \\ \text{time} &= t \\ \text{mass} &= m \\ \text{charge} &= q \end{aligned} \qquad \text{and} \qquad \begin{aligned} \text{distance} &= x \\ \text{charge} &= q \\ \text{potential} &= V \end{aligned}$$

TABLE 7-1. DIMENSIONS OF ELECTROSTATIC CONCEPTS

Concept	Dimensions	Potential-Theoretic Dimensions
Capacity	q^2t^2/mx^2	q/V
Charge	q	q
Energy	mx^2/t^2	Vq
Field	mx/t^2q	V/x
Force	mx/t^2	Vq/x
Potential	mx^2/t^2q	V

We give four examples to illustrate how concepts may be defined explicitly in terms of distance, charge, and potential.

(1) We can reasonably ask what the total amount of work required to assemble a charge distribution is if only an "infinitesimal" amount of charge is brought into position at one time. The way to compute this amount of work is to integrate the potential function against the charge distribution, provided the integral exists. Thus we define the **energy of a charge distribution** to be the integral of its potential with respect to the charge, provided the integral exists.

(2) The **total charge** of a charge distribution μ is $\mu(R^3)$.

(3) If a total amount of charge q is put on a piece of conducting metal in R^3, the charge will redistribute itself in such a way that the potential is a constant on the set where the metal is. The situation where the potential is constant on the metal is the one which minimizes energy

among all charges μ with total charge q and with μ vanishing away from the set where the metal is, and this situation is referred to as equilibrium. We define an **equilibrium potential** for a closed set E to be a potential which is 1 on E and which comes from a charge distribution which has all its charge on E. An **equilibrium set** is a set which has an equilibrium potential.

(4) The capacity of a conductor in R^3 is defined as the total amount of charge needed to produce a unit potential on the set where the conductor is. We thus define the **capacity** of any equilibrium set to be the total charge of a charge distribution producing an equilibrium potential.

To indicate the directions in which classical potential theory leads, we shall state without proof some of the theorems in the subject. The **support** of a charge is defined to be the complement of the union of all open sets U with the property that the charge vanishes on U and every measurable subset of U.

(1) *Uniqueness of charge:* A potential uniquely determines its charge.

(2) *Determination of potential:* A potential is completely determined by its values on the support of its charge.

(3) *Uniqueness of equilibrium potential:* A set E has at most one equilibrium potential. (This assertion is a corollary of (2).)

(4) *Characterization of equilibrium potential:* The equilibrium potential for an equilibrium set E is equivalent to the pointwise infimum of all potentials which have non-negative charges and which dominate the constant function 1 on E.

(5) *Principle of domination:* Let h and g be potentials arising from non-negative charges $\bar{\mu}$ and μ, respectively. If h dominates g on the support of μ, then h dominates g everywhere and $\bar{\mu}(R^3) \geq \mu(R^3)$.

(6) *Principle of balayage:* If g is a potential with a non-negative charge and if E is a closed set in R^3, then there is a unique potential $\bar{g}$ with a non-negative charge with support in E such that $\bar{g} = g$ on E. The potential $\bar{g}$ (called the balayage potential) satisfies $\bar{g} \leq g$ everywhere, and its total charge does not exceed the total charge of g. The balayage potential is equivalent to the pointwise infimum of all potentials which have non-negative charges and which dominate g on E. It is equivalent to the supremum of all potentials which are dominated by g on E and whose charges have support in E.

(7) *Principle of lower envelope:* The pointwise infimum of potentials with non-negative charges is equivalent to a potential with a non-negative charge.

(8) *Non-negative potentials:* The charge distribution of a non-negative potential has non-negative total charge.

(9) *Energy of equilibrium potential:* If E is an equilibrium set of finite energy, then the equilibrium potential minimizes energy among all potentials whose charges have support in E and whose total charge is equal to the capacity of E.

3. Equivalence of Brownian motion and potential theory

Kakutani [1944] observed that several of the basic quantities of potential theory, like equilibrium potential, had simple probabilistic interpretations in terms of Brownian motion. If E is an equilibrium set, the value of the equilibrium potential at x is the probability that a Brownian motion process started at x ever hits the set E. Doob and Hunt extended Kakutani's work, and it gradually became clear that in a certain sense Brownian motion and potential theory were really the same subject.

To say that they are exactly the same would be to say that every theorem about Brownian motion should interest a person studying potential theory, and conversely. Although it is doubtful that this situation is the case at present, it is certainly true that modern developments in the two subjects are moving more and more in this direction.

We shall now show that there is a natural way in terms of Brownian motion of obtaining the operator mapping charges into potentials, and that, conversely, from the potential operator it is possible to recover the family $\{P^t\}$ of transition operators for Brownian motion. These facts make it clear that in a technical sense the two theories are identical.

The proof in the first direction is easy and is completed by Proposition 7-4. We recall from the definition of μP^t that

$$(\mu P^t)(E) = \int_E f(x, t)dx$$

for a certain function $f(x, t)$.

Proposition 7-4: Every theorem about potentials can be formulated as a theorem about Brownian motion. Specifically, if μ is a charge, then the potential g of μ satisfies

$$g(x) = \lim_{T \to \infty} \int_0^T f(x, t)dt,$$

where

$$(\mu P^t)(E) = \int_E f(x, t)dx.$$

PROOF: We may assume that $\mu \geq 0$ without loss of generality. Then

$$\int_0^T f(x, t)dt = \int_0^T \frac{1}{(2\pi t)^{3/2}} \left[\int_{R^3} e^{-|x-y|^2/2t} d\mu(y)\right] dt.$$

By Fubini's Theorem we may interchange the order of integration. The above expression is

$$= \int_{R^3} \left[\int_0^T \frac{1}{(2\pi t)^{3/2}} e^{-|x-y|^2/2t} dt \right] d\mu(y).$$

We make the change of variable on t which sends $|x - y|^2/t$ into u^2. The expression becomes

$$= \int_{R^3} \left[\int_{|x-y|/\sqrt{T}}^{\infty} \frac{2}{(2\pi)^{3/2}} |x - y|^{-1} e^{-u^2/2} du \right] d\mu(y).$$

By the Monotone Convergence Theorem,

$$\begin{aligned}
\lim_{T\to\infty} \int_0^T f(x, t)dt &= \int_{R^3} \left[\int_0^{\infty} \frac{2}{(2\pi)^{3/2}} |x - y|^{-1} e^{-u^2/2} du \right] d\mu(y) \\
&= \int_{R^3} \frac{1}{|x - y|} d\mu(y) \left[\frac{1}{2\pi} \int_{-\infty}^{\infty} \frac{1}{\sqrt{2\pi}} e^{-u^2/2} du \right] \\
&= \frac{1}{2\pi} \int_{R^3} \frac{1}{|x - y|} d\mu(y) \\
&= g(x).
\end{aligned}$$

Proposition 7-4 is a precise statement of the connection between Brownian motion and potential theory in one direction. We see that formally the potential operator is

$$\lim_{T\to\infty} \int_0^T P^t dt.$$

Thus to complete the proof of the equivalence of the two theories, what we need to do essentially is recover a sequence from its limit. Of course, we cannot do so unless we know some other properties of the sequence, and it is the isolation of these properties that makes this half of the equivalence difficult. We shall not go into the details here, but we can indicate the general approach to the problem.

Let C_0 denote the set of continuous real-valued functions f on R^3 which vanish at infinity; that is, which are such that for any $\epsilon > 0$ there is a ball of finite radius in R^3 outside of which f is everywhere less than ϵ in absolute value. We define Q^t on C_0 by

$$(Q^t f)(y) = \int_{R^3} \frac{1}{(2\pi t)^{3/2}} e^{-|x-y|^2/2t} f(x)dx.$$

The following facts can be checked:

(1) If f is in C_0, then so is $Q^t f$.
(2) $\sup_y |(Q^t f)(y)| \leq \sup_y |f(y)|$.

(3) $Q^{t+s} = Q^t Q^s$.
(4) $(Q^t f)(y)$ converges uniformly to $f(y)$ as t decreases to 0.
(5) $(Q^t f)(y)$ converges uniformly to 0 as t increases to ∞.

Now any set function in the class M of differences of finite measures is completely determined by its effect on all the functions in C_0, and a direct calculation shows that

$$\int_{R^3} f d(\mu P^t) = \int_{R^3} (Q^t f) d\mu$$

for every $\mu \in M$ and $f \in C_0$. It follows that Q^t and this equation completely determine P^t. Therefore, it is enough to recover Q^t from the potential operator in order to prove our result.

For every $f \in C_0$ such that $(1/t)[(Q^t f)(y) - f(y)]$ converges uniformly as t decreases to 0, we define

$$Af = \lim_{t \downarrow 0} \frac{1}{t} [Q^t f - f].$$

It turns out that if A is known on its entire domain of definition, then Q^t is completely determined by the definition of A and by the first four properties of Q^t listed above. Thus if A could be defined within potential theory, then so could each of the operators Q^t: They are the unique family of operators such that the definition of A and properties (1), (2), (3), and (4) hold. The actual proof of this existence and uniqueness consists in writing down a concrete formula for Q^t in terms of A and t; we reproduce it in order to show that nothing appears in the formula except A, t, and the identity operator I:

$$Q^t f = \lim_{\lambda \to \infty} \sum_{k=0}^{\infty} \frac{1}{k!} t^k [\lambda^2 (\lambda I - A)^{-1} - \lambda I]^k f.$$

For every f in C_0 such that $\int_0^T (Q^t f)(y) dt$ converges uniformly as $T \to \infty$, we define

$$Gf = \lim_{T \to \infty} \int_0^T (Q^t f) dt.$$

It can be shown readily from the five properties of Q^t that G and $-A$ are inverse operators on their respective domains. Thus each uniquely determines the other. Finally (and here is where some work is required) G looks sufficiently like the potential operator when its definition is compared with the formulas of Proposition 7-4 that the potential operator determines G. Thus the potential operator determines G, G determines A, A determines Q^t, and Q^t determines P^t. Hence every theorem of Brownian motion theory is a theorem of potential theory.

4. Brownian motion and potential theory in n dimensions

In the mathematical formulation of Brownian motion and potential theory there is no need to restrict the underlying space to be three-dimensional. We can define an n-dimensional Brownian motion operator P^t by

$$(\mu P^t)(E) = \int_E \left[\frac{1}{(2\pi t)^{n/2}} \int_{R^n} e^{-|x-y|^2/2t} d\mu(y)\right] dx.$$

The potential operator differs in appearance from dimension to dimension more than the Brownian motion operator does, but its kernel is still a constant multiple of the integral of $|x|^{-(n-1)}$. The potential $g(x)$ of μ, the difference of two finite measures, is defined by

$$g(x) = \begin{cases} -\int_{R^1} |x - y| d\mu(y) & \text{in dimension 1} \\ 2\int_{R^2} \log |x - y| d\mu(y) & \text{in dimension 2} \\ c_n \int_{R^n} \frac{1}{|x - y|^{n-2}} d\mu(y) & \text{in dimension } n \geq 3, \end{cases}$$

where

$$c_n = \tfrac{1}{2}\pi^{-n/2}\Gamma(\tfrac{1}{2}(n - 2)).$$

In dimension $n \geq 3$, $g(x)$ is necessarily finite a.e., but in dimensions 1 and 2 we shall need to assume that g is finite a.e.

The fact that the kernel $1/|x|^{n-2}$ tends to zero at infinity in dimension $n \geq 3$ but the kernels $|x - y|$ and $\log |x - y|$ do not tend to zero in dimensions 1 and 2 gives us a clue that the potential theory or dimensions 1 and 2 will differ sharply from that in higher dimensions. We shall discuss the reason for this difference shortly.

In dimension $n \geq 3$, Brownian motion theory and potential theory are again equivalent. The formula

$$g(x) = \lim_{T\to\infty} \int_0^T \frac{1}{(2\pi t)^{n/2}} \int_{R^n} e^{-|x-y|^2/2t} d\mu(y) dt$$

generalized from Proposition 7-4 is still valid, and the discussion of Section 3 goes over with little change to establish the equivalence.

But in dimensions 1 and 2, it does not. The above formula is not true for these dimensions, and the argument in Section 3 fails after the operator G is introduced. The reason for this failure is the following. We recall that in dimension $n \geq 3$ the potential operator is formally $\lim_{T\to\infty} \int_0^T P^t dt$. In dimensions greater than or equal to 3, this quantity is finite, whereas in dimensions 1 and 2 it is infinite. Now

$\lim_{T\to\infty} \int_0^T P^t dt$ plays much the same role for Brownian motion that $\sum_{n=0}^{\infty} P^n$ plays for denumerable Markov chains. It is finite if the process is transient, and infinite if the process is recurrent. In fact, the distinction between transience and recurrence is what is relevant for Brownian motion here: In dimensions 3 and greater a Brownian motion particle after leaving the unit ball of R^n returns to it with probability less than one, whereas in dimensions 1 and 2 it returns with probability equal to one.

The potential operator in dimensions 1 and 2 arises in a different way. Specifically, the formula generalized from Proposition 7-4 is not valid in general, but it is valid if μ has total charge zero and if a mild additional condition is satisfied. The exact formulation of this result will interest us later, and we give it as the next proposition.

Proposition 7-5: In R^n for $n = 1$ or 2, let $\mu = \mu^+ - \mu^-$ be the difference of two finite measures, and suppose that

$$\int_{R^1} |x - y| d\mu^+(y) < \infty \quad \text{and} \quad \int_{R^1} |x - y| d\mu^-(y) < \infty \quad \text{a.e. if } n = 1$$

or

$$\int_{R^2} |\log |x - y|| d\mu^+(y) < \infty \quad \text{and} \quad \int_{R^2} |\log |x - y|| d\mu^-(y) < \infty \quad \text{a.e. if } n = 2.$$

If $\mu(R^n) \neq 0$, then

$$\lim_{T\to\infty} \int_0^T \frac{1}{(2\pi t)^{n/2}} \int_{R^n} e^{-|x-y|^2/2t} d\mu(y) dt = +\infty \quad \text{or} \quad -\infty \quad \text{a.e.}$$

If $\mu(R^n) = 0$, then

$$g(x) = \lim_{T\to\infty} \int_0^T \frac{1}{(2\pi t)^{n/2}} \int_{R^n} e^{-|x-y|^2/2t} d\mu(y) dt$$

exists, is finite a.e., and satisfies

$$g(x) = \begin{cases} -\int_{R^1} |x - y| d\mu(y) & \text{if } n = 1 \\ 2\int_{R^2} \log |x - y| d\mu(y) & \text{if } n = 2. \end{cases}$$

PROOF: We prove the result for $n = 1$; the ideas in the proof for $n = 2$ are similar. The same calculation as at the beginning of the proof of Proposition 7-4 shows that for $n = 1$

$$\int_0^T \frac{1}{\sqrt{2\pi t}} \int_{-\infty}^{\infty} e^{-|x-y|^2/2t} d\mu(y)dt$$
$$= \int_{-\infty}^{\infty} |x - y| \left[\int_{|x-y|/\sqrt{T}}^{\infty} \frac{2}{\sqrt{2\pi}} u^{-2} e^{-u^2/2} du \right] d\mu(y).$$

If we use integration by parts on the terms in the brackets, differentiating the exponential and integrating u^{-2}, we find that the right side is

$$= \int_{-\infty}^{\infty} |x - y| \frac{2}{\sqrt{2\pi}} \left[-u^{-1} e^{-u^2/2} \Big|_{\frac{|x-y|}{\sqrt{T}}}^{\infty} - \int_{|x-y|/\sqrt{T}}^{\infty} e^{-u^2/2} du \right] d\mu(y)$$
$$= -\int_{-\infty}^{\infty} |x - y| \left[\int_{|x-y|/\sqrt{T}}^{\infty} \frac{2}{\sqrt{2\pi}} e^{-u^2/2} du \right] d\mu(y)$$
$$+ \frac{2}{\sqrt{2\pi}} \int_{-\infty}^{\infty} \sqrt{T} e^{-|x-y|^2/2T} d\mu(y).$$

We let T tend to ∞ and consider each term separately. In the first term the expression in brackets increases to 1. If we write the integral as the difference of one with respect to μ^+ and one with respect to μ^- and use the fact that

$$\int_{-\infty}^{\infty} |x - y| d\mu^+(y) < \infty \quad \text{and} \quad \int_{-\infty}^{\infty} |x - y| d\mu^-(y) < \infty \qquad \text{a.e.,}$$

we see by the Dominated Convergence Theorem that the first term tends a.e. to

$$-\int_{-\infty}^{\infty} |x - y| d\mu(y).$$

Next we consider the second term. Suppose first that $\mu(R^1) \neq 0$. The second term tends, as $T \to \infty$, to

$$\lim_{T\to\infty} \sqrt{T} \lim_{T\to\infty} \frac{2}{\sqrt{2\pi}} \int_{-\infty}^{\infty} e^{-|x-y|^2/2T} d\mu(y),$$

and the integral and the second limit may be interchanged by dominated convergence to become

$$= \left(\lim_{T\to\infty} \sqrt{T}\right)\left(\frac{2}{\sqrt{2\pi}} \int_{-\infty}^{\infty} d\mu(y)\right)$$
$$= \pm\infty.$$

To complete the proof we shall show that if $\mu(R^1) = 0$, then

$$\int_{-\infty}^{\infty} \sqrt{T} e^{-|x-y|^2/2T} d\mu(y)$$

tends to zero a.e. as $T \to \infty$. Since $\mu(R^1) = 0$, we have

$$\int_{-\infty}^{\infty} \sqrt{T} e^{-|x-y|^2/2T} d\mu(y) = \int_{-\infty}^{\infty} \sqrt{T}(e^{-|x-y|^2/2T} - 1) d\mu(y).$$

We shall prove that the right side even tends to zero when μ is replaced by μ^+ or μ^-. Let us do so for μ^+. First we show that

$$|\sqrt{T}(e^{-|z|^2/2T} - 1)| \leq k|z|$$

for a fixed constant k and for all T. By differential calculus methods, we find that

$$\frac{e^{-|z|^2/2T} - 1}{|z|}$$

assumes its maximum value as a function of $|z|$ when $|z|$ satisfies

$$\frac{|z|^2}{T} = e^{|z|^2/2T} - 1.$$

The unique positive solution occurs for $2 \leq |z|^2/T \leq 3$. Let $b = |z|^2/T$ with $2 \leq b \leq 3$ be the point at which the solution occurs. Then

$$\left| \sqrt{T} \frac{e^{-|z|^2/2T} - 1}{|z|} \right| \leq \left| \sqrt{T} \frac{e^{-b/2} - 1}{\sqrt{bT}} \right| = \frac{1 - e^{-b/2}}{\sqrt{b}} = k$$

and

$$|\sqrt{T}(e^{-|z|^2/2T} - 1)| \leq k|z|.$$

Put $|z| = |x - y|$. Since

$$\int_{-\infty}^{\infty} k|x - y| d\mu^+(y) < \infty \qquad \text{a.e.,}$$

we have by dominated convergence

$$\lim_{T\to\infty} \int_{-\infty}^{\infty} \sqrt{T}(e^{-|x-y|^2/2T} - 1) d\mu^+(y)$$
$$= \int_{-\infty}^{\infty} \lim_{T\to\infty} [\sqrt{T}(e^{-|x-y|^2/2T} - 1)] d\mu^+(y)$$

for almost every x. The integrand on the right side is identically zero.

The hypotheses of Propositions 7-4 and 7-5 are worth reviewing and comparing. In the transient case, Proposition 7-4, we started with any element μ of M and we were able to conclude both that the potential operator was defined on μ and that its value was the Brownian motion limit. In the recurrent case, Proposition 7-5, we started with any element μ of M and we had to assume that the potential operator was defined on μ; we then concluded that the potential of μ was equal to the Brownian motion limit if and only if μ had total charge zero. We shall see that the same thing happens in potential theory defined for denumerable Markov chains.

5. Potential theory for denumerable Markov chains

We turn our attention now to those properties of Brownian motion which relate it to potential theory. In this section we shall answer the following questions:

(1) How can the connection between Brownian motion and classical potential theory be used to define a potential theory for denumerable Markov chains?

(2) How does potential theory differ in the transient and recurrent cases, and what form does the potential operator take?

(3) What is the nature of the inverse operator that transforms potentials into charges?

(4) What other Markov chain concepts play a role in potential theory?

For definiteness, let P be a denumerable Markov chain which either is recurrent or is transient with no absorbing states, and let α be a positive finite-valued P-superregular measure.

Before defining a potential theory for denumerable Markov chains, we should discuss some properties of the operators P^t and Q^t. The operators P^t and Q^t act respectively on differences of finite measures and on functions in C_0 according to the equations

$$(\mu P^t)(E) = \int_{R^n} \left[\int_E \frac{1}{(2\pi t)^{n/2}} e^{-|x-y|^2/2t} dy\right] d\mu(x)$$

and

$$(Q^t f)(y) = \int_{R^n} \frac{1}{(2\pi t)^{n/2}} e^{-|x-y|^2/2t} f(x)dx,$$

and they are related by the identity

$$\int_{R^n} (Q^t f) d\mu = \int_{R^n} fd(\mu P^t).$$

The linearity properties

$$(\mu + \nu)P^t = \mu P^t + \nu P^t$$

and

$$(c\mu)P^t = c(\mu P^t),$$

together with a certain property of continuity, imply that the action of P^t on differences of finite measures is analogous to the action of a matrix on row vectors. Similarly, the corresponding properties for Q^t imply that the action of Q^t on functions in C_0 is like that of a matrix on column vectors. But the real insight into P^t and Q^t comes in realizing that these matrices are identical. To be more specific, we must reformulate the assertion for a countable space.

Let P^* be a continuous linear operator on row vectors μ which have a finite sum, and let Q^* be a continuous linear operator on column vectors f whose components tend to 0. Suppose that P^* and Q^* are related by the identity

$$\mu \cdot (Q^* f) = (\mu P^*) \cdot f$$

for all μ and f of the type specified, where $\cdot$ stands for vector multiplication. Let $\delta^{(i)}$ and $d^{(i)}$ be the row and column vectors having ith component equal to 1 and all other components equal to 0. The vector $\delta^{(i)}$ is in the domain of P^*, and $d^{(i)}$ is in the domain of Q^*. If we define a square matrix $\{P^*_{ij}\}$ by

$$P^*_{ij} = (\delta^{(i)} P^*) \cdot d^{(j)} = (\delta^{(i)} P^*)_j,$$

then

$$(\mu P^*)_j = \sum_k \mu_k (\delta^{(k)} P^*)_j = \sum_k \mu_k P^*_{kj}.$$

Hence the operator P^* may be represented by the matrix $\{P^*_{ij}\}$. But by the identity relating P^* and Q^*,

$$P^*_{ij} = \delta^{(i)} \cdot (Q^* d^{(j)}).$$

Hence by a similar argument, the operator Q^* is also representable by the same matrix $\{P^*_{ij}\}$.

Thus the denumerable Markov chain analog of the pair of operators P^t and Q^t can be expected to be a single matrix depending on t. For $t = 1$, this matrix can be taken to be the transition matrix P of the Markov chain. Then the relations $P^{t+s} = P^t P^s$ and $Q^{t+s} = Q^t Q^s$ for integers t and s imply that the analog of P^t and Q^t for any other integer value of t is a power of the matrix P.

Lebesgue measure has a special property with respect to Brownian motion which is summed up in the equation

$$\int_E dx = \int_{R^n} \left[\int_E \frac{1}{(2\pi t)^{n/2}} e^{-|x-y|^2/2t} dy \right] dx.$$

If we call Lebesgue measure σ and use notation that earlier was reserved for finite measures, this equation becomes

$$\sigma = \sigma P^t.$$

That is, σ is regular for P^t. Thus the analog of σ for the Markov chain P should be a P-regular measure. But if P is transient, then P need not have a regular measure. We therefore relax our requirement and ask only that the analog of σ be a P-superregular measure. We can then decree that the specified measure α is to be the analog of σ.

The problem of defining potentials for Markov chains becomes a problem of translating notions about Brownian motion into notions about Markov chains. Following Propositions 7-4 and 7-5, we recall that a potential $g(x)$ is obtained from a charge μ in this way: If we abbreviate the equation

$$\begin{aligned}(\mu P^t)(E) &= \int_{R^n}\left[\int_E \frac{1}{(2\pi t)^{n/2}} e^{-|x-y|^2/2t} dy\right] d\mu(x) \\ &= \int_E \left[\int_{R^n} \frac{1}{(2\pi t)^{n/2}} e^{-|x-y|^2/2t} d\mu(x)\right] dy\end{aligned}$$

as

$$(\mu P^t)(E) = \int_E f(x, t)dx,$$

then

$$g(x) = \lim_{T\to\infty} \int_0^T f(x, t)dt.$$

Translating the relation for $(\mu P^t)(E)$ into notions about Markov chains, we write

$$\sum_{i\in E} (\mu P^n)_i = \sum_{i\in E} f_i^{(n)}\alpha_i.$$

If E is the one-point set $\{i\}$, we find that

$$f_i^{(n)} = \frac{1}{\alpha_i}(\mu P^n)_i$$

or

$$f^{(n)} = \text{dual}\ (\mu P^n).$$

The equation defining $g(x)$ translates into

$$g = \lim_{n\to\infty} [f^{(0)} + f^{(1)} + \cdots + f^{(n)}]$$

or

$$\boxed{g = \text{dual} \lim_{n\to\infty} [\mu(I + P + \cdots + P^n)]}$$

Classically, potentials are left as point functions and are never transformed into set functions because such a transformation is frequently impossible. In Markov chain potential theory, however, every column vector can be transformed into a row vector by the duality mapping.

If we take the dual of both sides of the boxed equation, we get

$$\text{dual } g = \lim_{n\to\infty} [\mu(I + P + \cdots + P^n)].$$

For simplicity in notation we shall adopt the convention that dual g, and not g, is the potential of μ. We can at last formulate a definition.

Definition 7-6: Any row vector μ with $\mu\mathbf{1}$ well defined and finite for which the limit

$$\nu = \lim_{n\to\infty} [\mu(I + P + \cdots + P^n)]$$

exists and is finite-valued is called a **left charge** with **potential measure** ν.

The condition that $\mu\mathbf{1}$ be well defined and finite is the analog of the condition that a charge in R^n be the difference of two finite measures.

The boxed equation for g yields an alternate possibility, namely

$$g = \lim_{n\to\infty} [(I + \hat{P} + \cdots + \hat{P}^n)(\text{dual } \mu)]$$

with $\alpha(\text{dual } \mu) = \mu\mathbf{1}$ finite. If we had gone through the same argument for the process $\hat{P}$, we would have obtained the same equation with the carets removed. We therefore complement Definition 7-6 as follows.

Definition 7-7: Any column vector f with αf well defined and finite for which the limit

$$g = \lim_{n\to\infty} [(I + P + \cdots + P^n)f]$$

exists and is finite-valued is called a **right charge** with **potential function** g.

From our knowledge of what happens with Brownian motion, we should expect that the Markov chain potential operators will arise in different ways in the transient and recurrent cases. Consequently, we shall treat the different kinds of processes separately, handling the transient case in Chapter 8 and the recurrent case in Chapter 9.

In the transient case of Brownian motion the operator was formally

$$\lim \int_0^T P^t dt = \int_0^\infty P^t dt,$$

and it is no surprise that for Markov chains it is the matrix $N = \sum_{n=0}^{\infty} P^n$ which turns out to be the potential operator both for left charges and right charges (see Theorem 8-3). Once we have the potential operator, it will not be difficult to develop a full theory in analogy with classical potential theory.

In the recurrent case, however, the problem of finding the potential operator is not so easy. The information that we will find, just as in

Proposition 7-5, is that for any charge of total charge zero on which the potential operator is defined the potential operator should agree with the operator which is formally

$$\lim \int_0^T P^t dt.$$

It will turn out that there are many possible potential operators for left charges and many others for right charges. Of these the matrices $-C$ and $-G$, respectively, will be representative (see Definition 9-24). But if we ask that the same matrix work for both left and right charges, then we shall see that there is a matrix K such that all such potential operators are of the form $-K + c1\alpha$, where c is a constant (see Theorem 9-84). With $-K$ as our operator, we have some hope of imitating classical potential theory if we redefine charge and potential in terms of K: The column vector f is a charge, for instance, with potential g if

$$g = -Kf.$$

From this new definition of charges and potentials, we shall be able, just as in the transient case, to prove theorems which are analogs of some of the main results of classical potential theory.

In discussing the relation of Brownian motion to potential theory in Section 3, we mentioned that the operator inverse to $-A$, where

$$Af = \lim_{t \downarrow 0} \frac{1}{t}(Q^t f - f),$$

was of the same form as the potential operator. It is thus quite believable that $-A$ should be essentially the inverse operator that transforms potentials into charges. Now the definition of A involves a derivative, and when concepts in Brownian motion are translated into concepts in Markov chains, derivatives transform into differences. Therefore, the proper analog of Af for Markov chains is $Pf - f = (P - I)f$. That is, $I - P$ plays the role of $-A$. In Theorems 8-4 and 9-15 we shall see that $I - P$ is indeed the operator that transforms potentials in the sense of Definitions 7-6 and 7-7 into charges.

With Brownian motion the operator A is a constant multiple of the Laplacian operator Δ for smooth enough functions, where

$$\Delta f = \left(\frac{\partial^2}{\partial x_1^{\,2}} + \cdots + \frac{\partial^2}{\partial x_n^{\,2}}\right) f.$$

If a function f satisfies the equation $\Delta f = 0$ in a neighborhood of a point x, then f is said to be **harmonic** in a neighborhood of x. The analog in the case of denumerable Markov chains is that if a column vector f satisfies $(P - I)f = 0$ at the point i, then f is **regular** at i. Thus we can expect that regular functions will have some of the same behavior for Markov chains that harmonic functions have classically. As an example, a function harmonic on a connected open set in R^n cannot assume its maximum value inside that open set unless the function is constant. We shall see in Corollary 8-44 that an analogous result holds for Markov chains.

A twice continuously differentiable function f is said to be **superharmonic** if $\Delta f \leq 0$. The analog of this property is the condition that $(P - I)f \leq 0$ or $f \geq Pf$. Thus the analog of a superharmonic function is a superregular function.

TABLE 7-2. MARKOV CHAIN ANALOGS OF POTENTIAL THEORY CONCEPTS

Classical Notion	Markov Chain Analog
R^n	State space S
P^t and Q^t	P^n
Lebesgue measure	α
Potential	$\lim \mu(I + P + \cdots + P^n)$ or $\lim (I + P + \cdots + P^n)f$
Total charge	$\mu 1$ or αf
Potential operator	Transient: N Recurrent: $-K$
Inverse operator	$I - P$
Harmonic function	Regular function
Superharmonic function	Superregular function
Connected set	Communicating set

6. Brownian motion as a limit of the symmetric random walk

The symmetric random walk in n dimensions was defined in Chapter 4 as a Markov chain obtained from sums of independent random variables on the integer lattice in R^n with the probability of going from any state to any of the $2n$ neighboring states equal to $1/(2n)$. In potential theory for Markov chains this process assumes the role of the "classical case," exhibiting in its potential theory much of the special behavior of the theory in Section 2. For instance, the matrix of the potential operator for this process has the same asymptotic behavior at infinity as the potential kernel has classically: $\log |x|$ in two dimensions, $1/|x|$ in three dimensions, and so on.

The reason for this coincidence is that Brownian motion is in a precise sense the limit of the symmetric random walk. Specifically if the random walk is considered first on the integer lattice, then on the half-integer lattice, then on the quarter-integer lattice, and so on, then the probabilities in the kth process of being in a fixed ball in R^n after time $4^k t$ converge to the probability in Brownian motion of being in that ball after time t. We shall prove this result only in the one-dimensional case, and we make use of the Central Limit Theorem (Theorem 1-68).

Consider for fixed k the random walk on the line having as states the points of the form $j2^{-k}$, for j any integer, and having transition probabilities $\frac{1}{2}$ from any state to each of its two neighbors. This process is the symmetric random walk with a change in scale. Let $x_n^{(k)}$ be the nth outcome function, and let $x_0^{(k)} = 0$. As in Section 1, we let x_t denote the position in Brownian motion.

Proposition 7-8: Brownian motion in one dimension is a limit of the symmetric random walk in this sense: If t is a diadic rational and if α and β are real numbers, then

$$\lim_{k\to\infty} \Pr_0[x_{4^k t}^{(k)} \in (\alpha, \beta)] = \Pr_0[x_t \in (\alpha, \beta)].$$

PROOF: The random variables $x_{n+1}^{(k)} - x_n^{(k)}$ are independent and identically distributed and have mean 0 and variance

$$\sigma^2 = \frac{1}{2}\left(\frac{1}{2^k}\right)^2 + \frac{1}{2}\left(-\frac{1}{2^k}\right)^2 = \frac{1}{4^k}.$$

Let $m = 4^k t$ be an integer. Since

$$x_m^{(k)} = \sum_{n=1}^{m} [x_n^{(k)} - x_{n-1}^{(k)}],$$

$x_m^{(k)}$ has mean 0 and variance $m/4^k = t$. Hence, by the Central Limit Theorem,

$$\lim_{k\to\infty} \Pr_0\left[\frac{\alpha}{\sqrt{t}} < \frac{x_m^{(k)}}{\sqrt{t}} < \frac{\beta}{\sqrt{t}}\right] = \Phi\left(\frac{\beta}{\sqrt{t}}\right) - \Phi\left(\frac{\alpha}{\sqrt{t}}\right)$$

or

$$\lim_{k\to\infty} \Pr_0[\alpha < x_m^{(k)} < \beta] = \Phi\left(\frac{\beta}{\sqrt{t}}\right) - \Phi\left(\frac{\alpha}{\sqrt{t}}\right).$$

On the other hand, by definition,

$$\begin{aligned}\Pr_0[x_t \in (\alpha, \beta)] &= \int_\alpha^\beta \frac{1}{\sqrt{2\pi t}} e^{-y^2/2t} dy \\ &= \int_{\alpha/\sqrt{t}}^{\beta/\sqrt{t}} \frac{1}{\sqrt{2\pi}} e^{-u^2/2} du \\ &= \Phi\left(\frac{\beta}{\sqrt{t}}\right) - \Phi\left(\frac{\alpha}{\sqrt{t}}\right).\end{aligned}$$

Therefore,

$$\lim_{k\to\infty} \Pr_0[x_{4^k t}^{(k)} \in (\alpha, \beta)] = \Pr_0[x_t \in (\alpha, \beta)].$$

7. Symmetric random walk in n dimensions

As was mentioned in Section 6, the n-dimensional symmetric random walk and n-dimensional Brownian motion share a number of properties because the second process is the limit of the first. Of these we shall prove for the random walk just two—that the process is recurrent in dimensions 1 and 2 and transient in dimension $n \geq 3$ and that in the transient case the columns of the N-matrix tend to zero. The second result is in analogy with the behavior of the potential kernel $1/|x|^{n-2}$ in Brownian motion.

For the first problem we note immediately that all states communicate in the random walks of all dimensions; hence each of them is either transient or recurrent. In one dimension the state space is the integers, and

$$P_{i,i+1} = P_{i,i-1} = \tfrac{1}{2}.$$

Since the mean step in zero, the process is recurrent by Proposition 5-22. A more direct proof of the recurrence proceeds as follows. It is impossible to get from state 2 to state 0 without going through 1. This fact, together with the translation invariance of the hitting probabilities, implies that

$$H_{20} = H_{21}H_{10} = (H_{10})^2.$$

But

$$\bar{H}_{00} = \tfrac{1}{2}H_{10} + \tfrac{1}{2}H_{-1,0} = \tfrac{1}{2}H_{10} + \tfrac{1}{2}H_{01} = H_{10}$$

since $H_{01} = H_{10}$. Therefore, the identity

$$H_{10} = \tfrac{1}{2}H_{00} + \tfrac{1}{2}H_{20} = \tfrac{1}{2} + \tfrac{1}{2}(H_{10})^2$$

implies that $H_{10} = 1$ and hence $\bar{H}_{00} = 1$. Consequently, the process is recurrent. Still a third proof can be based on a calculation of N_{00}. In fact, we have

$$P_{00}^{(2n)} = 2^{-2n}\binom{2n}{n}$$

because in order for the process to return in $2n$ steps to 0, it must make n steps to the right and n to the left; each such possibility has probability 2^{-2n}. By Lemma 1-59,

$$\binom{2n}{n} \sim c2^{2n}\frac{1}{\sqrt{n}}.$$

Hence the tail of the series $N_{00} = \sum P_{00}^{(2n)}$ dominates a constant multiple of the tail of the series $\sum 1/\sqrt{n}$. Therefore, N_{00} is infinite and the process is recurrent.

In the two-dimensional case there are two simultaneous processes going on in perpendicular directions, and for the process to return to the origin, both of them must return to their zero positions at the same time. Letting k be the number of steps to the right and $n - k$ the number of steps up in $2n$ steps, we have

$$P_{00}^{(2n)} = 4^{-2n} \sum_{k=0}^{n} \binom{2n}{k,\, k,\, n-k,\, n-k}.$$

If we multiply numerator and denominator of the multinomial coefficient by $(n!)^2$, we see that it equals $\binom{2n}{n}\binom{n}{k}^2$. Thus

$$P_{00}^{(2n)} = 4^{-2n}\binom{2n}{n} \sum_{k=0}^{n} \binom{n}{k}^2.$$

The identity

$$\sum_{k=0}^{n} \binom{n}{k}^2 = \binom{2n}{n}$$

shows that

$$P_{00}^{(2n)} = 4^{-2n}\binom{2n}{n}^2.$$

Since

$$\binom{2n}{n} \sim c2^{2n}\frac{1}{\sqrt{n}},$$

we have

$$P_{00}^{(2n)} \sim c^2\frac{1}{n}.$$

Thus the series $N_{00} = \sum P_{00}^{(2n)}$ dominates a multiple of $\sum 1/n$, N_{00} must be infinite, and the process is recurrent.

An alternate proof that the process is recurrent in two dimensions proceeds as follows. If we introduce the new coordinates

$$u = x + y$$
$$v = x - y,$$

then the two-dimensional symmetric random walk described in the coordinates (u, v) executes two one-dimensional symmetric random walks independently of each other. Hence $P_{(0,0)(0,0)}^{(2n)}$ is the probability that $u = 0$ and $v = 0$ after $2n$ steps, which is $(P_{00}^{(2n)})^2 \sim c^2/n$.

In three dimensions we calculate N_{00}. We have

$$\begin{aligned} P_{00}^{(2n)} &= 6^{-2n} \sum_{\substack{j,k \\ j+k \le n}} \binom{2n}{j,\, j,\, k,\, k,\, n-j-k,\, n-j-k} \\ &= 2^{-2n}\binom{2n}{n} \sum_{j,k} 3^{-2n}\binom{n}{j,\, k,\, n-j-k}^2. \end{aligned}$$

The coefficients $\binom{n}{j, k, n-j-k}$ are dominated by the central term

$$\binom{n}{n/3, n/3, n/3},$$

where the gamma function may be used for $(n/3)!$ if 3 does not divide n. This fact and the observation that the coefficients $\binom{n}{j, k, n-j-k}$ sum to 3^n implies that

$$P_{00}^{(2n)} \le 2^{-2n}\binom{2n}{n}3^{-n}\binom{n}{n/3, n/3, n/3}.$$

Summing on n and using the approximations of Section 1-6a, we see that the series $N_{00} = \sum P_{00}^{(2n)}$ is dominated by a multiple of $\sum 1/n^{3/2}$. Therefore N_{00} is finite, and the process is transient.

If any higher-dimensional random walk were recurrent, then the process projected to a three-dimensional set would be recurrent, and the latter process watched only when it changes state would also be recurrent. But this last process is exactly the three-dimensional symmetric walk. We conclude therefore that the random walk in all dimensions greater than three is transient, and we have completed the proof of the following proposition.

Proposition 7-9: The symmetric random walk is recurrent in dimensions one and two and is transient in all dimensions greater than or equal to three.

In the transient case of dimension $n \ge 3$, the jth entry in the 0th column of the N-matrix is of the order of a constant times $|j|^{-(n-2)}$. We conclude this chapter by proving the weaker result that the entries of that column tend to zero, but our proof will be for a more general situation.

Proposition 7-10: Let P be a Markov chain with an infinite state space such that

(1) P is transient.
(2) $P = P^T$.
(3) P has only finitely many non-zero entries in each row.

Then

$$\lim_{j\to\infty} H_{0j} = 0.$$

If, in addition, P represents sums of independent random variables, then

$$\lim_{j \to \infty} N_{0j} = 0.$$

In particular, the conclusions apply to the symmetric random walk in any dimension $n \geq 3$.

PROOF: We note that hypothesis (3) is equivalent to

(3′) For any state 0 and for any given m, there exist only finitely many states j for which $F_{0j}^{(k)} \neq 0$ for some $k \leq m$.

If the conclusion about H_{0j} is false, then for some $\epsilon > 0$ and for infinitely many j we have $H_{0j} > \epsilon$. By (1), N is finite-valued. Therefore,

$$\lim_{k \to \infty} P^k N = \lim_{k \to \infty} [N - (I + \cdots + P^{k-1})] = 0$$

and

$$\lim_{k \to \infty} P^k \bar{H} = 0$$

since $\bar{H} \leq H \leq N$. Choose m large enough so that $(P^m \bar{H})_{00} < \epsilon^2$. Since there are infinitely many j such that $H_{0j} > \epsilon$, we can find by (3′) such a j with $F_{0j}^{(k)} = 0$ for all $k \leq m$. Then

$$\begin{aligned} \mathrm{Pr}_0[\text{hit } 0 \text{ after time } m] &\geq \mathrm{Pr}_0[\text{ever hit } j \text{ and return to } 0] \\ &= H_{0j} H_{j0} \\ &= H_{0j} H_{0j} \quad \text{by (2)} \\ &> \epsilon^2. \end{aligned}$$

But

$$\mathrm{Pr}_0[\text{hit } 0 \text{ after time } m] = (P^m \bar{H})_{00} < \epsilon^2,$$

a contradiction. Therefore $H_{0j} \to 0$.

Finally if P represents sums of independent random variables,

$$N_{0j} = H_{0j} N_{jj} = H_{0j} N_{00}.$$

Hence $N_{0j} \to 0$. (Note that we really need only that N_{jj} is bounded.)

As Markov chains the symmetric random walks have some special properties, reflecting corresponding special properties of Brownian motion. For instance, α is a constant for the random walk, and $P = P^T$. Consequently $P = \hat{P}$. We shall see that although many results of classical potential theory generalize to all transient and most recurrent chains, some will require further assumptions which happen to be true for symmetric random walks.

CHAPTER 8

TRANSIENT POTENTIAL THEORY

1. Potentials

In this chapter P denotes a Markov chain all of whose states are transient—that is, a transient chain with no absorbing states. Every such chain has at least one (strictly) positive superregular measure, as we saw in Chapter 5; for example, the sum of 2^{-i} times the ith row of N is such a measure.

We select one such positive superregular measure, to be fixed throughout the chapter, and call it α. All of transient potential theory will be relative to the distinguished vector α.

Let $\hat{P}$ be the α-dual of P. Since all states are transient in $\hat{P}$ and since $\hat{\hat{P}} = P$, we see that $\hat{P}$ is the most general chain of the type we consider. The distinguished measure for $\hat{P}$ is taken to be *the same* α.

As an example, let P be a transient Markov chain whose states communicate. Then $P\mathbf{1} \leq \mathbf{1}$ and $0 < N = \sum_{k=0}^{\infty} P^k < \infty$. Every non-negative non-zero superregular row vector β is positive, for if $\beta_j \neq 0$, then for every state i and integer k

$$\beta_i \geq (\beta P^k)_i = \sum_m \beta_m P_{mi}^{(k)} \geq \beta_j P_{ji}^{(k)}.$$

The right side must be positive for some k, since j communicates with i. Thus in this special case any non-negative superregular row vector may be taken as α; in particular, α may be taken as a row of N.

In the general case, if $P\mathbf{1} \neq \mathbf{1}$, we have defined the enlarged chain $\bar{P}$ by adding an absorbing state a to P and by setting

$$\begin{aligned} \bar{P}_{ij} &= P_{ij} \qquad\qquad \text{if } i \neq a \text{ and } j \neq a \\ \bar{P}_{ia} &= 1 - \sum_k P_{ik} \\ P_{aj} &= \delta_{aj}. \end{aligned}$$

If $P\mathbf{1} = \mathbf{1}$, we shall agree that P is its own enlarged chain.

It will be convenient to say that the product of a row vector and a column vector is **finite** when we mean that it is well defined and finite. We recall the terminology **function** for column vector and **signed measure** for row vector.

Definition 8-1: If μ is a signed measure with $\mu 1$ finite and if

$$\nu = \lim_n [\mu(I + P + \cdots + P^{n-1})]$$

exists and is finite-valued, then μ is called a **left charge** with **potential measure** ν. If f is a function with αf finite and if

$$g = \lim_n [(I + P + \cdots + P^{n-1})f]$$

exists and is finite-valued, then f is called a **right charge** with **potential function** g. In either case a **pure potential** is a potential of a non-negative charge.

The condition that αf be finite is the natural analog of the classical theory as described in Chapter 7. It states that f is integrable with respect to the distinguished superregular measure α. Similarly, the condition on μ is that the distinguished superregular function 1 be integrable with respect to μ.

Potential functions have a simple probabilistic interpretation in terms of games. If f denotes a payment function in which f_j is the payment a player receives each time he is in state j, then $P^n f$ denotes the expected payment on the nth step. Thus $(I + P + \cdots + P^{n-1})f$ is the total payment before time n, and the potential g is the expected total payment in the long run. It is clear intuitively that g_i should equal $\sum_j N_{ij} f_j$, and we now prove this result.

Lemma 8-2: If $\mu 1$ is finite, then μN is finite-valued. If αf is finite, then Nf is finite-valued.

PROOF: We have

$$N_{ij} = H_{ij} N_{jj} \leq N_{jj}.$$

Thus

$$|(\mu N)_j| \leq \sum_i |\mu_i| N_{ij} \leq \sum_i |\mu_i| N_{jj} = (|\mu| 1) N_{jj} < \infty.$$

For the second half let $\mu = \text{dual} f$ and apply the first result to $\hat{P}$, noting that $\mu 1 = \alpha f$. Then

$$\infty > |\mu \hat{N}|_j = \left| \sum_i f_i \alpha_i (\alpha_j N_{ji} / \alpha_i) \right| = \left| \alpha_j \sum_i N_{ji} f_i \right| = \alpha_j |Nf|_j.$$

Since $\alpha_j \neq 0$, Nf is finite-valued.

Theorem 8-3: If αf is finite, then f is a charge and its potential is $g = Nf$.

PROOF: By Lemma 8-2, Nf is well defined and finite-valued. Hence so are both Nf^+ and Nf^-. By monotone convergence,

$$\lim_n [(I + P + \cdots + P^{n-1})f^+] = Nf^+$$

and

$$\lim_n [(I + P + \cdots + P^{n-1})f^-] = Nf^-.$$

Thus

$$\lim_n [(I + P + \cdots + P^{n-1})f] = Nf^+ - Nf^- = Nf.$$

Thus f is a charge if and only if it is integrable with respect to α, and N is the potential operator that transforms a charge into its potential. In particular, f is a charge for P if and only if it is a charge for $\hat{P}$. We shall now show that $I - P$ is the inverse operator.

Theorem 8-4: If g is a potential, then $(I - P)g$ is its charge.

PROOF: Let f be a charge with potential g. By Theorem 8-3, $g = Nf$. Hence, by Lemma 5-9, $(I - P)g = f$.

Therefore, there is a one-to-one correspondence between charges and potentials. Note that Theorem 8-4 implies that a potential is regular at all states where the charge is zero.

The method used to derive the second half of Lemma 8-2 is of general importance. We prove a result for all our P's for signed measures (or functions). We apply the result to $\hat{P}$ and obtain a corresponding result for functions (or signed measures). Then since $\hat{P}$ is the form of the most general transient chain being considered, the new result holds for all P's. Such results will loosely be described as **duals**.

The duals of Theorems 8-3 and 8-4 state that a signed measure μ is a charge if and only if $\mu 1$ is finite. Its potential is $\nu = \mu N$, and $\mu = \nu(I - P)$.

From now on we shall prove theorems only for functions; the dual results for signed measures can always be proved by the indicated method. The key to the success of the method is that the dual of a right charge for P is a left charge for $\hat{P}$, and the dual of a potential function for P is a potential measure for $\hat{P}$.

From Theorem 8-3 we see immediately that the class of potentials is quite extensive. We can even prove that there exists a strictly positive pure potential, a result we shall need later on.

Proposition 8-5: There exists a strictly positive pure potential.

PROOF: Number the states and let $f_j = 2^{-j}/\alpha_j$. Then

$$\alpha f = \sum_j \alpha_j(2^{-j}/\alpha_j) = 1,$$

so that f is the charge of a pure potential g, by Theorem 8-3. Furthermore,

$$g_i = \sum_j N_{ij}f_j \geq N_{ii}f_i \geq f_i > 0,$$

so that g is strictly positive.

For many purposes it is sufficient in studying potentials to consider only pure potentials. The reason for this simplification is the following.

Proposition 8-6: Any potential may be represented as the difference of two pure potentials.

PROOF: Write $g = Nf = Nf^+ - Nf^-$.

Note by Theorem 8-4 that a potential is superregular if and only if it is a pure potential.

We recall from Theorem 5-10 that a non-negative superregular function h is uniquely representable as $h = Nf + r$ with r regular. In the representation, $r = \lim_n P^n h \geq 0$ and $f = (I - P)h \geq 0$. (The dual of this result allows the unique representation of a non-negative superregular measure π as $\pi = \mu N + \rho$ with ρ regular. In this representation, $\rho = \lim_n \pi P^n \geq 0$ and $\mu = \pi(I - P)$.) This result is the analog of a classical theorem due to F. Riesz: In any open set of Euclidean space which corresponds to a transient version of Brownian motion, any non-negative superharmonic function is uniquely the sum of a pure potential for the region and a non-negative harmonic function. The pure potential may have infinite total charge. We now generalize the Markov chain result, and in so doing we obtain a useful necessary condition that potentials must satisfy.

Proposition 8-7: If $(I - P)h = f$, if h is finite-valued, and if Nf is finite-valued, then h has a representation in the form

$$h = Nf + r$$

with r regular. The vector r satisfies $r = \lim_n P^n h$.

PROOF: Set $r = h - Nf$. Then

$$\begin{aligned}(I - P)(h - Nf) &= (I - P)h - (I - P)(Nf)\\ &= (I - P)h - f \quad \text{by Lemma 5-9}\\ &= f - f = 0.\end{aligned}$$

Hence r is regular. Now

$$f = (I - P)h = h - Ph$$

or

$$h = Ph + f.$$

Since Nf is finite-valued and since $P^n f^+ \leq Nf^+$ and $P^n f^- \leq Nf^-$, $P^n f$ is finite-valued for every n. By induction we see that

$$P^{k-1}h = P^k h + P^{k-1}f$$

and that $P^k h$ is finite-valued. Summing for $k = 1, \ldots, n$, we obtain

$$h = P^n h + (I + P + \cdots + P^{n-1})f.$$

By dominated convergence the second term tends to Nf. Hence

$$h = \lim_n P^n h + Nf.$$

Corollary 8-8: If $(I - P)h = f$, if h is finite-valued, and if αf is finite, then h is a potential if and only if $\lim_n P^n h = 0$.

PROOF: By Theorem 8-3, f is a charge and Nf is its potential. Apply Proposition 8-7 and write

$$h = Nf + \lim P^n h.$$

If $\lim P^n h = 0$, then h is the potential of f. Conversely, if $\lim P^n h \neq 0$, then h cannot be a potential because, by Theorem 8-4, it would have to have f as its charge.

Corollary 8-9: If g is a potential, then $\lim_n P^n g = 0$.

PROOF: Take $h = g$ in Corollary 8-8.

In the discrete analog of the classical case—three-dimensional symmetric random walk—every potential g is bounded and satisfies $\lim_i g_i = 0$. In our theory we obtain only the weaker result, Corollary 8-9; that g may be unbounded will be shown in Section 7.

The stronger results of the classical theory are due to special features, as the next proposition shows. In the classical case α is chosen as $\mathbf{1}^T$

and N_{ii} is independent of i. Hence, $\alpha_i \geq kN_{ii}$ for some positive constant k. Furthermore, $\lim_i N_{ij} = 0$ by Proposition 7-10.

Proposition 8-10: If α is chosen so that $\alpha_i \geq kN_{ii}$ for all i with k a positive constant, then all potentials are bounded. If, in addition, $\lim_i N_{ij} = 0$, then $\lim_i g_i = 0$.

PROOF: The dual of $N_{ij} \leq N_{jj}$ is $N_{ij} \leq (N_{ii}/\alpha_i)\alpha_j$. If $\alpha_i \geq kN_{ii}$, then

$$|g_i| \leq \sum_j N_{ij}|f_j| \leq \frac{N_{ii}}{\alpha_i}\sum_j \alpha_j|f_j| \leq \frac{1}{k}\sum_j \alpha_j|f_j|.$$

Thus g is bounded. Now suppose $\lim_i N_{ij} = 0$. By Proposition 8-6 we may assume that g is a pure potential. Define a sequence of functions $h_j^{(i)} = N_{ij}/\alpha_j$ and a measure $\mu_j = \alpha_j f_j$. Then μ is a finite measure, and

$$h_j^{(i)} \leq \frac{N_{jj}}{\alpha_j} \leq \frac{1}{k}$$

is bounded independently of i and j. Hence, by dominated convergence,

$$\lim_i g_i = \lim_i \sum_j h_j^{(i)}\mu_j = \sum_j (\lim_i N_{ij})f_j = 0.$$

Both conditions of Proposition 8-10 hold for the basic example with α chosen as β. However, only the first condition holds for the reverse of the basic example (see Section 6). In Section 7 we shall see an example where both conditions fail.

2. The *h*-process and some applications

Duality is a transformation which interchanges the roles of row and column vectors. Our purpose now is to describe a useful transformation of transient chains into new transient chains in which row and column vectors are transformed into vectors of the same type.

Definition 8-11: Let h be a positive finite-valued superregular function for a transient chain P. The ***h*-process** is a Markov chain P^* with transition probabilities

$$P^*_{ij} = \frac{P_{ij}h_j}{h_i}.$$

It is left to the reader to verify that P^* is a transition matrix, that all states are transient, and that if the states of P communicate, the same is true for P^*. Let U be a diagonal matrix with diagonal entries $1/h_i$. Then $P^* = UPU^{-1}$.

Definition 8-12: The h-**process transformation** is a transformation defined on square matrices Y, row vectors π, and column vectors f by

$$Y^* = UYU^{-1}$$

$$\pi^* = \pi U^{-1}$$

$$f^* = Uf.$$

The h-process transformation yields results similar to those with duality. If P is a transient chain, then P^* also is transient. Moreover, powers of P transform to the P^* process the same way that P does, and the fundamental matrix for P^* is N^*. Sums and products are preserved in their given order; and equalities, inequalities, and limits are preserved entry-by-entry. Any superregular function (or signed measure) for P transforms into a superregular function (or signed measure) for P^*.

If α is the distinguished superregular measure for P, we select α^* as the distinguished superregular measure for P^*. Then $\alpha^* f^* = \alpha f$ and $N^* f^* = (Nf)^*$. Hence if f is a right charge with potential g in P, then f^* is a right charge for P^* with potential g^*.

If we decompose P as

$$P = \begin{array}{c} \\ E \\ \tilde{E} \end{array} \begin{array}{c} \begin{array}{cc} E & \tilde{E} \end{array} \\ \begin{pmatrix} T & U \\ R & Q \end{pmatrix} \end{array},$$

then

$$B^E = \begin{pmatrix} I & 0 \\ (\sum Q^n)R & 0 \end{pmatrix}.$$

From this decomposition we see that $(B^E)^*$ is the B^E-matrix of the P^*-chain because Q and R transform into Q^* and R^*, because products are preserved, and because $I^* = I$.

We shall now give some applications of the h-process.

Definition 8-13: The **support of a charge** is the set on which the charge is not 0; the **support of a potential** is the support of its charge. A charge or potential is said to have **support in** E if its support is a subset of E.

The function $\mathbf{1}$ is always superregular, and hence by the representation theorem $\mathbf{1} = Nf + r$, where $f = (I - P)\mathbf{1}$ and r is regular. That is,

$$f_i = 1 - (P\mathbf{1})_i = \bar{P}_{ia}$$

in the enlarged chain. Moreover,

$$(Nf)_i = \sum_j N_{ij}\bar{P}_{ja} = B_{ia}.$$

Thus $r_i = 1 - B_{ia}$ is the probability that the P-process, started at i, continues indefinitely. The enlarged chain is absorbing if and only if $\mathbf{1} = Nf$.

Proposition 8-14: Let $g > 0$ be a pure potential with support in E, and let P^* be the h-process for $h = g$. Then $g^* = \mathbf{1}$ is a potential, $\overline{P^*}$ is absorbing, and $(B^E)^*\mathbf{1} = \mathbf{1}$.

PROOF: Potentials transform into potentials; hence $g^* = \mathbf{1}$ is a potential. Since $\mathbf{1}$ is then of the form N^*f^*, $\overline{P^*}$ is absorbing. The absorbing state a can be reached only from a state i such that $(P^*\mathbf{1})_i < 1$; for such a state i,

$$0 < [(I - P^*)\mathbf{1}]_i = [(I - P^*)g^*]_i = f_i^*.$$

Hence $f_i > 0$ and i must be in E. Thus the P^*-process with probability one reaches E from all states, and $(B^E)^*\mathbf{1} = \mathbf{1}$.

What underlies Proposition 8-14 is this: The h-process tends to follow paths along which h is large. But since potentials tend to zero on the average ($P^n h \to 0$ for a potential), if h is a potential, then the paths in the h-process disappear. See Chapter 10 for details.

Proposition 8-15: If h is a non-negative finite-valued superregular function, then $B^E h \leq h$ for any set E.

PROOF: First suppose that $h > 0$. Form the h-process; then $h^* = \mathbf{1}$. Since $(B^E)^*\mathbf{1} \leq \mathbf{1}$, we have $B^E h \leq h$. (The conclusion that an inequality for the h-process implies an inequality for the original process is one we shall draw frequently. If it were false, then the inequality $B^E h \leq h$ would fail in some entry. But the h-process transformation preserves inequalities entry-by-entry.)

Now suppose that h has some zero entries. Apply the special case above to the function $h + \epsilon\mathbf{1}$. Then

$$B^E(h + \epsilon\mathbf{1}) \leq h + \epsilon\mathbf{1}.$$

Letting ϵ tend to zero, we obtain $B^E h \leq h$.

Proposition 8-16: If h is a non-negative superregular function and if E is any set of states, then $\bar{h} = B^E h$ satisfies the following:

(1) $\bar{h} \leq h$ and $\bar{h}_E = h_E$.

(2) $\bar{h}$ is the pointwise infimum of all non-negative superregular functions which dominate h on E.

(3) If $\bar{f} = (I - P)\bar{h}$, then

$$\bar{f}_E = (I - P^E)h_E \geq 0$$

and

$$\bar{f}_{\tilde{E}} = 0.$$

Therefore, $\bar{h}$ is regular on $\tilde{E}$ and is superregular everywhere.

(4) If $E \subset F$, then $B^E h \leq B^F h$.

PROOF: Statement (1) follows from Proposition 8-15 and the fact that $B^E_{ij} = \delta_{ij}$ for i in E. For (2) let x be a non-negative superregular function such that $x_E \geq h_E$. Since the $\tilde{E}$ columns of B^E are zero,

$$x \geq B^E x \geq B^E h = \bar{h},$$

and (2) holds provided we show in (3) that $\bar{h}$ is superregular. For (3), since $P\bar{h} \leq Ph \leq h$ is finite-valued,

$$\bar{f} = (I - P)(B^E h) = [(I - P)B^E]h = \begin{pmatrix} I - P^E & 0 \\ 0 & 0 \end{pmatrix}\begin{pmatrix} h_E \\ h_{\tilde{E}} \end{pmatrix}.$$

But h_E is P^E-superregular by the dual of Lemma 6-7, and (3) follows. Finally for (4), if $E \subset F$, then by conclusion (6) of Proposition 5-8,

$$B^E h = B^F(B^E h) \leq B^F h.$$

We now prove two lemmas and a proposition which conclude that a charge and its potential may both be computed from a knowledge of the values of the potential on the support. The first lemma is interesting in itself because of its game interpretation, which we shall discuss after proving the result.

Lemma 8-17: For any set of states E,

$$N = B^E N + {}^E N.$$

If g is a potential with charge f, then

$$g = B^E g + {}^E N f.$$

PROOF: In Theorem 4-11, let $\mathbf{f}_j$ be the number of times in j when and after E is reached (or 0 if E is not reached), and let $\mathbf{t}$ be the time when E is reached (or $+\infty$ if E is not reached). Then Theorem 4-11 yields

$$\begin{aligned} M_i[\mathbf{f}_j] &= \sum_k \Pr_i[x_t = k]\, M_k[\mathbf{n}_j] \\ &= \sum_k B^E_{ik} N_{kj}. \end{aligned}$$

But $\mathbf{f}_j$ is the difference of the total number of times the process is in j and the number of times it is in j before reaching E. Hence

$$\mathrm{M}_i[\mathbf{f}_j] = N_{ij} - {}^E N_{ij},$$

and the first equation follows. To get the second equation we multiply through by f; associativity in $B^E N f$ holds because $B^E N|f|$ is finite-valued.

In the game interpretation of potentials, f_j is a payment received each time the process is in state j, and g is the expected total gain if the process is started in i. The second equation of Lemma 8-17 states that g is the expected gain when and after E is entered *plus* the expected gain before E is entered. If all the payments are non-negative, then it is obvious from this interpretation that $g \geq B^E g$. If the support of f is in E, then all non-zero payments occur in E, and the expected gain before reaching E is zero. Hence, as we shall see formally in Proposition 8-19, $g = B^E g$.

Lemma 8-18: The fundamental matrix for P^E is N_E.

PROOF: The assertion is probabilistically clear because the number of times the process P is in a state of E when watched only in states of E is the same as the number of times the process P is in a state of E.

Proposition 8-19: If g is a potential with support in E, then g_E determines g, $g = B^E g$, $f_E = (I - P^E)g_E$, and $g_E = N_E f_E$.

PROOF: The fact that $g = B^E g$ is immediate from Lemma 8-17. Hence g_E determines g. Since $g = B^E g$, we have $f_E = (I - P^E)g_E$ by conclusion (3) of Proposition 8-16. Finally $g_E = N_E f_E$ either by Lemma 8-18 or by direct calculation:

$$\begin{pmatrix} g_E \\ g_{\tilde{E}} \end{pmatrix} = \begin{pmatrix} N_E & N_2 \\ N_3 & N_4 \end{pmatrix} \begin{pmatrix} f_E \\ 0 \end{pmatrix} = \begin{pmatrix} N_E f_E \\ N_3 f_E \end{pmatrix}.$$

Next we shall prove that the columns of B^E are always potentials.

Proposition 8-20: For any set of states E the columns of B^E are potentials with support in E, and

$$B^E = N \begin{pmatrix} I - P^E & 0 \\ 0 & 0 \end{pmatrix}.$$

PROOF: Since

$$\alpha\begin{pmatrix} I - P^E & 0 \\ 0 & 0 \end{pmatrix} = (\alpha_E(I - P^E) \quad 0)$$

is finite-valued, each column of $N\begin{pmatrix} I - P^E & 0 \\ 0 & 0 \end{pmatrix}$ is a potential with support in E. Thus by Proposition 8-19,

$$\begin{aligned} N\begin{pmatrix} I - P^E & 0 \\ 0 & 0 \end{pmatrix} &= B^E N\begin{pmatrix} I - P^E & 0 \\ 0 & 0 \end{pmatrix} \\ &= B^E\begin{pmatrix} N_E(I - P^E) & 0 \\ N_3(I - P^E) & 0 \end{pmatrix} \quad \text{with } N = \begin{pmatrix} N_E & N_2 \\ N_3 & N_4 \end{pmatrix} \\ &= B^E\begin{pmatrix} I & 0 \\ N_3(I - P^E) & 0 \end{pmatrix} \quad \text{by Lemma 8-18} \\ &= B^E. \end{aligned}$$

Corollary 8-21: For any set E, $\lim_n P^n B^E = 0$.

PROOF: Apply Proposition 8-20 and Corollary 8-9.

Finally we work toward a proof that a non-negative superregular function dominated by a potential is a potential, a result we state as Proposition 8-25.

Lemma 8-22: If E is a finite set and if h is non-negative superregular, then $B^E h$ is a pure potential of finite support.

PROOF: $B^E h$ is a finite linear combination of columns of B^E and is therefore by Proposition 8-20 a potential with support in the finite set E. Since h is non-negative superregular, $B^E h$ is non-negative superregular by conclusion (3) of Proposition 8-16. Hence, $B^E h$ is a pure potential.

Proposition 8-23: Every non-negative superregular function is the limit of an increasing sequence of pure potentials of finite support.

PROOF: Let $E_1 \subset E_2 \subset E_3 \subset \cdots$ be an increasing sequence of finite sets with union the set of all states S, and let $h^{(n)} = B^{E_n} h$. Then $h^{(n)}$ is an increasing sequence of pure potentials of finite support by

Lemma 8-22 and conclusion (4) of Proposition 8-16. If i is in E_n, then $h_i^{(n)} = h_i$, so that $\lim h^{(n)} = h$.

If g is a potential with charge f, then the **total charge** of g (or f) is defined to be αf (see Section 7-5).

Lemma 8-24: If $N\bar{f} \leq Nf$ with $\bar{f} \geq 0$ and αf finite, then

$$0 \leq \alpha \bar{f} \leq \alpha f < \infty.$$

PROOF: By the dual of Proposition 8-23, we may find a sequence of finite measures $\pi^{(n)}$ such that α is the monotone limit of $\pi^{(n)}N$. Since $N\bar{f} \leq Nf$, we have $\pi^{(n)}(N\bar{f}) \leq \pi^{(n)}(Nf)$ and

$$\lim_n \pi^{(n)}(N\bar{f}) \leq \lim_n \pi^{(n)}(Nf).$$

Since $\bar{f} \geq 0$,

$$\pi^{(n)}(N\bar{f}) = (\pi^{(n)}N)\bar{f},$$

and

$$\lim_n \pi^{(n)}(N\bar{f}) = (\lim_n \pi^{(n)}N)\bar{f} = \alpha\bar{f}$$

by monotone convergence with $\bar{f}$ as the measure. And since $\pi^{(n)}Nf^+ \leq \alpha f^+ < \infty$ and $\pi^{(n)}Nf^- \leq \alpha f^- < \infty$,

$$\pi^{(n)}(Nf) = \pi^{(n)}Nf^+ - \pi^{(n)}Nf^-.$$

Hence

$$\lim_n \pi^{(n)}(Nf) = \alpha f^+ - \alpha f^- = \alpha f$$

by monotone convergence for each term. Thus $\alpha\bar{f} \leq \alpha f$; $\alpha\bar{f} \geq 0$ since $\bar{f} \geq 0$, and $\alpha f < \infty$ by hypothesis.

Proposition 8-25: If h is a non-negative superregular function dominated by a potential g, then h is a potential and its total charge is no greater than the total charge of g.

PROOF: Let $g = Nf$. Write $h = N\bar{f} + \lim P^n h$ with $\bar{f} \geq 0$. Since $0 \leq h \leq g$, we have $0 \leq P^n h \leq P^n g$. But $P^n g \to 0$ by Corollary 8-9, so that $P^n h \to 0$ and $h = N\bar{f}$. Since $\alpha|f| < \infty$, we have, by Lemma 8-24; $0 \leq \alpha\bar{f} \leq \alpha f < \infty$. Hence h is a potential and $\alpha\bar{f} \leq \alpha f$.

Corollary 8-26: A non-negative potential $g = Nf$ has non-negative total charge.

PROOF: Let $g = Nf \geq 0$, and set $\bar{f} = 0$. Since αf is finite, $\alpha f \geq \alpha\bar{f} = 0$ by Lemma 8-24.

Proposition 8-25 has an interesting interpretation in terms of the enlarged chain. The jth column of N is a potential with charge $\{\delta_{ij}\}$. If the infimum of the column is positive, then the column dominates a constant function $k\mathbf{1}$, so that $\mathbf{1}$ is a potential by Proposition 8-25. Hence, the infimum of every column of N is zero unless the extended chain is absorbing. In particular, if $P\mathbf{1} = \mathbf{1}$, then the infimum of every column of N is 0.

In the case of the symmetric random walk in three dimensions, $P\mathbf{1} = \mathbf{1}$. Thus the infimum of every column of N is zero, and since P is symmetric, the infimum of every row is zero. This fact, although not providing a proof of Proposition 7-10, does give us more insight into that result.

3. Equilibrium sets and capacities

In proving analogs in the next section to the classical potential principles, we shall need to restrict the supports of the charges involved. The notion we shall need is that of an equilibrium set.

Definition 8-27: A set E is an **equilibrium set** for P if there is a pure potential which assumes the value 1 at every point of E and which has support in E. Such a potential is called an **equilibrium potential** for E. A set E is a **dual equilibrium set** for P if there is a pure potential measure with support in E which equals α on E.

We proceed to give two characterizations of equilibrium potentials.

Proposition 8-28: A set E is an equilibrium set if and only if both

(1) $\alpha e^E < \infty$ and
(2) for any starting distribution the set E is entered only finitely often a.e.

When E is an equilibrium set, the hitting vector h^E is the *unique* equilibrium potential and its charge is the escape vector e^E.

PROOF: Suppose E is an equilibrium set. If x is an equilibrium potential for E, then $B^E x = x$ by Proposition 8-19 and $x_E = \mathbf{1}$ by definition. Since $x_{\tilde{E}}$ does not affect the value of $B^E x$, we have

$$x = B^E x = B^E \mathbf{1} = h^E,$$

and h^E must be the equilibrium potential. Its charge is

$$(I - P)h^E = e^E$$

by Theorem 8-4 and Proposition 5-8. Thus E is an equilibrium set if and only if h^E is a potential or if and only if $\alpha e^E < \infty$ and $\lim P^n h^E = 0$. But $s^E = \lim P^n h^E$ by conclusion (7) of Proposition 5-8.

Corollary 8-29: All finite sets are equilibrium sets.

PROOF: Apply Propositions 4-28 and 8-28.

Proposition 8-30: If E is an equilibrium set, then the equilibrium potential is the pointwise infimum of all pure potentials which dominate $\mathbf{1}$ on E.

PROOF: Since $h^E = B^E\mathbf{1}$, the result follows from conclusion (2) of Proposition 8-16.

We shall use the notation η^E for dual e^E.

Definition 8-31: If E is an equilibrium set, the **capacity** of E is defined by $C(E) = \alpha e^E = \eta^E\mathbf{1}$.

In terms of total charge, Definition 8-31 states that the capacity of an equilibrium set is to be the total charge of the equilibrium potential.

Lemma 8-32: A set E is an equilibrium set if and only if both $(P^E)^n\mathbf{1} \to 0$ and $\alpha_E[(I - P^E)\mathbf{1}] < \infty$. If E is an equilibrium set, then $C(E) = \alpha_E[(I - P^E)\mathbf{1}] = [\alpha_E(I - \hat{P}^E)]\mathbf{1}$.

PROOF: We shall apply Proposition 8-28. $[(P^E)^n\mathbf{1}]_i$ is the probability starting in $i \in E$ of returning to E at least n times. Thus $(P^E)^n\mathbf{1} \to 0$ is a necessary and sufficient condition for being in E only finitely often a.e. for any starting distribution. Secondly $(I - P^E)\mathbf{1} = e^E_E$ and $\alpha_E e^E_E = \alpha e^E$. Hence αe^E is finite if and only if $\alpha_E[(I - P^E)\mathbf{1}] < \infty$. And if E is an equilibrium set, then

$$C(E) = \alpha_E[(I - P^E)\mathbf{1}].$$

Under duality a number is transformed into itself. Hence

$$\begin{aligned} C(E) &= [(\text{dual } \mathbf{1})(\text{dual } (I - P^E))](\text{dual } \alpha_E) \\ &= [\alpha_E(I - \hat{P}^E)]\mathbf{1}. \end{aligned}$$

Proposition 8-33: E is an equilibrium set if and only if $\mathbf{1}$ is a potential for P^E with α_E as the distinguished measure. Also $C(E)$ is the same computed for P as for P^E.

PROOF: 1 is always superregular for P^E. The two conditions given by Lemma 8-32 are precisely the conditions that 1 be a potential. Also $\alpha_E[(I - P^E)1]$ is the capacity of E in P^E.

Proposition 8-34: E is an equilibrium set for $\hat{P}$ if and only if it is a dual equilibrium set for P. When E is such a set, $\alpha_E = \hat{\eta}_E^E N_E$.

PROOF: E is an equilibrium set for $\hat{P}$ if and only if 1 is a potential for $\hat{P}^E$ with $1 = \hat{N}_E \hat{e}_E^E$. By duality, this condition is equivalent to the assertion that α_E is a potential measure for P^E with $\alpha_E = \hat{\eta}_E^E N_E$. The result then follows from the dual of Proposition 8-33.

We would like the result that $C(E) = \hat{C}(E)$. However, an equilibrium set for P need not be an equilibrium set for $\hat{P}$ (see Section 6). Therefore, the following is the best possible result:

Proposition 8-35: If E is an equilibrium set for both P and $\hat{P}$, then $C(E) = \hat{C}(E)$.

PROOF: By Proposition 8-34, we have $\alpha_E = \hat{\eta}_E^E N_E$, so that $\hat{\eta}_E^E = \alpha_E(I - P^E)$ by Lemma 8-18. By Lemma 8-32 applied to $\hat{P}$,

$$\begin{aligned} C(E) &= \alpha_E e_E^E = (\hat{\eta}_E^E N_E) e_E^E = \hat{\eta}_E^E (N_E e_E^E) = \hat{\eta}_E^E h_E^E \\ &= \hat{\eta}_E^E 1 = [\alpha_E(I - P^E)]1 = \hat{C}(E). \end{aligned}$$

Proposition 8-36: If F is a dual equilibrium set and $E \subset F$, then E is a dual equilibrium set and $\hat{C}(E) = \hat{\eta}^F h^E$.

PROOF: We shall use Proposition 8-28 to prove that E is a dual equilibrium set. By Proposition 8-34, $\alpha_F = (\hat{\eta}^F N)_F$ and $\hat{C}(F) = \hat{\eta}^F 1 < \infty$. Hence $\alpha_E = (\hat{\eta}^F N)_E$ and

$$\begin{aligned} \alpha_E(I - P^E) &= \left[\hat{\eta}^F N \begin{pmatrix} I - P^E & 0 \\ 0 & 0 \end{pmatrix}\right]_E \\ &= (\hat{\eta}^F B^E)_E \end{aligned}$$

by Proposition 8-20. Then

$$\alpha \hat{e}^E = \alpha_E[(I - \hat{P}^E)1] = [\alpha_E(I - P^E)]1 = (\hat{\eta}^F B^E)_E 1 = \hat{\eta}^F B^E 1 = \hat{\eta}^F h^E.$$

Since $\hat{\eta}^F h^E \leq \hat{\eta}^F 1 < \infty$, we have just verified the first condition of Proposition 8-28—that $\alpha \hat{e}^E < \infty$. The second condition is trivial for a subset of an equilibrium set. Hence E is a dual equilibrium set and $\hat{C}(E) = \alpha \hat{e}^E = \hat{\eta}^F h^E$.

The dual of this result states that any subset E of an equilibrium set F is an equilibrium set, and $C(E) = \eta^F \hat{h}^E$.

Proposition 8-37: The union of a finite number of equilibrium sets is an equilibrium set.

PROOF: Let $E_1, \ldots, E_n$ be equilibrium sets and let $E = \bigcup_{k=1}^n E_k$. Then

$$\begin{aligned}\sum_{i \in E} \alpha_i e_i^E &\leq \sum_{k=1}^n \sum_{i \in E_k} \alpha_i e_i^E \\ &\leq \sum_{k=1}^n \sum_{i \in E_k} \alpha_i e_i^{E_k} \\ &= \sum_{k=1}^n C(E_k) < \infty,\end{aligned}$$

and if the process is in each E_k only finitely often a.e., then it is in E only finitely often a.e. Hence, by Proposition 8-28, E is an equilibrium set.

Some of the classical results hold only if the support of a potential is a reasonably small set. It will always be satisfactory to have a finite set as support. A more general assumption is that the support is an equilibrium set. Since equilibrium sets include all finite sets, since a subset of an equilibrium set is an equilibrium set, and since finite unions of equilibrium sets are equilibrium sets, we may think of equilibrium sets as a class of "reasonably small" sets.

Choquet has introduced a generalized notion of capacity. In our case his definition takes the following form.

Definition 8-38: A **Choquet capacity** is a non-negative monotone increasing set function such that, for any sets $A_1, A_2, \ldots, A_n$,

$$\begin{aligned}C(A_1 \cap A_2 \cap \cdots \cap A_n) \leq \sum_i C(A_i) &- \sum_{i \neq j} C(A_i \cup A_j) \\ &+ \sum_{i \neq j \neq k} C(A_i \cup A_j \cup A_k) \\ &- \cdots - (-1)^n C(A_1 \cup \cdots \cup A_n).\end{aligned}$$

A simple way of constructing one of these capacities is to let π be a fixed starting distribution and to take $C(E)$ to be the probability of ever entering E. That is, $C(E) = \pi h^E$. This set function is monotone because h^E is. The right side of the inequality in the definition of capacity is the probability that all sets are entered. The left side is the probability that the intersection of the sets is entered, which is one way of entering all sets, though in general not the only way. Hence Choquet's definition is satisfied. Since we may clearly also replace $C(E)$ by $kC(E)$ with $k > 0$, π may be any non-negative finite measure.

We shall show from this construction that our Definition 8-31 yields a Choquet capacity on equilibrium sets. For convenience, we will give a proof for $\hat{P}$. For any fixed equilibrium set F of $\hat{P}$, Proposition 8-36 tells us that the capacity of any subset E is $\hat{\eta}^F h^E$. Hence the above argument applies with $\pi = \hat{\eta}^F$. Thus the Choquet conditions hold for all subsets of F and, since F is any equilibrium set, they hold for all equilibrium sets.

A more general method of obtaining a Choquet capacity within our framework is as follows. Let π be a measure, and let h be a strictly positive superregular function such that $\pi h < \infty$. Define $C(E) = \pi B^E h$. Forming the h-process, we see that $C(E) = \pi^*(B^E)^*1 = \pi^*(h^E)^*$ with $\pi^*1 = \pi h < \infty$. Thus by the special case $C(E)$ is a Choquet capacity for the h-process and hence satisfies the same axioms in the original chain. Moreover, we see that the situation in the earlier case is just the present case with $h = 1$. On the other hand, this more general method includes a second interesting case: If g is a pure potential for which πg is finite, then by Lemma 8-17,

$$\pi B^E g = \pi g - \pi\, {}^E N f;$$

$\pi B^E g$ is a Choquet capacity which assumes its maximum value πg on all sets E containing the support of f. If $\pi 1 = 1$, then $\pi B^E g$ in the game interpretation is the expected gain when and after E is reached.

Definition 8-31 is reasonable only for equilibrium sets, since otherwise it is possible to have $e^E = 0$. We could instead restrict the definition to finite sets and define the capacity of an infinite set as the supremum of the capacities of its finite subsets. We will show, under an additional assumption, that this new approach agrees with Definition 8-31 on equilibrium sets and assigns infinite capacity to all other sets.

Proposition 8-39: If $\hat{N}$ has columns which tend to zero, then a set E for which the supremum of the capacities of its finite subsets is finite is an equilibrium set, and the supremum is the capacity of the set.

PROOF: Let $E_1 \subset E_2 \subset \ldots$ be an increasing sequence of finite sets whose union is E. We must prove that if $\sup C(E_n)$ is finite, then h^E is a potential and $C(E) = \sup C(E_n)$. First we note that h^E is the monotone limit of h^{E_n}. If $i \in E_m$, then the ith component of e^{E_n} decreases for $n \geq m$. Thus $\lim e^{E_n} = \bar{e}$ exists. Since $\hat{N}$ has columns that tend to zero and since $\eta^{E_n}1 = C(E_n) \leq \sup C(E_n) < \infty$,

$$\eta^{E_n}\hat{N} \to (\text{dual } \bar{e})\hat{N}$$

by Proposition 1-58. By duality,

$$h^{E_n} = Ne^{E_n} \to N\bar{e},$$

and

$$\alpha\bar{e} = (\text{dual } \bar{e})1 = \sum_i \lim_n \eta_i^{E_n},$$

which by Fatou's Theorem is

$$\leq \lim_n (\eta^{E_n}1) = \lim C(E_n) = \sup C(E_n).$$

Hence $h^E = N\bar{e}$ and $\alpha\bar{e} < \infty$. Thus E is an equilibrium set. Also

$$C(E) = \alpha\bar{e} \leq \sup C(E_n).$$

But $C(E_n) \leq C(E)$ for every n, so that $\sup C(E_n) \leq C(E)$. Thus

$$C(E) = \sup C(E_n).$$

The converse, that an equilibrium set has the property that the supremum of the capacities of its finite subsets is finite, follows trivially from the monotonicity and the finiteness of capacity on equilibrium sets.

4. Potential principles

We shall now derive analogs of several of the fundamental theorems of classical potential theory. The first is the solution to the Dirichlet problem; in the uniqueness statement we shall need a lemma, for which we shall give two proofs.

Lemma 8-40: If $\bar{P}$ is an absorbing chain, then P has no bounded non-zero regular function.

PROOF 1: Suppose $Ph = h$ with $|h| \leq c1$. Since $h_i = \sum_i P_{ij}h_j$, we have $|h_i| \leq \sum_i P_{ij}|h_j|$ or $|h| \leq P|h|$. Therefore, $|h| \leq P^n|h| \leq P^n(c1)$. But P^n1 is the probability that the process continues at least until time n, which tends to zero as n tends to infinity because $\bar{P}$ is absorbing. Hence $h = 0$.

PROOF 2: Let a be the absorbing state of $\bar{P}$ and let $\mathbf{t} = \mathbf{a}$ be the time to absorption; $\mathbf{t}$ is a stopping time since $\bar{P}$ is absorbing. If $h = Ph$, set

$$h' = \begin{matrix} a \\ S \end{matrix} \begin{pmatrix} 0 \\ h \end{pmatrix}.$$

Then $h'(x_n(\omega))$ in the $\bar{P}$-process forms a bounded martingale. By Corollary 3-16 to the second martingale systems theorem,

$$0 = h'_a = \mathrm{M}[h'(x_t(\omega))] = \mathrm{M}[h'(x_0(\omega))].$$

But $x_0(\omega)$ is arbitrary, so that $h' = 0$ and $h = 0$.

The method in the second proof is of some importance, and we shall meet it again later.

Theorem 8-41: Let E be an arbitrary set of states, and suppose that ${}^EP$, the chain P with E made absorbing, is an absorbing chain. If h_E is any bounded function defined on E, then there exists a unique bounded function $\bar{h}$ whose restriction to E is h_E and which is regular on $\tilde{E}$. The function is

$$\bar{h} = B^E\begin{pmatrix} h_E \\ 0 \end{pmatrix}.$$

PROOF: For existence, set

$$\bar{h} = B^E\begin{pmatrix} h_E \\ 0 \end{pmatrix}.$$

The product is defined, since h_E is bounded and B^E has row sums one. Then the restriction of $\bar{h}$ to E is h_E because $(B^E)_{ij} = \delta_{ij}$ for i and j in E. Moreover,

$$(I - P)\bar{h} = (I - P)B^E\begin{pmatrix} h_E \\ 0 \end{pmatrix} = \begin{pmatrix} I - P^E & 0 \\ 0 & 0 \end{pmatrix}\begin{pmatrix} h_E \\ 0 \end{pmatrix} = \begin{pmatrix} (I - P^E)h_E \\ 0 \end{pmatrix},$$

so that $\bar{h}$ is regular outside of E; associativity is justified in the triple product because $(I + P)B^E\begin{pmatrix} |h_E| \\ 0 \end{pmatrix}$ is finite-valued.

For uniqueness, let

$$k = \begin{pmatrix} h_E \\ \bar{k} \end{pmatrix}$$

be another such bounded function. Then $\bar{h} - k$ is a bounded function which is zero on E and regular outside E. If Q is the transition matrix for the transient states of ${}^EP$, then $(\bar{h} - k)_{\tilde{E}}$ is a non-zero bounded Q-regular function, in contradiction to Lemma 8-40, since $\bar{Q}$ is absorbing.

Next we prove the Maximum Principle.

Theorem 8-42: Let E be an arbitrary set of states and suppose that h is a finite-valued function such that $h = B^E h$. Then the supremum of the values of h is equal to the supremum of the values of h on E. If the states of $\tilde{E}$ communicate in ${}^E P$, if ${}^E P$ is absorbing, and if h assumes its maximum on $\tilde{E}$, then h is constant on $\tilde{E}$.

REMARK: Corresponding results hold for infima by replacing h by $-h$.

PROOF:

$$h_i = \sum_{j \in E} B^E_{ij} h_j$$

$$\leq \sum_{j \in E} B^E_{ij} \left(\sup_{j \in E} h_j\right) \leq \sup_{j \in E} h_j.$$

Suppose that h assumes its maximum on $\tilde{E}$, that ${}^E P$ is absorbing, and that the states of $\tilde{E}$ communicate in ${}^E P$. Let i be a state where the maximum is assumed, and let k be any state of E that can be reached in ${}^E P$ from $\tilde{E}$. Since the transient states of ${}^E P$ communicate, we have $B^E_{ik} > 0$. Moreover,

$$\begin{aligned} h_i &= B^E_{ik} h_k + \sum_{j \neq k} B^E_{ij} h_j \\ &\leq B^E_{ik} h_k + h_i \sum_{j \neq k} B^E_{ij} \qquad \text{since } h_j \leq h_i \\ &= B^E_{ik} h_k + h_i (1 - B^E_{ik}) \quad \text{since } B^E 1 = 1 \\ &= h_i - B^E_{ik}(h_i - h_k). \end{aligned}$$

Therefore, $h_i = h_k$ for all such k. Then for any $m \in \tilde{E}$, $B^E_{mj} > 0$ precisely for those j for which $B^E_{ij} > 0$, and $h_j = h_i$ for those j. Thus

$$h_m = \sum_{j \in E} B^E_{mj} h_j = \sum_{j \in E} B^E_{mj} h_i = h_i.$$

Corollary 8-43: If g is a potential with support in a finite set, then g is bounded.

PROOF: Since $g = B^E g$ for any potential, we may apply Theorem 8-42. The supremum in E is over a finite number of values.

Corollary 8-44: Let E be an arbitrary set of states, and suppose that

(1) ${}^E P$ is an absorbing chain.
(2) the states of $\tilde{E}$ communicate in ${}^E P$.
(3) every state of E can be reached in ${}^E P$ from $\tilde{E}$.

If h is a bounded function regular outside of E, then h cannot assume its maximum on $\tilde{E}$ unless h is constant everywhere.

PROOF: By Theorem 8-41, h is the unique solution to the Dirichlet problem for the function h_E. Hence

$$h = B^E\begin{pmatrix} h_E \\ 0 \end{pmatrix}.$$

Multiplying through by B^E and applying Proposition 5-8, we have

$$B^E h = B^E B^E \begin{pmatrix} h_E \\ 0 \end{pmatrix} = B^E \begin{pmatrix} h_E \\ 0 \end{pmatrix} = h.$$

By Theorem 8-42, h is constant on $\tilde{E}$. As shown in the proof of that theorem, h assumes the same constant value at every state of E which can be reached in ${}^E P$ from $\tilde{E}$.

The result that follows is the Principle of Domination.

Theorem 8-45: Let h be a finite-valued non-negative superregular function, and let $g = Nf$ be a potential. If h dominates g on the support of f^+, then h dominates g everywhere. If, in addition, h is a potential $N\bar{f}$, then $\alpha f \leq \alpha \bar{f}$.

PROOF: If g is a pure potential supported in E, then $g = B^E g$ by Proposition 8-19. But by Proposition 8-16, $B^E g$ is the pointwise infimum of all non-negative superregular functions which dominate g on E. Thus the first half is proved if g is a pure potential. For arbitrary g, write $g = Nf^+ - Nf^-$. We have $Nf^+ - Nf^- \leq h$ on the support of f^+, so that

$$Nf^+ \leq h + Nf^-$$

on the support of f^+. Applying the special case to the superregular function $h + Nf^-$ and the potential Nf^+, we have

$$Nf^+ \leq h + Nf^- \quad \text{or} \quad g \leq h$$

everywhere. Finally, if $h = N\bar{f}$, then

$$Nf^+ \leq N(\bar{f} + f^-)$$

implies

$$\alpha f^+ \leq \alpha(\bar{f} + f^-)$$

by Lemma 8-24. Hence $\alpha f \leq \alpha \bar{f}$.

Next we prove the Principle of Balayage.

Theorem 8-46: If g is a pure potential and if E is any set of states, then there is a unique pure potential $\bar{g}$ with support in E such that $\bar{g} = g$ on E. The potential $\bar{g}$ satisfies $\bar{g} \leq g$ everywhere, and its total charge does not exceed the total charge of g.

PROOF: For existence, let $\bar{g} = B^E g$. Then $\bar{g} \leq g$, $\bar{g}_E = g_E$, and $\bar{g}$ is superregular by conclusions (1) and (3) of Proposition 8-16. By Proposition 8-25, $\bar{g}$ is a potential, and the total charge of $\bar{g}$ is less than or equal to that of g.

For uniqueness, if h were another such potential, we would have

$$\bar{g} = B^E \bar{g} = B^E h = h$$

by Proposition 8-15 and the fact that $\bar{g}_E = g_E = h_E$.

If E is any set and if g is a pure potential, we refer to the potential $\bar{g}$ of Theorem 8-46 as the **balayage potential** of g on E.

Corollary 8-47: The balayage potential $\bar{g} = B^E g$ of g on E is the pointwise infimum of all pure potentials which dominate g on E.

PROOF: Apply conclusion (2) of Proposition 8-16.

Corollary 8-48: The balayage potential of g on E is the supremum of all pure potentials with support in E which are dominated by g on E.

PROOF: Certainly the balayage potential does have the stated property. Thus let $\bar{g} = B^E g$ and let h be a potential with support in E and with $h_E \leq g_E$. Then by Proposition 8-19, $h = B^E h$ and

$$\bar{g} = B^E g \geq B^E h = h.$$

If g has support in E, then g itself is the balayage potential of g on E. In particular, h^E is the balayage potential of h^E on E for E an equilibrium set.

Next we prove the Principle of Lower Envelope.

Lemma 8-49: The pointwise infimum of non-negative superregular functions is non-negative superregular.

PROOF: It is clearly non-negative. If

$$h_\beta \geq P h_\beta$$

for all β, then

$$P(\inf h_\beta) \leq Ph_\beta \leq h_\beta$$

for all β, so that

$$P(\inf h_\beta) \leq \inf h_\beta.$$

Theorem 8-50: The pointwise infimum of pure potentials is a pure potential.

PROOF: Apply Lemma 8-49 and Proposition 8-25.

Finally we prove the Principle of Condensers. We are to think of two sets E and F as the two plates of a condenser with a positive charge placed on E and a negative charge placed on F in such a way as to produce a unit voltage drop. Since in equilibrium there should be a uniform voltage on each plate and since the 0-value of voltage is arbitrary, we will require that the potential be 1 on E and 0 on F. The theorem is proved for an equilibrium set E with **finite boundary** $\bar{E}$, that is, a set E that can be entered or left only through the finite set $\bar{E}$.

Theorem 8-51: Let E be an equilibrium set with finite boundary, and let F be any disjoint set of states. Then there is a potential $g = Nf$ which is 1 on E and 0 on F and which is such that f^+ has support in E, f^- has support in F, and $\alpha f \geq 0$.

PROOF: Let $g_i = {}^F H_{iE}$, the probability starting at i that E is reached before F. Clearly, g is 1 on E and 0 on F. Furthermore, $0 \leq g \leq h^E$. Since $P^n h^E \to 0$ for the equilibrium set E, we have $P^n g \to 0$.

Let $f = (I - P)g$. We are going to apply Corollary 8-8 to conclude that g is a potential with charge f, but to do so we must show that αf is finite. If i is in the complement of $E \cup F$, then $g_i = (Pg)_i$; hence f has support in $E \cup F$. Write

$$P^{E \cup F} = \begin{matrix} & \begin{matrix} E & F \end{matrix} \\ \begin{matrix} E \\ F \end{matrix} & \begin{pmatrix} X & Y \\ Z & W \end{pmatrix} \end{matrix}.$$

Noting that if $i \in E \cup F$, then $(Pg)_i$ is the probability that the next entry to $E \cup F$ is in E, we have

$$f_i = \begin{cases} g_i - (Pg)_i = 1 - (Pg)_i = 1 - (X\mathbf{1})_i & \text{if } i \in E \\ 0 - (Pg)_i = -(Z\mathbf{1})_i & \text{if } i \in F. \end{cases}$$

Hence

$$f_{E \cup F} = \begin{cases} 1 - X1 & \text{on } E. \\ -Z1 & \text{on } F. \end{cases}$$

Thus f^+ has support in E and f^- has support in F. Furthermore $X1 = 1$ except for the boundary of E; hence f^+ has finite support and $\alpha f^+ < \infty$. Moreover

$$\alpha f^- = \alpha_F Z1 = \sum_{j \in \tilde{E}} (\alpha_F Z)_j < \infty$$

since $Z_{ij} = 0$ except when j is on the boundary of E. Thus $\alpha|f| < \infty$. Hence g is a potential. Finally $\alpha f \geq 0$ by Corollary 8-26.

5. Energy

Classically, energy is the integral of the potential with respect to the charge, and we shall adopt the obvious analog of this definition. Throughout this section we shall write

$$\mu = \text{dual } f$$
$$\nu = \text{dual } g.$$

Definition 8-52: If $g = Nf$ is a potential and $|\mu|\ N\ |f| < \infty$, then its energy is defined to be $\mathbf{I}(g) = \mu g = \nu f$, and g is said to have **finite energy**.

If all potentials are bounded, then all of them have finite energy. For if $\bar{g} = N|f|$, then

$$\begin{aligned} |\mu|\ N\ |f| &= \sum_i \alpha_i |f_i| \bar{g}_i \\ &\leq \sup \bar{g}_i (\alpha |f|) \\ &< \infty. \end{aligned}$$

In any case a potential of finite support has finite energy.

We can write energy either purely in terms of the charge or purely in terms of the potential:

$$\mathbf{I}(g) = \mu(Nf) = \nu[(I - P)g].$$

Since the dual of a number is the same number, we also have

$$\mathbf{I}(g) = (\mu\hat{N})f = [\nu(I - \hat{P})]g.$$

If f is a charge for P, then, as noted after Theorem 8-3, f is also a charge for $\hat{P}$. In the two processes we have

$$\mathbf{I}(Nf) = \mu(Nf) = (\mu\hat{N})f$$

and

$$\mathbf{I}(\hat{N}f) = (\mu N)f = \mu(\hat{N}f),$$

and the energies are equal since the matrices associate by Corollary 1-5.

The expression for $\mathbf{I}(g)$ in Definition 8-52 disguises the fact that energy depends only on the values of the potential on the support E. We shall derive a simple dependence of $\mathbf{I}(g)$ on P^E.

Proposition 8-53: If g is a potential of finite energy with support in E, then

$$\mathbf{I}(g) = \nu_E[(I - P^E)g_E] = [\nu_E(I - \hat{P}^E)]g_E.$$

PROOF: Since f has support in E,

$$(I - P^E)g_E = f_E$$

by Proposition 8-19, and

$$\nu_E[(I - P^E)g_E] = \nu_E f_E = \nu f = \mathbf{I}(g).$$

The other half of the proposition is the dual of the first half.

Classically energy is non-negative. We shall prove shortly that the energy of a potential is non-negative provided it is finite. To do so, we first introduce a definition. If $g = Nf$ and $\bar{g} = N\bar{f}$ are potentials of finite energy, we define

$$(g, \bar{g}) = \tfrac{1}{2}(\mu\bar{g} + \bar{\mu}g),$$

provided the matrix products are well defined. (We shall show soon that this condition is always satisfied.)

Note that $(g, g) = \mathbf{I}(g)$. We wish to show that $(g, \bar{g})$ is an inner product. The reader should verify that $(g, \bar{g})$ satisfies (1), (2), and (4) in general and (3) when all the potentials have finite support. We shall prove (5) and the general case of (3) below.

(1) $(g, \bar{g}) = (\bar{g}, g)$.
(2) For every real number c, $(cg, \bar{g}) = c(g, \bar{g})$.
(3) $(g + g', \bar{g}) = (g, \bar{g}) + (g', \bar{g})$.
(4) If g is a pure potential for which $(g, g) = 0$, then $g = 0$.
(5) $(g, g) \geq 0$ for all g.

Lemma 8-54: If g has support in a finite set E, then

$$\mathbf{I}(g) = \tfrac{1}{2} \sum_{i \in E} \left[(\alpha_i m_i + \pi_i)g_i^2 + \sum_{j \in E} \alpha_i P_{ij}^E (g_i - g_j)^2\right] \geq 0,$$

where

$$m_i = 1 - \sum_{j \in E} P_{ij}^E \geq 0 \quad \text{and} \quad \pi_i = \alpha_i - \sum_{k \in E} \alpha_k P_{ki}^E \geq 0.$$

PROOF: We shall apply Proposition 8-53. The matrices involved are finite matrices so that distributivity and associativity hold. Hence

$$\begin{aligned}
\mathbf{I}(g) &= \nu_E g_E - \nu_E P^E g_E \\
&= \tfrac{1}{2} \sum_{i \in E} \left[\alpha_i g_i^2 + \alpha_i g_i^2 + \sum_{j \in E} (-2\alpha_i P_{ij}^E g_i g_j)\right] \\
&= \tfrac{1}{2} \sum_{i \in E} \left[\alpha_i\left(1 - \sum_{j \in E} P_{ij}^E\right) g_i^2 + \left(\alpha_i - \sum_{k \in E} \alpha_k P_{ki}^E\right) g_i^2 \right. \\
&\qquad \left. + \sum_{j \in E} (\alpha_i P_{ij}^E g_i^2 - 2\alpha_i P_{ij}^E g_i g_j + \alpha_i P_{ij}^E g_j^2)\right] \\
&= \tfrac{1}{2} \sum_{i \in E} \left[(\alpha_i m_i + \pi_i) g_i^2 + \sum_{j \in E} \alpha_i P_{ij}^E (g_i - g_j)^2\right].
\end{aligned}$$

Since P^E is a transition matrix and α_E is P^E-superregular, m and π are non-negative. Hence $\mathbf{I}(g) \geq 0$.

From properties (1), (2), and (3), we can prove that Schwarz's inequality holds for g and $\bar{g}$ whenever they have finite support.

Lemma 8-55: If $g = Nf$ and $\bar{g} = N\bar{f}$ are two potentials of finite support, then

$$(g, \bar{g})^2 \leq \mathbf{I}(g)\mathbf{I}(\bar{g}).$$

PROOF: By Lemma 8-54 we have

$$\mathbf{I}(xg - \bar{g}) = (xg - \bar{g}, xg - \bar{g}) \geq 0$$

for all real x. Hence by properties (1), (2), and (3), we find that

$$x^2(g, g) - 2x(g, \bar{g}) + (\bar{g}, \bar{g}) \geq 0$$

for all real x. If $(g, g) = 0$, then, for $-2x(g, \bar{g}) + (\bar{g}, \bar{g})$ to be non-negative for all x, it must be true that $(g, \bar{g}) = 0$, and the lemma is trivial. Otherwise, the discriminant of the quadratic equation in x must be non-positive, so that

$$4(g, \bar{g})^2 - 4(g, g)(\bar{g}, \bar{g}) \leq 0$$

or

$$\begin{aligned}
(g, \bar{g})^2 &\leq (g, g)(\bar{g}, \bar{g}) \\
&= \mathbf{I}(g)\mathbf{I}(\bar{g}).
\end{aligned}$$

Lemma 8-56: Let $g = Nf$ and $\bar{g} = N\bar{f}$ be pure potentials of finite energy, let

$$E_1 \subset E_2 \subset E_3 \cdots$$

be an increasing sequence of finite sets with union the set of all states S, and let

$$f^{(n)} = \begin{pmatrix} f_{E_n} \\ 0 \end{pmatrix}, \qquad \bar{f}^{(n)} = \begin{pmatrix} \bar{f}_{E_n} \\ 0 \end{pmatrix}.$$

$$g^{(n)} = Nf^{(n)}, \quad \text{and} \quad \bar{g}^{(n)} = N\bar{f}^{(n)}.$$

Then $(g, \bar{g}) = \lim_{n\to\infty} (g^{(n)}, \bar{g}^{(n)})$.

Proof: We have $(g, \bar{g}) = \frac{1}{2}(\mu\bar{g} + \bar{\mu}g)$, and by symmetry it is enough to show that $\bar{\mu}^{(n)}g^{(n)}$ converges to $\bar{\mu}g$. By monotone convergence, we have $\lim g^{(n)} = g$, since $0 \leq f^{(1)} \leq f^{(2)} \leq \cdots$. Let

$$h_i^{(n)} = \begin{cases} g_i^{(n)} & \text{if } i \in E_n \\ 0 & \text{otherwise.} \end{cases}$$

Then

$$\bar{\mu}^{(n)}g^{(n)} = \bar{\mu}h^{(n)}$$

and

$$\lim h^{(n)} = \lim g^{(n)} = g.$$

The functions $h^{(n)}$ are non-negative and increasing; also $\bar{\mu}$ is non-negative. Thus by monotone convergence,

$$\lim \bar{\mu}^{(n)}g^{(n)} = \lim \bar{\mu}h^{(n)} = \bar{\mu} \lim h^{(n)} = \bar{\mu}g.$$

Lemma 8-57: If g and $\bar{g}$ are pure potentials of finite energy, then

$$(g, \bar{g})^2 \leq (g, g)(\bar{g}, \bar{g}).$$

Consequently $(g, \bar{g}) < \infty$.

Proof: Form the approximations to g and $\bar{g}$ as in the statement of Lemma 8-56. By Lemmas 8-54 and 8-55,

$$(g^{(n)}, \bar{g}^{(n)})^2 \leq (g^{(n)}, g^{(n)})(\bar{g}^{(n)}, \bar{g}^{(n)}).$$

Applying Lemma 8-56 to each factor, we obtain

$$(g, \bar{g})^2 \leq (g, g)(\bar{g}, \bar{g}).$$

If g and $\bar{g}$ are any potentials of finite energy, then

$$|(g, \bar{g})| \leq (N|f|, N|\bar{f}|) \leq \sqrt{\mathbf{I}(N|f|)\mathbf{I}(N|\bar{f}|)} < \infty.$$

Therefore $(g, \bar{g})$ is always well defined. We can now prove (3) in general by breaking charges into positive and negative parts.

Proposition 8-58: If g has finite energy, then $\mathbf{I}(g) \geq 0$.

PROOF: Write $g = Nf = Nf^+ - Nf^-$, where Nf^+ and Nf^- are pure potentials. Then

$$\begin{aligned}
\mathbf{I}(g) &= (g, g) = (Nf^+ - Nf^-, Nf^+ - Nf^-) \\
&= \mathbf{I}(Nf^+) - 2(Nf^+, Nf^-) + \mathbf{I}(Nf^-) \\
&\geq \mathbf{I}(Nf^+) - 2\sqrt{\mathbf{I}(Nf^+)\mathbf{I}(Nf^-)} + \mathbf{I}(Nf^-) \quad \text{by Lemma 8-57} \\
&= (\sqrt{I(Nf^+)} - \sqrt{I(Nf^-)})^2 \\
&\geq 0.
\end{aligned}$$

We can finally prove Schwarz's inequality for all potentials of finite energy by proceeding just as in the proof of Lemma 8-55.

We now begin the proof of the fundamental result about energy, the theorem that justifies the name "equilibrium potential." Unfortunately the result fails for the most general transient chain, so that some extra hypothesis is needed. We shall prove the theorem—Theorem 8-61—under the hypothesis $P = \hat{P}$, a condition that *is* satisfied in the classical case of the three-dimensional symmetric random walk with $\alpha = \mathbf{1}^T$.

Lemma 8-59: The energy of the equilibrium potential on E is the capacity $C(E)$.

PROOF: $\mathbf{I}(h^E) = \eta^E h^E = \eta^E_E h^E_E = \eta^E_E \mathbf{1} = \eta^E \mathbf{1} = C(E)$.

Lemma 8-60: If E is an equilibrium set, if $g = Nf$ is a potential with support in E, and if $P = \hat{P}$, then $(g, h^E) = \alpha f$.

PROOF:

$$\begin{aligned}
2(g, h^E) &= \mu h^E + \eta^E g \\
&= \mu h^E + \eta^E Nf \\
&= \mu h^E + \mu \hat{N} e^E && \text{by duality} \\
&= \mu h^E + \mu N e^E && \text{since } N = \hat{N} \\
&= 2\mu h^E && \text{since } h^E = Ne^E \\
&= 2\mu \mathbf{1} && \text{since } f \text{ has support in } E \\
&= 2\alpha f.
\end{aligned}$$

Theorem 8-61: Suppose that P is a chain in which $P = \hat{P}$. If E is an equilibrium set, then the equilibrium potential for E minimizes energy among all potentials of finite energy whose support is in E and whose total charge is $C(E)$.

PROOF: Let $g = Nf$ be a potential with $\alpha f = C(E)$ and with support in E. By Lemma 8-60, $(g, h^E) = C(E)$, and by Lemma 8-59, $\mathbf{I}(h^E) = C(E)$. Furthermore, $\mathbf{I}(h^E) \neq 0$ by property (4). Therefore, by Schwarz's inequality, we have

$$\mathbf{I}(g) \geq \frac{(g, h^E)^2}{\mathbf{I}(h^E)} = \frac{C(E)^2}{C(E)} = C(E) = \mathbf{I}(h^E).$$

We shall see in the next section that the theorem need not be true if $P \neq \hat{P}$.

6. The basic example

In this section we shall work out what the results of the preceding five sections mean in terms of the basic example.

First we compute $\hat{P}$, the β-dual of P.

$$\hat{P}_{ij} = \frac{\beta_j P_{ji}}{\beta_i} = \begin{cases} \beta_{i-1} p_i / \beta_i = 1 & \text{if } j = i - 1 \\ \dfrac{\beta_j q_{j+1}}{\beta_0} = \beta_j - \beta_{j+1} & \text{if } i = 0, \end{cases}$$

Thus the reverse process proceeds deterministically a step at a time to the left until it reaches 0. From 0 it may step into any state and does so with probability

$$\hat{P}_{0j} = \beta_j - \beta_{j+1}.$$

Since $\sum_j \hat{P}_{0j} = 1 - \beta_\infty < 1$ and since 0 is reached from all states with probability one, the extended chain for $\hat{P}$ is absorbing. We saw in Section 5-10 that P has no non-zero regular measure; on the other hand, β is regular for $\hat{P}$ since

$$\sum_i \beta_i \hat{P}_{ij} = \beta_0(\beta_j - \beta_{j+1}) + \beta_{j+1} 1 = \beta_j.$$

From Section 5-10 we know that

$$N_{ij} = \begin{cases} \dfrac{\beta_j}{\beta_\infty} & \text{if } i \leq j \\ \dfrac{\beta_j}{\beta_\infty} - \dfrac{\beta_j}{\beta_i} & \text{if } i > j. \end{cases}$$

Hence

$$\hat{N}_{ij} = \frac{\beta_j N_{ji}}{\beta_i} = \begin{cases} \dfrac{\beta_j}{\beta_\infty} & \text{if } i \geq j \\ \dfrac{\beta_j}{\beta_\infty} - 1 & \text{if } i < j. \end{cases}$$

We note that N has columns tending to 0, but $\hat{N}$ has columns bounded away from 0. The latter fact by itself implies that the extended chain for $\hat{P}$ is absorbing.

Next we find the general form of potentials. In P if $g = Nf$, then

$$g_i = (Nf)_i = \frac{1}{\beta_\infty} \sum_{j=0}^{\infty} \beta_j f_j - \frac{1}{\beta_i} \sum_{j=0}^{i-1} \beta_j f_j. \tag{1}$$

In $\hat{P}$ if $g = \hat{N}f$, then

$$g_i = \frac{1}{\beta_\infty} \sum_{j=0}^{\infty} \beta_j f_j - \sum_{j=i+1}^{\infty} f_j. \tag{2}$$

In either case, g is finite-valued if $\beta|f| < \infty$, in agreement with Lemma 8-2. (For the reverse chain we have

$$\infty > \sum_j \beta_j |f_j| \geq \beta_\infty \sum_j |f_j|,$$

and hence $\sum_j |f_j| < \infty$.)

Let $\mu = \text{dual} f$ and $\nu = \text{dual } g$. Then as required by duality

$$\begin{aligned} \nu_i &= \frac{\beta_i}{\beta_\infty} \sum_{j=0}^{\infty} \beta_j f_j - \sum_{j=0}^{i-1} \beta_j f_j \\ &= \frac{\beta_i}{\beta_\infty} \sum_{j=0}^{\infty} \mu_j - \sum_{j=0}^{i-1} \mu_j = (\mu \hat{N})_i. \end{aligned}$$

Thus μ is a left charge with potential measure ν for $\hat{P}$.

Theorem 8-4 demands that $f = (I - P)g$ when g is a potential with charge f. We have from (1)

$$\begin{aligned} (Pg)_i &= p_{i+1} g_{i+1} + q_{i+1} g_0 \\ &= \frac{p_{i+1}}{\beta_\infty} (\beta f) - \frac{p_{i+1}}{\beta_{i+1}} \sum_{j=0}^{i} \beta_j f_j + \frac{q_{i+1}}{\beta_\infty} (\beta f) \\ &= \frac{1}{\beta_\infty} (\beta f) - \frac{1}{\beta_i} \sum_{j=0}^{i} \beta_j f_j. \end{aligned}$$

Hence $g_i - (Pg)_i = f_i$ and $(I - P)g = f$.

For both P and $\hat{P}$, we have

$$N_{ii} = \hat{N}_{ii} = \frac{\beta_i}{\beta_\infty},$$

and thus the condition $\beta_i \geq kN_{ii}$ of Proposition 8-10 is satisfied with $k = \beta_\infty$. Hence all potentials in both P and $\hat{P}$ are bounded. We can see directly the boundedness from (1) and (2). In (1) and (2) we have

$$|g_i| \leq \frac{2\beta|f|}{\beta_\infty}.$$

The estimate

$$|g_i| \le \frac{\beta|f|}{\beta_\infty}$$

of Proposition 8-10 is better; it takes into account the cancellation of the first and second terms in each expression. If $f \ge 0$ in (2), then the g_i increase monotonically to $\beta|f|/\beta_\infty$, so that the proposition gives the best possible bound. In (1) we have $\lim_i g_i = 0$, in agreement with the second half of Proposition 8-10 since the columns of N tend to zero. However, $\hat{N}$ does not have columns tending to zero, and $\lim_i g_i$ in (2) is not necessarily 0.

We determine the regular functions and signed measures for P and $\hat{P}$ as follows. If r is a P-regular function, then

$$r_i = (Pr)_i = p_{i+1}r_{i+1} + q_{i+1}r_0.$$

Thus

$$r_0 = p_1r_1 + q_1r_0, \quad p_1r_0 = p_1r_1,$$

and

$$r_0 = r_1.$$

Hence only the constant functions are P-regular. Dually only multiples of β are regular signed measures for $\hat{P}$. Since P has no regular signed measures, $\hat{P}$ has no regular functions. (Recall that $\hat{P}1 \ne 1$.) Therefore, the non-negative superregular functions of P are pure potentials plus non-negative constants, whereas only pure potentials are non-negative superregular functions for $\hat{P}$.

Next, we determine the equilibrium sets for P and $\hat{P}$. It is clear that P will be in any infinite set infinitely often a.e. Hence only finite sets are equilibrium sets. Let us verify this fact in terms of equilibrium potentials.

Let E be a finite set with m as last element. Then

$$e_i^E = \begin{cases} \dfrac{\beta_\infty}{\beta_m} & \text{if } i = m \\ 0 & \text{otherwise} \end{cases}$$

and

$$\beta e^E = \beta_m e_m^E = \beta_\infty.$$

By (1),

$$h_i^E = \frac{1}{\beta_\infty}(\beta e^E) - \frac{1}{\beta_i}\sum_{j=0}^{i-1} \beta_j e_j^E$$

or

$$h_i^E = \begin{cases} 1 & \text{if } i \le m \\ 1 - \dfrac{\beta_\infty}{\beta_i} & \text{if } i > m. \end{cases}$$

It is clear probabilistically that these are the correct values for h^E; when $i > m$ the only way to avoid hitting E from i is to march straight out to the right. The probability of doing so is β_∞/β_i. Thus we see that h^E satisfies the conditions of an equilibrium potential: All non-empty finite sets have capacity β_∞.

If E is an infinite set, then $h^E = 1$ and $e^E = (I - P)h^E = 0$. Hence h^E is not a potential, and E is not an equilibrium set. The fact that the supremum of the capacities of finite subsets of an infinite set E is β_∞ does not contradict Proposition 8-39, since $\hat{N}$ does not have columns tending to zero.

For the reverse chain $\hat{P}$ let E be any set (finite or infinite) with least element m. Then $\hat{h}_i^E = 1$ for $i \geq m$. For $i < m$,

$$\hat{h}_i^E = \hat{h}_0^E = \hat{H}_{0m} = \frac{\hat{N}_{0m}}{\hat{N}_{mm}} = \frac{(\beta_m/\beta_\infty) - 1}{\beta_m/\beta_\infty} = 1 - \frac{\beta_\infty}{\beta_m}.$$

The next to last equality follows from the fact that $m = 0$ is incompatible with $i < m$. The process can escape from E only via m, and for $m > 0$

$$\hat{e}_m^E = [(I - \hat{P})\hat{h}^E]_m = \hat{h}_m^E - \hat{h}_{m-1}^E = \frac{\beta_\infty}{\beta_m}.$$

If $m = 0$,

$$\hat{e}_0^E = 1 - \sum_j \hat{P}_{0j} = \beta_\infty = \frac{\beta_\infty}{\beta_0}.$$

Hence, in either case,

$$\beta e^E = \beta_m e_m^E = \beta_\infty.$$

Thus *all* sets are equilibrium sets, and every set has capacity β_∞. We note that only the finite sets are equilibrium sets for both P and $\hat{P}$, and their capacity β_∞ is the same in both, as predicted by Proposition 8-35.

The dual of Proposition 8-23 is that every non-negative superregular measure is the increasing limit of pure potential measures of finite support. We shall produce the charges for $\hat{P}$ which give rise to the potential measures which increase to β. Let E_m be the set $\{0, \ldots, m\}$. The functions $h^{E_m} = Ne^{E_m}$ are the functions which Proposition 8-23 gives as increasing to 1. Therefore, by duality the measures $\eta^{E_m}\hat{N}$ should increase to β, and the charges we seek in $\hat{P}$ are the η^{E_m}. From our above calculations we have

$$\eta_i^{E_m} = \begin{cases} \beta_\infty & \text{if } i = m \\ 0 & \text{otherwise,} \end{cases}$$

and

$$(\eta^{E_m}\hat{N})_i = \begin{cases} \beta_i & \text{if } i \leq m \\ \beta_i - \beta_\infty & \text{if } i > m. \end{cases}$$

Next suppose as in Lemma 8-24 that $N\bar{f} \leq Nf$ with $\bar{f} \geq 0$ and βf finite. Since by (1)

$$(N\bar{f})_0 = \frac{1}{\beta_\infty}(\beta\bar{f})$$

and

$$(Nf)_0 = \frac{1}{\beta_\infty}(\beta f),$$

we must have $\beta\bar{f} \leq \beta f$. The proof for the reverse chain is not so simple, however. If $\hat{N}\bar{f} \leq \hat{N}f$ with $\bar{f} \geq 0$ and βf finite, then

$$\frac{1}{\beta_\infty}(\beta\bar{f}) - \sum_{j=i+1}^{\infty} \bar{f}_j \leq \frac{1}{\beta_\infty}(\beta f) - \sum_{j=i+1}^{\infty} f_j$$

for all i. According to the proof of Lemma 8-24, we multiply through by η^{E_m} and take the limit on m. We have

$$\beta\bar{f} - \beta_\infty \sum_{j=m+1}^{\infty} \bar{f}_j \leq \beta f - \beta_\infty \sum_{j=m+1}^{\infty} f_j.$$

As $m \to \infty$, we obtain $\beta\bar{f} \leq \beta f$.

Turning to the potential principles, we shall first illustrate the Principle of Domination (Theorem 8-45). We do so in $\hat{P}$. Let $f \geq 0$ and $\bar{f} \geq 0$ be given, let E be the support of $\bar{f}$, and suppose that $g_i \geq \bar{g}_i$ for $i \in E$. For convenience, suppose $0 \in E$. From (2) we have

$$\frac{\beta f}{\beta_\infty} - \sum_{j>i} f_j \geq \frac{\beta\bar{f}}{\beta_\infty} - \sum_{j>i} \bar{f}_j \quad \text{for } i \in E.$$

Let k be in $\tilde{E}$, and let i be the largest state of E for which $i < k$ (i exists since $0 \in E$). Then

$$\begin{aligned} g_k = \frac{\beta f}{\beta_\infty} - \sum_{j>k} f_j \geq \frac{\beta f}{\beta_\infty} - \sum_{j>i} f_j \geq \frac{\beta\bar{f}}{\beta_\infty} - \sum_{j>i} \bar{f}_j \\ = \frac{\beta\bar{f}}{\beta_\infty} - \sum_{j>k} \bar{f}_j = \bar{g}_k, \end{aligned}$$

the next to last equality holding since $\bar{f}_j = 0$ for $i < j \leq k$. Hence $g \geq \bar{g}$.

Next we examine the Principle of Balayage (Theorem 8-46) for $\hat{P}$. Let $f \geq 0$ be given, and for convenience let E be an infinite set containing 0. We wish to choose $\bar{f}$ with support in E so that $\bar{g}_E = g_E$. For $i \in E$, we must have

$$\frac{\beta\bar{f}}{\beta_\infty} - \sum_{j>i} \bar{f}_j = \frac{\beta f}{\beta_\infty} - \sum_{j>i} f_j.$$

Let k be the next element of E greater than i. We must also have

$$\frac{\beta\bar{f}}{\beta_\infty} - \sum_{j>k} \bar{f}_j = \frac{\beta f}{\beta_\infty} - \sum_{j>k} f_j.$$

Subtracting, we find that $\bar{f}_k = \sum_{j=i+1}^{k} f_j$. This equation determines all of $\bar{f}$ except for $\bar{f}_0$. Adding the relations for $\bar{f}_k$ with $k > 0$, we obtain $\sum_{j>0} \bar{f}_j = \sum_{j>0} f_j$.

Since for $g_0 = \bar{g}_0$ we need

$$\frac{\beta\bar{f}}{\beta_\infty} - \sum_{j>0} \bar{f}_j = \frac{\beta f}{\beta_\infty} - \sum_{j>0} f_j,$$

we must choose $\beta\bar{f} = \beta f$. Thus set

$$\bar{f}_0 = f_0 + \sum_{j>0} \beta_j(f_j - \bar{f}_j)$$

$$\bar{f}_k = \sum_{j=i+1}^{k} f_j \quad \text{for } i, k \in E \text{ and } j \notin E \text{ when } i < j < k.$$

To see that we have actually chosen $\bar{f}_0 \geq 0$, we note that β_j decreases with j, so that if $i, k \in E$ with no $j \in E$ for $i < j < k$, then

$$\begin{aligned}
\beta_{i+1}(f_{i+1} - \bar{f}_{i+1}) + \beta_{i+2}(f_{i+2} - \bar{f}_{i+2}) + \cdots + \beta_k(f_k - \bar{f}_k) \\
&= \beta_{i+1}f_{i+1} + \beta_{i+2}f_{i+2} + \cdots + \beta_k f_k - \beta_k \bar{f}_k \\
&= \beta_{i+1}f_{i+1} + \cdots + \beta_k f_k - \beta_k(f_{i+1} + \cdots + f_k) \\
&= (\beta_{i+1} - \beta_k)f_{i+1} + \cdots + (\beta_{k-1} - \beta_k)f_{k-1} \\
&\geq 0.
\end{aligned}$$

We shall illustrate the Principle of Condensers (Theorem 8-51) for P in the case $E = \{0, 1, \ldots, a\}$ and $F = \{b, b+1, \ldots\}$ with $0 < a < b$. We have

$$g_i = {}^F H_{iE} = \begin{cases} 1 & \text{for } i \leq a \\ 1 - \dfrac{\beta_b}{\beta_i} & \text{for } a < i < b \\ 0 & \text{for } i \geq b. \end{cases}$$

Then

$$f_i = g_i - p_{i+1}g_{i+1} - q_{i+1}g_0 = \begin{cases} \dfrac{\beta_b}{\beta_a} & \text{if } i = a \\ -q_{i+1} & \text{if } i \geq b \\ 0 & \text{otherwise.} \end{cases}$$

Hence

$$\beta f = \beta_b - \sum_{i \geq b} \beta_i q_{i+1} = \beta_\infty.$$

We can verify from (1) that g is the potential of f, and we see that f^+ has support in E, f^- has support in F, and that $\beta f \geq 0$.

We can show by example that the Principle of Condensers does not hold for all equilibrium sets E. In $\hat{P}$ let E be the set of even states and let F be the set of odd states. All sets are equilibrium sets for $\hat{P}$, but E does not have a finite boundary. If

$$g_i = \begin{cases} 1 & \text{for } i \in E \\ 0 & \text{for } i \in F, \end{cases}$$

the theorem requires that g be a potential. But if $i \in F$,

$$f_i = 0 - (\hat{P}g)_i = -1.$$

Hence

$$(\hat{N}f^-)_0 = -\sum_{j \text{ odd}} \hat{N}_{0j} = -\sum_{j \text{ odd}} \left(\frac{\beta_j}{\beta_\infty} - 1\right).$$

The expression on the right side may be infinite if the β's are chosen properly. Let

$$\beta_i = \frac{1}{i+2} + \frac{1}{2}$$

$$\beta_\infty = \tfrac{1}{2}.$$

The β's determine the transition probabilities uniquely, and for this choice $(\hat{N}f^-)_0$ is infinite, a contradiction.

Equation (1) gives us the following relation for energy in the P-process.

$$\mathbf{I}(g) = \frac{1}{\beta_\infty}(\beta f)^2 - \sum_{i=0}^{\infty} f_i \sum_{j=0}^{i-1} \beta_j f_j.$$

It is not difficult to see that (2) yields the same value for the same f. But it is not easy to see that $\mathbf{I}(g) \geq 0$ if g is not a pure potential. We shall now show that Theorem 8-61 fails if the assumption $P = \hat{P}$ is dropped. In P let $E = \{0, 1, \ldots, m\}$. We have seen that $C(E) = \beta_\infty$. Thus any potential with total charge $\beta f = \beta_\infty$ equal to that of the equilibrium potential has energy

$$\beta_\infty - \sum_{i=0}^{\infty} f_i \sum_{j=0}^{i-1} \beta_j f_j.$$

The equilibrium potential has energy β_∞, which is a *maximum* (not a minimum) among pure potentials. For example, let

$$f_i = \begin{cases} \frac{1}{2}\beta_\infty & \text{if } i = 0 \\ \dfrac{1}{2\beta_1}\beta_\infty & \text{if } i = 1 \\ 0 & \text{otherwise.} \end{cases}$$

Then $\beta f = \beta_\infty$ and

$$\mathbf{I}(g) = \beta_\infty - f_0(\beta_0 f_1) = \beta_\infty - \frac{1}{4\beta_1}\beta_\infty^2 < \beta_\infty.$$

7. An unbounded potential

In Proposition 8-10 we saw that a sufficient condition for all potentials to be bounded is that α be chosen so that $\alpha_i \geq kN_{ii}$ for all i. The purpose of this section is to show that unbounded potentials may exist when this hypothesis is not satisfied; the potential we exhibit will have a bounded charge whose support is at the same time an equilibrium set and a dual equilibrium set.

The chain P will be a modification of sums of independent random variables on the line with $p_1 = \frac{1}{3}$ and $p_{-1} = \frac{2}{3}$. Let

$$\left.\begin{aligned} P_{i,i-1} &= \tfrac{2}{3} \\ P_{i,i+1} &= \tfrac{1}{3} \end{aligned}\right\} \quad \text{for } i \leq 0$$

and

$$\left.\begin{aligned} P_{ii} &= 1 - \frac{1}{i} \\ P_{i,i-1} &= \frac{2}{3i} \\ P_{i,i+1} &= \frac{1}{3i} \end{aligned}\right\} \quad \text{for } i > 0.$$

If the process is watched only when it changes states, it becomes the $p_1 = \frac{1}{3}$, $p_{-1} = \frac{2}{3}$ process, so that we may compute H from the latter chain.

$$H_{ij} = \begin{cases} 1 & \text{if } i \geq j \\ (\frac{1}{2})^{j-i} & \text{if } i < j. \end{cases}$$

Therefore,

$$\bar{H}_{ii} = P_{i,i+1}\cdot 1 + P_{i,i-1}\cdot\tfrac{1}{2} + P_{ii}\cdot 1 = 1 - \tfrac{1}{2}P_{i,i-1},$$

and

$$N_{jj} = \frac{1}{1 - \bar{H}_{jj}} = \frac{2}{P_{j,j-1}}.$$

Hence

$$N_{ij} = \begin{cases} 3 & \text{if } j \leq 0, \quad i \geq j \\ 3(\frac{1}{2})^{j-i} & \text{if } j \leq 0, \quad i < j \\ 3j & \text{if } j > 0, \quad i \geq j \\ 3j(\frac{1}{2})^{j-i} & \text{if } j > 0, \quad i < j. \end{cases}$$

Let $E = \{1, 2, 3, \ldots\}$. Since the process goes toward $-\infty$ with probability one, it can be in E only finitely often a.e. Moreover, $e_i^E = 0$ unless $i = 1$, so that $\alpha e^E < \infty$ for any choice of α. Thus E is an equilibrium set.

We shall take α to be the zeroth row of N. Then

$$\alpha_j = N_{0j} = \begin{cases} 3 & \text{if } j \leq 0 \\ 3j(\frac{1}{2})^j & \text{if } j > 0. \end{cases}$$

If we calculate $\hat{P}$, we find that

$$\hat{P}_{ij} = P_{ij} \quad \text{for } i > 0,$$

$$\left.\begin{aligned} \hat{P}_{i,i+1} &= \tfrac{2}{3} \\ \hat{P}_{i,i-1} &= \tfrac{1}{3} \end{aligned}\right\} \quad \text{for } i < 0,$$

and

$$\hat{P}_{01} = \hat{P}_{0,-1} = \tfrac{1}{3}.$$

With probability one the $\hat{P}$ process reaches 0 from all states, and from there it can disappear. Hence the extended chain for $\hat{P}$ is absorbing, and $\hat{P}$ is in any set only finitely often a.e. As before, $\alpha\hat{e}^E < \infty$, and E is therefore an equilibrium set for $\hat{P}$.

Thus E is both an equilibrium set and a dual equilibrium set. We shall choose a bounded charge with support in E. Let

$$f_i = \begin{cases} 1 & \text{if } i \in E \\ 0 & \text{otherwise.} \end{cases}$$

Then

$$\alpha f = \sum_{j>0} 3j(\tfrac{1}{2})^j = \frac{3(\frac{1}{2})}{(1 - \frac{1}{2})^2} = 6.$$

Thus f is a charge. Its potential is

$$g_i = \sum_{j>0} N_{ij} = \begin{cases} \sum_{j>0} 3j(\frac{1}{2})^{j-i} & \text{if } i \leq 0 \\ \sum_{j=1}^{i} 3j + \sum_{j>i} 3j(\frac{1}{2})^{j-i} & \text{if } i > 0. \end{cases}$$

Summing these expressions, we find that

$$g_i = \begin{cases} 6(\tfrac{1}{2})^{|i|} & \text{for } i \leq 0 \\ \tfrac{3}{2}(i^2 + 3i + 4) & \text{for } i > 0. \end{cases}$$

Thus $\lim_{i\to+\infty} g_i = +\infty$, and g is unbounded.

We note that $P^n g$ for large n is a weighted sum of g values along paths that the process is likely to take. Thus the fact that $\lim_{i\to+\infty} g_i = +\infty$ does not contradict $P^n g \to 0$ since the process moves toward $+\infty$ with probability zero. On the other hand, in the direction that the process does go, namely $-\infty$, we do have $\lim_{i\to-\infty} g_i = 0$.

8. Applications of potential-theoretic methods

Many useful quantities for transient chains arise as means of non-negative random variables: $h_i = \mathrm{M}_i[\mathbf{z}]$. To compute h we can often use a systems theorem argument (Theorem 4-11 with the random time identically one) to obtain an expression of the form $h = Ph + f$. Under appropriate circumstances we may write $(I - P)h = f$, and if h can be shown to be a potential, we conclude $h = Nf$. The purpose of this section is to give some sufficient conditions under which all these steps are valid and to apply the results.

We first restrict our attention to the case of a bounded non-negative random variable $\mathbf{z}$. Later in this section we extend our results to obtain Theorem 8-67, which is a powerful tool applicable even if the vector h is not necessarily finite-valued.

To maximize the number of potentials, we choose a row of N as α. Then all finite-valued functions of the form Nf are potentials, and a set E is an equilibrium set if and only if the process is in E only finitely often a.e. Let $\mathbf{z}$ be a bounded non-negative random variable, and let $\mathbf{z}^{(n)}(\omega) = \mathbf{z}(\omega_n)$. We shall assume that $\mathbf{z}^{(1)} \leq \mathbf{z}$. Then $\mathrm{M}_i[\mathbf{z}]$ is finite for all i, and $\mathrm{M}_i[\mathbf{z}^{(1)}] \leq \mathrm{M}_i[\mathbf{z}]$. Define

$$h_i = \mathrm{M}_i[\mathbf{z}]$$

and

$$f_i = \mathrm{M}_i[\mathbf{z} - \mathbf{z}^{(1)}] \geq 0.$$

Lemma 8-62: The column vector h is superregular and satisfies $(I - P)h = f$. Furthermore, $\mathbf{z}^{(\infty)} = \lim_{n\to\infty} \mathbf{z}^{(n)}$ exists and is finite.

PROOF: By Theorem 4-11 with the random time taken to be identically one, we have

$$\mathrm{M}_i[\mathbf{z}^{(1)}] = \sum_k P_{ik}\mathrm{M}_k[\mathbf{z}] = (Ph)_i.$$

Therefore,

$$[(I - P)h]_i = \mathrm{M}_i[\mathbf{z}] - \mathrm{M}_i[\mathbf{z}^{(1)}] = \mathrm{M}_i[\mathbf{z} - \mathbf{z}^{(1)}]$$
$$= f_i.$$

Since $f \geq 0$, h is superregular; and since $\mathbf{z}^{(1)} \leq \mathbf{z}$, we have

$$\mathbf{z} \geq \mathbf{z}^{(1)} \geq \mathbf{z}^{(2)} \geq \cdots \geq 0.$$

Therefore $\mathbf{z}^{(\infty)}$ exists and is finite.

Lemma 8-63: The function h satisfies $h = Nf$ if and only if $\mathbf{z}^{(\infty)} = 0$ almost everywhere.

PROOF: By dominated convergence,

$$\mathrm{M}_i[\mathbf{z}^{(\infty)}] = \lim \mathrm{M}_i[\mathbf{z}^{(n)}],$$

and by Theorem 4-11 with the random time n,

$$\{\mathrm{M}_i[\mathbf{z}^{(n)}]\} = P^n\{\mathrm{M}_i[\mathbf{z}]\} = P^n h.$$

Hence

$$\mathrm{M}_i[\mathbf{z}^{(\infty)}] = \lim P^n h.$$

By Theorem 5-10,

$$h_i = (Nf)_i + \mathrm{M}_i[\mathbf{z}^{(\infty)}].$$

Thus $h = Nf$ if and only if $\mathrm{M}_i[\mathbf{z}^{(\infty)}] = 0$ for every i. Since $\mathbf{z}^{(\infty)} \geq 0$, $\mathrm{M}_i[\mathbf{z}^{(\infty)}] = 0$ for all i if and only if $\mathbf{z}^{(\infty)} = 0$ a.e.

Lemma 8-64: If either of these conditions is satisfied, then $h = Nf$:

(1) There exists an equilibrium set E such that $\mathbf{z}(\omega) = 0$ for every path ω which does not go through E.
(2) The enlarged chain is absorbing and $\mathbf{z}(\omega) = 0$ for every path ω which begins in the absorbing state.

PROOF: The second condition is just the first for E, the set of all transient states. For the first condition, on every path which does not pass through E, $\mathbf{z}(\omega) = 0$, so that $\mathbf{z}^{(n)}(\omega) = 0$ for every $n \geq 1$ on such paths. On almost all paths which do pass through E, there is an n which is a function of the path and which denotes the last time the process passes through E on that path. Therefore, on almost all paths there is an n depending on the path such that $\mathbf{z}^{(n)}(\omega) = 0$. Hence $\mathbf{z}^{(\infty)} = 0$ a.e., and the result follows by Lemma 8-63.

We shall now generalize our considerations. Let $\mathbf{z}(\omega)$ be a non-negative random variable with $h_i = M_i[\mathbf{z}]$ not necessarily finite. Suppose $\mathbf{z} \geq \mathbf{z}^{(1)}$, and define ${}^m\mathbf{z}(\omega) = \min(m, \mathbf{z}(\omega))$. We agree to set $\mathbf{z} - \mathbf{z}^{(1)} = 0$ at all points where $\mathbf{z}^{(1)}$ is infinite. Then

$$\mathbf{z}(\omega) = \lim_{m \to \infty} {}^m\mathbf{z}(\omega).$$

The crucial property of the functions ${}^m\mathbf{z}$ is that

$$({}^m\mathbf{z})^{(1)} = {}^m(\mathbf{z}^{(1)}).$$

We denote the common value of $({}^m\mathbf{z})^{(1)}$ and ${}^m(\mathbf{z}^{(1)})$ by ${}^m\mathbf{z}^{(1)}$, and we define ${}^m\mathbf{z}^{(n)}$ analogously.

Lemma 8-65: If $\mathbf{z} \geq \mathbf{z}^{(1)} \geq 0$, then $\mathbf{z} - \mathbf{z}^{(1)} \geq {}^m\mathbf{z} - {}^m\mathbf{z}^{(1)} \geq 0$.

PROOF:

$${}^m\mathbf{z} - {}^m\mathbf{z}^{(1)} = \begin{cases} \mathbf{z} - \mathbf{z}^{(1)} & \text{if } \mathbf{z} \leq m \\ 0 & \text{if } \mathbf{z}^{(1)} > m \\ m - \mathbf{z}^{(1)} & \text{if } \mathbf{z} > m \geq \mathbf{z}^{(1)}. \end{cases}$$

Define vectors ${}^mh$ and ${}^mf$ by

$${}^mh_i = M_i[{}^m\mathbf{z}]$$

and

$${}^mf_i = M_i[{}^m\mathbf{z} - {}^m\mathbf{z}^{(1)}].$$

Lemma 8-66: ${}^{m+1}f \geq {}^mf$ and $\lim_m {}^mf = f$.

PROOF: We note that ${}^m({}^{m+1}\mathbf{z}) = {}^m\mathbf{z}$. Applying Lemma 8-65 to ${}^{m+1}\mathbf{z}$, we find that ${}^{m+1}\mathbf{z} - {}^{m+1}\mathbf{z}^{(1)} \geq {}^m\mathbf{z} - {}^m\mathbf{z}^{(1)}$. Then ${}^{m+1}f_i \geq {}^mf_i$. Since $\lim ({}^m\mathbf{z} - {}^m\mathbf{z}^{(1)}) = \mathbf{z} - \mathbf{z}^{(1)}$ a.e., $\lim {}^mf = f$ by the Monotone Convergence Theorem.

Theorem 8-67: Let $\mathbf{z}$ be a non-negative random variable. Define

$$\mathbf{z}^{(1)}(\omega) = \mathbf{z}(\omega_1)$$
$$\mathbf{z}(\omega) - \mathbf{z}^{(1)}(\omega) = 0 \quad \text{when } \mathbf{z}^{(1)}(\omega) = +\infty$$
$$h_i = M_i[\mathbf{z}]$$
$$f_i = M_i[\mathbf{z} - \mathbf{z}^{(1)}],$$

and suppose that $\mathbf{z} \geq \mathbf{z}^{(1)}$. If $\mathbf{z}$ satisfies either one of the following conditions, then $h = Nf$:

(1) There exists an equilibrium set E such that $\mathbf{z}(\omega) = 0$ for every path ω which does not go through E.

(2) The extended chain is absorbing and $\mathbf{z}(\omega) = 0$ for every path ω which begins in the absorbing state.

PROOF: For each m, ${}^m\mathbf{z}$ is bounded. If one of the conditions applies to $\mathbf{z}$, then it applies to ${}^m\mathbf{z}$ because $0 \leq {}^m\mathbf{z} \leq \mathbf{z}$. Hence by Lemma 8-64,

$$ {}^m h = N\, {}^m f. $$

By Lemma 8-66, ${}^m f$ increases to f, and by monotone convergence ${}^m h$ increases to h. Therefore $h = Nf$ by the Monotone Convergence Theorem.

We now apply Theorem 8-67 in four special cases.

First let P be an absorbing chain with fundamental matrix N. We define $a_i^{(r)} = \mathrm{M}_i[\mathbf{a}(\omega)^r]$, where $\mathbf{a}(\omega)$ is the absorption time defined in Chapter 5. The column vector $a^{(r)}$ is indexed by the transient states of P.

Proposition 8-68: If P is an absorbing chain, then

$$ a^{(r)} = \sum_{m=1}^{r-1} \binom{r}{m} NQa^{(m)} + N\mathbf{1}. $$

PROOF: Start the process in a transient state i, and let $\mathbf{a}$ be the absorption time. Since i is transient,

$$ \mathbf{a}^r(\omega) = (\mathbf{a} + 1)^r(\omega_1) $$

so that

$$ \mathbf{a}^r(\omega) - \mathbf{a}^r(\omega_1) = (\mathbf{a} + 1)^r(\omega_1) - \mathbf{a}^r(\omega_1) $$

or

$$ (\mathbf{a}^r - (\mathbf{a}^r)^{(1)})(\omega) = ((\mathbf{a} + 1)^r - \mathbf{a}^r)(\omega_1). $$

By Theorem 4-11 with the random time identically one,

$$ \begin{aligned} \mathrm{M}_i[\mathbf{a}^r - (\mathbf{a}^r)^{(1)}] &= \sum_k P_{ik}\mathrm{M}_k[(\mathbf{a} + 1)^r - \mathbf{a}^r] \\ &= \sum_{k\ \text{abs.}} R_{ik}\mathrm{M}_k[1] + \sum_{k\ \text{trans.}} Q_{ik}\mathrm{M}_k\left[\sum_{m=0}^{r-1} \binom{r}{m} \mathbf{a}^m\right] \\ &= \sum_{m=1}^{r-1} \binom{r}{m} \sum_k Q_{ik} a_k^{(m)} + 1 \end{aligned} $$

since $\sum R_{ik} + \sum Q_{ik} = 1$. In Theorem 8-67 let $\mathbf{z} = \mathbf{a}^r$. Then $\mathbf{z}$

satisfies $\mathbf{z} \geq \mathbf{z}^{(1)}$, and condition (2) of the theorem holds for $\mathbf{z}$. As we have just shown,

$$f = \sum_{m=1}^{r-1} \binom{r}{m} Qa^{(m)} + 1,$$

and we also have $h = a^{(r)}$. Hence $h = Nf$ by the theorem.

Corollary 8-69: Let P be an absorbing chain. Then there exist real numbers c and d such that

$$a^{(r)} \leq cNa^{(r-1)} \leq da^{(r)}.$$

In particular, $a^{(r)}$ is finite-valued if and only if $Na^{(r-1)}$ is finite-valued.

PROOF: By Proposition 8-68,

$$a^{(r)} = N \sum_{m=1}^{r-1} \binom{r}{m} Qa^{(m)} + N1,$$

even if both sides are infinite. Hence

$$\begin{aligned} a^{(r)} &\leq (2^r - 2)NQa^{(r-1)} + N1 && \text{since } a^{(m)} \leq a^{(r-1)} \\ &= (2^r - 2)(N - I)a^{(r-1)} + N1 && \text{since } N - I = NQ \\ &\leq (2^r - 2)Na^{(r-1)} + N1 \\ &\leq (2^r - 1)Na^{(r-1)} && \text{since } 1 \leq a^{(r-1)}. \end{aligned}$$

For the other inequality,

$$\begin{aligned} a^{(r)} &\geq NQa^{(r-1)} \\ &= Na^{(r-1)} - a^{(r-1)} \\ &\geq Na^{(r-1)} - a^{(r)} \quad \text{since } a^{(r-1)} \leq a^{(r)}. \end{aligned}$$

Hence $Na^{(r-1)} \leq 2a^{(r)}$.

As a second example, let P be a transient chain, and for any two transient states i and j define

$$W_{ij} = \mathrm{M}_i[\mathbf{n}_j^2].$$

The reader may verify with the aid of Theorem 4-11 that if $\mathbf{z} = \mathbf{n}_j^2$, then

$$\mathrm{M}_i[\mathbf{z} - \mathbf{z}^{(1)}] = (2N_{dg} - I)_{ij}.$$

Now $\{j\}$ is an equilibrium set, and $\mathbf{n}_j^2 = 0$ for all paths not going through this set. Hence, for fixed j, the column vector h with $h_i = W_{ij}$ satisfies condition (1) of Theorem 8-67. Therefore,

$$W = N(2N_{dg} - I).$$

We note that W is finite-valued; similarly one shows that $M_i[\mathbf{n}_j^r] < \infty$ for $r > 2$.

Next let E be an equilibrium set and let $\mathbf{z}$ be the time at which the process is in E for the last time (or 0 if E is never reached). Set $v^E = M_i[\mathbf{z}]$. Then

$$\mathbf{z} - \mathbf{z}^{(1)} = \begin{cases} 1 & \text{if } E \text{ is ever reached after time 0} \\ 0 & \text{otherwise.} \end{cases}$$

Hence $M_i[\mathbf{z} - \mathbf{z}^{(1)}] = \bar{h}_i^E$, the probability that, from i, E is reached after time 0. The random variable $\mathbf{z}$ satisfies condition (1) of the theorem. Hence

$$v^E = N\bar{h}^E.$$

Finally, let E be an equilibrium set and let j be any state. Let $\mathbf{z}_j$ be the number of times in j before E is left for the last time (or 0 if E is never reached). Define

$$N_{ij}^E = M_i[\mathbf{z}_j].$$

Then $\mathbf{z}_j$ satisfies condition (1) of the theorem, and we have

$$\mathbf{z} - \mathbf{z}^{(1)} = \begin{cases} 1 & \text{if } x_0(\omega) = j \text{ and } E \text{ is ever reached} \\ 0 & \text{otherwise.} \end{cases}$$

Hence

$$M_i[\mathbf{z} - \mathbf{z}^{(1)}] = \delta_{ij}h_j^E,$$

and

$$N_{ij}^E = N\{\delta_{ij}h_j^E\} = N_{ij}h_j^E.$$

9. General denumerable stochastic processes

We shall show that any denumerable stochastic process can be represented within a transient Markov chain in such a manner that potential theory applied to the chain yields corresponding results for the stochastic process.

Throughout this section we shall deal with a probability space Ω with measure μ, and a fixed sequence $\{\mathscr{R}_i\}$ of partitions of Ω. Each $\mathscr{R}_i$ has a denumerable number of cells, and $\mathscr{R}_i \subset \mathscr{R}_{i+1}$. For convenience, we assume that $\mathscr{R}_0 = \{\Omega\}$. We recall that $(f_n, \mathscr{R}_n)$ is a stochastic process if f_n is constant on each cell of $\mathscr{R}_n$. (This condition is Definition 2-5 expressed in terms of partitions.) If $U \in \mathscr{R}_n$, define $f_n(U)$ to be this constant value.

Definition 8-70: If $\{\mathscr{R}_i\}$ is a sequence of partitions, the **space-time Markov chain** for $\{\mathscr{R}_i\}$ is defined to be a Markov chain whose states are all ordered pairs $\langle U, n\rangle$, where $U \in \mathscr{R}_n$ and $\mu(U) > 0$, and whose transition probabilities are

$$P_{\langle U,n\rangle,\langle V,m\rangle} = \begin{cases} \dfrac{\mu(V)}{\mu(U)} & \text{if } m = n + 1 \text{ and } V \subset U \\ 0 & \text{otherwise.} \end{cases}$$

The chain is started in the state $\langle \Omega, 0\rangle$, which will be called state 0.

Proposition 8-71: The space-time chain is transient, and if $i = \langle U, n\rangle$, then

$$N_{0i} = H_{0i} = P_{0i}^{(n)} = \mu(U).$$

PROOF: State i can be entered only on the nth step, as is clear from Definition 8-70. Hence the chain is in i at most once, and $N_{0i} = H_{0i} = P_{0i}^{(n)}$. Along the path from 0 to i there is a unique sequence of cells

$$U \subset U_{n-1} \subset U_{n-2} \subset \cdots \subset U_1 \subset \Omega \quad \text{with } U_k \in \mathscr{R}_k.$$

Then

$$\begin{aligned} P_{0i}^{(n)} &= P_{0\langle U_1,1\rangle} P_{\langle U_1,1\rangle,\langle U_2,2\rangle} \cdots P_{\langle U_{n-1},n-1\rangle,\langle U,n\rangle} \\ &= \frac{\mu(U_1)}{\mu(\Omega)} \frac{\mu(U_2)}{\mu(U_1)} \cdots \frac{\mu(U)}{\mu(U_{n-1})} = \mu(U). \end{aligned}$$

For example, let Ω be a sequence space with some probability measure μ, and let $\mathscr{R}_n$ be the partition such that $\mathscr{R}_n^* = \mathscr{F}_n$. As usual, we may think of a cell of $\mathscr{F}_n$ as a path $\langle i_1, i_2, \ldots, i_n\rangle$ of length n in Ω. A state of the space-time chain may also be thought of as such a path, and the chain moves from $\langle i_1, i_2, \ldots, i_n\rangle$ to $\langle j_1, j_2, \ldots, j_{n+1}\rangle$ only if $j_k = i_k$ for $1 \leq k \leq n$. The probability of such a transition is

$$\Pr[x_{n+1} = j_{n+1} \mid x_1 = i_1 \wedge \cdots \wedge x_n = i_n].$$

The starting state 0 may be thought of as the empty sequence.

Definition 8-72: If $(f_n, \mathscr{R}_n)$ is a stochastic process, then the function f defined on the states of a space-time chain by

$$f(\langle U, n\rangle) = f_n(U)$$

is said to **correspond to** the process $(f_n, \mathscr{R}_n)$.

We write $f \sim (f_n, \mathscr{R}_n)$ when f corresponds to $(f_n, \mathscr{R}_n)$. If we identify two stochastic processes $(f_n, \mathscr{R}_n)$ and $(g_n, \mathscr{R}_n)$ when $f_n = g_n$ a.e. for every n, then the correspondence between all stochastic processes on $\{\mathscr{R}_n\}$ and all functions on the states of the space-time chain is one-one. We now restrict ourselves to the case where the f_n are real-valued functions. Under the following definition the correspondence preserves inequalities, linear combinations, and limits.

Definition 8-73: Operations on stochastic processes are defined by

(1) $(f_n, \mathscr{R}_n) \leq (g_n, \mathscr{R}_n)$ if $f_n \leq g_n$ a.e. for all n.
(2) $a(f_n, \mathscr{R}_n) + b(g_n, \mathscr{R}_n) = (af_n + bg_n, \mathscr{R}_n)$.
(3) $\lim_k (f_n{}^{(k)}, \mathscr{R}_n) = (f_n, \mathscr{R}_n)$ if $\lim f_n{}^{(k)} = f_n$ a.e. for all n.

Lemma 8-74: If $(f_n, \mathscr{R}_n) \sim f$, then

$$(\mathrm{M}[f_{n+k} \mid \mathscr{R}_n], \mathscr{R}_n) \sim P^k f$$

in the sense that if either quantity is well defined, then so is the other; and if they are both well-defined, then they correspond.

PROOF: We shall proceed by induction on k. If $k = 0$, the result is trivial. Suppose that both quantities exist for some k and that they correspond. Then

$$(P^{k+1}f)_i = \sum_j P_{ij}(P^k f)_j$$

or

$$(P^{k+1}f)_{\langle U,n\rangle} = \sum_{\substack{V \subset U \\ V \in \mathscr{R}_{n+1}}} \frac{\mu(V)}{\mu(U)} (P^k f)_{\langle V,n+1\rangle}.$$

By inductive hypothesis, $(\mathrm{M}[f_{n+k} \mid \mathscr{R}_n], \mathscr{R}_n) \sim P^k f$. Hence, by definition of the correspondence,

$$\begin{aligned}(P^k f)_{\langle V,n+1\rangle} &= \mathrm{M}[f_{n+1+k} \mid \mathscr{R}_{n+1}](V) \\ &= \frac{1}{\mu(V)} \sum_{\substack{W \subset V \\ W \in \mathscr{R}_{n+1+k}}} \mu(W) f_{n+1+k}(W),\end{aligned}$$

and

$$\begin{aligned}(P^{k+1}f)_{\langle U,n\rangle} &= \sum_{\substack{V \subset U \\ V \in \mathscr{R}_{n+1}}} \frac{\mu(V)}{\mu(U)} \sum_{\substack{W \subset V \\ W \in \mathscr{R}_{n+1+k}}} \frac{\mu(W)}{\mu(V)} f_{n+1+k}(W) \\ &= \frac{1}{\mu(U)} \sum_{\substack{W \subset U \\ W \in \mathscr{R}_{n+1+k}}} \mu(W) f_{n+1+k}(W) \\ &= \mathrm{M}[f_{n+1+k} \mid \mathscr{R}_n](U).\end{aligned}$$

That is, $P^{k+1}f$ exists if and only if $\mathrm{M}[f_{n+(k+1)} \mid \mathscr{R}_n]$ does, and if they exist, then they correspond.

Proposition 8-75: If $(h_n, \mathscr{R}_n) \sim h$, then $(h_n, \mathscr{R}_n)$ is a supermartingale (martingale) if and only if h is superregular (regular) and $P^k h$ is finite-valued for all k.

PROOF: If h is superregular, then $Ph \leq h$. Hence, by Lemma 8-74, $\mathrm{M}[h_{n+1} \mid \mathscr{R}_n] \leq h_n$ a.e. for all n. If $(P^k h)$ is finite-valued, then

$$(P^k h)_0 = \sum_{U \in \mathscr{R}_k} \mu(U) h_k(U) = \mathrm{M}[h_k]$$

is finite. Since h_n is constant on cells of $\mathscr{R}_n$, Definition 3-5 is satisfied.

Conversely, if $(h_n, \mathscr{R}_n)$ is a supermartingale, then $\mathrm{M}[h_{n+1} \mid \mathscr{R}_n] \leq h_n$, and hence $Ph \leq h$ by Lemma 8-74. Moreover, $\mathrm{M}[h_{k+n}] = (P^{k+n}h)_0$ is finite. If $i = \langle U, n \rangle$, then $P_{0i}^{(n)} = \mu(U) > 0$. From

$$(P^{k+n}h)_0 = \sum_j P_{0j}^{(n)} (P^k h)_j,$$

we see that $(P^k h)_i$ must be finite. The proof for martingales simply replaces $Ph \leq h$ by $Ph = h$.

Definition 8-76: $(f_n, \mathscr{R}_n)$ is a **stochastic process charge** with **potential** $(g_n, \mathscr{R}_n)$ if $\sum_n \mathrm{M}[|f_n|] < \infty$ and if $(g_n, \mathscr{R}_n) = \sum_k (\mathrm{M}[f_{n+k} \mid \mathscr{R}_n], \mathscr{R}_n)$. If, in addition, $f_n \geq 0$ then the potential is called a **pure potential.**

We shall make use of potential theory results for the space-time chain P. As the distinguished measure α, we select $\alpha_j = N_{0j} > 0$. As usual, if αf is finite, then f is a charge.

Proposition 8-77: Charge functions correspond to charge stochastic processes, and their potentials also correspond.

PROOF: If $(f_n, \mathscr{R}_n) \sim f$, then by Proposition 8-71

$$\begin{aligned}
\alpha|f| &= \sum_{\langle U,n \rangle} \mu(U) \cdot \mathrm{M}[|f_n| \mid \mathscr{R}_n](U) \\
&= \sum_n \sum_{U \in \mathscr{R}_n} \mu(U) \cdot |f_n(U)| \\
&= \sum_n \mathrm{M}[|f_n|].
\end{aligned}$$

Since sums and limits are preserved, we have by Proposition 8-74

$$g = Nf = \sum_k P^k f \sim \sum_k (\mathrm{M}[f_{n+k} \mid \mathscr{R}_n], \mathscr{R}_n).$$

Thus g is the potential of f if and only if the two conditions of Definition 8-76 are fulfilled by $(f_n, \mathscr{R}_n)$ and $(g_n, \mathscr{R}_n)$.

We give two applications of the correspondence. The first is a decomposition of non-negative supermartingales.

Proposition 8-78: If $(h_n, \mathscr{R}_n)$ is a non-negative supermartingale, then there is a unique representation

$$(h_n, \mathscr{R}_n) = (r_n, \mathscr{R}_n) + (g_n, \mathscr{R}_n)$$

of $(h_n, \mathscr{R}_n)$ as the sum of a martingale and a potential. In the representation the martingale is a non-negative martingale, the potential is a pure potential, and r_n satisfies $r_n = \lim_k \mathrm{M}[h_{n+k} \mid \mathscr{R}_n]$. Moreover, $(g_n, \mathscr{R}_n)$ is the difference between a martingale and a process consisting of an increasing sequence of random variables.

PROOF: Existence and uniqueness of the representation follows immediately from Theorem 5-10 and Propositions 8-75 and 8-77. Then $r_n = \lim_k \mathrm{M}[h_{n+k} \mid \mathscr{R}_n]$ by Lemma 8-74. For the last part, let $(f_n, \mathscr{R}_n)$ be the charge of $(g_n, \mathscr{R}_n)$. Set

$$s_n = f_0 + \cdots + f_{n-1}$$

and

$$s = \lim s_n.$$

Then s_n increases monotonically to s, and

$$\mathrm{M}[s] \le \sum_n \mathrm{M}[f_n \mid \mathscr{R}_n] < \infty$$

by monotone convergence. Since $f_n \ge 0$,

$$\begin{aligned} g_n &= \sum_k \mathrm{M}[f_{n+k} \mid \mathscr{R}_n] = \mathrm{M}\Big[\sum_k f_{n+k} \mid \mathscr{R}_n\Big] \\ &= \mathrm{M}[s - s_n \mid \mathscr{R}_n] = \mathrm{M}[s \mid \mathscr{R}_n] - s_n. \end{aligned}$$

Hence $\{g_n\}$ is the difference between the martingale $\{\mathrm{M}[s \mid \mathscr{R}_n]\}$ and the increasing sequence s_n.

As the second application, we give a proof of the Upcrossing Lemma, Proposition 3-11, as it applies to non-negative supermartingales. The present estimate is better than the one in Chapter 3.

Proposition 8-79: Let r and s be real numbers with $0 \le r < s$. Let $\boldsymbol{\beta}(\omega)$ be the number of upcrossings on ω of $[r, s]$ by the non-negative supermartingale $(\mathbf{f}_k, \mathscr{R}_k)$ up to time n. Then

$$\mathrm{M}[\boldsymbol{\beta}] \le \frac{r}{s - r}.$$

PROOF: Let $f \sim (\mathbf{f}_k, \mathscr{R}_n)$; f is non-negative superregular. Let E and F be the sets of states in the space-time chain defined by

$$E = \{\langle U, m\rangle \mid m \leq n \quad \text{and} \quad \mathbf{f}_m(U) \leq r\}$$
$$F = \{\langle U, m\rangle \mid m \leq n \quad \text{and} \quad \mathbf{f}_m(U) \geq s\}.$$

Hence $f_i \leq r$ for $i \in E$ and $f_j \geq s$ for $j \in F$. For any other state, $i = \langle U, m\rangle$ with $m \leq n$, $r < f_i < s$. Now $h_0^F = (B^F 1)_0$ is the probability that the random variables $\mathbf{f}_0, \ldots, \mathbf{f}_n$ ever take on a value greater than or equal to s. Similarly, $(B^E B^F 1)_0$ is the probability of at least one upcrossing, and in general $[(B^E B^F)^k 1]_0$ is the probability of at least k upcrossings by $\mathbf{f}_0, \mathbf{f}_1, \ldots, \mathbf{f}_n$. Hence

$$\mathrm{M}[\boldsymbol{\beta}] = \sum_{k=1}^{\infty} \Pr[\boldsymbol{\beta} \geq k] = \sum_{k=1}^{\infty} [(B^E B^F)^k 1]_0.$$

Since the chain cannot be in F after time n, F is an equilibrium set, and $h^F = B^F 1$ is a potential. For $i \in F$

$$(B^F 1)_i = 1 \leq \frac{1}{s} f_i.$$

By the Principle of Domination (Theorem 8-45),

$$(B^F 1) \leq \frac{1}{s} f$$

everywhere. Hence $B^E B^F 1 \leq (1/s) B^E f$. For $i \in E$,

$$(B^E f)_i \leq r.$$

Since f is superregular and since $r1$ is superregular,

$$B^E f \leq r1$$

everywhere by conclusion (2) of Proposition 8-16. Thus

$$B^E B^F 1 \leq \frac{r}{s} 1.$$

By induction,

$$(B^E B^F)^k 1 \leq \left(\frac{r}{s}\right)^k 1,$$

and hence

$$\mathrm{M}[\boldsymbol{\beta}] \leq \sum_{k=1}^{\infty} \left(\frac{r}{s}\right)^k = \frac{r}{s - r}.$$

10. Problems

1. If P is a transient chain whose states communicate and if $\alpha > 0$ is superregular, show that
   (a) $\hat{P}$ stops (disappears) a.e. if and only if α is a potential measure.
   (b) The mean stopping time is finite if and only if α is a charge. [*Hint:* Adjoin an absorbing state to $\hat{P}$.]

2. Let P be a recurrent chain, and let E be a finite set. Show that for any specified values of h_E there is a unique bounded function h with the specified values which is regular on $\tilde{E}$. Show that h takes on its maximum on E.

3. Illustrate the result of the previous problem for the symmetric random walk on the integers with the specified values $h_0 = 0$ and $h_1 = 1$.

4. Let P have only transient states, let π be a specified probability vector, and let $\alpha = \pi N$. Let $f \geq 0$ be a charge, and define the random variable

$$\mathbf{s} = \sum_{n=0}^{\infty} f(x_n).$$

   Show that $\mathbf{s}$ is finite a.e. and that $\mathrm{M}_\pi[\mathbf{s}] = \alpha f$. What is the value of $\mathrm{M}_i[\mathbf{s}]$? Give a game-interpretation.

5. In the framework of Problem 4, introduce a second charge $\bar{f}$, its potential $\bar{g}$, and $\bar{\mathbf{s}}$. Prove that if $\mu = \text{dual}\, f$, then

$$(g, \bar{g}) = \tfrac{1}{2}\mathrm{M}_\pi[\mathbf{s}\bar{\mathbf{s}}] + \tfrac{1}{2}\mu\bar{f}.$$

   Find the corresponding expression for $\mathbf{I}(g)$.

6. If P is a chain with only transient states, let Var_{ij} be the variance of $\mathbf{n}_j$ for the process started at i. Show that

$$\mathrm{Var}_{ij} = N_{ij}(2N_{jj} - N_{ij} - 1).$$

7. In the framework of Problem 6, if $P\mathbf{1} = \mathbf{1}$, prove that for each j there is an i such that $\mathrm{Var}_{ij} \geq N_{ij}$.

Problems 8 to 19 refer to the following Markov chain: The states are the non-negative integers. From state i either the process moves one step to the right with probability $p_i > 0$, or it remains at i.

8. Find H and N.

9. Give a *simple* characterization of
   (a) the regular functions,
   (b) the non-negative superregular functions,
   (c) the pure potentials, where $\alpha_j = N_{0j}$,
   (d) the potentials, where $\alpha_j = N_{0j}$.

10. What does Theorem 5-10 say about this chain?

11. If g is a potential with charge f, give a simple characterization in terms of g of the support of f. Of the support of f^+.

12. Use Problems 9 and 11 to verify that Theorem 8-45 holds for this chain. [*Hint:* Distinguish the cases where the support of f^+ is finite and where it is infinite.]

13. Use Problems 9 and 11 to construct a counterexample to Theorem 8-45 if the assumption $h \geq 0$ is omitted.
14. If $\alpha_j = N_{0j}$, form $\hat{P}$ and compute $\hat{N}$.
15. For $E = \{0, 1, 2\}$, find $\bar{h}^E$. [See the end of Section 8.] Show that $v^E = N\bar{h}^E$ has the desired interpretation. [Remember that the chain starts at time 0, not at time 1.]
16. Show that $C(E) = 1$ for all equilibrium sets. Show that both the hypothesis and the conclusion of Proposition 8-39 are false for P.
17. Show that $\hat{C}(E) = 1$ for all sets E. Show that both the hypothesis and the conclusion of Proposition 8-39 hold for $\hat{P}$.
18. Let $E = \{0, 1, \ldots, n\}$. Find B^E. If g is a potential, what is the form of its balayage potential on E? Show that Theorem 8-46 is satisfied.
19. Show that if $\mu =$ dual f, then

$$\mathbf{I}(g) = \frac{1}{2}\Big(\sum_i \mu_i\Big)^2 + \frac{1}{2}\sum_i \mu_i^2.$$

Problems 20 to 30 refer to sums of independent random variables on the integers with $p_{-1} = \frac{1}{3}$ and $p_1 = \frac{2}{3}$. Use the results of Problems 12 to 19 in Chapter 5.

20. Show that there are two essentially different positive regular measures and that all regular measures are linear combinations of the two basic measures.
21. Show that if $f \geq 0$ and $\alpha f < +\infty$ for either α, then $g = Nf$ is finite-valued. Show also that $\lim P^n g = 0$.
22. Let $E = \{0, 1, \ldots, n\}$. Compute B^E. Choose a non-negative super-regular function h (not a constant), and verify the various parts of Proposition 8-16.
23. For E as above, compute e^E, and verify that $Ne^E = h^E$.
24. For E as above, compute $C(E) = \alpha e^E$ for each of the two basic measures. What happens as n increases?
25. Form $\hat{P}$ and compute $\hat{N}$ for each of the two basic measures. In each case, does $\hat{N}$ have columns tending to 0?
26. Use the results of the last two problems to show that each assignment of capacities is consistent with Proposition 8-39, even though $\lim_n C(E)$ is finite in one case and not in the other.
27. Show that there are infinite equilibrium sets for this chain.
28. Choose $\alpha = 1^T$. Prove that if $\lim_{n \to +\infty} h_n = 0$ and $\sum |h_i - h_{i-1}| < \infty$, then h is a potential.
29. Choose a function h satisfying the conditions of Problem 28, compute its charge f, and check that $h = Nf$.
30. Let $E = \{0, 1, 2\}$. Compute P^E and the fundamental matrix of this finite chain. Verify that the latter is N_E.

CHAPTER 9

RECURRENT POTENTIAL THEORY

1. Potentials

Throughout this chapter P is a recurrent chain which is either null or noncyclic ergodic. For such a chain, $\lim_n P^n$ always exists; we let $L = \lim P^n$.

In a recurrent chain the non-negative finite-valued superregular measures are uniquely determined up to multiplication by a constant, and the non-zero ones are positive and regular. We choose one such non-zero regular measure and call it α.

If P is noncyclic ergodic, then $L_{ij} = \alpha_j / \sum_k \alpha_k$; whereas if P is null, then $L_{ij} = 0$. In either case, $L_{ik} - L_{ij}\alpha_k/\alpha_j = 0$.

Duality for P is defined with respect to the regular measure α. The dual $\hat{P}$ of a null chain is null, and the dual of a noncyclic ergodic chain is noncyclic ergodic. In general, if two results are duals, we shall prove only one of the pair. As usual, the key to the proof by duality of the second result is that $\hat{P}$ is the most general chain of the type we consider in this chapter.

As Definition 9-1 suggests, we define charges and potentials in the same way as in transient potential theory.

Definition 9-1: If μ is a signed measure with $\mu 1$ finite and if

$$\nu = \lim_n \left[\mu(I + P + \cdots + P^{n-1})\right]$$

exists and is finite-valued, then μ is called a **left charge** with **potential measure** ν and **total charge** $\mu 1$. If f is a function with αf finite and if $g = \lim_n \left[(I + P + \cdots + P^{n-1})f\right]$ exists and is finite-valued, then f is called a **right charge** with **potential function** g and **total charge** αf. The **support** of a charge is the set on which the charge is not zero; the support of a potential is the support of its charge.

The definition of the support of a potential is not justified until we prove a uniqueness theorem for the charge of a potential, but such a result will follow directly from conclusion (2) of Theorem 9-15.

We note that the dual of a left (right) charge for P is a right (left) charge for $\hat{P}$ and that their total charges are the same (since the dual of a number is the same number).

Although we adopt the same definitions as with transient chains, the results are sometimes significantly different. For example, the only pure potential is the zero potential: If $f \geq 0$, then by monotone convergence $\lim_n [(I + P + \cdots + P^{n-1})f]_i = \sum_j N_{ij} f_j$, where $N_{ij} = +\infty$ for every i and j. Thus the limit is finite-valued only if $f = 0$.

On the other hand, every row (or column) of $I - P$ is a charge, and the potential of the ith row (column) of $I - P$ is the ith row (column) of $I - L$. For if μ is the ith row of $I - P$, then $|\mu|1 \leq 2 < \infty$ and

$$\begin{aligned} \nu_j &= \lim_n [\mu(I + P + \cdots + P^{n-1})]_j = \lim_n (I - P^n)_{ij} \\ &= (I - L)_{ij}; \end{aligned}$$

the assertion for columns is dual.

Our first potential theory result will be that every charge has total charge zero. To prove this fact, we require the Doeblin Ratio Limit Theorem, Theorem 9-4. We recall that $H_{ij}^{(n)}$ is the probability starting in i of reaching j before or at time n and that $N_{ij}^{(n)}$ is the mean number of times the process started at i is in j up to and including time n. Hence $H_{ij}^{(n)} = \sum_{k=0}^{n} F_{ij}^{(k)}$ and $N_{ij}^{(n)} = \sum_{k=0}^{n} (P^k)_{ij}$; from the latter relation we see that $\hat{N}_{ij}^{(n)} = (\alpha_j/\alpha_i) N_{ji}^{(n)}$. In terms of this notation the Doeblin Ratio Limit Theorem states that in any Markov chain with a positive superregular measure

$$\lim_n \frac{N_{ij}^{(n)}}{N_{i'j'}^{(n)}}$$

exists and is finite for any states i, j, i', and j' which communicate. We shall give a simple proof of this important result.

Lemma 9-2: Let P be any Markov chain with a positive superregular measure α. If i and j communicate, then the quantities

$$N_{jj}^{(n)} - N_{ij}^{(n)}$$

and

$$N_{ii}^{(n)} \frac{\alpha_j}{\alpha_i} - N_{ij}^{(n)}$$

are non-negative and bounded. In particular, $|N_{jj}^{(n)} - N_{ij}^{(n)}| \leq {}^iN_{jj}$.

Proof: If i and j communicate, then ${}^i\bar{H}_{jj} < 1$, so that ${}^iN_{jj} = 1/(1 - {}^i\bar{H}_{jj}) < \infty$. It is clear that

$$N_{jj}^{(n)} - N_{ij}^{(n)} \geq 0$$

and that

$$N_{jj}^{(n)} \leq {}^iN_{jj} + N_{ij}^{(n)},$$

and hence the first expression is non-negative and bounded by ${}^iN_{jj} < \infty$. By duality, $N_{jj}^{(n)} - N_{ji}^{(n)}(\alpha_j/\alpha_i)$ is non-negative and bounded; if we multiply by α_i/α_j and interchange j and i, we obtain the second result.

Lemma 9-3: Let P be any Markov chain with a positive superregular measure α. If i and j communicate, then for all n

$$\frac{N_{ij}^{(n)}}{N_{jj}^{(n)}} \leq 1 \quad \text{and} \quad \frac{N_{ij}^{(n)}}{N_{ii}^{(n)}} \leq \frac{\alpha_j}{\alpha_i},$$

and

$$\lim_n \frac{N_{ij}^{(n)}}{N_{jj}^{(n)}} = H_{ij} \quad \text{and} \quad \lim_n \frac{N_{ij}^{(n)}}{N_{ii}^{(n)}} = \frac{\alpha_j}{\alpha_i}\hat{H}_{ji} = {}^i\bar{N}_{ij}.$$

Proof: By Lemma 9-2,

$$0 \leq N_{jj}^{(n)} - N_{ij}^{(n)} \leq c \quad \text{for all } n.$$

Hence

$$0 \leq 1 - \frac{N_{ij}^{(n)}}{N_{jj}^{(n)}} \leq \frac{c}{N_{jj}^{(n)}}.$$

Therefore $N_{ij}^{(n)}/N_{jj}^{(n)} \leq 1$ for all n. If j is recurrent, then $N_{jj}^{(n)} \to +\infty$, and the ratio $N_{ij}^{(n)}/N_{jj}^{(n)}$ must tend to 1; we have $H_{ij} = 1$ since i must be recurrent if i and j communicate. Hence

$$\frac{N_{ij}^{(n)}}{N_{jj}^{(n)}} \to H_{ij}.$$

If j is transient, then

$$\frac{N_{ij}^{(n)}}{N_{jj}^{(n)}} \to \frac{N_{ij}}{N_{jj}} = \frac{H_{ij}N_{jj}}{N_{jj}} = H_{ij}.$$

The other results are duals, and the assertion about ${}^i\bar{N}_{ij}$ follows from Corollary 6-20.

The following is the Ratio Limit Theorem.

Theorem 9-4: Let P be any Markov chain with a positive superregular measure α, and let i, j, i', and j' be any states which communicate. Then

$$\lim_n \frac{N_{ij}^{(n)}}{N_{i'j'}^{(n)}}$$

exists. If all four states are recurrent, the limit is $\alpha_j/\alpha_{j'}$. If the states are transient, the limit is

$$\frac{H_{ij}\hat{H}_{jj'}\alpha_j}{H_{i'j'}H_{j'j}\alpha_{j'}}.$$

REMARK: Since the states in question communicate, they must be either all recurrent or all transient.

PROOF: Write

$$\frac{N_{ij}^{(n)}}{N_{i'j'}^{(n)}} = \left[\frac{N_{i'j'}^{(n)}}{N_{j'j'}^{(n)}}\right]^{-1}\left[\frac{N_{j'j}^{(n)}}{N_{j'j'}^{(n)}}\right]\left[\frac{N_{j'j}^{(n)}}{N_{jj}^{(n)}}\right]^{-1}\left[\frac{N_{ij}^{(n)}}{N_{jj}^{(n)}}\right],$$

and apply Lemma 9-3 to each factor.

Proposition 9-5: Every charge has total charge zero.

PROOF: If μ is a left charge, then $\sum |\mu_k| < \infty$ and

$$\nu_j = \lim_n \sum_k \mu_k N_{kj}^{(n)}$$

is finite. Therefore

$$\sum_k \mu_k \left(\frac{N_{kj}^{(n)}}{N_{jj}^{(n)}}\right) \to 0.$$

Since by Lemma 9-3, $N_{kj}^{(n)}/N_{jj}^{(n)} \leq 1$, dominated convergence gives

$$0 = \lim \sum_k \mu_k \left(\frac{N_{kj}^{(n)}}{N_{jj}^{(n)}}\right) = \sum_k \mu_k \lim \left(\frac{N_{kj}^{(n)}}{N_{jj}^{(n)}}\right) = \sum \mu_k = \mu 1.$$

The result for functions is dual.

The condition that a function f satisfy $\alpha f = 0$ is a strong necessary condition for it to be a charge, but it is by no means sufficient. In fact, it is not even sufficient in general if f also has finite support. We shall return to discuss this point at length in Section 2.

We now establish as Theorem 9-7 an identity which will play a fundamental role when we develop an operator which transforms charges into potentials.

Lemma 9-6: Let $\{a_n\}$ and $\{b_n\}$ be two sequences of real numbers such that $a_n \geq 0$, $\sum a_n = a < \infty$, $|b_n| < B$, and $|b_n - b_{n-1}| \to 0$. Then

$$\lim_{n\to\infty} \sum_{k=0}^{n} a_k(b_n - b_{n-k}) = 0.$$

PROOF: Let $\epsilon > 0$ be given. Choose N sufficiently large that $\sum_{k>N} a_k < \epsilon/(4B)$ and pick N' large enough so that for all $n \geq N'$

$$|b_n - b_{n-1}| \leq \frac{\epsilon}{2aN}.$$

Then for $n \geq N + N'$, we have

$$\begin{aligned}\left|\sum_{k=0}^{n} a_k(b_n - b_{n-k})\right| &\leq \sum_{k=0}^{N} a_k|b_n - b_{n-k}| + \sum_{k=N+1}^{n} a_k(|b_n| + |b_{n-k}|) \\ &\leq \sum_{k=0}^{N} a_k \sum_{j=0}^{k-1} |b_{n-j} - b_{n-j-1}| + 2B \sum_{k=N+1}^{n} a_k \\ &\leq \sum_{k=0}^{N} a_k \sum_{j=0}^{k-1} \frac{\epsilon}{2aN} + \frac{\epsilon}{4B}\cdot 2B, \quad \text{since } n - j \geq N' \\ &\leq \frac{\epsilon}{2a}\sum_{k=0}^{N} a_k + \frac{\epsilon}{2} \\ &\leq \epsilon.\end{aligned}$$

Theorem 9-7: Let i, j, and k be arbitrary states in a recurrent Markov chain which is either null or noncyclic ergodic. Then

$$\lim_{n\to\infty} [(N_{kk}^{(n)} - N_{ik}^{(n)})\alpha_j/\alpha_k + N_{ij}^{(n)} - N_{kj}^{(n)}] = {}^kN_{ij}.$$

PROOF: We may assume that neither i nor j equals k, since otherwise both sides are clearly zero. We begin by establishing four equations:

$$(1) \qquad N_{kk}^{(n)} = \sum_{\nu=0}^{\infty} F_{ik}^{(\nu)} N_{kk}^{(n)}$$

$$(2) \qquad N_{ik}^{(n)} = \sum_{\nu=0}^{n} F_{ik}^{(\nu)} N_{kk}^{(n-\nu)}$$

$$(3) \qquad N_{kj}^{(n)} = \sum_{\nu=0}^{\infty} F_{ik}^{(\nu)} N_{kj}^{(n)}$$

$$(4) \qquad N_{ij}^{(n)} = \sum_{\nu=0}^{n} F_{ik}^{(\nu)} N_{kj}^{(n-\nu)} + {}^kN_{ij}^{(n)}.$$

Equations (1) and (3) follow from the fact that $\sum F_{ik}^{(\nu)} = H_{ik} = 1$. Equation (2) comes from Theorem 4-11 with the random time $\mathbf{t} = \min(\mathbf{t}_k, n)$, and equation (4) is a similar result, except that the sum has been broken into two parts representing what happens after and before state k is reached for the first time.

Multiply (1) by α_j/α_k, (2) by $-\alpha_j/\alpha_k$, (3) by -1, and (4) by 1, and add. We obtain

$$(N_{kk}^{(n)} - N_{ik}^{(n)})\alpha_j/\alpha_k + (N_{ij}^{(n)} - N_{kj}^{(n)})$$
$$= {}^kN_{ij}^{(n)} + \sum_{\nu=0}^{n} F_{ik}^{(\nu)}(b_n - b_{n-\nu}) + \sum_{\nu=n+1}^{\infty} F_{ik}^{(\nu)}b_n,$$

where $b_n = N_{kk}^{(n)}\alpha_j/\alpha_k - N_{kj}^{(n)}$, and $\{b_n\}$ is a bounded sequence by Lemma 9-2. The first term ${}^kN_{ij}^{(n)}$ on the right side tends to ${}^kN_{ij}$, and the third term tends to zero since $\{b_n\}$ is bounded and $\sum F_{ik}^{(\nu)}$ is finite. It is thus sufficient to show that

$$\lim_n \sum_{\nu=0}^{n} a_\nu(b_n - b_{n-\nu}) = 0,$$

where $a_\nu = F_{ik}^{(\nu)}$. Since $a_n \geq 0$, $\sum a_n = 1$, and $\{b_n\}$ is bounded, we need show only that $(b_n - b_{n-1}) \to 0$ to apply Lemma 9-6. But

$$\begin{aligned} b_n - b_{n-1} &= (N_{kk}^{(n)} - N_{kk}^{(n-1)})\frac{\alpha_j}{\alpha_k} - (N_{kj}^{(n)} - N_{kj}^{(n-1)}) \\ &= P_{kk}^{(n)}\frac{\alpha_j}{\alpha_k} - P_{kj}^{(n)} \\ &\to L_{kk}\frac{\alpha_j}{\alpha_k} - L_{kj} \\ &= 0. \end{aligned}$$

If in the recurrent chain P we make a set of states E absorbing, then ${}^EP$ is an absorbing chain and results about transient chains may be applied to it. For example, the result $N_{ij} = H_{ij}N_{jj}$ yields ${}^EN_{ij} = {}^EH_{ij}\,{}^EN_{jj}$. We shall also make frequent use of the fact noted in Section 6-2 that ${}^E\hat{N}_{ij} = (\alpha_j/\alpha_i)\,{}^EN_{ji}$.

At this point we begin developing the machinery needed for the main result of this section, Theorem 9-15. We first need two preliminary identities, which we establish as Propositions 9-12 and 9-13.

Lemma 9-8: For any pair of states i and j,

$$\frac{{}^jN_{ii}}{\alpha_i} = \frac{{}^iN_{jj}}{\alpha_j}.$$

PROOF 1: If $i = j$, both sides are zero. If $i \neq j$, then from $\bar{H}_{jj} = 1$ and (3) of Lemma 4-19, we find

$${}^i\bar{H}_{jj} + {}^j\bar{H}_{ji}H_{ij} = \bar{H}_{jj} = 1,$$

so that

$$1 - {}^i\bar{H}_{jj} = {}^j\bar{H}_{ji}.$$

Hence

$$ {}^iN_{jj} = 1/(1 - {}^i\bar{H}_{jj}) = 1/{}^j\bar{H}_{ji}. $$

Therefore, by Proposition 5-4 and Corollary 6-21,

$$ \frac{{}^jN_{ii}}{{}^iN_{jj}} = {}^j\bar{H}_{ji}\,{}^jN_{ii} = {}^j\bar{N}_{ji} = \frac{\alpha_i}{\alpha_j}. $$

PROOF 2: Set $i = j$ in Theorem 9-7. Then

$$ \lim\,[(N_{kk}^{(n)} - N_{ik}^{(n)})\alpha_i + (N_{ii}^{(n)} - N_{ki}^{(n)})\alpha_k] = \alpha_k\,{}^kN_{ii}. $$

Interchange i and k, and the left side stays the same. The right side becomes $\alpha_i\,{}^iN_{kk}$.

Lemma 9-9: For any states i, j, k with $i \neq j$,

$$ {}^jN_{ki} + {}^iN_{kj}\,\frac{\alpha_i}{\alpha_j} = {}^jN_{ii}. $$

PROOF:

$$ \begin{aligned} {}^jN_{ki} + {}^iN_{kj}\,\frac{\alpha_i}{\alpha_j} &= {}^jH_{ki}\,{}^jN_{ii} + {}^iH_{kj}\,{}^iN_{jj}\,\frac{\alpha_i}{\alpha_j} \\ &= ({}^jH_{ki} + {}^iH_{kj})\,{}^jN_{ii} \quad \text{by Lemma 9-8} \\ &= {}^jN_{ii}. \end{aligned} $$

Lemma 9-10:

$$ \left| N_{ki}^{(n)} - N_{k0}^{(n)}\,\frac{\alpha_i}{\alpha_0} \right| \leq {}^0N_{ii}. $$

PROOF: If $i = 0$, both sides are zero. Otherwise we have

$$ \hat{N}_{ik}^{(n)} \leq {}^0\hat{N}_{ik} + \hat{N}_{0k}^{(n)}. $$

If we multiply through by α_i/α_k, we obtain

$$ N_{ki}^{(n)} \leq {}^0N_{ki} + \frac{\alpha_i}{\alpha_0}\left(\frac{\alpha_0}{\alpha_k}\,\hat{N}_{0k}^{(n)}\right) = {}^0N_{ki} + \frac{\alpha_i}{\alpha_0}\,N_{k0}^{(n)}. $$

Hence

$$ N_{ki}^{(n)} - \frac{\alpha_i}{\alpha_0}\,N_{k0}^{(n)} \leq {}^0N_{ki}. $$

Interchanging i and 0 and multiplying by α_i/α_0 gives

$$ \frac{\alpha_i}{\alpha_0}\,N_{k0}^{(n)} - N_{ki}^{(n)} \leq {}^iN_{k0}\,\frac{\alpha_i}{\alpha_0}. $$

Therefore,

$$\left| N_{ki}^{(n)} - \frac{\alpha_i}{\alpha_0} N_{k0}^{(n)} \right| \leq {}^0N_{ki} + {}^iN_{k0} \frac{\alpha_i}{\alpha_0} = {}^0N_{ii} \quad \text{by Lemma 9-9.}$$

As usual, we agree that ${}^EN$, and in particular ${}^0N$, is a square matrix indexed by the set of all states; the entries of ${}^EN$ on the rows or columns indexed by E are all zeros.

Lemma 9-11: If $\mu 1$ is finite, then $\mu\, {}^0N$ is finite-valued. Dually, if αf is finite, then ${}^0Nf$ is finite-valued.

PROOF: If $\mu 1$ is finite, then

$$\sum_i |\mu_i|\, {}^0N_{ij} = \sum_i |\mu_i|\, {}^0H_{ij}\, {}^0N_{jj} \leq {}^0N_{jj} \sum_i |\mu_i| < \infty.$$

Proposition 9-12: If $\mu 1 = 0$, then

$$\lim_{n \to \infty} \sum_k \mu_k \left[N_{ki}^{(n)} - \frac{\alpha_i}{\alpha_0} N_{k0}^{(n)} \right] = \sum_k \mu_k\, {}^0N_{ki}.$$

Dually, if $\alpha f = 0$, then

$$\lim_{n \to \infty} \sum_k [N_{ik}^{(n)} - N_{0k}^{(n)}] f_k = \sum_k {}^0N_{ik} f_k.$$

PROOF: Let

$$S_k^{(n)} = \left[(N_{00}^{(n)} - N_{k0}^{(n)}) \frac{\alpha_i}{\alpha_0} + (N_{ki}^{(n)} - N_{0i}^{(n)}) \right].$$

Since $\mu 1 = 0$, we have

$$\sum_k \mu_k \left[N_{ki}^{(n)} - \frac{\alpha_i}{\alpha_0} N_{k0}^{(n)} \right] = \sum_k \mu_k S_k^{(n)}.$$

By Theorem 9-7, $\lim_n S_k^{(n)} = {}^0N_{ki}$. Hence if we can prove that $S_k^{(n)}$ is bounded independently of n and k, then, since $\mu 1$ is finite, the result of the proposition follows by dominated convergence. (Note that $\sum_k \mu_k\, {}^0N_{ki}$ is absolutely convergent by Lemma 9-11.) We have

$$|S_k^{(n)}| \leq \left| N_{00}^{(n)} \frac{\alpha_i}{\alpha_0} - N_{0i}^{(n)} \right| + \left| N_{ki}^{(n)} - N_{k0}^{(n)} \frac{\alpha_i}{\alpha_0} \right|.$$

The first term on the right is bounded according to Lemma 9-2, and the second term is bounded according to Lemma 9-10.

Proposition 9-13: If $\alpha f = 0$, then f is a charge if and only if

$$\lim_n [P^n({}^0Nf)]$$

exists and is finite-valued. If f is a charge, then its potential g satisfies

$$g = {}^0Nf - \lim_n [P^n({}^0Nf)].$$

PROOF: We have

$$(P\,{}^0N)_{ij} = \sum_k P_{ik}\,{}^0N_{kj} = \begin{cases} 0 & \text{if } j = 0 \text{ since } {}^0N_{k0} = 0 \\ {}^0\bar{N}_{0j} = \dfrac{\alpha_j}{\alpha_0} & \text{if } i = 0 \text{ and } j \neq 0 \\ {}^0N_{ij} - \delta_{ij} & \text{if } i \neq 0 \text{ and } j \neq 0 \end{cases}$$

$$((I - P)\,{}^0N)_{ij} = \begin{cases} 0 & \text{if } j = 0 \\ -\alpha_j/\alpha_0 & \text{if } i = 0 \text{ and } j \neq 0 \\ \delta_{ij} & \text{if } i \neq 0 \text{ and } j \neq 0. \end{cases}$$

By Lemma 9-11, ${}^0Nf$ is finite-valued, and hence associativity holds in the triple product $(I - P)\,{}^0Nf$. We find

$$((I - P)\,{}^0Nf)_i = \begin{cases} \sum_{j \neq 0} \delta_{ij} f_j = f_i & \text{for } i \neq 0 \\ \sum_{j \neq 0} \left(-\dfrac{\alpha_j}{\alpha_0}\right) f_j = \dfrac{1}{\alpha_0}(\alpha_0 f_0) = f_0 & \text{for } i = 0. \end{cases}$$

The next to last equality uses the fact that $\alpha f = 0$. We see therefore that $(I - P)\,{}^0Nf = f$, so that

$$\begin{aligned} \lim_{n \to \infty} (I + P + \cdots + P^{n-1})f &= \lim_{n \to \infty} (I + P + \cdots + P^{n-1})[(I - P)\,{}^0Nf] \\ &= \lim_{n \to \infty} (I - P^n)\,{}^0Nf. \end{aligned}$$

Associativity where required in the last equation follows from the distributive property, which holds because ${}^0Nf$ is finite-valued.

In the discussion that follows, E and F denote non-empty subsets of states.

Lemma 9-14: $B^E\,{}^FN = {}^FN - {}^EN$, provided $F \subset E$.

PROOF: By Theorem 4-11 with the time to reach E as the random time, we find that $(B^E\,{}^FN)_{ij} = \sum_k B^E_{ik}\,{}^FN_{kj}$ is the mean of the total number of times that the process is in state j before F, counting from the time when E is first entered. This mean is the difference of the number of times in j before F and the number of times in j before E, since F cannot be entered unless E is entered.

Theorem 9-15: If f is a function with $\alpha f = 0$ and if

$$\lim\,[(I + P + \cdots + P^{n-1})f]_0$$

exists and is finite for some state 0, then f is a charge. If g is its potential, then

(1) $g = {}^0Nf + g_0\mathbf{1}$.
(2) $f = (I - P)g$.
(3) $P^n g \to 0$.
(4) $B^E g = g$ if the support is contained in E.
(5) $f_E = (I - P^E)g_E$ if the support is contained in E.

PROOF: By Proposition 9-12, if $\lim_n \sum_k N_{0k}^{(n)} f_k$ exists, then

$$\lim_n \sum_k [N_{ik}^{(n)} - N_{0k}^{(n)}]f_k = \sum_n {}^0N_{ik}f_k$$

or

$$\left(\lim_n \sum_k N_{ik}^{(n)} f_k\right) - \left(\lim_n \sum_k N_{0k}^{(n)} f_k\right) = ({}^0Nf)_i.$$

Hence $\lim_n \sum_k N_{ik}^{(n)} f_k$ exists, and f is a charge.

(1) Therefore $g_i - g_0 = ({}^0Nf)_i$ and (1) follows.

(2) From the proof of Proposition 9-13 we have $(I - P)\,{}^0Nf = f$, since $\alpha f = 0$. Thus if we multiply (1) by $I - P$, we get (2).

(3) By (2) we have $g = Pg + f$, and, since $P^k f$ is finite-valued, we see by induction that $P^k g$ is finite-valued and that

$$P^{k-1}g = P^k g + P^{k-1}f.$$

Adding these relations for $k = 1, \ldots, n$, we obtain

$$g = P^n g + (I + P + \cdots + P^{n-1})f.$$

Hence $P^n g \to 0$.

(4) Let $0 \in E$. By Lemma 9-14 with $F = \{0\}$,

$$B^E\,{}^0N = {}^0N - {}^EN,$$

and by (1)

$$\begin{aligned} B^E g &= B^E\,{}^0Nf + B^E g_0\mathbf{1} \\ &= {}^0Nf - {}^ENf + g_0\mathbf{1} \quad \text{since } B^E\mathbf{1} = \mathbf{1} \\ &= g - {}^ENf \quad \text{by (1)}. \end{aligned}$$

Since f has support in E, ${}^{E}Nf = 0$ because ${}^{E}N_{ij} = 0$ when $j \in E$.

(5) $f = (I - P)g = (I - P)(B^E g) = [(I - P)B^E]g$

$$= \begin{pmatrix} I - P^E & 0 \\ 0 & 0 \end{pmatrix} \begin{pmatrix} g_E \\ g_{\tilde{E}} \end{pmatrix} \quad \text{by Proposition 6-17}$$

$$= \begin{pmatrix} (I - P^E)g_E \\ 0 \end{pmatrix}.$$

We note from conclusion (2) that a potential uniquely determines its charge.

Corollary 9-16: If $(I - P)g = f$, then g is a potential if and only if αf is finite and $P^n g \to 0$.

PROOF: If g is a potential, then $P^n g \to 0$ by conclusion (3) of Theorem 9-15 and f is the charge by conclusion (2). Hence αf is finite by definition. Conversely, if $(I - P)g = f$, we have by induction

$$g = P^n g + (I + \cdots + P^{n-1})f.$$

If $P^n g \to 0$, then $g = \lim\,[(I + P + \cdots + P^{n-1})f]$; if αf is finite, then g is a potential by definition.

We already know that the columns of $I - L$ are potentials whose charges are the corresponding columns of $I - P$. Corollary 9-16 allows us to enlarge this result as follows.

Corollary 9-17: If P is a null chain and if g is a function for which αg is finite, then g is a potential.

PROOF: We shall apply Corollary 9-16. By writing $g = g^+ - g^-$, we may assume that $g \geq 0$. Then

$$(P^n g)_i = \sum_j P_{ij}^{(n)} g_j = \frac{1}{\alpha_i} \sum_j (\alpha_j g_j) \hat{P}_{ji}^{(n)}.$$

Since αg is finite and since $\hat{P}_{ji}^{(n)}$ is bounded by one, we have by dominated convergence

$$\lim_n\, (P^n g)_i = \frac{1}{\alpha_i} \sum_j (\alpha_j g_j) \lim_n \hat{P}_{ji}^{(n)} = 0.$$

Therefore $P^n g \to 0$. Set $f = (I - P)g$. Then

$$\alpha|f| \leq \alpha(g + Pg) = \alpha g + \alpha(Pg) = \alpha g + (\alpha P)g = \alpha g + \alpha g < \infty.$$

A corresponding result for noncyclic ergodic chains will be proved in Section 3.

2. Normal chains

The condition at the end of Section 1 that g is a potential in a null chain if αg is finite does not provide a sufficiently large class of potentials for a satisfactory potential theory. In this section we shall impose the condition on P that any function f of finite support with $\alpha f = 0$ be a charge and that the corresponding result for signed measures hold; such a chain will be called **normal**. The justification for considering normal chains will consist in our showing that the class of normal chains is quite extensive; we shall see in Section 3, for example, that all noncyclic ergodic chains are normal.

Our procedure in introducing normal chains will be not to define them as above but to give a definition which is computationally simpler to check. The point of departure is the identity

$$g = {}^0Nf - \lim_n [P^n({}^0Nf)]$$

of Proposition 9-13.

Proposition 9-18: If for each j there is some i such that $\lim (P^n\, {}^0N)_{ij}$ exists, then $\lim (P^n\, {}^0N)$ exists and has constant columns which are finite-valued.

PROOF: Let j be a fixed state and let

$$f_k = \begin{cases} \alpha_j/\alpha_0 & \text{for } k = 0 \\ -1 & \text{for } k = j \\ 0 & \text{otherwise.} \end{cases}$$

Since $\alpha f = 0$, we have $(I - P)\, {}^0Nf = f$. Hence

$$\begin{aligned} \lim\, [(I + P + \cdots + P^{n-1})f]_i & \\ &= \lim\, [(I + P + \cdots + P^{n-1})(I - P)\, {}^0Nf]_i \\ &= \lim\, [(I - P^n)\, {}^0Nf]_i \\ &= ({}^0Nf)_i + \lim\, (P^n\, {}^0N)_{ij}. \end{aligned}$$

This limit exists for some i by hypothesis, and hence, by Theorem 9-15, f is a charge and $\lim (P^n\, {}^0N)_{kj}$ exists for all k. Hence $\lim (P^n\, {}^0N)$ exists. By Fatou's Theorem,

$$P \lim (P^n\, {}^0N) \le \lim P\, P^n\, {}^0N = \lim (P^n\, {}^0N),$$

so that each column of the limit is non-negative superregular. Hence the columns of the limit are constants by Proposition 6-3.

Definition 9-19: If the indicated limits exist, then

$$ {}^i\nu_j = \lim_n \sum_k (P^n)_{mk} \, {}^iN_{kj} $$

and

$$ {}^i\lambda_j = {}^i\nu_j / {}^iN_{jj} = \lim_n \sum_k (P^n)_{mk} \, {}^iH_{kj}. $$

Notation independent of m in Definition 9-19 is justified by Proposition 9-18. We note that ${}^i\nu_j$ exists if and only if ${}^i\lambda_j$ exists, that ${}^i\nu_j$ is finite, and that $0 \leq {}^i\lambda_j \leq 1$. Furthermore, ${}^i\lambda_j$ is the probability that j is entered in the long run before i, and ${}^i\nu_j$ is the mean number of times in j, in the long run, before reaching i.

Definition 9-20: A chain is **normal** if for some fixed state 0 and for all j, ${}^0\lambda_j$ and ${}^0\hat{\lambda}_j$ both exist.

We note the important fact that the dual of a normal chain is normal.

Proposition 9-21: If P is a normal chain and if, for a given function f with $\alpha f = 0$, $\sum_k {}^0N_{kk} f_k$ is finite, then its potential g exists, is bounded, and satisfies $g = [{}^0N - \mathbf{1}\,{}^0\nu]f$.

PROOF: Since P is normal, $P^n\,{}^0N \to \mathbf{1}\,{}^0\nu$. Furthermore,

$$ \sum_k (P^n)_{ik} \, {}^0N_{kj} \leq \sum_k (P^n)_{ik} \, {}^0N_{jj} = {}^0N_{jj} $$

and ${}^0Nf^+$ and ${}^0Nf^-$ are both finite-valued. Thus we have dominated convergence in Proposition 9-13, f is a charge, and $g = [{}^0N - \mathbf{1}\,{}^0\nu]f$. Since by Theorem 9-15 $|g_i - g_0| = |\sum_k {}^0N_{ik} f_k| \leq \sum_k {}^0N_{kk} |f_k|$, g is bounded.

Corollary 9-22: If P is normal and if f is a function of finite support with $\alpha f = 0$, then f is a charge.

PROOF: We have $\sum_k {}^0N_{kk}|f_k| < \infty$. Apply Proposition 9-21.

The converse to Corollary 9-22 is the following.

Lemma 9-23: If the function f defined by

$$ f_i = \begin{cases} \alpha_j/\alpha_0 & \text{if } i = 0 \\ -1 & \text{if } i = j \\ 0 & \text{otherwise} \end{cases} $$

is a charge with potential g, then ${}^0\lambda_j$ exists and $g_0 = {}^0\nu_j$. If all functions f of support $\{0, j\}$ with $\alpha f = 0$ are charges and if all signed measures μ of support $\{0, j\}$ with $\mu 1 = 0$ are charges, then P is normal.

PROOF: For the first assertion we have, by Proposition 9-13,

$$\begin{aligned} g_0 &= ({}^0Nf)_0 - \lim (P^n\, {}^0Nf)_0 \\ &= -\lim (P^n\, {}^0Nf)_0 \quad \text{since } {}^0N_{0k} = 0 \text{ for all } k \\ &= \lim (P^n\, {}^0N)_{0j} \qquad \text{since } {}^0N_{k0} = 0 \text{ for all } k \\ &= {}^0\nu_j. \end{aligned}$$

Hence ${}^0\lambda_j$ exists. For the second assertion we see from the hypothesis about functions that ${}^0\lambda_j$ exists for all j. Dually we obtain from the hypothesis about signed measures that ${}^0\hat{\lambda}_j$ exists for all j. Hence P is normal.

Definition 9-24: Matrices C and G are defined by

$$C_{ij} = \lim_n (N_{jj}^{(n)} - N_{ij}^{(n)})$$

$$G_{ij} = \lim_n \left(N_{ii}^{(n)} \frac{\alpha_j}{\alpha_i} - N_{ij}^{(n)}\right)$$

whenever the indicated limits exist.

According to Lemma 9-2, the quantities defining C_{ij} and G_{ij} are bounded and non-negative. Thus all entries of C and G which exist are finite and non-negative. We have further that $C_{ii} = G_{ii} = 0$ for every i and that

$$\hat{G}_{ij} = \frac{\alpha_j C_{ji}}{\alpha_i}.$$

Hence $\hat{G} = \text{dual}\, C$ and $\hat{C} = \text{dual}\, G$.

Lemma 9-25: G_{0j} exists if and only if ${}^0\nu_j$ exists. If they both exist, then $G_{0j} = {}^0\nu_j$. Dually $C_{j0} = (\alpha_0/\alpha_j)\, {}^0\hat{\nu}_j$.

PROOF: We need only note that for the potential defined in Lemma 9-23

$$g_0 = \lim_n \left[N_{00}^{(n)} \frac{\alpha_j}{\alpha_0} - N_{0j}^{(n)}\right] = G_{0j}.$$

Hence G_{0j} exists if and only if $g_0 = {}^0\nu_j$ exists, and then they are equal. The other result follows by duality.

Theorem 9-26: If for some fixed state 0 and all other states j,

(1) ${}^0\lambda_j$ and ${}^0\hat{\lambda}_j$ both exist,
or (2) G_{0j} and C_{j0} both exist,
or (3) all functions and signed measures with support $\{0, j\}$ and total charge 0 are potential charges,

then P is normal. Conversely, if P is normal, then C, G, ${}^i\lambda_j$, ${}^i\nu_j$ all exist (for both P and $\hat{P}$), $G_{ij} = {}^i\nu_j = {}^i\lambda_j\,{}^iN_{jj}$ and

$$C_{ij} = \frac{\alpha_j}{\alpha_i}\,{}^j\hat{\nu}_i = {}^j\hat{\lambda}_i\,{}^iN_{jj},$$

and all functions and signed measures of finite support and total charge 0 are potential charges.

PROOF: (1) is the definition of normality. If (2) holds, then ${}^0\nu_j$ and ${}^0\hat{\nu}_j$ exist by Lemma 9-25, and (1) holds. The sufficiency of (3) was shown in Lemma 9-23.

Conversely, suppose that P is normal. Then Corollary 9-22 assures that all f with finite support and $\alpha f = 0$ are potential charges. Consider the charge

$$f_k = \begin{cases} \alpha_j/\alpha_i & \text{if } k = i \\ -1 & \text{if } k = j \\ 0 & \text{otherwise.} \end{cases}$$

By Lemmas 9-23 and 9-25, its potential has as ith component $G_{ij} = {}^i\nu_j$, and ${}^i\nu_j = {}^i\lambda_j\,{}^iN_{jj}$ by definition. The remaining assertions follow by duality.

Corollary 9-27: If P is normal, then

$${}^0N_{ij} - {}^0\nu_j = -[G_{ij} - G_{i0}(\alpha_j/\alpha_0)].$$

PROOF: The potential of Lemma 9-23 has 0th component $g_0 = {}^0\nu_j$. By Theorem 9-15,

$$g_i = g_0 + ({}^0Nf)_i = {}^0\nu_j - {}^0N_{ij}.$$

By definition,

$$g_i = \lim\left[N_{i0}^{(n)}\frac{\alpha_j}{\alpha_0} - N_{ij}^{(n)}\right] = G_{ij} - G_{i0}\frac{\alpha_j}{\alpha_0}.$$

Corollary 9-28: If P is a normal chain and if f is a function with $\alpha f = 0$ and $\sum_k {}^0N_{kk}f_k$ finite, then f is a charge and its potential g satisfies $g = -Gf$.

PROOF: By Proposition 9-21,

$$\begin{aligned} g &= [{}^0N - \mathbf{1}\,{}^0\nu]f \\ &= \left[-G + \frac{1}{\alpha_0}\{G_{i0}\alpha_j\}\right]f \quad \text{by Corollary 9-27} \\ &= -Gf \qquad\qquad\qquad\quad \text{since } \alpha f = 0. \end{aligned}$$

The dual of Corollary 9-28 is that in a normal chain if μ is a signed measure with $\mu\mathbf{1} = 0$ and $\sum_k \mu_k\,{}^0N_{kk}$ finite, then μ is a charge and its potential ν satisfies $\nu = -\mu C$.

We can use Theorem 9-26 to conclude that all symmetric sums of independent random variables processes are normal; in particular, the one-dimensional and two-dimensional symmetric random walks are normal.

Corollary 9-29: If P is a null or noncyclic ergodic sums of independent random variables process with $P = P^T$, then P is normal,

$$C_{ij} = G_{ij} = \tfrac{1}{2}\,{}^jN_{ii} \quad \text{and} \quad {}^i\lambda_j = \tfrac{1}{2}.$$

PROOF: If we put j in for k and i in for i and j in Theorem 9-7, we obtain

$$\lim_{n\to\infty}\left[(N_{jj}^{(n)} - N_{ij}^{(n)})\frac{\alpha_i}{\alpha_j} + (N_{ii}^{(n)} - N_{ji}^{(n)})\right] = {}^jN_{ii}.$$

Now $\alpha_i = \alpha_j$ and $N_{ii}^{(n)} = N_{jj}^{(n)}$ in sums of independent random variables, and $N_{ij}^{(n)} = N_{ji}^{(n)}$ in a symmetric process. Hence

$$\lim\,[2(N_{jj}^{(n)} - N_{ij}^{(n)})] = {}^jN_{ii}$$

or $C_{ij} = \frac{1}{2}\,{}^jN_{ii}$. Alternatively

$$\lim\left[2(N_{ii}^{(n)}\frac{\alpha_j}{\alpha_i} - N_{ij}^{(n)})\right] = {}^jN_{ii}$$

or $G_{ij} = \frac{1}{2}\,{}^jN_{ii} = \frac{1}{2}\,{}^iN_{jj}$. Hence ${}^i\lambda_j = \frac{1}{2}$.

The strongest known result for concluding that a function f is in the domain of the potential operator G is Corollary 9-28, but that result involves a condition that is hard to check. The definition and theorem to follow give a more useful condition.

Definition 9-30: A function f in a normal chain is said to be a **weak charge** if $\alpha f = 0$ and if Gf and $\hat{C}f$ are both finite-valued. A signed measure μ is called a weak charge if $\mu\mathbf{1} = 0$ and if μC and $\mu\hat{G}$ are both finite-valued.

The dual of a weak charge for P is a weak charge for $\hat{P}$.

Theorem 9-31: If P is a normal chain and if f is a weak charge, then f is a charge and its potential g is bounded and satisfies $g = -Gf$.

PROOF: Set i and j both equal to k in Corollary 9-27. Then

$$^0N_{kk} - {}^0\nu_k = -[G_{kk} - G_{k0}(\alpha_k/\alpha_0)] = G_{k0}(\alpha_k/\alpha_0).$$

Since ${}^0\nu_k = G_{0k}$, ${}^0N_{kk} = G_{k0}(\alpha_k/\alpha_0) + G_{0k}$. Thus

$$\sum_k {}^0N_{kk}f_k = \sum_k \left(\frac{\alpha_k G_{k0}}{\alpha_0} + G_{0k}\right) f_k = (\hat{C}f)_0 + (Gf)_0,$$

and $\sum_k {}^0N_{kk}f_k$ must be finite. The result follows by Proposition 9-21 and Corollary 9-28.

Dually if P is normal and μ is a weak charge, then ν exists, is bounded by a multiple of α, and satisfies $\nu = -\mu C$.

For the symmetric random walks in one and two dimensions, we have $C = G = \hat{C} = \hat{G}$ by Corollary 9-29. Therefore, Theorem 9-31 states for these cases that if $\alpha f = 0$ and if Gf is finite-valued, then f is a charge and its potential is bounded and satisfies $g = -Gf$; this is the analog of the Brownian motion result.

We shall now introduce the recurrent analog of the equilibrium set for transient chains—namely, the small ergodic set. Potentials with supports in small ergodic sets will be found to have special properties.

Lemma 9-32: If h is a non-negative bounded column vector such that $\lim_n P^n h$ exists, then the limit is a constant vector.

PROOF: By Fatou's Theorem

$$P \lim (P^n h) \leq \lim P^{n+1}h.$$

Thus the limit is finite-valued, non-negative, and superregular. Hence it is constant.

In particular, if $\lim P^n B^E$ exists, then it has constant columns.

Definition 9-33: If the indicated limit exists, then

$$\lim_{n\to\infty} P^n B^E = 1\lambda^E.$$

The entry λ_i^E is the probability in the long run of entering E at i. In the special case where E is a two-point set, we have ${}^0\lambda_i = \lambda_i^{\{i,0\}}$.

Proposition 9-34: If λ^E exists, then $\lambda^E \geq 0$, $\lambda^E_{\tilde{E}} = 0$, and $\lambda^E 1 \leq 1$.

Proof: The first two assertions follow from the facts that $P^n B^E \geq 0$ for every n and that $(P^n B^E)_{ij} = 0$ if j is not in E. Also,

$$(\lambda^E 1)1 = (1\lambda^E)1 = (\lim P^n B^E)1 \leq \lim (P^n B^E 1) = 1,$$

so that $\lambda^E 1 \leq 1$.

Definition 9-35: If λ^E exists and is such that $\lambda^E 1 = 1$, then E is said to be a **small set**.

To justify the name small set and to prove existence of small sets, we shall show that subsets of small sets are small and that in a normal chain all finite sets are small.

Proposition 9-36: If E is a subset of a small set F, then E is small and $\lambda^E = \lambda^F B^E$.

Proof: By Proposition 5-8, we have $B^E = B^F B^E$. Since each row of $P^n B^F$ tends to the probability vector λ^F, it follows from Proposition 1-57 that

$$P^n B^E = P^n B^F B^E \rightarrow 1\lambda^F B^E.$$

Thus λ^E exists and $\lambda^E = \lambda^F B^E$. In addition, $\lambda^E 1 = \lambda^F B^E 1 = \lambda^F 1 = 1$.

Lemma 9-37: If P is a normal chain and if $\sum_{k \in E} {}^0N_{kk} P^E_{kj}$ is finite for every $j \in E$, then λ^E exists and the columns of $B^E - 1\lambda^E$ are potentials with support in E.

Proof:

$$(I - P)B^E = \begin{pmatrix} I - P^E & 0 \\ 0 & 0 \end{pmatrix},$$

and the columns on the right are charges for a bounded potential by Proposition 9-21. Thus the limits

$$\lim_n (I + P + \cdots + P^{n-1}) \begin{pmatrix} I - P^E & 0 \\ 0 & 0 \end{pmatrix} = \lim_n (I - P^n)B^E$$

exist, and the resulting potentials are the columns of $B^E - 1\lambda^E$.

Proposition 9-38: In a normal chain all finite sets are small.

PROOF: Let E be a finite set. By Lemma 9-37, λ^E exists. Moreover

$$\begin{aligned}\lambda^E 1 &= \sum_{j \in E} \lambda_j^E = \sum_{j \in E} \lim (P^n B^E)_{ij} \\ &= \lim \sum_{j \in E} (P^n B^E)_{ij} \\ &= \lim (P^n B^E 1)_i \\ &= 1.\end{aligned}$$

Corollary 9-39: In a normal chain ${}^i\lambda_j + {}^j\lambda_i = 1$.

PROOF: Since ${}^i\lambda_j = \lambda_j^{\{i,j\}}$ and ${}^j\lambda_i = \lambda_i^{\{i,j\}}$, we may apply Proposition 9-38.

Definition 9-40: A set E of states is an **ergodic set** if P^E is an ergodic chain.

Proposition 9-41: E is ergodic if and only if $\alpha_E 1 < \infty$. All finite sets are ergodic, a subset of an ergodic set is ergodic, and the union of two ergodic sets is ergodic.

PROOF: $\alpha_E P^E = \alpha_E$. Hence P^E is ergodic if and only if $\alpha_E 1 < \infty$. Thus the ergodic sets are the sets of finite α-measure; the remaining statements follow from this observation.

Small sets and ergodic sets might seem to be related notions, but we shall see later that they are actually independent.

Proposition 9-42: If E is a small set and if g is a bounded potential with support in E, then g is regular at points of $\tilde{E}$ and $\lambda^E g = 0$. Conversely, if E is small and ergodic and if g is a bounded function which is regular at points of $\tilde{E}$, and which satisfies $\lambda^E g = 0$, then g is a bounded potential with support in E.

PROOF: Let g be a bounded potential with support in E. Since $(I - P)g$ is the charge, g is regular in $\tilde{E}$. By Theorem 9-15, $B^E g = g$ and hence $P^n B^E g = P^n g$. Since $P^n g \to 0$, $P^n B^E g \to 0$. But by Proposition 1-57, since E is small, $P^n B^E g \to 1(\lambda^E g)$. Hence $\lambda^E g = 0$.

For the converse $g(x_n)$ is a bounded martingale at points of $\tilde{E}$, since g is bounded and regular in $\tilde{E}$. By Corollary 3-16 with the stopping time the time to reach E, we have $g = B^E g$. Now, since E is small,

$$P^n g = P^n B^E g \to 1(\lambda^E g) = 0 \quad \text{by Proposition 1-57.}$$

In addition,

$$\begin{aligned}\alpha[(I - P)g] &= \alpha[(I - P)B^E g] \\ &= \alpha\begin{pmatrix}(I - P^E)g_E \\ 0\end{pmatrix} \\ &= \alpha_E(g_E - P^E g_E).\end{aligned}$$

Since E is ergodic and g is bounded, $\alpha_E g_E$ is finite and hence

$$\alpha_E(g_E - P^E g_E) = \alpha_E g_E - (\alpha_E P^E) g_E = 0.$$

Therefore g is a potential by Corollary 9-16, and it is bounded by hypothesis. Its charge $\begin{pmatrix}(I - P^E)g_E \\ 0\end{pmatrix}$ has support in E.

For potentials of total charge zero in recurrent chains, the Principle of Balayage takes the following form.

Proposition 9-43: Let E be a small ergodic set. If x is a function bounded on E, then there is a unique potential g with support in E which differs from x on E by a constant function. The potential is $g = B^E x - (\lambda^E x)\mathbf{1}$.

PROOF: Let $g = B^E x - (\lambda^E x)\mathbf{1}$. Then g is bounded, and

$$(I - P)g = \begin{pmatrix}(I - P^E)x_E \\ 0\end{pmatrix};$$

therefore g is regular on $\tilde{E}$. Since $\lambda^E g = 0$, g is a potential with support in E, by Proposition 9-42; and g differs from x by the constant $\lambda^E x$ on E.

For uniqueness, let g' be another such potential. Then $g_E - g'_E = k\mathbf{1}$ and $g - g' = B^E g - B^E g'$. Then $g - g' = k\mathbf{1}$. Since $P^n(g - g') \to 0$, we must have $k = 0$. Hence $g = g'$.

The result to follow is the total-charge-zero version of the Principle of Condensers.

Proposition 9-44: Suppose that E and F are disjoint sets such that $E \cup F$ is small and ergodic. Then there exists a function h such that:

(1) $h_i = 1$ if $i \in E$ and $h_j = 0$ if $j \in F$.
(2) $f = (I - P)h$ has its positive values in E and its negative values in F.
(3) h is the sum of a potential and a constant function.

PROOF: For existence, let x be a function which is 1 on E and 0 on F, and set $h = B^{E \cup F}x$. By Proposition 9-43, (1) and (3) are satisfied and f has support in $E \cup F$. For (2) we note that, if $i \in E$, then f_i is the probability of return to F before E and, if $i \in F$, then $-f_i$ is the probability of return to E before F.

For uniqueness, let x be a function satisfying (1), (2), and (3). Then $x = g + c\mathbf{1}$, where g is a potential, by (3). By (2),

$$(I - P)g = (I - P)x$$

vanishes outside of $E \cup F$ so that g has support in $E \cup F$. Hence $g = B^{E \cup F}g$. Therefore,

$$B^{E \cup F}x = B^{E \cup F}g + B^{E \cup F}(c\mathbf{1}) = g + c\mathbf{1} = x,$$

and x is uniquely determined by its values on $E \cup F$, which are fixed by (1).

We conclude this section with some results about normal chains which will be needed later.

Proposition 9-45: In a normal chain

$$G_{ik}\frac{\alpha_j}{\alpha_k} + G_{kj} - G_{ij} = {}^kN_{ij}$$

$$C_{ik}\frac{\alpha_j}{\alpha_k} + C_{kj} - C_{ij} = {}^kN_{ij}$$

$$G_{ik}\frac{\alpha_i}{\alpha_k} + G_{ki} = {}^kN_{ii}$$

$$C_{ik}\frac{\alpha_i}{\alpha_k} + C_{ki} = {}^kN_{ii}.$$

PROOF: The first two expressions follow from Theorem 9-7 and Definition 9-24. For the other two, set $i = j$.

By Lemma 9-32, if $\lim\,(P^n\,{}^EN)$ exists, then it is finite-valued and has constant columns.

Definition 9-46: ${}^E\nu_j = \lim_n \sum_k (P^n)_{ik}\,{}^EN_{kj}$, provided the limit exists.

Proposition 9-47: ${}^E\nu_j$ exists if and only if $\lambda_j^{E\cup\{j\}}$ exists. If they both exist, then ${}^E\nu_j = \lambda_j^{E\cup\{j\}}\,{}^EN_{jj}$. Hence in a normal chain ${}^E\nu$ exists for all finite sets E.

PROOF: ${}^E\nu_j = \lim \sum_k (P^n)_{ik}\, {}^EN_{kj}$ and

$$\begin{aligned}\lambda_j^{E\cup\{j\}}\, {}^EN_{jj} &= \lim \sum_k (P^n)_{ik} B_{kj}^{E\cup\{j\}}\, {}^EN_{jj} \\ &= \lim \sum_k (P^n)_{ik}\, {}^EH_{kj}\, {}^EN_{jj} \\ &= \lim \sum_k (P^n)_{ik}\, {}^EN_{kj}.\end{aligned}$$

REMARK: Actually, it can be shown that if E is small, then $E \cup \{j\}$ is small. Hence ${}^E\nu$ exists for all small sets. (See the problems.)

3. Ergodic chains

For this section let P be a noncyclic ergodic chain, and choose α so that $\alpha\mathbf{1} = 1$. Then $L = \lim P^n = A = \mathbf{1}\alpha$. The dual of P, namely $\hat{P}$, is also noncyclic ergodic, and the mean first passage time matrix for P has all finite entries (see Proposition 6-42).

We begin by proving that all noncyclic ergodic chains are normal and by giving an existence theorem for potentials.

Lemma 9-48: For every i and j,

$$0 \le [N_{jj}^{(n)} - N_{ij}^{(n)}] \le M_{ij} < \infty \quad \text{and} \quad \lim\, [N_{jj}^{(n)} - N_{ij}^{(n)}] = M_{ij}\alpha_j.$$

PROOF: Summing over the powers of P in Lemma 6-34, we obtain

$$N_{ij}^{(n)} = \sum_{k=0}^{\infty} F_{ij}^{(k)} N_{jj}^{(n-k)},$$

where we use the convention $N_{jj}^{(m)} = 0$ if $m < 0$. Since $\sum_{k=0}^{\infty} F_{ij}^{(k)} = H_{ij} = 1$, we have

$$N_{jj}^{(n)} - N_{ij}^{(n)} = \sum_{k=0}^{\infty} F_{ij}^{(k)}[N_{jj}^{(n)} - N_{jj}^{(n-k)}].$$

As $n \to \infty$, we obtain

$$\begin{aligned}\lim_n\, [N_{jj}^{(n)} - N_{ij}^{(n)}] &= \lim_n \sum_{k=0}^{\infty} F_{ij}^{(k)}[N_{jj}^{(n)} - N_{jj}^{(n-k)}] \\ &= \sum_{k=0}^{\infty} F_{ij}^{(k)} \lim_n\, [N_{jj}^{(n)} - N_{jj}^{(n-k)}]\end{aligned}$$

by dominated convergence, since $N_{jj}^{(n)} - N_{jj}^{(n-k)} \le k$ and $\sum F_{ij}^{(k)}k = M_{ij} < \infty$. Thus

$$\begin{aligned}\lim\, [N_{jj}^{(n)} - N_{ij}^{(n)}] &= \sum_{k=0}^{\infty} F_{ij}^{(k)} \lim_n \sum_{m=n-k+1}^{n} P_{jj}^{(m)} \\ &= \alpha_j \sum_{k=0}^{\infty} kF_{ij}^{(k)} \\ &= M_{ij}\alpha_j.\end{aligned}$$

Theorem 9-49: Every noncyclic ergodic chain is normal, and $C_{ij} = M_{ij}\alpha_j$. In matrix form $C = MD^{-1}$.

PROOF: By Lemma 9-48, C exists and satisfies $C_{ij} = M_{ij}\alpha_j$. Since $\hat{P}$ is noncyclic ergodic, $\hat{C}$ exists and hence G exists. Thus P is normal by Theorem 9-26.

Dually, $G_{ij} = \alpha_j \hat{M}_{ji}$ or $G = \hat{M}^T D^{-1}$.

For noncyclic ergodic chains we can prove a stronger result than Theorem 9-31 about the existence of potentials.

Theorem 9-50: If f is a function with $\alpha f = 0$ and Gf finite-valued, then f is a charge and its potential g is such that $g = -Gf$ and αg is finite. If, in addition, $\hat{C}f$ is finite-valued, then g is bounded. Dually, if μ is a signed measure with $\mu 1 = 0$ and μC finite-valued, then μ is a charge and its potential ν is such that $\nu = -\mu C$ and $\nu 1$ is finite; if, in addition, $\mu \hat{G}$ is finite-valued, then ν is bounded by a multiple of α.

PROOF: We shall prove the dual statements.

$$[\mu N^{(n)}]_i = \sum_k \mu_k N_{ki}^{(n)} = -\sum_k \mu_k [N_{ii}^{(n)} - N_{ki}^{(n)}]$$

since $\sum_k \mu_k N_{ii}^{(n)} = N_{ii}^{(n)}(\mu 1) = 0$. Now $N_{ii}^{(n)} - N_{ki}^{(n)} \le M_{ki}$ by Lemma 9-48, and $\alpha_i \sum_k \mu_k M_{ki} = (\mu C)_i$ is finite by hypothesis. Hence, by dominated convergence, μ is a charge and

$$\nu_i = -\sum_k \mu_k C_{ki}.$$

By the dual of Theorem 9-15,

$$\nu = \mu\, {}^0N + \frac{\nu_0}{\alpha_0}\alpha.$$

To show that $\nu 1$ is finite, it suffices to show that $|\mu|\, {}^0N1 < \infty$. But

$$\begin{aligned} |\mu|\, {}^0N1 &= \sum_{i,j} |\mu_i|\, {}^0N_{ij} \\ &= \sum_i |\mu_i| M_{i0} \\ &= \frac{1}{\alpha_0} \sum |\mu_i| C_{i0} < \infty \quad \text{by hypothesis.} \end{aligned}$$

If $\mu \hat{G}$ is finite-valued, then ν is bounded by a multiple of α, according to Theorem 9-31.

Corollary 9-51:

$$(I - P)(-C) = I - A$$

and

$$(-G)(I - P) = I - A.$$

PROOF: A row of $I - P$ is a charge whose potential is the corresponding row of $I - A$. Since

$$\begin{aligned} 0 \le (PC)_{ij} &= \sum_k P_{ik}C_{kj} \\ &= \alpha_j \sum_k P_{ik}M_{kj} \\ &= \alpha_j(\bar{M}_{ij} - 1) \quad \text{by Proposition 6-41} \\ &< \infty, \end{aligned}$$

$(I - P)(-C) = PC - C$ is finite-valued. Hence $(I - P)(-C) = I - A$ by Theorem 9-50. The second result is dual.

From Theorem 9-50 we have a sufficient condition on charges for their potentials to exist. We turn now to conditions on functions to ensure that they are potentials, the same problem that we touched on for null chains at the end of Section 1. We shall prove as Theorem 9-53 the result for noncyclic ergodic chains that corresponds to Corollary 9-17 for null chains.

Lemma 9-52: If h is a function for which αh is finite, then $P^n h \to Ah$.

PROOF:

$$\begin{aligned} \sum_j P_{ij}^{(n)} h_j &= \frac{1}{\alpha_i} \sum_j (\alpha_j h_j) \hat{P}_{ji}^{(n)} \\ &\to \frac{1}{\alpha_i} \sum_j (\alpha_j h_j)\alpha_i \quad \text{by dominated convergence} \\ &= \alpha h. \end{aligned}$$

Theorem 9-53: If g is a function for which αg is finite, then g is a potential if and only if $\alpha g = 0$. If g is such a potential and if f is its charge, then $g = (I - A)\ {}^0Nf$.

PROOF: Set $f = (I - P)g$. Then $\alpha f = \alpha[(I - P)g] = \alpha g - \alpha Pg = 0$. By Lemma 9-52, $P^n g \to Ag = 1(\alpha g)$. Therefore, by Corollary 9-16, g is a potential if and only if $\alpha g = 0$. If g is such a potential, then by Theorem 9-15

$$g = {}^0Nf + g_0 1.$$

If we multiply through by α, we obtain

$$0 = \alpha\, {}^0Nf + g_0.$$

Hence

$$g = {}^0Nf - A\, {}^0Nf = (I - A)\, {}^0Nf.$$

Thus integrable potentials, those for which αg is finite, are precisely those integrable functions whose integral is zero. In particular, every bounded potential has integral zero.

Examples show that associativity of $A({}^0Nf)$ in Theorem 9-53 may fail. However, if it does hold, then Gf is finite-valued and $g = -Gf$. In fact, since the columns of ${}^0N$ are bounded, $\alpha\, {}^0N$ is finite-valued. Hence, by Lemma 9-52,

$$\lim P^n\, {}^0N = A\, {}^0N.$$

On the other hand, by definition

$$\lim P^n\, {}^0N = \mathbf{1}\, {}^0\nu.$$

Hence by associativity

$$g = {}^0Nf - \mathbf{1}\, {}^0\nu f = [{}^0N - \mathbf{1}\, {}^0\nu]f.$$

Therefore, by Corollary 9-27,

$$\begin{aligned} g_i &= -(Gf)_i + G_{i0}\frac{\alpha f}{\alpha_0} \\ &= -(Gf)_i \quad \text{since } \alpha f = 0. \end{aligned}$$

For a further discussion of this point, see the Additional Notes.

The existence of small sets for noncyclic ergodic chains is settled by the following proposition. Obviously all sets in such chains are ergodic sets.

Proposition 9-55: In a noncyclic ergodic chain all sets are small, and $\lambda^E = \alpha B^E$ and ${}^E\nu = \alpha\, {}^EN$. In particular, for the set of all states, $\lambda^S = \alpha$.

PROOF: By Proposition 1-57, $P^nB^E \to AB^E$. Hence $\lambda^E = \alpha B^E$ and $\lambda^E\mathbf{1} = \alpha B^E\mathbf{1} = \alpha\mathbf{1} = 1$, so that every set E is small. For $E = S$, $B^S = I$ and thus $\lambda^S = \alpha$. The assertion about ${}^E\nu$ is proved similarly.

Corollary 9-56: If αh is finite and if $(I - P)h = f$, then $B^E h$ is finite-valued and $B^E h = h - {}^E N f$. If f has support in E, then $B^E h = h$. If in addition h is bounded, then $\lambda^E h = \alpha h$.

PROOF: Let $g = h - (\alpha h)\mathbf{1}$. Then g is a potential by Theorem 9-53, and $(I - P)g = (I - P)h = f$. The proof of conclusion (4) of Theorem 9-15 shows that $B^E g$ is finite-valued and $B^E g = g - {}^E N f$. Therefore $B^E h$ is finite-valued and $B^E h = h - {}^E N f$. If f has support in E, then ${}^E N f = 0$ since ${}^E N_{ij} = 0$ for j in E. Hence $B^E h = h$ and $\alpha(B^E h) = \alpha h$. If h is bounded, associativity holds, and we conclude from Proposition 9-55 that $\lambda^E h = \alpha h$.

For total-charge-zero potentials in noncyclic ergodic chains, the following is the form that the Principle of Balayage takes.

Proposition 9-57: If αh is finite and $\lambda^E h$ is finite, then there is a unique potential g with support in E which differs from h by a constant on E.

PROOF: For existence $B^E h$ is finite-valued by Corollary 9-56, and we let

$$g = B^E h - (\lambda^E h)\mathbf{1}.$$

Then g differs from h by the constant $\lambda^E h$ on E. Since $\alpha g = 0$, g is a potential by Theorem 9-53. Moreover,

$$f = (I - P)g = (I - P)B^E h = \begin{pmatrix} (I - P^E)h_E \\ 0 \end{pmatrix},$$

so that the support of f is in E. For uniqueness, let g' be another such potential. Then $g - g'$ is constant on E. Since $B^E(g - g') = g - g'$, $g - g'$ is constant everywhere. But if $P^n(g - g') \to 0$, then $g - g'$ must be 0. Hence g is unique.

The following summary of the results for ergodic chains may be helpful. We consider the set of all states as a denumerable measure space of finite total measure $\alpha\mathbf{1}$; the measure assigned to state i is α_i. A function h is integrable if αh is finite; we restrict our attention to integrable functions.

We know that an integrable function is uniquely representable as the sum of a constant and a potential. The constant is the integral of the function. Hence an integrable function is a potential if and only if its integral is zero.

We conclude this section with some results about M and λ^E which hold for ergodic chains.

Proposition 9-58:

$$M_{ik} + M_{kj} - M_{ij} = \frac{{}^kN_{ij}}{\alpha_j},$$

$$M_{ik} + M_{ki} = \frac{{}^kN_{ii}}{\alpha_i}.$$

PROOF: In Proposition 9-45, substitute $M_{ij}\alpha_j$ for C_{ij} in the second and fourth equations.

Lemma 9-59: In any infinite ergodic chain, for fixed i and k

$$\lim_j {}^iN_{kj} = \lim_j {}^iH_{kj} = 0.$$

PROOF: Since ${}^iH_{kj} \leq {}^iN_{kj}$, it suffices to prove the result for ${}^iN_{kj}$. But $\sum_j {}^iN_{kj} = M_{ki} < \infty$, so that $\lim_j {}^iN_{kj} = 0$.

Proposition 9-60: In any infinite noncyclic ergodic chain, for fixed i

$$\lim_j {}^i\lambda_j = 0.$$

PROOF:

$$\begin{aligned} \lim_j {}^i\lambda_j &= \lim_j \sum_k \alpha_k\, {}^iH_{kj} && \text{by Proposition 9-55} \\ &= \sum_k \alpha_k \lim_j {}^iH_{kj} && \text{by dominated convergence} \\ &= 0 && \text{by Lemma 9-59.} \end{aligned}$$

Proposition 9-61: $\hat{\lambda}_i^E = \alpha_i \bar{M}_{iE}$ for i in E.

PROOF: By Proposition 6-16,

$$\alpha_j \hat{B}_{ji}^E = \alpha_i\, {}^E\bar{N}_{ij} \quad \text{for } i \text{ in } E.$$

Summing on j gives

$$\hat{\lambda}_i^E = \sum_j \alpha_j \hat{B}_{ji}^E = \alpha_i \sum_j {}^E\bar{N}_{ij} = \alpha_i \bar{M}_{iE}.$$

Thus Proposition 6-24 is the assertion that all sets are small in an ergodic chain.

Lemma 9-62: If $S_{ij} = M_{ij} + M_{ji}$, then

$$S_{ij} = \frac{{}^iN_{jj}}{\alpha_j},$$

$$S_{ij} = S_{ji} = \hat{S}_{ij} = \hat{S}_{ji},$$

$$M_{ij} = {}^j\hat{\lambda}_i S_{ji},$$

and

$$\frac{M_{ji}}{M_{ij}} = \frac{{}^i\hat{\lambda}_j}{{}^j\hat{\lambda}_i}.$$

PROOF: The first result is a restatement of Proposition 9-58. Since ${}^i\hat{N}_{jj} = {}^iN_{jj}$, we have $S_{ij} = \hat{S}_{ij}$; $S_{ij} = S_{ji}$ by definition. Finally, the last two results follow from the identities

$$M_{ij} = \frac{C_{ij}}{\alpha_j} = \frac{1}{\alpha_i}\,{}^j\hat{\nu}_i = {}^j\hat{\lambda}_i S_{ji},$$

which are consequences of Theorem 9-49 and Lemma 9-25.

From this lemma we see that

$$\hat{M}_{ij} = {}^j\lambda_i S_{ji} = {}^j\lambda_i \hat{S}_{ij} = {}^j\lambda_i(M_{ij} + M_{ji}),$$

a formula which gives a means of computing $\hat{M}$ from quantities in P. Moreover, from Proposition 9-61 we have

$${}^j\hat{\lambda}_i = \alpha_i \bar{M}_{i,\{i,j\}},$$

so that the lemma gives, on multiplication by S_{ji},

$$M_{ij} = {}^j\hat{\lambda}_i S_{ji} = \alpha_i \frac{{}^jN_{ii}}{\alpha_i} \bar{M}_{i,\{i,j\}}$$

or

$$M_{ij} = {}^jN_{ii}\bar{M}_{i,\{i,j\}}.$$

Proposition 9-63: In any infinite noncyclic ergodic chain

$$\lim_j \frac{M_{ji}}{M_{ij}} = 0,$$

$$\lim_j M_{ij} = +\infty$$

and

$$\lim_j \frac{C_{ij} - G_{ij}}{\alpha_j} = +\infty.$$

PROOF: By Lemma 9-62,

$$\frac{M_{ji}}{M_{ij}} = \frac{{}^i\hat{\lambda}_j}{{}^j\hat{\lambda}_i}.$$

By Proposition 9-60 the numerator tends to 0 and the denominator tends to 1. Hence the first assertion follows. Therefore, for all but finite many states j,

$$2M_{ij} \geq M_{ji} + M_{ij} \geq \bar{M}_{jj} = \frac{1}{\alpha_j}.$$

Since $\sum \alpha_j < \infty$, $\alpha_j \to 0$ and $M_{ij} \to \infty$. Finally

$$\begin{aligned}(C_{ij} - G_{ij})/\alpha_j &= M_{ij} - \frac{{}^i\nu_j}{\alpha_j} = M_{ij} - \frac{1}{\alpha_j}\,{}^i\lambda_j\,{}^iN_{jj} \\ &= M_{ij} - {}^i\lambda_j S_{ij} = M_{ij} - {}^i\lambda_j(M_{ij} + M_{ji}) \\ &= M_{ij}\left[1 - {}^i\lambda_j\left(1 + \frac{M_{ji}}{M_{ij}}\right)\right].\end{aligned}$$

The factor in brackets tends to 1 since ${}^i\lambda_j \to 0$ and $M_{ji}/M_{ij} \to 0$. Thus the third assertion of the proposition follows from the fact that $M_{ij} \to \infty$.

Corollary 9-64: In any infinite noncyclic ergodic chain, $C \neq G$.

PROOF: If $C = G$, then $(C_{ij} - G_{ij})/\alpha_j = 0$ for every j.

On the other hand, there are many finite ergodic chains with $C = G$. An example is

$$P = \begin{pmatrix} 1 - a & a \\ a & 1 - a \end{pmatrix}$$

for $0 < a < 1$.

4. Classes of ergodic chains

Let P be a recurrent chain. In this section we shall investigate the finiteness of the rth moments of certain random times and obtain formulas for these moments. Let

$$M_{ij}^{(r)} = \mathrm{M}_i[\mathbf{t}_j{}^r],$$

$$b_j^{(r)} = \mathrm{M}_j[\bar{\mathbf{t}}_j{}^r],$$

$$c_j^{(r)} = \sum_k \alpha_k \,\mathrm{M}_k[\mathbf{t}_j{}^r] \quad \text{or} \quad c^{(r)} = \alpha M^{(r)}.$$

These quantities are rth moments of the first passage times, the return times, and the equilibrium first passage times, respectively.

Let 0 be a distinguished state and write

$$P = \begin{matrix} & \{0\} & \widetilde{\{0\}} \\ \{0\} & P_{00} & U \\ \widetilde{\{0\}} & R & Q \end{matrix} \quad \text{and} \quad \alpha = \begin{matrix} \{0\} & \widetilde{\{0\}} \\ (\alpha_0 & \bar{\alpha}). \end{matrix}$$

The chain ${}^0P$ obtained from P by making 0 absorbing is an absorbing chain and has $N = \sum Q^k$ as its fundamental matrix. The time to absorption $\mathbf{a}$ in the chain ${}^0P$ is the same as the time $\mathbf{t}_0$ to reach 0 in the chain P. As in Section 8-8 we let $a^{(r)}$ be a column vector indexed by the transient states of ${}^0P$ and satisfying

$$a_i^{(r)} = \mathrm{M}_i[\mathbf{t}_0{}^r] = M_{i0}^{(r)}.$$

Since $M_{00}^{(r)} = 0$, Proposition 8-68 enables us to compute any column of $M^{(r)}$. From the relation $c_0^{(r)} = \bar{\alpha}a^{(r)}$ we also have a formula for the computation of $c^{(r)}$. An easy calculation gives $b_0^{(r)}$ in terms of $a^{(m)}$ with $m \leq r$:

$$\begin{aligned} b_0^{(r)} = \mathrm{M}_0[\bar{\mathbf{t}}_0{}^r] &= \sum_k P_{0k}\,\mathrm{M}_k[(\mathbf{t}_0 + 1)^r] \\ &= P_{00} + \sum_{k \neq 0} P_{0k} \sum_{m=0}^{r} \binom{r}{m} \mathrm{M}_k[\mathbf{t}_0{}^m] \\ &= P_{00} + \sum_{m=0}^{r} \binom{r}{m} U a^{(m)}. \end{aligned}$$

The first three propositions to follow give conditions for the finiteness of $M^{(r)}$, $b^{(r)}$, and $c^{(r)}$.

Proposition 9-65: If $r > 0$, then $b_0^{(r)} < \infty$ if and only if $c_0^{(r-1)} < \infty$.

Proof: Since $a^{(m)} \leq a^{(r)}$ for $m \leq r$, $b_0^{(r)} < \infty$ if and only if $Ua^{(r)} < \infty$. Multiplying the inequalities of Corollary 8-69 through by U gives

$$Ua^{(r)} \leq cUNa^{(r-1)} \leq dUa^{(r)}.$$

For $j \neq 0$,

$$(UN)_j = \sum_{k \neq 0} P_{0k}\,{}^0N_{kj} = {}^0\bar{N}_{0j} = \frac{\alpha_j}{\alpha_0}.$$

Thus

$$UNa^{(r-1)} = \frac{1}{\alpha_0} \sum_{j \neq 0} \alpha_j\,\mathrm{M}_j[\mathbf{t}_0^{r-1}] = \frac{1}{\alpha_0} c_0^{(r-1)},$$

and $c_0^{(r-1)} < \infty$ if and only if $b_0^{(r)} < \infty$.

Lemma 9-66: For any two distinct states i and j, let $\mathbf{t}_{i,j}(\omega) = \mathbf{t}_j(\omega_{\mathbf{t}_i})$ be the additional time needed to reach j after i. Then for any powers r and s

$$\mathrm{M}_\pi[\mathbf{t}_i{}^r\, \mathbf{t}_{i,j}^s] = \mathrm{M}_\pi[\mathbf{t}_i{}^r]\, \mathrm{M}_i[\mathbf{t}_j{}^s].$$

PROOF:

$$\begin{aligned}
\mathrm{M}_\pi[\mathbf{t}_i{}^r\, \mathbf{t}_{i,j}^s] &= \sum_{m,n} \mathrm{Pr}_\pi[\mathbf{t}_i = m \wedge \mathbf{t}_{i,j} = n]\, m^r n^s \\
&= \sum_{m,n} \mathrm{Pr}_\pi[\mathbf{t}_i = m]\, \mathrm{Pr}_\pi[\mathbf{t}_{i,j} = n \mid \mathbf{t}_i = m]\, m^r n^s \\
&= \sum_{m,n} \mathrm{Pr}_\pi[\mathbf{t}_i = m]\, \mathrm{Pr}_i[\mathbf{t}_j = n]\, m^r n^s \\
&\qquad\qquad \text{by the strong Markov property} \\
&= \sum_{m} \mathrm{Pr}_\pi[\mathbf{t}_i = m]\, m^r \Big(\sum_n \mathrm{Pr}_i[\mathbf{t}_j = n] n^s\Big) \\
&= \mathrm{M}_\pi[\mathbf{t}_i{}^r]\, \mathrm{M}_i[\mathbf{t}_j{}^s].
\end{aligned}$$

Proposition 9-67: The vector $b^{(r)}$ is finite-valued if and only if $M^{(r)}$ is finite-valued.

PROOF: By definition $b_j^{(r)} = \mathrm{M}_j[\bar{\mathbf{t}}_j{}^r]$. Let $\mathbf{t} = \min(\mathbf{t}_i, \bar{\mathbf{t}}_j)$ for $i \neq j$, and let $\mathbf{u} = \bar{\mathbf{t}}_j - \mathbf{t}$ (or 0 if $\mathbf{t} = \infty$). Then $\bar{\mathbf{t}}_j = \mathbf{t} + \mathbf{u} \geq \mathbf{u}$, so that

$$\begin{aligned}
b_j^{(r)} &\geq \mathrm{M}_j[\mathbf{u}^r] \\
&= \sum_k \mathrm{Pr}_j[x_{\mathbf{t}} = k]\, \mathrm{M}_k[\bar{\mathbf{t}}_j{}^r] \quad \text{by Theorem 4-11} \\
&\geq \mathrm{Pr}_j[x_{\mathbf{t}} = i]\, \mathrm{M}_i[\bar{\mathbf{t}}_j{}^r] \\
&= {}^j\bar{H}_{ji} M_{ij}^{(r)}.
\end{aligned}$$

Since ${}^j\bar{H}_{ji} > 0$, if $b_j^{(r)} < \infty$, then $M_{ij}^{(r)} < \infty$ for all i.

Conversely, suppose that $M^{(r)}$ is finite-valued. Since $\bar{\mathbf{t}}_j \leq \mathbf{t}_i + \mathbf{t}_{i,j}$ for any state $i \neq j$,

$$\begin{aligned}
b_j^{(r)} &\leq \mathrm{M}_j[(\mathbf{t}_i + \mathbf{t}_{i,j})^r] \\
&= \sum_{m=0}^{r} \binom{r}{m} \mathrm{M}_j[\mathbf{t}_i{}^m\, \mathbf{t}_{i,j}^{r-m}] \\
&= \sum_{m=0}^{r} \binom{r}{m} M_{ji}^{(m)}\, M_{ij}^{(r-m)} \quad \text{by Lemma 9-66.}
\end{aligned}$$

Hence $b^{(r)}$ is finite-valued.

Proposition 9-68: If $b_0^{(r)} < \infty$ for some state 0, then $b_j^{(r)} < \infty$ for all j.

PROOF: The proof is by induction on r. For $r = 0$, $b_j^{(0)} = 1$, and the result is trivial. Suppose that the result holds for $r \leq n$ and that $b_0^{(n+1)} < \infty$. Then $b_0^{(r)} < \infty$ for $r \leq n$, so that $b_j^{(r)} < \infty$ for $r \leq n$ by inductive assumption. By Proposition 9-67, $M^{(r)}$ is finite-valued for $r \leq n$, and by Proposition 9-65, $c_0^{(r)} < \infty$ for $r \leq n$. Thus for $j \neq 0$

$$\begin{aligned} c_j^{(n)} &= \sum_k \alpha_k \, \mathrm{M}_k[\mathbf{t}_j{}^n] \\ &\leq \sum_k \alpha_k \, \mathrm{M}_k[(\mathbf{t}_0 + \mathbf{t}_{0,j})^n] \quad \text{since } \mathbf{t}_j \leq \mathbf{t}_0 + \mathbf{t}_{0,j} \\ &= \sum_k \alpha_k \sum_{r=0}^{n} \binom{n}{r} \mathrm{M}_k[\mathbf{t}_0{}^r \, \mathbf{t}_{0,j}^{n-r}] \\ &= \sum_k \alpha_k \sum_{r=0}^{n} \binom{n}{r} M_{k0}^{(r)} M_{0j}^{(n-r)} \\ &= \sum_{r=0}^{n} \binom{n}{r} c_0^{(r)} M_{0j}^{(n-r)} < \infty. \end{aligned}$$

Therefore $b^{(n+1)} < \infty$ by Proposition 9-65.

Finally we show the connection between the moments in P and those in $\hat{P}$.

Lemma 9-69: $\bar{F}_{00}^{(n)} = \hat{\bar{F}}_{00}^{(n)}$ in a recurrent chain.

PROOF: By Lemma 6-34, for $n \geq 1$,

$$P_{00}^{(n)} = \sum_{k=1}^{n} \bar{F}_{00}^{(k)} P_{00}^{(n-k)}.$$

By induction we see that $\bar{F}_{00}^{(n)}$ is a function of $P_{00}^{(k)}$ for $k \leq n$; similarly $\hat{\bar{F}}_{00}^{(n)}$ is the same function of $\hat{P}_{00}^{(k)}$. Since $P_{00}^{(k)} = \hat{P}_{00}^{(k)}$, $\bar{F}_{00}^{(n)} = \hat{\bar{F}}_{00}^{(n)}$.

Proposition 9-70: $b^{(r)} = \hat{b}^{(r)}$ and $c^{(r)} = \hat{c}^{(r)}$.

PROOF: By Lemma 9-69,

$$b_0^{(r)} = \mathrm{M}_0[\bar{\mathbf{t}}_0{}^r] = \sum_n n^r \bar{F}_{00}^{(n)} = \sum_n n^r \hat{\bar{F}}_{00}^{(n)} = \mathrm{M}_0[\hat{\bar{\mathbf{t}}}_0{}^r] = \hat{b}_0^{(r)}.$$

For the second assertion,

$$\begin{aligned}\alpha_i \Pr_i[\mathbf{t}_0 = m] &= \sum \alpha_i P_{ik_1} P_{k_1 k_2} \ldots P_{k_{m-1}0} \\ &= \sum \alpha_0 \hat{P}_{0k_{m-1}} \ldots \hat{P}_{k_2 k_1} \hat{P}_{k_1 i},\end{aligned}$$

where each of the sums is taken over the denumerable number of sequences $k_1, \ldots, k_{m-1}$ with each k_j different from 0. Hence for $m > 0$,

$$\begin{aligned}\sum_i \alpha_i \Pr_i[\mathbf{t}_0 = m] &= \alpha_0 \sum \hat{P}_{0k_{n-1}} \ldots \hat{P}_{k_2 k_1} \\ &= \alpha_0(1 - \hat{\bar{H}}_{00}^{(m-1)}) \\ &= \alpha_0(1 - \bar{H}_{00}^{(m-1)}) \qquad \text{by Lemma 9-69} \\ &= \sum_i \alpha_i \Pr_i[\hat{\mathbf{t}}_0 = m] \quad \text{by symmetry.}\end{aligned}$$

If we multiply through by m^r and sum on m, we obtain $c_0^{(r)} = \hat{c}_0^{(r)}$.

We thus arrive at a hierarchy of recurrent chains:

Definition 9-71: A recurrent chain P has **ergodic degree** r if $b_0^{(r)} < \infty$ but $b_0^{(r+1)} = \infty$. If $b_0^{(r)} < \infty$ for every r, then P is said to have **infinite ergodic degree**. [It should be noted that $b_0^{(0)} = 1$; hence the degree is always defined.]

We may summarize our previous results as follows:

(1) Ergodic degree does not depend on the choice of the state 0.
(2) P is of ergodic degree $r > 0$ if and only if $c_0^{(r-1)} < \infty$ but $c_0^{(r)} = \infty$. (The choice of 0 is immaterial.)
(3) P is of ergodic degree r if and only if $M^{(r)}$ is finite-valued but $\alpha M^{(r)}$ is infinite-valued.
(4) $\hat{P}$ has the same ergodic degree as P.
(5) If P is of infinite ergodic degree, then $M^{(r)}$, $b^{(r)}$, and $c^{(r)}$ are finite-valued for all r.

For example, null chains have ergodic degree 0 and ergodic chains have ergodic degree at least 1. We shall see in Section 6 that the basic example may be of any degree $r = 0, 1, 2, \ldots, \infty$.

Proposition 9-72: Every finite recurrent chain has infinite ergodic degree.

PROOF: For a fixed state 0 the result follows by induction on r from Propositions 9-68, 9-67, and 9-65.

5. Strong ergodic chains

A **strong ergodic chain** is a recurrent Markov chain of ergodic degree 2 or greater. Every strong ergodic chain is ergodic, and by Proposition 9-72 all finite recurrent chains are strong ergodic. By Proposition 9-65, P is strong ergodic if and only if αM is finite-valued. If P is strong ergodic, then so is $\hat{P}$. By Proposition 9-70, $\alpha M = \alpha \hat{M}$.

In this section P is a noncyclic strong ergodic chain and α is chosen so that $\alpha 1 = 1$.

Proposition 9-73: If P is strong ergodic and if f is a bounded function, then Gf is finite-valued. If, in addition, $\alpha f = 0$, then f is a charge and its potential g satisfies $g = -Gf$.

PROOF: If $|f| \leq k1$, then

$$G|f| \leq kG1 = k(\hat{M}^T D^{-1})1 = k\hat{M}^T \alpha^T = k(\alpha \hat{M})^T < \infty.$$

Hence $G|f|$ is finite-valued. If $\alpha f = 0$, then f is a charge with potential $-Gf$ by Theorem 9-50.

Definition 9-74: For any noncyclic ergodic chain P, a matrix Z is defined by $Z = \sum_{n=0}^{\infty} (P - A)^n$ whenever that sum exists *and is finite-valued.*

Proposition 9-75: Z exists if and only if P is strong ergodic. If P is strong ergodic, then $Z = A - G(I - A)$.

PROOF: Suppose P is strong ergodic. Since $\alpha(I - A) = 0$, each column of $I - A$ is a charge with potential

$$-G(I - A) = \sum_{n=0}^{\infty} P^n(I - A)$$

by Proposition 9-73. By induction we verify that for $n > 0$

$$P^n - A = (P - A)^n$$

and hence $P^n(I - A) = P^n - A = (P - A)^n$. Therefore

$$-G(I - A) = I - A + \sum_{n=1}^{\infty} P^n(I - A) = I - A + \sum_{n=1}^{\infty} (P - A)^n$$
$$= -A + \sum_{n=0}^{\infty} (P - A)^n.$$

Hence Z exists and equals $A - G(I - A)$.

Conversely, suppose Z exists. Then

$$\begin{aligned}(\alpha M)_j\alpha_j &= (\alpha M D^{-1})_j = (\alpha C)_j \\ &\leq \liminf_n \sum_k [\alpha_k(N_{jj}^{(n-1)} - N_{kj}^{(n-1)})] \quad \text{by Fatou's Theorem} \\ &= \liminf_n (N_{jj}^{(n-1)} - n\alpha_j) \\ &= \liminf_n \sum_{m=0}^{n-1} (P^m - A)_{jj} \\ &= \liminf_n \sum_{m=0}^{n-1} [(P - A)^m]_{jj} - A_{jj} \\ &= Z_{jj} - \alpha_j.\end{aligned}$$

Hence αM is finite-valued, and P is strong ergodic.

Proposition 9-76: If P is strong ergodic, then

(1) dual $Z = \hat{Z}$.
(2) $Z\mathbf{1} = \mathbf{1}$ and $\alpha Z = \alpha$.
(3) $Z(I - P) = I - A = (I - P)Z$.
(4) $Z(I - P + A) = I = (I - P + A)Z$.
(5) $Z = A - G(I - A) = A - (I - A)C$.

PROOF: For (1) we have

$$\begin{aligned}\text{dual } Z &= \text{dual}\left(\sum_{n=0}^{\infty} (P - A)^n\right) = \sum_{n=0}^{\infty} [\text{dual } (P - A)]^n \\ &= \sum_{n=0}^{\infty} (\hat{P} - A)^n = \hat{Z}.\end{aligned}$$

Hence in conclusions (2) through (5) the second result is always the dual of the first, and we need verify only the first. Conclusion (5) comes from Proposition 9-75. For (2), we have

$$\begin{aligned}Z\mathbf{1} &= [A - G(I - A)]\mathbf{1} = A\mathbf{1} - (G - GA)\mathbf{1} \\ &= \mathbf{1} - G\mathbf{1} + GA\mathbf{1} \quad \text{since } G\mathbf{1} < \infty \text{ by Proposition 9-73} \\ &= \mathbf{1} - G\mathbf{1} + G\mathbf{1} \\ &= \mathbf{1}.\end{aligned}$$

For (3),

$$\begin{aligned}Z(I - P) &= [A - G(I - A)](I - P) \\ &= (-G + GA)(I - P) \\ &= -G + GA + GP - GAP,\end{aligned}$$

since by Proposition 9-73 all terms are finite-valued,

$$\begin{aligned} &= -G + GA + GP - GA \\ &= -G(I - P) \\ &= I - A \quad \text{by Corollary 9-51.} \end{aligned}$$

Conclusion (4) follows directly from (2) and (3).

Proposition 9-77: In a strong ergodic chain

$$C = EZ_{dg} - Z$$

and

$$M = (EZ_{dg} - Z)D.$$

PROOF: The second assertion follows from the first, since $C = MD^{-1}$. For the first we have

$$\begin{aligned} Z_{ij} - \alpha_j &= \lim_n \sum_{m=0}^{n} (P^m - A)_{ij} \\ &= \lim [N_{ij}^{(n)} - (n + 1)\alpha_j] \end{aligned}$$

and similarly

$$Z_{jj} - \alpha_j = \lim [N_{jj}^{(n)} - (n + 1)\alpha_j]$$

Hence

$$Z_{jj} - Z_{ij} = \lim [N_{jj}^{(n)} - N_{ij}^{(n)}] = C_{ij}.$$

The next proposition shows that Z may be used as a single potential operator in place of both $-C$ and $-G$. Since (dual Z) $= \hat{Z}$, duality takes as simple a form as for transient potentials. Beginning in Section 8 we shall develop an operator $-K$ which exists for all normal chains and which has properties similar to those of Z.

Proposition 9-78: In a strong ergodic chain, Z may be used as either a right or a left potential operator.

PROOF: If $g = -Gf$ and $\alpha f = 0$, then

$$g = [A - G(I - A)]f = Zf$$

by Proposition 9-75. The result for signed measures is dual.

The operator Z is used in Kemeny and Snell [1960] in the analysis of finite recurrent chains, which are all strong ergodic. A number of quantities associated with recurrent chains are computed in that book in terms of Z. The proposition to follow is a sample.

Proposition 9-79: If P is strong ergodic, then

(1) $M_{\alpha j} = \dfrac{Z_{jj}}{\alpha_j} - 1$,
(2) $Z_{jj} = \bar{M}_{\alpha j}\alpha_j$,
(3) $M_{\alpha j} - M_{ij} = \hat{M}_{\alpha i} - \hat{M}_{ji}$.

PROOF: From Proposition 9-77 and 9-76,

$$\alpha M = (\alpha E Z_{dg} - \alpha Z)D = (E Z_{dg} - \alpha)D = E Z_{dg} D - 1^T,$$

and (1) follows. For (2),

$$\bar{M}_{\alpha j} = M_{\alpha j} + \alpha_j \bar{M}_{jj} = M_{\alpha j} + 1 = \frac{Z_{jj}}{\alpha_j}.$$

Finally by (1) and Proposition 9-77,

$$M_{\alpha j} - M_{ij} = \frac{Z_{ij}}{\alpha_j} - 1 = \frac{\hat{Z}_{ji}}{\alpha_i} - 1 = \hat{M}_{\alpha i} - \hat{M}_{ji}.$$

6. The basic example

The basic example P is recurrent if and only if $\beta_n \to 0$. Then β is regular, and we choose $\alpha = \beta$. First we consider the case where P is null, that is, where $\sum_i \beta_i = +\infty$.

Since a null chain with high probability will be outside a given finite set after a long time, the finite set must be re-entered from the left. Hence if E is finite,

$$\lambda_i^E = \begin{cases} 1 & \text{if } i \text{ is the first state of } E \\ 0 & \text{otherwise.} \end{cases}$$

In particular,

$${}^i\lambda_j = \begin{cases} 1 & \text{if } j < i \\ 0 & \text{if } j \geq i. \end{cases}$$

Since

$${}^iN_{jj} = \frac{1}{1 - {}^i\bar{H}_{jj}} = \frac{1}{{}^j\bar{H}_{ji}} = \begin{cases} \dfrac{\beta_j}{\beta_i} & \text{if } j < i \\ 1 & \text{if } j \geq i, \end{cases}$$

we have

$$G_{ij} = {}^i\nu_j = {}^i\lambda_j\, {}^iN_{jj} = \begin{cases} \dfrac{\beta_j}{\beta_i} & \text{if } j < i \\ 0 & \text{if } j \geq i. \end{cases}$$

To compute C, we note that the reverse chain $\hat{P}$ enters finite sets from the right. If E is finite,

$$\hat{\lambda}_i^E = \begin{cases} 1 & \text{if } i \text{ is the last state of } E \\ 0 & \text{otherwise.} \end{cases}$$

Hence

$${}^i\hat{\lambda}_j = \begin{cases} 1 & \text{if } j > i \\ 0 & \text{if } j \leq i. \end{cases}$$

Since ${}^i\hat{N}_{jj} = {}^iN_{jj}$,

$$\hat{G}_{ij} = {}^i\hat{\lambda}_j \, {}^iN_{jj} = \begin{cases} 1 & \text{if } j > i \\ 0 & \text{if } j \leq i \end{cases}$$

and

$$C_{ij} = \frac{\beta_j}{\beta_i} \hat{G}_{ji} = \begin{cases} \dfrac{\beta_j}{\beta_i} & \text{if } j < i \\ 0 & \text{if } j \geq i. \end{cases}$$

Note that $C = G$ and that $-C$ or $-G$ is obtained if, in the transient case, we let $\beta_\infty = +\infty$ in the formula for N.

Let us consider λ^E for infinite sets. For convenience we assume that 0 is in E. The probability of entering E in the long run at a state $j > 0$ is no greater than the probability of being between 0 and j in the long run, and hence $\lambda_j^E = 0$ for $j > 0$. For any state k, let k' be the next state in E (with the convention $k = k'$ if $k \in E$). Then

$$\lambda_0^E = \lim_n \sum_k P_{0k}^{(n)} B_{k0}^E = \lim_n \left[\sum_{k > 0} P_{0k}^{(n)} \left(1 - \frac{\beta_{k'}}{\beta_k}\right) + P_{00}^{(n)} \right]$$

$$= 1 - \lim_n \sum_k P_{0k}^{(n)} \frac{\beta_{k'}}{\beta_k}.$$

Therefore the last term must be 0 for a small set.

We shall use this criterion to give an example of an ergodic set which is not a small set. Let $\frac{1}{2} \leq p < 1$ and

$$P_{i,i+1} = \begin{cases} p & \text{if } i \text{ is a power of 2} \\ 1 & \text{otherwise,} \end{cases}$$

and let E be the set of all powers of 2 together with 0. Then

$$\beta_i = \begin{cases} 1 & \text{if } i = 0 \text{ or } 1 \\ p^n & \text{if } i > 1 \text{ and } 2^{n-1} < i \leq 2^n. \end{cases}$$

Since $\sum \beta_i = 2 + \sum_{n=1}^{\infty} 2^{n-1} p^n = \infty$ if $p \geq \frac{1}{2}$, the process is null.

On the other hand, $\sum_{i \in E} \beta_i = 1 + \sum_{k=0}^{\infty} p^k < \infty$, since $p < 1$; hence E is an ergodic set. Now

$$\sum_k P_{0k}^{(n)} \frac{\beta_{k'}}{\beta_k} = \sum_k P_{0k}^{(n)} \cdot 1 = 1,$$

so that $\lambda^E = 0$ and E is not small.

For this example the bounded function $g = \mathbf{1}$ is regular outside of E and has $\lambda^E g = 0$, but $P^n g$ does not tend to 0 and hence g is not a potential. Thus Proposition 9-42 fails if we require that E be ergodic but not necessarily small.

Moreover, the same process and the same set give an example of a function f with support in E and having $\alpha f = 0$ and Gf finite-valued which is such that $-Gf$ is not a potential. For any function f,

$$(-Gf)_i = -\frac{1}{\beta_i} \sum_{j=0}^{i-1} \beta_j f_j.$$

If we let

$$f_i = \begin{cases} 0 & \text{for } i = 0 \text{ and for } i \notin E. \\ -1 & \text{for } i = 1 \\ 1 & \text{for other } i \in E, \end{cases}$$

and specialize to the case $p = \frac{1}{2}$, then

$$\alpha f = -1 + \sum_{n>0} 2^{-n} = 0$$

and

$$(-Gf)_i = \begin{cases} 0 & \text{for } i = 0 \text{ or } 1 \\ 2 & \text{otherwise.} \end{cases}$$

By Corollary 9-17 the function

$$g_i = \begin{cases} 2 & \text{for } i = 0 \text{ or } 1 \\ 0 & \text{otherwise} \end{cases}$$

is a potential. Since the sum of $-Gf$ and the potential g is a constant vector and not a potential, $-Gf$ cannot be a potential. We should note that f is not a weak charge since

$$(\hat{C}f)_i = (\hat{G}f)_i = \sum_{j>i} f_j = +\infty.$$

With minor modifications of the above example, we can make $\lambda^E \mathbf{1} = \lambda_0^E$ assume any value between 0 and 1. For example, to get $\lambda_0^E = p$ with $\frac{1}{2} \le p < 1$, redefine E to consist of 0 and all states of the form $2^k + 1$. To get values of $\lambda_0^E < \frac{1}{2}$, redefine the process (and the set) so that the process can move from i to 0 only when i is a power of n.

Now we consider the case where P is ergodic. Let

$$\alpha_j = \beta_j \Big/ \sum_k \beta_k$$

and define

$$\sigma_i = \sum_{k=0}^{i-1} \beta_k.$$

Then $\sigma_0 = 0$ and $\sigma_\infty = \sum_k \beta_k$; σ_∞ is finite for all ergodic basic examples, and $\alpha_j = \beta_j/\sigma_\infty$.

The formula for ${}^iN_{jj}$ still applies, but $\lambda^E = \alpha B^E$ and λ^E is positive on all of E. In particular, if $i < j$,

$${}^i\lambda_j = \sum_k \alpha_k \, {}^iH_{kj} = \sum_{k=i+1}^{j} \alpha_k \frac{\beta_j}{\beta_k} = (j - i)\alpha_j.$$

If $i > j$, then

$${}^i\lambda_j = 1 - {}^j\lambda_i = 1 - (i - j)\alpha_i.$$

Therefore

$$G_{ij} = {}^i\lambda_j \, {}^iN_{jj} = \begin{cases} (j - i)\alpha_j & \text{if } j \geq i \\ \dfrac{\alpha_j}{\alpha_i} + (j - i)\alpha_j & \text{if } j < i. \end{cases}$$

To get C, we first compute $\hat{G}$. If $i > j$,

$$\begin{aligned} {}^i\hat{\lambda}_j &= \sum_k \alpha_k \, {}^i\hat{H}_{kj} = \sum_{k=j}^{i-1} \alpha_k + \sum_{k<j} \alpha_k \, {}^i\hat{H}_{0j} \\ &= \frac{1}{\sigma_\infty}\left[\sum_{k=j}^{i-1} \beta_k + \sum_{k<j} \beta_k\left(1 - \frac{\alpha_i}{\alpha_j}\right)\right] \\ &= \frac{1}{\sigma_\infty}\left[(\sigma_i - \sigma_j) + \sigma_j\left(1 - \frac{\alpha_i}{\alpha_j}\right)\right] \\ &= \frac{\sigma_i}{\sigma_\infty} - \frac{\alpha_i}{\alpha_j}\frac{\sigma_j}{\sigma_\infty}. \end{aligned}$$

If $i < j$,

$${}^i\hat{\lambda}_j = 1 - {}^j\hat{\lambda}_i = 1 - \frac{\sigma_j}{\sigma_\infty} + \frac{\alpha_j}{\alpha_i}\frac{\sigma_i}{\sigma_\infty}.$$

Thus

$$\hat{G}_{ij} = \begin{cases} \dfrac{\alpha_j}{\alpha_i}\dfrac{\sigma_i}{\sigma_\infty} - \dfrac{\sigma_j}{\sigma_\infty} & \text{if } j \leq i \\ 1 + \dfrac{\alpha_j}{\alpha_i}\dfrac{\sigma_i}{\sigma_\infty} - \dfrac{\sigma_j}{\sigma_\infty} & \text{if } j > i, \end{cases}$$

and

$$C_{ij} = \frac{\alpha_j}{\alpha_i}\hat{G}_{ji} = \begin{cases} \dfrac{\sigma_j}{\sigma_\infty} - \dfrac{\alpha_j}{\alpha_i}\dfrac{\sigma_i}{\sigma_\infty} & \text{if } j \geq i \\[2ex] \dfrac{\sigma_j}{\sigma_\infty} - \dfrac{\alpha_j}{\alpha_i}\dfrac{\sigma_i}{\sigma_\infty} + \dfrac{\alpha_j}{\alpha_i} & \text{if } j < i. \end{cases}$$

Also

$$\hat{C}_{ij} = \frac{\alpha_j}{\alpha_i} G_{ji} = \begin{cases} (i - j)\alpha_j & \text{if } j \leq i \\ (i - j)\alpha_j + 1 & \text{if } j > i. \end{cases}$$

In an ergodic basic example, C is never equal to G. For suppose $C_{0j} = G_{0j}$ for all j. Then

$$\frac{\sigma_j}{\sigma_\infty} = j\alpha_j$$

or

$$\sigma_j = j\beta_j$$

for all j. By induction, $\beta_j = \beta_{j+1}$ for every j, in contradiction to the fact that β_j must tend to 0 in a recurrent basic example.

In an ergodic basic example, we have

$$M_{ij} = \begin{cases} \dfrac{1}{\alpha_j}\dfrac{\sigma_j}{\sigma_\infty} - \dfrac{1}{\alpha_i}\dfrac{\sigma_i}{\sigma_\infty} & \text{if } j \geq i \\[2ex] \dfrac{1}{\alpha_j}\dfrac{\sigma_j}{\sigma_\infty} - \dfrac{1}{\alpha_i}\dfrac{\sigma_i}{\sigma_\infty} + \dfrac{1}{\alpha_i} & \text{if } j < i, \end{cases}$$

since $M_{ij} = C_{ij}/\alpha_j$.

It is clear that $\mathrm{M}_i[\bar{\mathfrak{t}}_i{}^n]$ is finite if and only if $\mathrm{M}_0[\bar{\mathfrak{t}}_0{}^n]$ is finite. In other words, ergodic degree is independent of the state. On the other hand, $\mathrm{M}_0[\bar{\mathfrak{t}}_0{}^n]$ is finite if and only if $\mathrm{M}_0[\hat{\bar{\mathfrak{t}}}_0{}^n]$ is finite and if and only if $\mathrm{M}_\alpha[\hat{\mathfrak{t}}_0{}^{n-1}]$ is finite. But trivially, $\mathrm{M}_\alpha[\hat{\mathfrak{t}}_0{}^{n-1}] = \sum_k \alpha_k k^{n-1} = (1/\sigma_\infty)\sum_k \beta_k k^{n-1}$. Hence P and $\hat{P}$ are of ergodic degree at least n if and only if $\sum_k \beta_k k^{n-1}$ is finite. The chain with $p_i = (i/(i+1))^{n+1}$ has $\beta_k = (k+1)^{-n-1}$; for this chain $\sum_k k^{n-1}\beta_k < \infty$ while $\sum_k k^n\beta_k = \infty$, and the chain is of degree n. To obtain a chain of infinite degree, let $p_i = p$ for every $i > 0$; such a chain represents "repetition of a single task." Then $\beta_k = p^k$ and $\sum_k k^n p^k < \infty$ for all n.

Turning to ergodic potentials, we know that if $\alpha f = 0$ and if Gf is finite-valued, then f is a charge with potential g and

$$g_i = -(Gf)_i = -\sum_j j\alpha_j f_j - \frac{1}{\alpha_i}\sum_{j<i}\alpha_j f_j.$$

Hence if $\alpha f = 0$ and if $\sum_j j\alpha_j f_j$ is finite, then f is a charge and its potential g satisfies $g = -Gf$.

We can show as follows that $\alpha f = 0$ is not a sufficient condition for f to be a charge, even in an ergodic chain. Let h be a function such that $h \geq 0$, $\alpha h = +\infty$, $f = (I - P)h$, and $\alpha f = 0$. For example, choose $p_i = (i/(i + 1))^{3/2}$, so that $\beta_i = (i + 1)^{-3/2}$ and the chain is ergodic; then $h_i = \sqrt{i + 1}$ has all the required properties. For any such function h

$$(I + P + \cdots + P^{n-1})f = (I - P^n)h$$

and, by Fatou's Theorem,

$$\liminf_n (P^n h) \geq Ah = +\infty.$$

Hence f is not a charge.

For ergodic chains, potentials among all integrable functions are characterized by the fact that they have integral zero—that is, $\alpha g = 0$ or $\nu 1 = 0$. We shall conclude this section by constructing a non-integrable potential. The chain we use will be the reverse of the fixed-task example with $p = \frac{1}{2}$. Then $\alpha_k = (\frac{1}{2})^{k+1}$ and

$$P_{i+1,i} = 1$$

$$P_{0i} = (\tfrac{1}{2})^{i+1} = \alpha_i.$$

Let ν be a row vector; the necessary and sufficient condition on ν that ν be a potential is that $\nu P^n \to 0$ and that $[\nu(I - P)]1 = 0$. Now

$$P_{0j}^{(0)} = \delta_{0j} \quad \text{and} \quad P_{0j}^{(n)} = \sum_k P_{0k} P_{kj}^{(n-1)} = \sum_k \alpha_k P_{kj}^{(n-1)}$$
$$= \alpha_j.$$

If $i \geq n$, then $P_{ij}^{(n)} = \delta_{i,j+n}$, and if $i < n$, then

$$P_{ij}^{(n)} = \sum_k P_{ik}^{(i)} P_{kj}^{(n-i)} = P_{0j}^{(n-i)} = \alpha_j.$$

Therefore

$$P_{ij}^{(n)} = \begin{cases} \delta_{i,j+n} & \text{if } i \geq n \\ \alpha_j & \text{if } i < n. \end{cases}$$

Now $(\nu P^n)_j = \sum_i \nu_i P_{ij}^{(n)} = \alpha_j \sum_{i<n} \nu_i + \nu_{j+n}$. Thus $\nu P^n \to 0$ if and only if

$$\lim_{n\to\infty} \sum_{i=0}^{n} \nu_i = 0.$$

On the other hand, we are trying to construct ν so that $|\nu|1 = +\infty$, and hence the series $\sum_{i=0}^{\infty} \nu_i$ must converge conditionally to 0.

The condition $[\nu(I - P)]1 = 0$ imposes more restrictions on ν. We have

$$[\nu(I - P)]_j = (\nu_j - \nu_{j+1}) - \alpha_j \nu_0,$$

so that $[\nu(I - P)]1$ is well defined if $\sum |\nu_j - \nu_{j+1}| < +\infty$. And if $\sum |\nu_j - \nu_{j+1}| < +\infty$, then

$$[\nu(I - P)]1 = \nu_0 - \lim_j \nu_j - \sum_j \alpha_j \nu_0 = -\lim_j \nu_j.$$

We conclude that ν is a nonintegrable potential if:

(1) $\sum_{i=0}^{\infty} \nu_i$ converges conditionally to 0.
(2) $\sum |\nu_j - \nu_{j+1}| < +\infty$.

If $\nu_0 = 0$, these conditions are necessary and sufficient. It is easily verified that the sequence

$$0, 1, -1, \tfrac{1}{4}, \tfrac{1}{4}, -\tfrac{1}{4}, -\tfrac{1}{4}, \tfrac{1}{9}, \tfrac{1}{9}, \tfrac{1}{9}, -\tfrac{1}{9}, -\tfrac{1}{9}, -\tfrac{1}{9}, \tfrac{1}{16}, \tfrac{1}{16}, \tfrac{1}{16}, \tfrac{1}{16}, -\tfrac{1}{16}, \ldots$$

satisfies conditions (1) and (2).

7. Further examples

EXAMPLE 1: Independent trials process.

In an independent trials process $P_{ij} = p_j$ independently of i, where $p_j > 0$ and $\sum p_j = 1$. In such a chain $P^n = P$ since

$$P_{ij}^{(n)} = \sum_k P_{ik}^{(n-1)} P_{kj} = \sum_k P_{ik}^{(n-1)} p_j = p_j = P_{ij}.$$

It follows that P is recurrent and that $A = \lim P^n = P$; the chain is noncyclic ergodic and $\alpha_j = p_j$. In addition,

$$Z = I + \sum_{n=1}^{\infty} (P^n - A) = I,$$

so that P is strong ergodic.

If $\alpha f = 0$, then f is a charge with potential $g = Zf = f$. If $\mu 1 = 0$, then μ is a charge with potential $\nu = \mu Z = \mu$.

We have

$$\Pr_0[\bar{\mathbf{t}}_0 = k] = (1 - p_0)^{k-1} p_0.$$

Hence P is of infinite ergodic degree. To compute the M-matrix, we note that

$$\Pr_i[\bar{\mathbf{t}}_j = k] = (1 - p_j)^{k-1} p_j$$

and therefore

$$\bar{M}_{ij} = p_j \sum_k k(1 - p_j)^{k-1} = p_j\left(\frac{1}{p_j^2}\right) = \frac{1}{p_j}.$$

EXAMPLE 2: Reflecting random walk.

We return to the reflecting p-q random walk of Example 1 of Section 6-7. We first determine its ergodic degree. From 0 the chain returns to 0 either in one step or in an even number of steps, say $2n$, and

$$\Pr_0[\bar{\mathbf{t}}_0 = 1] = q.$$

Each path returning to 0 in $2n$ steps has n steps to the right and n to the left. Thus the probability of such a path is $(pq)^n$. From Feller [1957], p. 71, we see that the number of such paths is $\binom{2n-2}{n-1}\frac{1}{n}$. Therefore

$$\Pr_0[\bar{\mathbf{t}}_0 = 2n] = \binom{2n-2}{n-1}\frac{1}{n}(pq)^n.$$

We know from earlier results that the chain is recurrent if and only if $p \leq \frac{1}{2}$. The moments of the return time to 0 in this case are

$$\mathrm{M}_0[\bar{\mathbf{t}}_0^r] = q + \sum_{n=1}^{\infty} (2n)^r \binom{2n-2}{n-1}\frac{1}{n}(pq)^n.$$

By Stirling's formula, for large n

$$(2n)^r \binom{2n-2}{n-1}\frac{1}{n}(pq)^n \sim c(4pq)^n \cdot n^{r-3/2}$$

where c is a constant. If $p = \frac{1}{2}$, then $4pq = 1$, the series converges only for $r = 0$, and the chain is null. But if $p < \frac{1}{2}$, then $4pq < 1$ and all moments are finite. We may summarize as follows:

$$\text{If} \begin{Bmatrix} p > \frac{1}{2} \\ p = \frac{1}{2} \\ p < \frac{1}{2} \end{Bmatrix}, \quad \text{then } P \text{ is} \begin{cases} \text{transient} \\ \text{null} \\ \text{ergodic of infinite degree.} \end{cases}$$

If $p < \frac{1}{2}$, we can find $C_{ij} = M_{ij}\alpha_j$ from the calculation of M_{ij} in Section 6-7. The reflecting random walk satisfies $P = \hat{P}$, and hence

$$G_{ij} = \hat{G}_{ij} = \frac{\alpha_j C_{ji}}{\alpha_i} = \alpha_j M_{ji}$$

If $p = \frac{1}{2}$, the process is with high probability far from state 0 after a long time. Hence ${}^i\lambda_j = 1$ if $j > i$. Thus

$$G_{ij} = \begin{cases} {}^iN_{jj} & \text{if } j > i \\ 0 & \text{if } j \leq i. \end{cases}$$

We may compute ${}^iN_{jj}$ for $j > i$ as follows. For $j > 0$,

$$
\begin{aligned}
{}^0\bar{H}_{jj} &= \tfrac{1}{2}\,{}^0H_{j+1,j} + \tfrac{1}{2}\,{}^0H_{j-1,j} \\
&= \tfrac{1}{2} + \tfrac{1}{2}\,{}^0H_{j-1,j} \\
&= \frac{1}{2} + \frac{1}{2}\frac{j-1}{j} \quad \text{from Section 5-4} \\
&= 1 - \frac{1}{2j}.
\end{aligned}
$$

Therefore, ${}^0N_{jj} = 2j$ and ${}^iN_{jj} = {}^0N_{j-i,j-i} = 2(j - i)$. Thus

$$
G_{ij} = \begin{cases} 2(j-i) & \text{if } j > i \\ 0 & \text{if } j \le i. \end{cases}
$$

To find the C matrix, we note that

$$
\begin{aligned}
C_{ij} &= \frac{\alpha_j}{\alpha_i} G_{ji} \quad \text{since } P = \hat{P} \\
&= G_{ji} \qquad \text{since } \alpha = \mathbf{1}^T.
\end{aligned}
$$

In the case $p = \frac{1}{2}$, the condition $\alpha f = 0$ is the condition $\sum_j f_j = 0$. Then

$$
\begin{aligned}
(Gf)_i &= 2 \sum_{j>i} (j - i) f_j \\
&= 2 \sum_j (j - i) f_j + 2 \sum_{j \le i} (i - j) f_j \\
&= 2 \sum_j j f_j + 2 \sum_{j<i} (i - j) f_j.
\end{aligned}
$$

Now

$$
(\hat{C}f)_j = (G^T f)_j = (f^T G)_j = 2 \sum_{i<j} f_i (j - i),
$$

and the right side is always finite. Thus if Gf is finite-valued and $\alpha f = 0$, f is a weak charge. That is, if $\sum_j f_j = 0$ and if $\sum_j j\,|f_j| < \infty$, then

$$
g_i = -2 \sum_j j f_j - 2 \sum_{j<i} (i - j) f_j
$$

is a bounded potential.

EXAMPLE 3: Sums of independent random variables on the line.

We consider one of the chains covered in Proposition 5-22; they represent sums of independent random variables on the line with finitely many k-values. If the mean of the k-values is zero, the process is recurrent. And since $\alpha = \mathbf{1}^T$, such a chain must be null.

In a recurrent process of this type

$$\left[N_{ii}^{(n)}\frac{\alpha_j}{\alpha_i} - N_{ij}^{(n)}\right] = [N_{ii}^{(n)} - N_{ij}^{(n)}] = [N_{jj}^{(n)} - N_{ij}^{(n)}].$$

It follows that G exists if and only if C exists, and if they both exist, then $C = G$. We are going to show that for this process ${}^i\lambda_j$ does indeed exist for all i and j and hence the process is normal.

Noting that

$${}^i\lambda_j = \lim_n \sum_k P_{0k}^{(n)}\,{}^iH_{kj},$$

we write for fixed N

$$\sum_k P_{0k}^{(n)}\,{}^iH_{kj} = \sum_{k=-N}^{N} P_{0k}^{(n)}\,{}^iH_{kj} + \sum_{k<-N} P_{0k}^{(n)}\,{}^iH_{kj} + \sum_{k>N} P_{0k}^{(n)}\,{}^iH_{kj}.$$

Suppose for the moment that the two limits

$${}^iH_{+\infty,j} = \lim_{k\to+\infty} {}^iH_{kj}$$

and

$${}^iH_{-\infty,j} = \lim_{k\to-\infty} {}^iH_{kj}$$

exist. Choose N large enough so that

$$|{}^iH_{kj} - {}^iH_{+\infty,j}| < \epsilon$$

and

$$|{}^iH_{-k,j} - {}^iH_{-\infty,j}| < \epsilon$$

for $k > N$. Then

$$\begin{aligned}\sum_k P_{0k}^{(n)}\,{}^iH_{kj} &= \sum_{k=-N}^{N} P_{0k}^{(n)}\,{}^iH_{kj} + {}^iH_{-\infty,j}\Big(\sum_{k<-N} P_{0k}^{(n)}\Big)\\ &\quad + {}^iH_{+\infty,j}\Big(\sum_{k>N} P_{0k}^{(n)}\Big) + \sum_k P_{0k}^{(n)}\epsilon_k,\end{aligned}$$

where

$$\epsilon_k = \begin{cases} 0 & \text{for } -N \le k \le N \\ {}^iH_{kj} - {}^iH_{\pm\infty,j} & \text{otherwise.}\end{cases}$$

Since $|\epsilon_k| < \epsilon$, the last term on the right is less than ϵ in absolute value for every n. Moreover, the first term on the right tends to zero with n, since P is null. Finally by the Central Limit Theorem (Theorem 1-68)

$$\lim_n \sum_{k<-N} P_{0k}^{(n)} = \lim_n \sum_{k>N} P_{0k}^{(n)} = \frac{1}{2}.$$

Therefore ${}^i\lambda_j$ exists and satisfies

$${}^i\lambda_j = \tfrac{1}{2}\,{}^iH_{-\infty,j} + \tfrac{1}{2}\,{}^iH_{+\infty,j}.$$

The missing step in the argument is the proof that ${}^iH_{-\infty,j}$ and ${}^iH_{+\infty,j}$ exist. We shall now show that ${}^iH_{-\infty,j}$ exists; the other proof is similar. Since ${}^i\lambda_j = {}^0\lambda_{j-i}$ and ${}^0\lambda_{-j} = {}^j\lambda_0 = 1 - {}^0\lambda_j$, it suffices to prove that ${}^0H_{-\infty,j}$ exists for $j > 0$. If v is the largest k-value and $E = \{0, \ldots, v\}$, then

$$ {}^0H_{-k,j} = \sum_{s\in E} B^E_{-k,s}\,{}^0H_{sj}, $$

since neither 0 nor j can be reached from the negative side except through E. If we can show that $B^E_{-\infty,s} = \lim_{k\to-\infty} B^E_{k,s}$ exists, then since E is finite, we will have

$$ {}^0H_{-\infty,j} = \sum_{s\in E} B^E_{-\infty,s}\,{}^0H_{sj}. $$

Thus form the ladder process P^+. Then $p_j^+ = 0$ unless $0 < j \le v$. By the special choice of E we have $B^E_{-k,s} = (B^+)^E_{-k,s}$. Letting $f_j = p_j^+$ and $u_n = H^+_{-n,0}$, we see that $u_0 = 1$ and

$$ u_n = \sum_{k=1}^{n} p_k^+ H^+_{-(n-k),0} = \sum_{k=0}^{n-1} p_{n-k}^+ H_{-k,0} = \sum_{k=0}^{n-1} f_{n-k}u_k. $$

Hence by the Renewal Theorem (Theorem 1-67), we have

$$ \lim_{n\to\infty} H^+_{-n,0} = \frac{1}{\mu^+}, $$

where $\mu^+ = \sum jp_j^+$. But $(B^+)^E_{-k,0} = H^+_{-k,0}$, and

$$ H^+_{-k,s} = (B^+)^E_{-k,s} + \sum_{j<s} (B^+)^E_{-k,j}H^+_{j,s}. $$

Since $\lim_{k\to\infty} H^+_{-k,s} = 1/\mu^+$, we can prove by induction on s that $(B^+)^E_{-\infty,s}$ exists. Therefore ${}^0H_{-\infty,j}$ exists, and P is normal.

We conclude this section by treating the case of the one-dimensional symmetric random walk, in which $p_{-1} = p_{+1} = \frac{1}{2}$. For $j > 0$, it is clear that ${}^0H_{-\infty,j} = 0$ and ${}^0H_{+\infty,j} = 1$. Hence ${}^i\lambda_j = \frac{1}{2}$ and $G_{ij} = \frac{1}{2}\,{}^iN_{jj}$, in agreement with Corollary 9-29. For $j > 0$, ${}^0N_{jj}$ is the same as for Example 2 with $p = \frac{1}{2}$, since the two processes stopped at 0 are identical. Hence ${}^0N_{jj} = 2j$. If $j > i$, then

$$ {}^iN_{jj} = {}^0N_{j-i,j-i} = 2(j - i), $$

whereas if $j < i$, then ${}^iN_{jj} = {}^jN_{ii} = 2(i - j)$. Hence

$$ G_{ij} = \tfrac{1}{2}\,{}^iN_{jj} = |i - j|. $$

Thus the potential operator is the absolute value of the distance, just as in classical one-dimensional potential theory. The conditions

$\alpha f = 0$ and Gf finite are the conditions $\sum_j f_j = 0$ and $\sum_j j|f_j| < \infty$. Since $\hat{C}f = Gf$, we see that if $\sum f_j = 0$ and $\sum j|f_j| < \infty$, then

$$g_i = -\sum_j |i - j| f_j$$

is a bounded potential.

8. The operator *K*

We introduce a new matrix to serve as potential operator for recurrent chains. The operator K will combine many of the properties of C and G and will have the single drawback that some of its entries may be negative. Our procedure will be first to tie the K-operator into our present notion of potential, then to define in terms of K the recurrent version of capacity, and finally to introduce so-called generalized potentials, in which we do not require total charge zero. With the generalized potentials we shall be able to prove analogs of the classical potential principles.

Let a distinguished state 0 be specified.

Definition 9-80: The K-matrix is defined by

$$K_{ij} = \lim_{n \to \infty} \left[N_{00}^{(n)} \frac{\alpha_j}{\alpha_0} - N_{ij}^{(n)} \right],$$

whenever the limit exists.

Lemma 9-81: If the indicated entries of C and G exist, then

$$K_{ij} = C_{ij} + (G_{0j} - C_{0j})$$

and

$$K_{ij} = G_{ij} + (C_{i0} - G_{i0}) \frac{\alpha_j}{\alpha_0}.$$

PROOF:

$$\begin{aligned} K_{ij} &= \lim \left[N_{00}^{(n)} \frac{\alpha_j}{\alpha_0} - N_{ij}^{(n)} \right] \\ &= \lim \left[N_{00}^{(n)} \frac{\alpha_j}{\alpha_0} - N_{0j}^{(n)} \right] - \lim [N_{jj}^{(n)} - N_{0j}^{(n)}] + \lim [N_{jj}^{(n)} - N_{ij}^{(n)}] \\ &= (G_{0j} - C_{0j}) + C_{ij}. \end{aligned}$$

By Proposition 9-45,

$$G_{i0} \frac{\alpha_j}{\alpha_0} + G_{0j} - G_{ij} = {}^0N_{ij} = C_{i0} \frac{\alpha_j}{\alpha_0} + C_{0j} - C_{ij},$$

and hence

$$C_{ij} + (G_{0j} - C_{0j}) = G_{ij} + (C_{i0} - G_{i0}) \frac{\alpha_j}{\alpha_0}.$$

Lemma 9-82: K exists if and only if P is normal.

PROOF: If P is normal, then C and G exist, and hence K exists by Lemma 9-81. Conversely, suppose K exists. Since $K_{i0} = C_{i0}$ and $K_{0j} = G_{0j}$, C_{i0} and G_{0j} exist for all i and j. Thus P is normal by Theorem 9-26.

Lemma 9-83: $\hat{K}$ = dual K.

PROOF: Since $\hat{G}$ = dual C and $\hat{C}$ = dual G, we have

$$\begin{aligned}(\alpha_j/\alpha_i)\hat{K}_{ji} &= (\alpha_j/\alpha_i)\left[\hat{G}_{ji} + (\hat{C}_{j0} - \hat{G}_{j0})\frac{\alpha_i}{\alpha_0}\right] \\ &= C_{ij} + (G_{0j} - C_{0j}) = K_{ij}.\end{aligned}$$

The fact that $\hat{K}$ = dual K is the key property that the K matrix has and the C and G matrices lack. It is what is behind our first important result.

Theorem 9-84: Let

$$\nu = \lim\,[\mu(I + P + \cdots + P^n)]$$

and

$$g = \lim\,[(I + P + \cdots + P^n)f]$$

be potentials of weak charges in a normal chain P. Then $\nu = -\mu K$ and $g = -Kf$. If Y is any matrix such that for all potentials ν and g of finite support $\nu = \mu Y$ and $g = Yf$, then $Y = -(K + k(\mathbf{1}\alpha))$, where k is a constant.

PROOF: By Theorem 9-31, $\nu = -\mu C$ and $g = -Gf$. But from Lemma 9-81 and the fact that $\mu\mathbf{1} = \alpha f = 0$, we see that $-\mu K = -\mu C$ and $-Kf = -Gf$. Hence K has the desired property. If Y is an operator that serves for charges μ, then $\mu Y = -\mu K$, so that $\mu(Y + K) = 0$. Taking μ to be the row vector with 1 in the ith entry and -1 in the jth entry, we see that μ is a charge and that the ith and jth rows of $Y + K$ are equal. Hence $Y + K$ has constant columns. A similar argument with potential functions shows that $Y + K$ has rows proportional to α. Therefore $Y + K = -k(\mathbf{1}\alpha)$ for some k.

Actually at any time when $\nu = -\mu C$ or $g = -Gf$, we may use K in place of C and G. Thus K serves for both functions and measures. And the theorem shows that the only other two-sided potential operators differ from $-K$ by a multiple of $\mathbf{1}\alpha$.

We note from the definition of K that if $C = G$, then $K = C = G$. Thus in the classical cases (the symmetric random walks), the three operators coincide.

We introduce the notation

$$^E d_i = {}^E\hat{\nu}_i/\alpha_i, \quad l_i^E = \hat{\lambda}_i^E/\alpha_i \quad \text{and} \quad W^E = \begin{pmatrix} I - P^E & 0 \\ 0 & 0 \end{pmatrix}.$$

Thus $^E d = \text{dual } {}^E\hat{\nu}$ and $l^E = \text{dual } \hat{\lambda}^E$.

Proposition 9-85: If P is normal and E is a finite set, then

$$(-K)W^E = B^E - 1\lambda^E$$

and

$$W^E(-K) = \begin{pmatrix} {}^E\bar{N} \\ 0 \end{pmatrix} - l^E\alpha.$$

PROOF: Lemma 9-37 proves that a column of $B^E - 1\lambda^E$ is a potential with charge the corresponding column of W^E. The first result then follows from Theorem 9-84. The second result follows by duality from Proposition 6-16.

Corollary 9-86: If P is normal and E is a finite set, then

$$-K_E(I - P^E) = I - 1\lambda_E^E$$

and

$$(I - P^E)(-K_E) = I - l_E^E\alpha_E.$$

PROOF: Restrict the equations of Proposition 9-85 to square matrices indexed by the states of E.

We shall see shortly how Corollary 9-86 may be used to compute P^E from K_E. Although we have the formula $P^E = T + UNR$ for P^E, this expression is not of practical value for infinite chains, since N is indexed by the states of $\tilde{E}$. On the other hand, K_E *can* be computed without finding all of K, and hence P^E can be calculated from K_E for finite sets E by using only finite matrices.

From the fact that $-K$ is a two-sided potential operator, we see from the proofs of Proposition 9-85 and Corollary 9-86 that

$$(I - P^E)C_E = -I + l_E^E\alpha_E$$

and

$$G_E(I - P^E) = -I + 1\lambda^E.$$

We turn now to a discussion of capacity. Throughout the remainder of the section we shall assume that P is a normal chain.

Proposition 9-87: For any finite set E there is a constant $k(E)$ such that

$$\lambda_E^E K_E = k(E)\alpha_E \quad \text{and} \quad K_E l_E^E = k(E)\mathbf{1}.$$

Furthermore,

$$k(E) = \lambda_E^E K_E l_E^E \quad \text{and} \quad k(E) = \hat{k}(E).$$

PROOF: In Corollary 9-86, multiply the first equation by K_E on the right and the second by K_E on the left and equate the results. Then

$$\mathbf{1}(\lambda_E^E K_E) = (K_E l_E^E)\alpha_E.$$

Thus for some constant $k(E)$, we have $K_E l_E^E = k(E)\mathbf{1}$ and $\lambda_E^E K_E = k(E)\alpha_E$. Multiplication of $\lambda_E^E K_E = k(E)\alpha_E$ on the right by l_E^E gives $k(E) = \lambda_E^E K_E l_E^E$, since $\alpha_E l_E^E = 1$. The dual of this equation is $k(E) = \hat{\lambda}_E^E \hat{K}_E \hat{l}_E^E$, and thus $k(E) = \hat{k}(E)$.

Definition 9-88: For a finite set E the constant $k(E)$ such that $K_E l_E^E = k(E)\mathbf{1}$ is called the **capacity** of E.

Just as the K-matrix in general depends upon the state 0 selected, so capacity in general is a function of the distinguished state. If $E = \{i\}$, then $\lambda_j^E = \delta_{ij}$, and from Proposition 9-87 we see that

$$K_{ii} = k(\{i\})\alpha_i$$

or

$$k(\{i\}) = (G_{0i} - C_{0i})/\alpha_i.$$

In particular, $k(\{0\}) = 0$.

If we form a K' matrix by using a distinguished state $0'$, then

$$\begin{aligned} K'_{ij} - K_{ij} &= (G_{0'j} - C_{0'j}) - (G_{0j} - C_{0j}) \\ &= (C_{00'} - G_{00'})\alpha_j/\alpha_{0'} \end{aligned}$$

by Proposition 9-45. But $K = K' + k\mathbf{1}\alpha$ by Theorem 9-84. Thus

$$k = \frac{G_{00'} - C_{00'}}{\alpha_{0'}} = k(\{0'\})$$

and

$$K = K' + k(\{0'\})\mathbf{1}\alpha.$$

Therefore, from Proposition 9-87 we see that

$$k(E) = k'(E) + k(\{0'\})$$

for every finite set E. Thus capacity is determined up to an additive constant. If we let $E = \{0\}$, we find that $0 = k'(\{0\}) + k(\{0'\})$ or that $k'(\{0\}) = -k(\{0'\})$. Note that since $k(\{0'\}) = (G_{00'} - C_{00'})/\alpha_{0'}$, when $C = G$ the capacity does not depend upon the choice of 0.

In general, there is no reason why $k(E)$ should be positive. However, if 0 is in E, then

$$k(E) = (Kl^E)_0 = (Gl^E)_0 \geq 0,$$

since Kl^E is constant on E. The dual expression is

$$\alpha_0 k(E) = (\lambda^E C)_0.$$

In the following discussions we shall restrict ourselves to sets which contain the reference point. We accordingly denote by $\mathscr{L}_0$ the collection of all finite sets which contain 0.

Proposition 9-89: If E is in $\mathscr{L}_0$, then $k(E) > 0$ if and only if there is a j in E such that ${}^0\lambda_j > 0$ and $\hat{\lambda}_j^E > 0$ (or, equivalently, there is a j in E such that ${}^0\hat{\lambda}_j > 0$ and $\lambda_j^E > 0$).

PROOF: For $E \in \mathscr{L}_0$,

$$\begin{aligned} k(E) &= \sum_{j \in E} G_{0j} l_j^E \\ &= \sum_{j \in E} \frac{{}^0\nu_j \hat{\lambda}_j^E}{\alpha_j} \\ &= \sum_{j \in E} {}^0\lambda_j \left(\frac{{}^0N_{jj}}{\alpha_j}\right) \hat{\lambda}_j^E. \end{aligned}$$

Thus $k(E) > 0$ if and only if there is a non-zero term. Since ${}^0N_{jj}/\alpha_j > 0$, the first assertion of the proposition follows. The equivalent condition follows by duality.

Proposition 9-90: If P is noncyclic ergodic or if P is a symmetric sums of independent random variables process, then $k(E) > 0$ for all sets E in $\mathscr{L}_0$ with more than one point.

PROOF: If P is ergodic, then $\lambda^E = \alpha B^E$ by Proposition 9-55. Hence $\lambda_j^E \geq \alpha_j > 0$ for $j \in E$. Since $\hat{P}$ is also ergodic, ${}^0\hat{\lambda}_j > 0$ for $j \neq 0$. Therefore, if E contains a state other than 0, then $k(E) > 0$ by Proposition 9-89.

If P is a symmetric sums of independent random variables process, then $K_{ij} = \frac{1}{2}\,{}^jN_{ii} \geq \frac{1}{2}$ for $j \neq i$, by Corollary 9-29. If $l_0^E > 0$, then $k(E) \geq K_{j0} l_0^E > 0$ for any other state $j \in E$. If $l_0^E = 0$, then $l_j^E > 0$ for some $j \in E$, and $k(E) \geq K_{0j} l_j^E > 0$.

Proposition 9-91: For any set E in $\mathscr{L}_0$, K_E^{-1} exists if and only if $k(E) > 0$. If $k(E) > 0$, then

$$K_E = \left(P^E + \frac{1}{k(E)} l_E^E \lambda_E^E - I\right)^{-1}.$$

PROOF: If $k(E) = 0$, then $\lambda_E^E K_E = 0$ and K_E^{-1} does not exist. If $k(E) > 0$, then by Corollary 9-86,

$$K_E\left(P^E + \frac{1}{k(E)} l_E^E \lambda_E^E - I\right) = I - \mathbf{1}\lambda_E^E + K_E\left(\frac{1}{k(E)} l_E^E \lambda_E^E\right)$$

$$= I - \mathbf{1}\lambda_E^E + \mathbf{1}\lambda_E^E \quad \text{by Proposition 9-87}$$

$$= I.$$

From Proposition 9-91 we obtain the following expression for P^E in terms of K_E^{-1} and α_E.

Corollary 9-92: If E is a set in $\mathscr{L}_0$ with $k(E) > 0$, then

$$P^E = I + K_E^{-1} - \frac{(K_E^{-1}\mathbf{1})(\alpha_E K_E^{-1})}{\alpha_E K_E^{-1}\mathbf{1}}.$$

PROOF: By Proposition 9-91,

$$P^E = I + K_E^{-1} - \frac{1}{k(E)} l_E^E \lambda_E^E.$$

Multiplying through on the right by $\mathbf{1}$ gives

$$K_E^{-1}\mathbf{1} = \frac{1}{k(E)} l_E^E.$$

If we multiply both sides by α_E, we get

$$k(E) = \frac{1}{\alpha_E K_E^{-1}\mathbf{1}}.$$

Hence

$$l_E^E = \frac{K_E^{-1}\mathbf{1}}{\alpha_E K_E^{-1}\mathbf{1}}.$$

By duality,

$$\lambda_E^E = \frac{\alpha_E K_E^{-1}}{\alpha_E K_E^{-1}\mathbf{1}},$$

and substitution for $k(E)^{-1} l_E^E \lambda_E^E$ gives the desired result.

We now begin a series of lemmas which lead to the fact that $k(E)$ is a Choquet capacity in the sense of Definition 8-38.

Lemma 9-93: If E is a finite set, then

$$\lim_n (B^E N^{(n)} - N^{(n)}) = {}^E d\alpha - {}^E N.$$

PROOF: By Proposition 6-17,

$$B^E N^{(n)} - N^{(n)} = {}^E N P^{n+1} - {}^E N.$$

Now

$$\text{dual } ({}^E N P^{n+1}) = \hat{P}^{n+1}\, {}^E\hat{N} \to 1\, {}^E\hat{\nu} \quad \text{by Proposition 9-47,}$$

so that

$${}^E N P^{n+1} \to {}^E d\alpha.$$

Lemma 9-94: If E is a finite set, then

$$B^E K = K + {}^E N - {}^E d\alpha$$

and

$$K\begin{pmatrix} {}^E\bar{N} \\ 0 \end{pmatrix} = K + {}^E N - 1\, {}^E\nu.$$

PROOF: Each row of $B^E - I$ is a weak charge. By Theorem 9-84 and Lemma 9-93,

$$(B^E - I)K = -\lim_n (B^E - I)N^{(n)} = {}^E N - {}^E d\alpha.$$

The second result is dual.

Lemma 9-95: If E is a finite set, then

$$\lambda^E K = k(E)\alpha + {}^E\nu$$

and

$$K l^E = k(E)1 + {}^E d.$$

PROOF: Multiply the equation of Lemma 9-94 on the right by l^E to obtain

$$\begin{aligned} B^E(K l^E) &= K l^E + {}^E N l^E - {}^E d(\alpha l^E) \\ &= K l^E - {}^E d. \end{aligned}$$

But $(K l^E)_E = K_E l^E_E$, which by Proposition 9-87 equals $k(E)1$. Hence

$$K l^E = k(E)1 + {}^E d.$$

The other result is dual.

Lemma 9-96: If E, F, and L are finite sets with E and F both contained in L, then

$$[k(E) - k(F)] = \lambda^L({}^F d - {}^E d) = ({}^F\nu - {}^E\nu)l^L.$$

PROOF: Multiplying the equation of Lemma 9-95 through by λ^L, we obtain

$$\lambda^L K l^E = k(E) + \lambda^{L\ E} d$$

and similarly

$$\lambda^L K l^F = k(F) + \lambda^{L\ F} d.$$

Subtracting, we see that it is sufficient to prove that $\lambda^L K l^E = \lambda^L K l^F$. But

$$\lambda^L K l^E = (\lambda^L_L K_L) l^E_L = k(L) \alpha_L l^E_L = k(L)$$

since the support of l^E is in L. Similarly, $\lambda^L K l^F = k(L)$. The other equality is dual.

Lemma 9-97: For any sets $A_1, A_2, \ldots, A_r$ which have at least one point in common,

$$\begin{aligned} {}^{A_1 \cap A_2 \cap \cdots \cap A_r}N_{ij} \geq \sum_s {}^{A_s}N_{ij} - \sum_{s \neq t} {}^{A_s \cup A_t}N_{ij} \\ + \sum_{s \neq t \neq u} {}^{A_s \cup A_t \cup A_u}N_{ij} - \cdots + (-1)^{r+1}\ {}^{A_1 \cup A_2 \cup \cdots \cup A_r}N_{ij}. \end{aligned}$$

PROOF: The left side is the mean number of times in state j, starting in i, before the intersection of sets is reached. We shall show that the right side is the mean number of times in j before all of the sets have been reached. The former is clearly at least as large as the latter. Let $\mathbf{n}_j(\omega)$ be the number of times on ω that the process is in j before all sets are entered. On the path ω let S_k be the set of times at which the process is in j before A_k is entered. Then $\mathbf{n}_j(\omega)$ is the cardinal number of $S_1 \cup S_2 \cup \cdots \cup S_r$ and equals

$$\sum_s n(S_s) - \sum_{s \neq t} n(S_s \cap S_t) \pm \cdots,$$

where $n(A)$ is the cardinal number of A. Since $n(S_s \cap \cdots \cap S_t)$ is the number of times in j before $A_s \cup \cdots \cup A_t$ is entered, the result follows.

Proposition 9-98: Capacity is a monotone increasing set function, and for any sets $A_1, A_2, \ldots, A_r$ in $\mathscr{L}_0$,

$$\begin{aligned} k(A_1 \cap \cdots \cap A_r) \leq \sum_i k(A_i) - \sum_{i \neq j} k(A_i \cup A_j) \\ + \sum_{i \neq j \neq k} k(A_i \cup A_j \cup A_k) - \cdots \\ + (-1)^{r+1} k(A_1 \cup \cdots \cup A_r). \end{aligned}$$

PROOF: We first prove monotonicity. Let $A \subset B$; then

$$^{A}N \geq {}^{B}N.$$

Thus

$$P^n\, {}^{A}N \geq P^n\, {}^{B}N.$$

As $n \to \infty$,

$$^{A}\nu \geq {}^{B}\nu.$$

Therefore

$$({}^{A}\nu - {}^{B}\nu)l^L \geq 0.$$

If L contains A and B, then by Lemma 9-96, $k(B) - k(A) \geq 0$ or

$$k(B) \geq k(A).$$

Therefore k is monotone. The other inequality of the proposition follows in exactly the same manner starting from the result of Lemma 9-97.

In the next section we shall prove potential principles for sets of positive capacity. It is therefore particularly annoying when all sets in a chain have capacity zero, and we treat this case now.

Definition 9-99: A normal chain is **degenerate** if, for every choice of the reference state 0, $k(E) = 0$ for all finite sets E.

Lemma 9-100: If, for a single reference point 0, $k(E) = 0$ for all $E \in \mathscr{L}_0$ and for all one-point sets, then the chain is degenerate.

PROOF: If $0'$ is any other reference point, then

$$k(E) = k'(E) + k(\{0'\}) = k'(E).$$

Hence it suffices to show that $k(E) = 0$. If $0 \in E$, $k(E) = 0$ by hypothesis. If $0 \notin E$, let $0'$ be any point of E, and $0 = k'(\{0'\}) \leq k'(E) = k(E) \leq k(E \cup \{0\}) = 0$. Both inequalities follow by the monotonicity proved in Proposition 9-98.

Lemma 9-101: A normal chain is degenerate if and only if for every pair of states i and j either ${}^{i}\lambda_j = 0$ and ${}^{i}\hat{\lambda}_j = 1$ *or* ${}^{i}\lambda_j = 1$ and ${}^{i}\hat{\lambda}_j = 0$.

PROOF: If the chain is degenerate, then $k(\{i, j\}) = 0$ if i is chosen as the reference point. Hence ${}^{i}\lambda_j = 0$ or ${}^{i}\hat{\lambda}_j = 0$ by Proposition 9-89. By symmetry ${}^{j}\lambda_i = 0$ or ${}^{j}\hat{\lambda}_i = 0$, and hence ${}^{i}\lambda_j = 1$ or ${}^{i}\hat{\lambda}_j = 1$.

Conversely, if ${}^0\lambda_j = 0$ or ${}^0\hat{\lambda}_j = 0$ for all states j, then $k(E) = 0$ for all $E \in \mathscr{L}_0$ (by Proposition 9-89 since $\lambda_j^E \leq {}^0\lambda_j$). And

$$k(\{j\}) = \frac{1}{\alpha_j}(G_{0j} - C_{0j}) = \frac{{}^0N_{jj}}{\alpha_0}({}^0\lambda_j - {}^j\hat{\lambda}_0).$$

But ${}^0\lambda_j$ and ${}^j\hat{\lambda}_0$ are either both 0 or both 1. Hence $k(\{j\}) = 0$. Thus the converse follows from Lemma 9-100.

The basic example and its reverse are degenerate when null. It can be shown that degenerate chains are all of a type similar to the basic example. (See the problems.)

Proposition 9-102: If P is not degenerate, then there are two states 0 and 1 such that, with 0 as reference point, $k(E) > 0$ for all sets containing both states.

PROOF: Choose an arbitrary point as the reference state 0. Then, by Lemma 9-100, there is either a one-point set or a set in $\mathscr{L}_0$ which has non-zero capacity.

If $k(\{1\}) > 0$ for some state, then we may choose 0 and 1 as indicated, since $k(E) \geq k(\{1\}) > 0$ if $1 \in E$. If $k(\{1\}) < 0$ for some state, then use 1 as a new reference point, and $k'(\{0\}) = -k(\{1\}) > 0$. Hence we simply interchange the roles of 0 and 1.

Otherwise $k(E) \neq 0$ for some $E \in \mathscr{L}_0$. Then $k(E) > 0$. Let E be a set in $\mathscr{L}_0$ with positive capacity and containing as few states as possible, let 1 be a state of E other than 0, and let $F = E - \{1\}$. By minimality, $k(F) = 0$. Thus, by Proposition 9-98,

$$0 = k(\{0\}) = k(F \cap \{0, 1\}) \leq k(F) + k(\{0, 1\}) - k(E),$$

and

$$k(\{0, 1\}) \geq k(E) > 0.$$

Hence any set containing 0 and 1 also has positive capacity.

We conclude this section by applying our results to ergodic chains.

Proposition 9-103: If P is ergodic, $\sum_{k \in E} \lambda_k^E M_{k0} = k(E)$.

PROOF: Since $\alpha_0 k(E) = (\lambda^E C)_0$, we have $k(E) = (\lambda^E M)_0$.

This proposition enables us to give an interpretation to $k(E)$. Since $\lambda^E = \alpha B^E$, $k(E) = \alpha(B^E M)_0 = \mathrm{M}_\alpha$[time to reach 0 after E is entered] $= \mathrm{M}_\alpha[\mathbf{t}_0 - \mathbf{t}_E]$. This function is monotone in E since $\mathbf{t}_0 - \mathbf{t}_E$ is monotone on each path. The capacity inequalities follow from the fact that the time to enter all of n sets is no greater than the time to enter the intersection; hence the time to reach 0 from the intersection is no greater than the time to reach 0 after all n sets are entered. We also see why $k(E) > 0$, if E has more than one state.

Corollary 9-104: $(\lambda^E M)_j = (\hat{\lambda}^E \hat{M})_j$ for all $j \in E$.

PROOF: Choose $j \in E$ as reference point. Then

$$(\lambda^E M)_j = k(E) = \hat{k}(E) = (\hat{\lambda}^E \hat{M})_j.$$

Proposition 9-105: In an infinite ergodic chain $K + k1\alpha$ has negative entries for each k.

PROOF:

$$\begin{aligned}(K + k1\alpha)_{jj} &= G_{0j} - C_{0j} + k\alpha_j \\ &= \left(-\frac{C_{0j} - G_{0j}}{\alpha_j} + k\right)\alpha_j.\end{aligned}$$

Hence

$$\lim_j \frac{(K + k1\alpha)_{jj}}{\alpha_j} = -\infty \quad \text{by Proposition 9-63.}$$

Proposition 9-106: Let $\{E_n\}$ be an increasing sequence of finite sets with union the set of all states S. Then $\lim_{n\to\infty} k(E_n) < +\infty$ if and only if the chain is strong ergodic. In a strong ergodic chain, $k(E) = M_{\alpha 0} - M_{\alpha E}$, and hence $\hat{M}_{\alpha E} = M_{\alpha E}$.

PROOF: $k(E) = \mathrm{M}_\alpha[\mathbf{t}_0 - \mathbf{t}_E]$ and therefore

$$\sum_{i \in E} \alpha_i M_{i0} \le k(E) \le M_{\alpha 0}.$$

Hence as $E_n \to S$, $k(E_n) \to M_{\alpha 0}$, which is finite only in a strong ergodic chain. Then

$$M_{\alpha 0} - M_{\alpha E} = k(E) = \hat{k}(E) = \hat{M}_{\alpha 0} - \hat{M}_{\alpha E}.$$

Hence $\hat{M}_{\alpha E} = M_{\alpha E}$, since $\alpha M = \alpha \hat{M}$.

For strong ergodic chains the concept of capacity can be extended to infinite sets, and Proposition 9-103, Corollary 9-104, and Proposition 9-106 hold for all sets E.

9. Potential principles

In this section we shall assume that P is a nondegenerate normal chain. We shall assume that 0 and 1 are chosen according to Proposition 9-102, and we denote the collection of all finite sets containing both 0 and 1 as $\mathscr{L}$. Thus capacity is uniquely defined, $k(E) > 0$ for $E \in \mathscr{L}$ and K_E^{-1} exists by Proposition 9-91.

Definition 9-107: Let f be any function whose non-zero components are on a finite set E. Then f is the **charge** of the **generalized potential** $g = -Kf$, with **support** E and **total charge** αf. If $f \geq 0$, then g is a **pure potential.** Let μ be any signed measure whose non-zero components are on E. Then μ is the **charge** of the **generalized potential measure** $\nu = -\mu K$, with **support** E and **total charge** $\mu 1$.

In this section g and f will be used only for generalized potentials and their charges.

Proposition 9-108: g determines f, and if the support is in $E \in \mathscr{L}$, then g_E determines g.

PROOF: Since $E \in \mathscr{L}$, $k(E) > 0$, K_E^{-1} exists, and

$$f_E = -K_E^{-1} g_E, \qquad g = -K\begin{pmatrix} f_E \\ 0 \end{pmatrix}.$$

To see that g determines f even if the support F fails to contain 0 or 1, we simply let $E = F \cup \{0, 1\}$.

Proposition 9-109: For any g with support in E, $E \in \mathscr{L}$,

$$\lambda^E g = -(\alpha f)k(E)$$
$$g = B^E g - (\alpha f)\, {}^E d$$
$$(I - P^E)g_E = f_E - (\alpha f)l_E^E.$$

PROOF: Since $\lambda_E^E K_E = k(E)\alpha_E$ by Proposition 9-87, $-\lambda^E g = (\alpha f)k(E)$. Since $B^E K = K + {}^E N - {}^E d\alpha$ by Lemma 9-94, $-B^E g = -g - {}^E d(\alpha f)$. Since $(I - P^E)(-K_E) = I - l_E^E \alpha_E$ by Corollary 9-86,

$$(I - P^E)g_E = f_E - l_E^E(\alpha f).$$

Proposition 9-110: If g is a pure potential and E is any set in $\mathscr{L}$, not necessarily the support of g, then

$$-\lambda^E g \geq (\alpha f)k(E)$$
$$g \geq B^E g - (\alpha f)\, {}^E d$$
$$(I - P^E)g_E \geq f_E - (\alpha f)l_E^E.$$

PROOF:

$$\begin{aligned}
-\lambda^E g &= \lambda^E K f \\
&= (k(E)\alpha + {}^E\nu)f \quad \text{by Lemma 9-95} \\
&\geq k(E)(\alpha f) \qquad \text{since } {}^E\nu \geq 0 \text{ and } f \geq 0. \\
-B^E g &= B^E K f \\
&= Kf + {}^E N f - (\alpha f)\, {}^E d \quad \text{by Lemma 9-94} \\
&\geq Kf - (\alpha f)\, {}^E d
\end{aligned}$$

or

$$B^E g \leq g + (\alpha f)\, {}^E d.$$

By Proposition 9-85,

$$W^E(-K) = \begin{pmatrix} {}^E\bar{N} \\ 0 \end{pmatrix} - l^E \alpha.$$

Hence

$$\begin{aligned}
(I - P^E) g_E &= {}^E\bar{N} f - (\alpha f) l_E^E \\
&\geq f_E - (\alpha f) l_E^E,
\end{aligned}$$

since ${}^E\bar{N}_{ii} = 1$ for $i \in E$ and $f \geq 0$.

The recurrent Maximum Principle is the following corollary.

Corollary 9-111: A generalized potential of non-negative total charge assumes its maximum on the support.

PROOF: By Proposition 9-109, $g = B^E g - (\alpha f)\, {}^E d$. But αf and ${}^E d$ are both non-negative. Hence $g \leq B^E g$, and $g_i \leq \max_{k \in E} g_k$.

Lemma 9-112: If g' is a generalized potential with support in $E \in \mathscr{L}$ and if g is any pure potential such that $g_E \leq g'_E$, then $\alpha f' \geq \alpha f$.

PROOF: By hypothesis and by Propositions 9-109 and 9-110,

$$0 \geq \lambda^E(g' - g) \geq (-\alpha f')k(E) + (\alpha f)k(E).$$

From this it follows that $\alpha f' \geq \alpha f$, since $k(E) > 0$.

Definition 9-113: An **equilibrium potential** for E is a potential with support in E which has total charge one and has constant values on E.

This definition of equilibrium potential agrees with the one for transient chains (Definition 8-27) only up to a constant multiple, but it will be a more convenient definition for recurrent chains. We shall discuss the effect of the change after proving that equilibrium potentials exist.

Theorem 9-114: Any set E in $\mathscr{L}$ has a unique equilibrium potential. The equilibrium charge is l^E and the value of the equilibrium potential on the set is $-k(E)$. For j not in E it has the value $-k(E) - {}^E d_j$.

PROOF: Lemma 9-95, together with the fact that $\alpha l^E = 1$, shows that $-Kl^E$ is an equilibrium potential and has the values specified in the theorem.

Conversely, if f is an equilibrium charge, then by Proposition 9-109, $f_E = (I - P^E)g_E + (\alpha f)l_E^E = (I - P^E)g_E + l_E^E$. Since g_E is constant and $P^E 1 = 1$, $(I - P^E)g_E = 0$. Thus $f_E = l_E^E$, and the equilibrium potential is unique.

If we renormalize the equilibrium potential for E so that its value is one on E, then its total charge becomes $-1/k(E)$. Thus if we were following the definitions of transient potential theory, we would define $-1/k(E)$ to be the capacity of E. The set function $-1/k(E)$ is a monotone function, as is $k(E)$, but it does not have as nice a probabilistic interpretation as our choice.

We shall now prove the recurrent Principle of Balayage.

Theorem 9-115: If g is a pure potential and if $E \in \mathscr{L}$ then there is a unique pure potential g' with support in E such that

(1) $g'_E = g_E$,
(2) $g' \leq g$,
(3) $\alpha f' \geq \alpha f$,
(4) $\alpha f' = -k(E)^{-1}\lambda^E g$,

and this unique potential is $B^E g + (\lambda^E g/k(E))\, {}^E d$.

PROOF: Since g' is determined by its values on E, there is only one potential satisfying (1), and its charge is

$$\begin{aligned} f'_E &= -K_E{}^{-1} g_E \\ &= (I - P^E)g_E - \frac{\lambda^E g}{k(E)}\, l_E^E \quad \text{by Proposition 9-91} \\ &\geq (I - P^E)g_E + (\alpha f) l_E^E \\ &\geq f_E. \end{aligned}$$

The last two steps are by Proposition 9-110. Thus g' is pure. Hence

$\alpha f' \geq \alpha f$, by Lemma 9-112, and we have (3). Result (4) follows by multiplying the second equation above by α. Furthermore

$$\begin{aligned} g' &= B^E g' - (\alpha f')^E d && \text{by Proposition 9-109} \\ &\leq B^E g - (\alpha f)^E d && \text{by (1) and (3)} \\ &\leq g && \text{by Proposition 9-110.} \end{aligned}$$

Hence we have (2). The formula for g' follows from (4) and the second part of Proposition 9-109.

The potential g' is, as in the transient case, referred to as the **balayage potential** of g on E.

Next, we prove the recurrent Principle of Domination.

Theorem 9-116: If g and g' are pure potentials, if g' has its support in $E \in \mathscr{L}$, and if $g_E \geq g'_E$, then $g \geq g'$.

PROOF:

$$\begin{aligned} g &\geq B^E g - (\alpha f)^E d && \text{by Proposition 9-110} \\ &\geq B^E g' - (\alpha f')^E d && \text{by hypothesis and by Lemma 9-112} \\ &= g' && \text{by Proposition 9-109.} \end{aligned}$$

Corollary 9-117: The balayage potential of g on E is the infimum of all pure potentials that dominate g on E.

PROOF: If $\bar{g}$ is a pure potential and $\bar{g}_E \geq g_E$, then $\bar{g}_E \geq g'_E$; hence $\bar{g} \geq g'$ by domination.

Proposition 9-118: If g is a pure potential with support in $E \in \mathscr{L}$ and total charge 1, then

$$\min_{i \in E} g_i \leq -k(E) \leq \max_{i \in E} g_i.$$

PROOF: Assume that the first inequality is false. Then $-(Kf)_i > -k(E)$ for all $i \in E$. Hence we may choose a $c > 1$ such that

$$-K(cf) \geq -k(E)1 = -Kl^E$$

on E. Thus $c = c(\alpha f) \leq \alpha l^E = 1$, by Lemma 9-112, which is a contradiction. The other inequality is proved similarly.

We retain the same definition of energy as in transient theory.

Definition 9-119: If $g = -Kf$ is a potential, the **energy** of g, denoted $\mathbf{I}(g)$, is

$$\mathbf{I}(g) = \mu g = \nu f.$$

Lemma 9-120: For any potential g with support in a set $E \in \mathscr{L}$, $\mathbf{I}(g) = \nu_E(I - P^E)g_E + (\alpha f)(\hat{\lambda}^E g)$. If $P = \hat{P}$, then

$$\mathbf{I}(g) = \nu_E(I - P^E)g_E - k(E)(\alpha f)^2.$$

PROOF: By Proposition 9-109,

$$(I - P^E)g_E = f_E - (\alpha f)l_E^E.$$

Hence

$$\nu_E(I - P^E)g_E = \mathbf{I}(g) - (\alpha f)(\nu_E l_E^E) = \mathbf{I}(g) - (\alpha f)(\hat{\lambda}^E g).$$

If $P = \hat{P}$, then $\hat{\lambda}^E g = \lambda^E g = -(\alpha f)k(E)$ by Proposition 9-109.

Lemma 9-121: $\nu_E(I - P^E)g_E = \frac{1}{2}\sum_{i,j\in E} \alpha_i P_{ij}^E(g_i - g_j)^2 \geq 0$, and the value is 0 only if g is constant on E.

PROOF: The proof proceeds as in Lemma 8-54, but $P^E\mathbf{1} = \mathbf{1}$ and $\alpha_E P^E = \alpha_E$; hence $m_i = 0$ and $\pi_i = 0$. Let F be the subset of all states of E on which $g = g_0$. If $F \neq E$, there are states $i \in F$ and $j \in \tilde{F}$ such that $P_{ij}^E > 0$, since the states of E communicate. Then

$$\nu_E(I - P^E)g_E \geq \tfrac{1}{2}\alpha_i P_{ij}^E(g_i - g_j)^2 > 0.$$

Lemma 9-122: If $E \in \mathscr{L}$, the energy of the equilibrium potential of E is $-k(E)$.

PROOF: Since $g = -Kl^E$ is constant on E, $(I - P^E)g_E = 0$. Since $\alpha f = 1$, $\hat{\lambda}^E g = \hat{\lambda}^E(-k(E)\mathbf{1}) = -k(E)$. Hence by Lemma 9-120, the result follows.

Theorem 9-123: If $P = \hat{P}$ and $E \in \mathscr{L}$, then among all potentials with support in E and total charge 1 the equilibrium potential alone minimizes energy.

PROOF: If g has support in E and $\alpha f = 1$, $\mathbf{I}(g) \geq -k(E)$, by Lemmas 9-120 and 9-121. Equality holds only if g is constant on E, in which case g is the equilibrium potential.

10. A model for potential theory

By an **electric circuit** we shall mean a denumerable number of terminals, some of which are connected by wires. The wire connecting terminals i and j has resistance r_{ij} and conductance $c_{ij} = 1/r_{ij}$. If there is no connection between i and j, we let $c_{ij} = 0$; also we define $c_{ii} = 0$. We shall assume that:

(1) For each i, $\sum_j c_{ij} < \infty$. This condition is satisfied, for example, if each terminal is connected to only finitely many other terminals.
(2) The circuit is connected.

From physics we have the following two laws. In the present context the first one may be taken as a definition of current in terms of voltage:

(1) (Ohm) If i is at voltage v_i and j is at voltage v_j, the current flowing from j to i is $(v_i - v_j)c_{ij}$.
(2) (Kirchhoff) The sum of all currents flowing into a given terminal is 0.

If an outside source is attached to a certain terminal, then the Kirchhoff Law (2) does not apply unless account is taken of current flowing from the outside source. But for all terminals i which are not attached to any outside sources, Ohm's Law and Kirchhoff's Law imply:

(3) $\sum_k (v_i - v_k)c_{ik} = 0$.

If a finite set E of terminals is kept at prescribed non-zero voltages by an outside source and if there is a finite set F with $E \subset F$ such that all points in the set $\tilde{F}$ are grounded (kept at 0 voltage), then we shall call the problem of finding voltages at the points i of $F - E$ in such a way that (3) holds a **standard voltage problem**. Note that for finite circuits the voltages may be prescribed at an arbitrary subset E of terminals.

Definition 9-124: A Markov chain P with $P\mathbf{1} = \mathbf{1}$ is said to **represent** some given electric circuit if each state corresponds to a terminal and if any standard voltage problem can be solved in such a way that the voltage vector is P-regular at points of $F - E$.

It follows from Theorem 8-41 that if P represents a circuit, then the solution v to a standard voltage problem is unique and satisfies

$$v = B^{E \cup \tilde{F}} v.$$

Thus the voltage at a point i of $F - E$ can be interpreted as the expected final payment if the chain is started at i and stopped at $E \cup \tilde{F}$ and if a payment of v_j is received if the process reaches state j of E.

Proposition 9-125: For any electric circuit there is a unique Markov chain P such that $P_{ii} = 0$ and P represents the circuit.

PROOF: We first prove uniqueness; let P represent the circuit. Let i and j be any two distinct states, and let $E = \{j\}$ and $F = \{i, j\}$. Put a unit voltage at j and ground $\tilde{F}$. Then by (3),

$$\sum_k (v_i - v_k)c_{ik} = 0.$$

Since $v_k = 0$ except when k is i or j and since $c_{ii} = 0$, we have

$$v_i \sum_k c_{ik} = v_i c_{ii} + v_j c_{ij} = c_{ij}.$$

Hence

$$v_i = \frac{c_{ij}}{\sum_k c_{ik}}.$$

(The denominator is not zero since the circuit is connected.) Now since P represents the circuit,

$$v_i = (Pv)_i = \sum_k P_{ik} v_k = P_{ii} v_i + P_{ij} v_j.$$

Since $P_{ii} = 0$ and since $v_j = 1$, we have

$$v_i = P_{ij}.$$

Therefore

$$P_{ij} = \frac{c_{ij}}{\sum_k c_{ik}},$$

and P is unique.

Next we prove existence. Let the circuit be given, and define

$$P_{ij} = \frac{c_{ij}}{\sum_k c_{ik}}.$$

Then $P_{ij} \geq 0$, $P_{ii} = 0$, and

$$\sum_j P_{ij} = \frac{\sum_j c_{ij}}{\sum_k c_{ik}} = 1.$$

Hence P is a transition matrix with $P_{ii} = 0$ and $P\mathbf{1} = \mathbf{1}$. Thus let E and F be finite sets with $E \subset F$, let v_E be specified, and let $v_{\tilde{F}} = 0$. We are to show the standard voltage problem has a solution. Define v by

$$v = B^{E \cup \tilde{F}} \begin{pmatrix} v_E \\ v_{\tilde{F}} \\ 0 \end{pmatrix}.$$

Then v is regular on $F - E$, or

$$v_i - \sum_k P_{ik} v_k = [(I - P)v]_i = 0$$

for i in $F - E$. Since $P\mathbf{1} = \mathbf{1}$,

$$\sum_k (v_i - v_k) P_{ik} = 0,$$

or

$$\sum_k (v_i - v_k) \frac{c_{ik}}{\sum_m c_{im}} = 0.$$

Thus

$$\sum_k (v_i - v_k) c_{ik} = 0,$$

and v is a solution to the standard voltage problem.

Corollary 9-126: Every standard voltage problem has a solution, and that solution is unique.

PROOF: Existence follows from Proposition 9-125 and uniqueness follows from Theorem 8-41.

Shortly we shall show exactly how general the class of chains that represent circuits is. But first we shall exhibit the connection between the currents and voltages of this section and the charges and potentials of Markov chain potential theory. In so doing, what we will be showing is that electric circuits provide a model for the discrete potential theory associated with the class of chains that represent circuits.

In physics, current is the time rate at which charge flows past a point—that is, the derivative of charge with respect to time. Markov chains, however, have a time scale that is discrete and not continuous, and the proper analog of the time rate at which charge flows past a point is the amount of charge that moves past the point in unit time. With discrete time the charge moves to some point, stays for unit time, and then moves to another point. Hence the magnitude of the current at a point is equal to the magnitude of the accumulated charge at that point.

Now in a standard voltage problem, current flows in and out of the circuit through the terminals which are attached to the outside source. The above considerations lead us to define the **charge at terminal** i to be the current μ_i which flows into the circuit; μ_i is taken to be negative if the current flows out. By Kirchhoff's Law (2), the charge will be zero on the set $F - E$. For i in $E \cup \tilde{F}$ the charge is given by

$$\mu_i = \sum_k (v_i - v_k)c_{ik}.$$

Before we can connect μ and v in terms of the representing chain, we need one preliminary result.

Proposition 9-127: Let P be a Markov chain which represents an electric circuit with conductances c_{ij}. Then the row vector α defined by

$$\alpha_i = \sum_k c_{ik}$$

is P-regular, and the α-dual of P is P.

PROOF: We have

$$\alpha_i P_{ij} = \alpha_i \frac{c_{ij}}{\sum_k c_{ik}} = \alpha_i \frac{c_{ij}}{\alpha_i} = c_{ij} = c_{ji} = \alpha_j P_{ji},$$

and hence

$$\sum_i \alpha_i P_{ij} = \sum_i \alpha_j P_{ji} = \alpha_j.$$

Therefore α is regular. Since $P_{ij} = (\alpha_j P_{ji})/\alpha_i$, the α-dual of P is P.

In terms of Proposition 9-127 we can transform the equation $\mu_i = \sum_k (v_i - v_k)c_{ik}$ as follows:

$$\begin{aligned}\mu_i &= v_i\alpha_i - \sum_k \alpha_i P_{ik} v_k \\ &= v_i\alpha_i - \sum_k v_k \alpha_k P_{ki} \\ &= [(\text{dual } v)(I - P)]_i.\end{aligned}$$

Hence $\mu = (\text{dual } v)(I - P)$. Since $P = \hat{P}$,

$$\text{dual } \mu = (I - P)v.$$

Let $f_i = \mu_i/\alpha_i$, that is, f is the dual of the vector of charges at the various terminals. Then

$$f = (I - P)v.$$

We note that all pairs of states communicate in P since the circuit is connected; hence P is either transient or recurrent. But v has only finitely many non-zero entries, so that αv is finite. It follows that if P

is transient or null, then v is a potential in the Markov chain sense. And if P is ergodic, then $v - (\alpha f)1$ is a potential. In any case, f is the charge.

Conversely, if, in a chain which represents a circuit, g is a potential vanishing outside a finite set F with a charge f such that f has total charge 0 and f has support in $E \cup \tilde{F}$, where $E \subset F$, then g solves the standard voltage problem for E and F with the specified values g_E on E. It is in this sense that electric circuits form a model for potential theory.

We turn to the problem of classifying all Markov chains which represent circuits. A chain P is said to be **α-reversible** if $\hat{P}$, the α-dual of P, equals P.

Proposition 9-128: A Markov chain with $P_{ii} = 0$ represents a circuit if and only if its states communicate and it has a positive regular measure α with respect to which it is α-reversible.

PROOF: If P represents a circuit, then it has a regular measure α and is α-reversible by Proposition 9-127. Its states communicate since the circuit is connected.

Conversely, suppose that P is a transition matrix with the stated properties. Introduce the electric circuit with the states of P as terminals and with $c_{ij} = \alpha_i P_{ij}$. The circuit is well defined since

$$c_{ii} = \alpha_i P_{ii} = 0$$

and since

$$c_{ij} = \alpha_i P_{ij} = \alpha_j P_{ji} = c_{ji}.$$

Since the states communicate, the circuit is connected. To see that P represents the circuit, we note that

$$\alpha_i = \sum_k \alpha_i P_{ik} = \sum_k c_{ik}.$$

Thus

$$P_{ij} = \frac{c_{ij}}{\alpha_i} = \frac{c_{ij}}{\sum_k c_{ik}}.$$

Finally we consider the problem of when the chain representing a circuit is recurrent and when it is transient.

Lemma 9-129: Let P be a chain which represents an electric circuit. Let a unit voltage be put at 0, let F be a finite set containing 0, and let $\tilde{F}$ be kept at voltage 0. The charge at the terminal 0 is

$$\mu_0 = \alpha_0 \cdot {}^0\bar{H}_{0\tilde{F}}.$$

PROOF:

$$
\begin{aligned}
{}^{0}\bar{H}_{0\tilde{F}} &= \sum_k P_{0k}\,{}^{0}H_{k\tilde{F}} \\
&= \sum_k P_{0k}(1 - {}^{\tilde{F}}H_{k0}) \quad \text{since } {}^{\tilde{F}}P \text{ is absorbing} \\
&= \sum_k P_{0k}(1 - B_{k0}^{\{0\}\cup\tilde{F}}) \\
&= \sum_k P_{0k}(v_0 - v_k) \\
&= \frac{1}{\alpha_0}\sum_k c_{0k}(v_0 - v_k) \\
&= \frac{\mu_0}{\alpha_0}.
\end{aligned}
$$

Lemma 9-130: In any Markov chain a state 0 is recurrent if and only if ${}^{0}\bar{H}_{0\tilde{F}} \to 0$ for some (or every) increasing sequence of finite sets F with union the set of all states.

PROOF: If 0 is transient, then there is a positive probability $1 - \bar{H}_{00}$ that the chain never returns to 0. Hence

$$
{}^{0}\bar{H}_{0\tilde{F}} \geq 1 - \bar{H}_{00} > 0
$$

for every finite set F, and ${}^{0}\bar{H}_{0\tilde{F}}$ cannot approach 0.

Conversely, let 0 be recurrent. Choose N sufficiently large that $\bar{H}_{00}^{(N)} > 1 - \epsilon$. Choose δ close enough to 1 so that

$$
1 - \epsilon < \delta^N < 1.
$$

Then construct an increasing sequence of finite sets $A_0, A_1, \ldots, A_N$ such that $A_0 = \{0\}$ and such that the probability of stepping from any state of A_k to a state of A_{k+1} is greater than δ. Let F be any finite set containing A_N. The probability that the process started at 0 is, for every $n \leq N$, in A_n after n steps is greater than δ^N. Hence

$$
{}^{0}\bar{H}_{0\tilde{F}}^{(N)} < 1 - \delta^N < \epsilon.
$$

Since

$$
1 - \epsilon < \bar{H}_{00}^{(N)} \leq {}^{\tilde{F}}\bar{H}_{00}^{(N)} + {}^{0}\bar{H}_{0\tilde{F}}^{(N)},
$$

we have

$$
{}^{\tilde{F}}\bar{H}_{00} \geq {}^{\tilde{F}}\bar{H}_{00}^{(N)} > 1 - 2\epsilon.
$$

But

$$
{}^{\tilde{F}}\bar{H}_{00} + {}^{0}\bar{H}_{0\tilde{F}} = 1,
$$

so that

$$^0\bar{H}_{0\tilde{F}} < 2\epsilon$$

for all F containing A_N.

Proposition 9-131: Let P be a chain which represents an electric circuit. Then the following is a necessary and sufficient condition for P to be recurrent: If terminal 0 is kept at a unit voltage, if all terminals outside a finite set F are grounded, and if F is allowed to increase to S, then the current at terminal 0 tends to zero. Furthermore a necessary and sufficient condition for P to be ergodic is that $\sum_{i,j} c_{ij} < \infty$.

PROOF: The first assertion follows directly from Lemmas 9-129 and 9-130. The second assertion follows from the fact that P is ergodic if and only if $\alpha 1 < \infty$, where $\alpha_i = \sum_j c_{ij}$.

11. A nonnormal chain and other examples

We shall show by examples in this section that all three of the following conjectures are false:

(1) Small sets and ergodic sets are identical concepts, or at least one of the notions includes the other.
(2) All null chains are normal.
(3) The existence of either C or G implies the existence of both.

First, we settle the independence of the notions of small sets and ergodic sets. We saw in Section 6 an example of an ergodic set which is not small, and we shall now produce a small set which is not ergodic.

Let P be a chain with states the non-negative integers and with transition probabilities

$$P_{0i} = p_i = P_{i0} \quad \text{for } i > 0$$

$$P_{ii} = q_i = 1 - p_i.$$

All other entries of P are 0. We impose no requirements on the p_i yet except that $p_i > 0$ and $\sum p_i = 1$. It is clear that $\bar{H}_{00} = 1$ and hence P is recurrent. Since $P = P^T$, $\alpha = 1^T$ is regular and P is null. Only finite sets are ergodic, and thus it is sufficient to exhibit an infinite small set. For any set E containing 0,

$$\lambda_j^E = \lim_n \sum_k P_{0k}^{(n)} B_{kj}^E = \lim_n P_{0j}^{(n)} = 0 \quad \text{for } j > 0,$$

whereas

$$\begin{aligned}\lambda_0^E &= \lim_n \sum_k P_{0k}^{(n)} B_{k0}^E \\ &= \lim_n P_{00}^{(n)} + \lim_n \sum_{k \in \tilde{E}} P_{0k}^{(n)} \\ &= 1 - \lim_n P_{0E}^{(n)}.\end{aligned}$$

Thus any set containing 0 such that $P_{0E}^{(n)} \to 0$ satisfies $\lambda_0^E = 1$ and is therefore a small set. Let

$$p_i = 4^{-k} \quad \text{for } 2^k - 1 \le i \le 2^{k+1} - 2.$$

The p_i assume the constant value 4^{-k} on a block of length 2^k. Thus

$$\sum p_i = \sum_{k=1}^{\infty} 2^k \cdot 4^{-k} = 1.$$

Let E consist of 0 and one state, such as 2^k, from each block, and label the representative of the kth block in E as e_k. Then

$$P_{0E}^{(n)} = P_{00}^{(n)} + \sum_{k=1}^{N-1} P_{0e_k}^{(n)} + P_{0T_N}^{(n)},$$

where $T_N = \{e_N, e_{N+1}, e_{N+2}, \ldots\}$. We can form 2^N disjoint sets like T_N each differing from it in every representative selected from the Nth block on. By symmetry $P_{0T'_N}^{(n)} = P_{0T_N}^{(n)}$ for all such sets T'_N. Hence $P_{0T_N}^{(n)} \le 1/2^N$ for all n. Since $P_{0j}^{(n)} \to 0$ in a null chain,

$$\limsup_n P_{0E}^{(n)} \le \frac{1}{2^N},$$

and we must have $P_{0E}^{(n)} \to 0$.

Second, we shall construct a nonnormal null chain. We shall use generating functions and require the following facts:

(1) If $F(t) = \sum_n f_n t^n$ and $G(t) = \sum_n g_n t^n$, then

$$F(t) \cdot G(t) = \sum_n \left(\sum_{k=0}^{n} f_k g_{n-k} \right) t^n.$$

That is, the coefficients of the series for $F(t) \cdot G(t)$ are the convolutions $\sum f_k g_{n-k}$.

(2) The Abel sum of the series $\sum_n f_n$ is $\lim_{t \to 1^-} F(t)$. If the series $\sum f_n$ converges, then its Abel sum exists and has the same value (Proposition 1-62).

Let P be any recurrent chain; we begin by deriving a necessary and sufficient condition for the series $\sum_n (P_{11}^{(n)} - P_{01}^{(n)})$ defining C_{01} to be

Abel summable. Let $E = \{0, 1\}$ and define generating functions as follows:

$$A_{00}(t) = \sum_n {}^1\bar{F}_{00}^{(n)}t^n, \qquad A_{01}(t) = \sum_n {}^0\bar{F}_{01}^{(n)}t^n$$

$$A_{10}(t) = \sum_n {}^1\bar{F}_{10}^{(n)}t^n, \qquad A_{11}(t) = \sum_n {}^0\bar{F}_{11}^{(n)}t^n$$

$$P_{ij}(t) = \sum_n P_{ij}^{(n)}t^n$$

$$H_0(t) = A_{00}(t) + A_{01}(t) = \sum_n \bar{F}_{0E}^{(n)}t^n$$

$$H_1(t) = A_{10}(t) + A_{11}(t) = \sum_n \bar{F}_{1E}^{(n)}t^n$$

$$Q(t) = \sum_n (P_{11}^{(n)} - P_{01}^{(n)})t^n = P_{11}(t) - P_{01}(t)$$

$$R(t) = \frac{1 - H_1(t)}{1 - H_0(t)}.$$

We note that the series defining C_{01} is Abel summable if and only if $\lim_{t\to 1^-} Q(t)$ exists. We shall prove that this limit exists if and only if $\lim_{t \to 1^-} R(t)$ exists. We have

$$P_{01}^{(n)} = \sum_k ({}^1\bar{F}_{00}^{(k)}P_{01}^{(n-k)} + {}^0\bar{F}_{01}^{(k)}P_{11}^{(n-k)})$$

or

$$P_{01}(t) = A_{00}(t)P_{01}(t) + A_{01}(t)P_{11}(t).$$

Also

$$P_{11}^{(n)} = \delta_{n0} + \sum_k ({}^1\bar{F}_{10}^{(k)}P_{01}^{(n-k)} + {}^0\bar{F}_{11}^{(k)}P_{11}^{(n-k)})$$

or

$$P_{11}(t) = 1 + A_{10}(t)P_{01}(t) + A_{11}(t)P_{11}(t).$$

Solving these equations for $P_{01}(t)$ and $P_{11}(t)$, we find

$$\begin{aligned} Q(t) &= P_{11}(t) - P_{01}(t) \\ &= \frac{1 - A_{00}(t) - A_{01}(t)}{(1 - A_{00}(t))(1 - A_{11}(t)) - A_{01}(t)A_{10}(t)} \\ &= \frac{1}{(1 - A_{00}(t))R(t) + A_{10}(t)}. \end{aligned}$$

Since $A_{00}(1) = {}^1\bar{H}_{00} < 1$ and since $A_{10}(1) = {}^1\bar{H}_{10} > 0$, $\lim_{t\to 1^-} Q(t)$ exists if and only if $\lim_{t\to 1^-} R(t)$ exists.

The example of a nonnormal chain will be like the earlier example in this section, only "doubled." The states are

$$0, a_1, a_2, a_3, \ldots$$

$$1, b_1, b_2, b_3, \ldots,$$

and if $p_i > 0$, $p_i' > 0$, $\sum p_i = 1$, and $\sum p_i' = 1$, then the transition probabilities are

$$\begin{aligned} P_{0a_i} &= p_i = P_{a_i 1} \\ P_{a_i a_i} &= q_i = 1 - p_i \\ P_{1b_i} &= p_i' = P_{b_i 0} \\ P_{b_i b_i} &= q_i' = 1 - p_i'. \end{aligned}$$

All other entries are 0. We see easily that $\bar{H}_{00} = 1$ and that $\alpha = 1^T$ is regular, hence the chain is recurrent and null. For $E = \{0, 1\}$,

$$\bar{F}_{0E}^{(n)} = F_{01}^{(n)} = \sum_i p_i q_i^{n-2} p_i \quad \text{for } n \geq 2.$$

Therefore

$$H_0(t) = \sum_i p_i^2 \sum_{n=2}^{\infty} q_i^{n-2} t^n = \sum_i \frac{p_i^2 t^2}{1 - q_i t}.$$

Similarly,

$$H_1(t) = \sum_i \frac{(p_i')^2 t^2}{1 - q_i' t},$$

and we have defined $R(t)$ by

$$R(t) = \frac{1 - H_1(t)}{1 - H_0(t)}.$$

We again choose the p_i's in blocks as follows. Let $\{n_k\}$ and $\{n_k'\}$ be two rapidly increasing sequences (with magnitude specified later) such that $n_k < n_k' < n_{k+1}$. Let there be n_k consecutive p_i's equal to $\epsilon_k = 1/(2^k n_k)$, and let there be n_k' consecutive (p_i')'s equal to $\epsilon_k' = 1/(2^k n_k')$. The remainder of the proof consists in showing that for suitably chosen $\{n_k\}$ and $\{n_k'\}$

$$\lim_{n \to \infty} R(1 - \epsilon_n) \neq \lim_{n \to \infty} R(1 - \epsilon_n').$$

We shall only sketch the argument.

We begin by estimating $R(1 - \epsilon_n)$ for large n. We have

$$\begin{aligned} H_0(1 - \epsilon_n) &= \sum_i \frac{p_i^2 (1 - \epsilon_n)^2}{1 - (1 - p_i)(1 - \epsilon_n)} \sim \sum_i \frac{p_i^2}{p_i + \epsilon_n} \\ &= \sum_{k=1}^{\infty} n_k \frac{\epsilon_k^2}{\epsilon_k + \epsilon_n} = \sum_{k=1}^{\infty} \frac{1}{2^k} \frac{\epsilon_k}{\epsilon_k + \epsilon_n}. \end{aligned}$$

For $k = n$, $\epsilon_k/(\epsilon_k + \epsilon_n) = \frac{1}{2}$. Choose the sequence n_k so that ϵ_n is negligible compared to ϵ_k when $k < n$. Then

$$H_0(1 - \epsilon_n) \sim \sum_{k=1}^{n-1} 2^{-k} \cdot 1 + 2^{-n} \cdot 2^{-1} = 1 - 2^{-n+1} + 2^{-n-1}.$$

Similarly,

$$H_1(1 - \epsilon_n) \sim \sum_{k=1}^{\infty} 2^{-k} \frac{\epsilon_k'}{\epsilon_k' + \epsilon_n} \sim \sum_{k=0}^{n-1} 2^{-k} = 1 - 2^{-n+1},$$

which differs from $H_0(1 - \epsilon_n)$ if ϵ_n is chosen to be negligible compared with ϵ'_n but much bigger than ϵ'_{n-1}.

$$R(1 - \epsilon_n) \sim \frac{2^{-n+1}}{2^{-n+1} - 2^{-n-1}} = \frac{4}{3}.$$

If $t = 1 - \epsilon'_n$ and n is large we obtain, similarly,

$$H_0(1 - \epsilon'_n) \sim \sum_{k=1}^{\infty} 2^{-k} \frac{\epsilon_k}{\epsilon_k + \epsilon'_n} \sim \sum_{k=1}^{n} 2^{-k} = 1 - 2^{-n}$$

and

$$H_1(1 - \epsilon'_n) \sim \sum_{k=1}^{\infty} 2^{-k} \frac{\epsilon'_k}{\epsilon'_k + \epsilon'_n} \sim \sum_{k=1}^{n-1} 2^{-k} + 2^{-n}.2^{-1}$$
$$= 1 - 2^{-n+1} + 2^{-n-1}.$$

The asymmetry in $H_0(1 - \epsilon'_n)$ and $H_1(1 - \epsilon_n)$ arises because the condition $n_k < n'_k < n_{k+1}$ is not symmetric. We thus find

$$R(1 - \epsilon'_n) \sim \frac{2^{-n+1} - 2^{-n-1}}{2^{-n}} = \frac{3}{2}.$$

Therefore, $\lim_{t \to 1^-} R(t)$ does not exist and C_{01} cannot exist. In particular, P is not normal.

This example has the property that neither C nor G exists. The reverse process has transition probabilities $\hat{P}_{0b_i} = P_{b_i 0} = p'_i = \hat{P}_{b_i 1}$ and $\hat{P}_{1a_i} = p_i = \hat{P}_{a_i 0}$. Thus $\hat{P}$ is the same as P except that the roles of 0 and 1 have been interchanged. The above argument therefore shows that $\hat{C}_{10}$ does not exist, and since $\hat{C}_{10} = G_{01}$ if either exists, G cannot exist.

Not even reversibility ($P = \hat{P}$) is a sufficient condition for a null chain to be normal. A slight modification of the above example provides a counterexample. Let the states be as before, and define p_i and p'_i as above. Let

$$P_{0a_i} = P_{a_i 0} = \tfrac{1}{2}p_i, \qquad P_{1b_1} = P_{b_1 1} = \tfrac{1}{2}p'_i,$$
$$P_{a_i a_i} = 1 - \tfrac{1}{2}p_i, \qquad P_{b_i b_i} = 1 - \tfrac{1}{2}p'_i,$$

and

$$P_{01} = P_{10} = \tfrac{1}{2}.$$

Set all other entries of P equal to 0. Then $P = P^T$, so that $\alpha = 1^T$ and $\hat{P} = IP^TI = P$. But the same kind of computation as for the preceding example shows that C_{01} does not exist. Thus we see that even a symmetric P need not be normal.

Third, we show that the existence of one of C and G does not imply the existence of the other. Again we modify the first example of a nonnormal chain. Let

$$P_{0a_i} = p_i, \quad P_{a_i a_i} = q_i, \quad P_{1b_i} = p_i', \quad \text{and} \quad P_{b_i b_i} = q_i'$$

as before, and use the same p's. But set

$$P_{a_i 0} = P_{a_i 1} = \tfrac{1}{2}p_i \quad \text{and} \quad P_{b_i 0} = P_{b_i 1} = \tfrac{1}{2}p_i'.$$

It is clear that $\bar{F}^{(n)}_{0,\{0,1\}}$ and $\bar{F}^{(n)}_{1,\{0,1\}}$ are the same as before, so that $\lim_{t\to 1^-} R(t)$ does not exist and C_{01} does not exist. On the other hand, the reverse chain is no longer of the same type and the argument for the nonexistence of G fails. In fact,

$$\begin{aligned} {}^0\lambda_1 &= \lim_n \left[P^{(n)}_{01} + \sum_i P^{(n)}_{0a_i}\, {}^0H_{a_i 1} + \sum_i P^{(n)}_{0b_i}\, {}^0H_{b_i 1}\right] \\ &= \lim_n \left[P^{(n)}_{01} + \frac{1}{2}\sum_i (P^{(n)}_{0a_i} + P^{(n)}_{0b_i})\right] \\ &= \frac{1}{2}. \end{aligned}$$

Hence $G_{01} = \frac{1}{2}\, {}^0N_{11}$ exists. Moreover, if x and y are any two states other than 0 and 1,

$${}^x\lambda_y = \tfrac{1}{2}\, {}^xH_{0y} + \tfrac{1}{2}\, {}^xH_{1y}.$$

Therefore all of G exists. The reverse chain is an example in which C exists but G does not.

12. Two-dimensional symmetric random walk

The purpose of this section is to show how the results of Section 8 may be used to work out some of the matrices associated with the two-dimensional symmetric random walk.

First we shall find the operator K. In this example, $K = C = G$ and

$$K_{(x,y),(x',y')} = K_{(x-x',y-y'),(0,0)}.$$

Hence it suffices to compute one column of K. We let

$$k(x, y) = K_{(x,y),(0,0)}.$$

A row of $I - P$ is a charge whose potential is the corresponding row of I. For this process a row of $I - P$ has only finitely many non-zero entries and is therefore a weak charge. By Theorem 9-84, $(I - P)K = -I$. Thus $k(x, y)$ is the average of its values at the four neighboring points, except that at the origin the average is one larger. We know also that $k(0, 0) = 0$. The high degree of symmetry of the chain implies that

$$k(x, y) = k(-x, y) = k(y, x).$$

In particular, the values at the four points neighboring the origin must be the same. Since their average is one, $k(0, 1) = 1$.

We shall need one more result. It can be shown (see Spitzer [1964]) that

$$k(x, x) = \frac{4}{\pi}\left(\frac{1}{1} + \frac{1}{3} + \cdots + \frac{1}{2x - 1}\right) \quad \text{for } x > 0.$$

This identity, together with the properties above, determines the function k.

In fact, it suffices to restrict attention to $0 \le x \le y$. We know that $k(0, 0) = 0$, $k(0, 1) = 1$, and $k(1, 1) = 4/\pi$. If we know the values of $k(x, y)$ up to a given y_0 for all x such that $0 \le x \le y$, then we can find the values of $k(x, y_0 + 1)$ for $0 \le x \le y_0 + 1$. The averaging and symmetry properties give

$$k(0, y_0 + 1) = 4k(0, y_0) - 2k(1, y_0) - k(0, y_0 - 1)$$

$$k(x, y_0 + 1) = 4k(x, y_0) - k(x + 1, y_0) - k(x - 1, y_0) - k(x, y_0 - 1) \quad \text{for } 0 < x < y_0$$

$$k(y_0, y_0 + 1) = 2k(y_0, y_0) - k(y_0 - 1, y_0).$$

And $k(y_0 + 1, y_0 + 1)$ is given by the identity for $k(x, x)$.

The equations above thus are recursion equations for $k(x, y)$. Actually these equations are so simple that we apparently have a very rapid method of computing K_E for large finite sets E. Unfortunately the recursion is highly sensitive to rounding errors.

In Table 9-1 we give values of $k(x, y)$ for a wedge in the plane. The computations were carried out to 9-place accuracy, but by $y = 10$ the effect of rounding errors was noticeable. Any larger table would require much more accuracy of computation.

TABLE 9-1. $k(x, y)$ FOR A WEDGE-SHAPED REGION

$y = 9$	2.429	2.431	2.444	2.461	2.486	2.514	2.546	2.579	2.614	2.649
$y = 8$	2.353	2.357	2.372	2.395	2.424	2.459	2.496	2.535	2.574	
$y = 7$	2.267	2.274	2.293	2.322	2.359	2.400	2.444	2.489		
$y = 6$	2.168	2.178	2.203	2.241	2.288	2.339	2.391			
$y = 5$	2.052	2.065	2.101	2.153	2.213	2.276				
$y = 4$	1.908	1.930	1.984	2.056	2.134					
$y = 3$	1.721	1.762	1.849	1.952						
$y = 2$	1.454	1.546	1.698							
$y = 1$	1.000	1.273								
$y = 0$	0									
	$x = 0$	$x = 1$	$x = 2$	$x = 3$	$x = 4$	$x = 5$	$x = 6$	$x = 7$	$x = 8$	$x = 9$

If we wish to find $k(E)$, λ^E, and P^E for a finite set of points E, we first construct K_E and then compute the inverse K_E^{-1}. Let $z = K_E^{-1}\mathbf{1}$. Since $\alpha = \mathbf{1}^T$ and since K is symmetric, the results in the statement and proof of Corollary 9-92 simplify to

$$k(E) = 1/(\mathbf{1}^T z)$$

$$\lambda_E^E = k(E) z^T$$

$$P^E = I + K_E^{-1} - k(E) z z^T.$$

Calculations for various sets E appear in Tables 9-2, 9-3, and 9-4.

TABLE 9-2. $k(E)$, λ^E, AND P^E WHEN E CONSISTS OF THREE POINTS ON A LINE; $E = \{(0, 0), (a, 0), (2a, 0)\}$

$a = 1$				$a = 2$			
$k(E) = 0.785$				$k(E) = 1.082$			
λ^E	0.393	0.215	0.393	λ^E	0.372	0.256	0.372
P^E	0.460	0.393	0.148	P^E	0.610	0.256	0.134
	0.393	0.215	0.393		0.256	0.488	0.256
	0.148	0.393	0.460		0.134	0.256	0.610
$a = 3$				$a = 4$			
$k(E) = 1.256$				$k(E) = 1.379$			
λ^E	0.365	0.270	0.365	λ^E	0.361	0.277	0.361
P^E	0.663	0.212	0.125	P^E	0.693	0.189	0.118
	0.212	0.576	0.212		0.189	0.621	0.189
	0.125	0.212	0.663		0.118	0.189	0.693

TABLE 9-3. $k(E)$, λ^E, AND P^E WHEN E CONSISTS OF THREE POINTS ON A DIAGONAL; $E = \{(0, 0), (a, a), (2a, 2a)\}$

$a = 1$				$a = 3$			
$k(E) = 0.955$				$k(E) = 1.407$			
λ^E	0.375	0.250	0.375	λ^E	0.360	0.279	0.360
P^E	0.558	0.295	0.147	P^E	0.699	0.185	0.117
	0.295	0.411	0.295		0.185	0.631	0.185
	0.147	0.295	0.558		0.117	0.185	0.699

$a = 5$			
$k(E) = 1.622$			
λ^E	0.356	0.287	0.356
P^E	0.738	0.157	0.106
	0.157	0.687	0.157
	0.106	0.157	0.738

TABLE 9-4. $k(E)$ AND λ^E FOR THREE SETS E

21 points arranged in an isosceles right triangle of base 6

$k(E) = 1.611$

λ^E

0.162					
0.046	0.059				
0.041	0	0.051			
0.041	0	0	0.051		
0.044	0	0	0	0.059	
0.109	0.044	0.041	0.041	0.046	0.162

13 points arranged in a figure ×

$k(E) = 1.778$

λ^E

0.153						0.153
	0.064				0.064	
		0.030		0.030		
			0.015			
		0.030		0.030		
	0.064				0.064	
0.153						0.153

30 points in a 5-by-6 rectangle

$k(E) = 1.670$

λ^E

0.105	0.042	0.038	0.038	0.042	0.105
0.044	0	0	0	0	0.044
0.041	0	0	0	0	0.041
0.044	0	0	0	0	0.044
0.105	0.042	0.038	0.038	0.042	0.105

It is hard to acquire an intuition for the capacities of sets aside from their monotonicity. However, the values of P^E and λ^E are quite intuitive. The latter, in this random walk, may be thought of as the entrance probabilities to E if the chain is started near ∞. For example, in the case of the 5-by-6 rectangle in Table 9-4, it is clear that the corner positions should be considerably more probable than the points on the side. Points on the short sides are more probable than points on the long ones, and the rectangle cannot be entered at an interior point. Equally instructive are the values of λ^E for three-point sets in Tables 9-2 and 9-3. The middle point is least likely to be hit first, but the difference decreases as the points are spread farther apart.

13. Problems

1. Prove that ${}^iN_{jj} = {}^jN_{ii}$ for recurrent sums of independent random variables.

2. Prove that for any null chain and for any states i and j there is a finite set E such that

$$\limsup_n \frac{N_{ij}^{(n)}}{N_{iE}^{(n)}} < \epsilon.$$

3. Let a recurrent chain P be started in state 0. Let α_i^E be the mean number of times in state i in the time required to reach the set E and then return to 0. Thus, for example, if E is the set of all states, then $\alpha^E = (1/\alpha_0)\alpha$.
   (a) Prove that for any set E there is a constant k_E such that $\alpha^E = k_E\alpha$.
   (b) Let Q be the transition matrix for the transient states when 0 is made absorbing and let $\bar{\alpha}$ be the restriction of α to the transient states. Show that if E is a set which does not contain 0, then

$$1 - {}^E\overline{H}_{00} = \frac{1}{\alpha_0}\bar{\alpha}[(I - Q)B^E\mathbf{1}],$$

   where B^E is restricted to the transient states.
   (c) Conclude that if E is a set which does not contain 0, then $1/k_E$ is the transient capacity of the set E in the chain Q, provided the distinguished superregular measure is taken as $\bar{\alpha}$.

4. Prove that for a recurrent chain

$${}^jN_{ik} + {}^iN_{jk} = {}^jN_{ii}\frac{\alpha_k}{\alpha_i}.$$

5. For the symmetric random walk in two dimensions, verify that the function whose (x, y)th entry is $(|x| + |y| + 1)^{-1}$ is a potential, using only Corollary 9-16. Show that its charge f satisfies $\alpha f = 0$.

6. Let P be the one-dimensional symmetric random walk, and let $E = \{0, 1, 2, 3\}$.
   (a) Find B^E, λ^E, P^E.
   (b) Find all potential functions with support in E, and find their charges.
   (c) Compute αf and αg for each.

7. Let P be the symmetric random walk in two dimensions, and let a, b, and c be three distinguished states. We play a game as follows: The process is started in 0. Each time it is in a or b we win a dollar, and each time it is in c we lose two dollars.
   (a) Let $g_0^{(n)} = \mathrm{M}_0$ [expected gain to time n]. Prove that $\lim g_0^{(n)}$ exists, and find a *computable* expression for it.
   (b) What happens if the game is changed so that we lose only one dollar when the process is in state c?

Problems 8 to 10 lead to the fact that in a normal chain the union of a small set and a finite set is small.

8. Let E be a small set, and let $F = E \cup \{k\}$ have one more point. Show that

$$B_{ij}^E = B_{ij}^F + B_{ik}^F B_{kj}^E \quad \text{for all } j \in E.$$

   From this equation show that if λ_k^F exists, then λ^F exists and F is a small set.

9. In Problem 8 show that

$$^kH_{ij} = \sum_{m \in E} B^F_{im}\, {}^kH_{mj} \quad \text{for all } j \in E.$$

Use the identity of Problem 8 to eliminate the factors B^F_{im}, and solve the resulting identity for B^F_{ik}. Prove from this result that λ^F_k exists, provided P is normal.

10. Use the results of the two previous problems to prove that the union of a small set and a finite set is small in a normal chain.

Problems 11 to 22 develop a new null example and use it to illustrate results in the chapter. The state space consists of all points $z = (x, y)$ in the plane with integer coordinates ≥ 1. It will be convenient to let $n = x + y$. We let (1, 1) be our state 0. Define

$$P_{(x,y),(x+1,y)} = \frac{x}{x+y+1}, \quad P_{(x,y),(x,y+1)} = \frac{y}{x+y+1},$$

and

$$P_{(x,y),0} = \frac{1}{x+y+1}.$$

11. Verify that P is null recurrent and that $\alpha_z = 2/[(n-1)n]$.
12. Compute $\hat{P}$.
13. Prove that $P^{(m)}_{0z}$ is the same for all z with $n = x + y$ fixed. Do the same for $\hat{P}^{(m)}_{0z}$.
14. Let $E = \{z \mid x + y \leq n_0\}$. Find λ^E and $\hat{\lambda}^E$.
15. Show that

$$^0N_{zz'} = \begin{cases} \binom{x'-1}{x-1}\binom{y'-1}{y-1}\Big/\binom{n'}{n}, & \text{if } x \leq x',\ y \leq y' \\ 0 & \text{otherwise.} \end{cases}$$

16. Let f be defined by $f_0 = -1$, $f_{(3,2)} = 10$, and $f_z = 0$ otherwise. Show that f is a charge, and use parts (1) and (3) of Theorem 9-15 to find the potential g.
17. Check that $f = (I - P)g$ for the functions of Problem 16. Does $\alpha g = 0$? Verify that $\lambda^E g = 0$ for all finite sets containing the support.
18. Show that $G_{0z} = 0$ and $G_{z0} = (n-1)n/2$.
19. Use Problem 18 and Proposition 9-45 to show that

$$G_{zz'} = \frac{(n-1)n}{(n'-1)n'} - {}^0N_{zz'}.$$

20. Verify that the potential g found in Problem 16 is $-Gf$.
21. Find C, and compute a potential measure of finite support in two different ways (in analogy with Problems 16 and 20).
22. Let E be the triangular set of Problem 14, and let x be its characteristic function. Verify Proposition 9-43.

Problems 23 to 26 develop some theoretical results for an ergodic chain in terms of the operator K.

23. Express K in terms of M and $\hat{M}$.
24. Prove that $M_{ij} = (K_{ij} - K_{jj})/\alpha_j$.
25. Show that $\hat{M}_{ij} - M_{ji} = k(\{i\}) - k(\{j\})$.
26. What happens to the formulas in Problems 24 and 25 if the reference point 0 is changed?

Problems 27 to 32 carry this development further for finite recurrent chains.

27. Show that $k(S) = M_{\alpha 0} = \hat{M}_{\alpha 0}$.
28. Prove that $K\mathbf{1} = k(S)\mathbf{1}$ and $\alpha K = k(S)\alpha$.
29. Prove that $M\alpha^T = c\mathbf{1}$, where $c = k(S) - \sum_i k(\{i\})\alpha_i$.
30. Prove that
$$K = \left(\frac{1}{k(S)}\mathbf{1}\alpha + P - I\right)^{-1}.$$
31. Prove that the set of charges is the same as the set of potentials.
32. To what extent do the results of Problems 27 to 31 generalize to strong ergodic chains?

Problems 33 to 39 are intended to illustrate Problems 23 to 32 for the Land of Oz example, which was defined in Chapter 4. [See also Chapter 6, Problem 1.] We choose the middle state (nice weather) to be the distinguished state 0.

33. Show that $P = \hat{P}$ (the chain is reversible).
34. Find M.
35. Find K, using the result of Problem 23.
36. Find $k(S)$, using the result of Problem 27.
37. Find K, using the result of Problem 30, and compare with the value of K found in Problem 35.
38. Check the results given in Problems 24, 25, 28, and 29.
39. Find the most general charge and the most general potential function. Verify that the set of charges is the same as the set of potentials.

Problems 40 to 48 work out the probabilistic solution of the so-called Second Boundary Value Problem in the sense that Theorem 8-41 presented the solution of the First Boundary Value Problem. Let P be an absorbing chain whose transient states communicate and whose absorbing states form a finite set B. B is thought of as our "boundary." To each state k in B we associate a "neighboring" transient state k'. For a given function h, we define its normal derivative d_k at k to be $h_k - h_{k'}$. The problem is to find a function h which is regular on the transient states and has specified normal derivatives.

40. Prove that $h_{\bar{B}} = NRh_B$ for any solution.
41. Modify the original chain so that instead of stopping at an absorbing state k, it moves to the neighboring k' with probability 1. Show that this new chain is recurrent.

42. Let P^* be the transition matrix of the modified chain watched in B. Prove that this is an ergodic chain.
43. Show that the requirement that h have the specified normal derivatives can be written as $(I - P^*)h_B = d$, where d has the given values d_k as components.
44. Prove that $\alpha^* d = 0$ is a necessary condition for a solution to exist.
45. Show that $\alpha^* d = 0$ is also sufficient by showing that d is a charge, that its potential will serve as h_B, and that the h supplied by Problem 40 is a solution.
46. Prove that the most general solution differs from the given one only by a constant.
47. Prove that if the modified chain (indexed on all states) is a normal chain, then the most general solution is

$$h = -K\begin{pmatrix} d \\ 0 \end{pmatrix} + c\mathbf{1}.$$

48. Show that if the transient states are the lattice points in a bounded convex set in n-dimensional Euclidean space and if the process moves as a symmetric random walk which is stopped when it moves out of the convex set, then we can apply the previous results.

Problems 49 to 53 give a complete characterization of degenerate chains.

49. Prove that if P is degenerate, so is $\hat{P}$.
50. Show that if P is degenerate and if we let $i < j$ stand for ${}^j\lambda_i = 1$, then $<$ is a simple ordering.
51. Prove that if $k < i < j$, then ${}^jH_{ki} = 1$. Deduce from this fact that, in moving to the right, the process can move at most one step at a time. [*Hint:* Consider λ^E for $E = \{k, i, j\}$.]
52. Prove that the ordering of states must be that of the integers, the positive integers, or the negative integers.
53. Show that the basic example and its reverse illustrate two of the possible orderings, and construct an example of the third.

CHAPTER 10

TRANSIENT BOUNDARY THEORY

1. Motivation for Martin boundary theory

For purposes of motivation it is convenient to think of the state space of a Markov chain P with only transient states as being similar to the open unit disk of two-dimensional Euclidean space. In two-space the boundary of the disk—namely the circle S^1—has the property that there is a one-one correspondence between the non-negative harmonic functions $h(re^{i\theta})$ in the disk and the non-negative Borel measures μ^h on the circle. The correspondence is

$$h(re^{i\theta}) = \int_{S^1} P(re^{i\theta}, t) d\mu^h(t), \tag{*}$$

where $P(re^{i\theta}, t)$ is the Poisson kernel

$$\frac{1 - r^2}{1 - 2r\cos(\theta - t) + r^2}.$$

Transient boundary theory seeks an analogous representation theorem for all non-negative P-regular functions defined on the state space.

The first problem that arises is to find what the analogs of the circle (the boundary) and the kernel should be. We would like a representation

$$h(i) = \int K(i, x) d\mu^h(x).$$

In the case of the disk, a calculation with Green's identities shows that any kernel $P(re^{i\theta}, t)$ giving rise to the correspondence (*) and satisfying

$$\frac{1}{2\pi} \int_0^{2\pi} P(re^{i\theta}, t) dt = 1$$

must be the normal derivative at t of the Green's function for the disk relative to the point $re^{i\theta}$. That is,

$$P(re^{i\theta}, t) = \left[\frac{\partial}{\partial n} G(\cdot, re^{i\theta})\right]_t.$$

Application of l'Hospital's rule shows that

$$\lim_{\substack{z\to t\\ \text{radially}}} \frac{G(z, re^{i\theta})}{G(z, p)} = \left[\frac{\partial}{\partial n} G(\cdot, re^{i\theta}) \Big/ \frac{\partial}{\partial n} G(\cdot, p)\right]_t$$
$$= P(re^{i\theta}, t)/P(p, t),$$

where p is any fixed reference point in the disk. Hence, except for the positive factor $P(p, t)$ which depends on t but not on $re^{i\theta}$, the Poisson kernel is equal to

$$\lim_{\substack{z\to t\\ \text{radially}}} \frac{G(z, re^{i\theta})}{G(z, p)}. \tag{**}$$

Therefore this last function may be used in the representation (*) in place of the Poisson kernel; the distinction between the kernels is just a normalizing factor (depending on t) which can be absorbed by changing the measures.

Two comments are in order. First, the limit in (**) need not be taken radially. Any method of approach of z to t, as long as z stays in the interior of the disk, will give the same value. Second, the considerations above apply equally well to any domain in n-dimensional space with a sufficiently smooth boundary. Although the explicit form of the kernel will vary from region to region, it will always be connected to the Green's function in the way we have just described.

R. S. Martin [1941] made use of these observations to define an ideal boundary for an *arbitrary* domain in Euclidean space. If the Green's function for the region is denoted $G(z, y)$, he noted that points t on the ordinary topological boundary of the region did not necessarily have the property that

$$\lim_{z\to t} \frac{G(z, y)}{G(z, p)}$$

exists. He suggested that distinct ideal boundary points u should be associated to subsequences $\{z_n\}$ which yield distinct values for the limits

$$\lim_{z_n\to t} \frac{G(z_n, y)}{G(z_n, p)} = K(y, u).$$

He went on to show that the desired representation theorem is indeed obtained in terms of this boundary and the kernel $K(y, u)$.

Doob [1959], taking advantage of the fact that the N-matrix for a transient Markov chain is the analog of the Green's function (see Proposition 7-4), showed that Martin's approach could be used to obtain a boundary for Markov chains. (We remark that N_{ij} corresponds to $G(j, i)$ with the indices in the reversed order.) As the analog of Martin's kernel he used limits on j of expressions of the form N_{ij}/N_{0j}.

There is a minor restriction imposed by the Doob approach, namely that N_{0j} is assumed non-zero for all j. For a more general chain in which it is not possible to get from state 0 to every other state, what Doob did was to consider only those states that could be reached from 0. We shall not follow him in this respect. We simply use limits of $N_{ij}/(\pi N)_j$ instead, where π is a probability vector such that πN is strictly positive. In terms of this kernel there is a natural space to try as the one corresponding to the closed unit disk. The space should consist of one point for each possible limit of $N_{ij}/N_{\pi j}$. Actually we shall find that this space is too large—that the space has to be cut down a bit for the representation to be unique. The price of uniqueness is that the cut-down space is not compact.

The introduction of π in place of 0 itself leads to a problem. The representation will have to be restricted to π-integrable regular functions h, those for which πh is finite. This requirement evaporated in Martin's or Doob's treatment because for them π assigned unit mass to a point 0 and $h(0)$ was automatically finite.

Hunt [1960] gave a new approach to Martin boundary theory for Markov chains which was more probabilistic in nature than Doob's. We follow Hunt's probabilistic approach, except that we use a different metric and get a boundary which is more like Doob's.

2. Extended chains

We begin by introducing the machinery which we shall use in later sections to develop Martin boundary theory for Markov chains. We are going to use a broader notion of Markov chain than we have been considering so far—namely, a process whose time index starts at $-\infty$ and whose behavior is Markovian only after it has entered certain sets.

That is, we extend the concept of Markov chain in two ways. First, we shall allow any finite measure π as a starting distribution. This is only a minor modification in the theory and is a convenience in that it removes the necessity of normalization in certain constructions. Second, we shall extend the Markov chain to the past, that is, to a stochastic process $\{x_n\}$ where n runs through all the integers (including negative integers). This second extension is an essential one and will be the main topic of this section.

First we must extend the concept of a stochastic process. The **state space** will be a denumerable set S (with at least two elements) and two distinguished other states a and b. The underlying set Ω of the measure space will consist of all doubly infinite sequences

$$\omega = (\ldots, c_{-2}, c_{-1}, c_0, c_1, c_2, \ldots)$$

such that

(1) $c_n \in S$ or $c_n = a$ or $c_n = b$.
(2) If $c_n = a$, then $c_m = a$ for all $m < n$.
(3) If $c_n = b$, then $c_m = b$ for all $m > n$.
(4) $c_n \in S$ for at least one n.

The interpretation of (2) and (3) is that state a stands for "not yet started" and state b stands for "stopped." Thus (4) has the meaning that each path in Ω represents some nontrivial possibility for the process. We shall refer to Ω as a **double sequence space**.

We define the outcome functions x_n as usual except that n may be any integer. A **basic cylinder set** is any truth set in Ω of a statement of the form

$$x_m = c_m \wedge x_{m+1} = c_{m+1} \wedge \cdots \wedge x_n = c_n,$$

where at least one c_i is an element of S. The field generated by the basic cylinder sets is denoted $\mathscr{F}$, and the smallest Borel field containing $\mathscr{F}$ is $\mathscr{G}$.

Definition 10-1: An extended stochastic process $\{x_n\}$ is the set of outcome functions for a measure space $(\Omega, \mathscr{G}, \Pr)$ such that

(1) Ω is a double sequence space with state space $S \cup \{a, b\}$.
(2) $\mathscr{G}$ is the smallest Borel field containing the field of cylinder sets of Ω.
(3) $\Pr[\{\omega \mid x_n = i\}] < \infty$ for every integer n and every i in S.

Note that we do not augment the measure space $(\Omega, \mathscr{G}, \Pr)$ by allowing all subsets of sets of measure zero to be measurable.

We shall use interchangeably the notations $\Pr[P]$ and $\Pr[p]$, where P is the truth set of the statement p. Thus the third condition may be replaced by the condition $\Pr[x_n = i] < \infty$ for all n and for all i in S. From it and from condition (4) in the definition of Ω, we find that $\Pr$ must be sigma-finite. However, the measure $\Pr$ need not be finite.

As promised, Definition 10-1 extends the definition of stochastic process in two ways: The time index n runs through all the integers, and the measure $\Pr$ need not be a probability measure or even a finite measure.

In examples it is necessary to have a method of constructing the measures for extended stochastic processes. If the measure Pr has already been defined consistently on basic cylinder sets, we need a version of the Kolmogorov Theorem to prove that Pr is completely additive on $\mathscr{F}$. Theorem 1-19 then will give the extension of Pr to all of $\mathscr{G}$. At this point, therefore, we stop to outline a proof that Pr is completely additive on $\mathscr{F}$.

Now the argument of Lemma 2-1 easily shows that Pr is non-negative and additive. Extend Ω to include the set of *all* doubly infinite sequences of a's, b's, and elements of S, and define Pr to be zero on all cylinder subsets of the set added. Then if Pr were a finite measure, we could temporarily rearrange the time scale and then conclude complete additivity by Theorem 2-4. But, in general, Pr is merely sigma-finite and therefore we shall write it as the countable sum of totally finite non-negative additive set functions, each of which is a measure on cylinder sets depending on a bounded time interval. Each of the summands is completely additive by the above argument, and therefore Pr is completely additive by Lemma 1-3. Thus all we need to do is decompose Pr as such a sum. The countable family of statements indexed by $i \in S$ and by $n \geq 0$, consisting of

$$x_0 = i$$

$$x_{-n-1} = i \wedge x_{-n} = b$$

$$x_n = a \wedge x_{n+1} = i,$$

is a disjoint exhaustive family in the original double sequence space, and each statement is assigned finite Pr-measure. For each of these statements q, define

$$\Pr^{(q)}(E) = \Pr[E \cap \{\omega \mid q\}]$$

for E in the field of cylinder sets. Then the family $\{\Pr^{(q)}\}$ is the required family of set functions.

We now fix our attention on a single extended stochastic process $\{x_n\}$.

Although we may be dealing with an infinite measure space, the conditional probability

$$\Pr[p \mid q] = \Pr[p \wedge q]/\Pr[q]$$

is still well defined as long as $0 < \Pr[q] < \infty$. We define $\Pr[p \mid q]$ to be zero if $\Pr[q] = 0$.

Definition 10-2: For $E \subset S$ and for any $\omega \in \Omega$ such that $x_n(\omega) \in E$ for some n, let $\mathbf{u}_E(\omega)$ be the infimum of all n such that $x_n(\omega) \in E$ and let $\mathbf{v}_E(\omega)$ be the supremum of all such n. Define $\mathbf{u} = \mathbf{u}_S$ and $\mathbf{v} = \mathbf{v}_S$; $\mathbf{u}$ and $\mathbf{v}$ are called the initial time and the final time, respectively.

By condition (4) of the definition of Ω, we see that $\mathbf{u}(\omega)$ and $\mathbf{v}(\omega)$ are defined for all ω. Moreover, $\mathbf{u}_E(\omega) \leq \mathbf{v}_E(\omega)$ whenever $\mathbf{u}_E(\omega)$ and $\mathbf{v}_E(\omega)$ are defined. The values $\mathbf{u}_E(\omega) = -\infty$ and $\mathbf{v}_E(\omega) = +\infty$ are possible. If $x_n(\omega) \in E$ for some n, we have

$$x_n(\omega) = \begin{cases} a & \text{if } n < \mathbf{u}(\omega) \\ \text{element of } S - E & \text{if } \mathbf{u}(\omega) \leq n < \mathbf{u}_E(\omega) \\ \text{element of } S - E & \text{if } \mathbf{v}_E(\omega) < n \leq \mathbf{v}(\omega) \\ b & \text{if } \mathbf{v}(\omega) < n. \end{cases}$$

Proposition 10-3: The functions $\mathbf{u}_E$ and $\mathbf{v}_E$ have a $\mathscr{G}$-measurable subset of Ω as domain and are each $\mathscr{G}$-measurable.

PROOF: We prove the result for $\mathbf{u}_E$. We have

$$\{\omega \mid \mathbf{u}_E(\omega) \leq k\} = \bigcup_{i \in E} \bigcup_{n=-\infty}^{k} \{\omega \mid x_n(\omega) = i\},$$

and the union of these sets on k is the domain of $\mathbf{u}_E$.

Definition 10-4: Let E be a subset of S and define

$$y_n(\omega) = x_{(n + \mathbf{u}_E(\omega))}(\omega)$$

for all $n \geq 0$ and for all ω such that $\mathbf{u}_E(\omega) > -\infty$. Let $\bar{\Omega}$ be the ordinary sequence space with state space $S \cup \{b\}$ with the measure of each measurable set $A \subset \bar{\Omega}$ defined to be

$$\Pr[\{\omega \mid (y_0(\omega), y_1(\omega), \ldots) \in A\}].$$

The measure space $\bar{\Omega}$ and its outcome functions together are called the **process watched starting in** E.

The process watched starting in a set E is an ordinary stochastic process, except that the starting distribution need not be a probability measure and can possibly be infinite.

Let $j \in S$. The mean number ν_j of times that the process $\{x_n\}$ is in j can, as usual, be computed by

$$\nu_j = \sum_n \Pr[x_n = j],$$

except the sum is over all integers n. By Definition 10-1, each summand is finite, but the sum may be infinite.

Definition 10-5: An extended stochastic process $\{x_n\}$ is an **extended chain** with transition matrix P if the following conditions hold for each finite subset E of S:

(1) The domain of $\mathbf{u}_E$ has positive measure.
(2) $\Pr[\mathbf{u}_E = -\infty] = 0$.
(3) The process watched starting in E is a Markov chain with transition matrix P and finite starting measure (but not necessarily a probability measure).
(4) For all $j \in S$, $\nu_j < \infty$.

Note that the transition matrix P of an extended chain necessarily satisfies $P\mathbf{1} = \mathbf{1}$. The state space of P never needs to be any bigger than $S \cup \{b\}$, but as we shall see shortly it must contain S. If b is in the state space, it clearly must be an absorbing state.

From the definition of the process watched starting in E, we see that the total starting measure for the process is equal to the measure of the set of paths on which there is a first entry to E. That is, it is the measure of the set where $\mathbf{u}_E > -\infty$. By conditions (1) and (2), this measure is positive. Hence the process watched starting in E has a starting measure which is not identically zero.

Applying this observation to the one-point set $\{j\}$, we see that j must be included in the state space of P.

Let $\{x_n\}$ be an extended stochastic process satisfying (1), (2), and (3). We shall derive as Proposition 10-6 a necessary and sufficient condition for (4) to hold. Let E be a finite set of S. We introduce the abbreviations

$$\mu_i^{E,m} = \Pr[\mathbf{u}_E = m \wedge x_m = i]$$

and

$$\mu^E = \sum_m \mu^{E,m}.$$

Then μ^E is the starting measure of the process watched starting in E and is a finite measure with support in E by (3). Our remarks above showed that μ^E is not identically zero.

Since the process watched starting in E is a Markov chain and since (2) holds, the following computation is justified: For $j \in E$,

$$\Pr[x_n = j \wedge x_{n+1} = k \wedge x_{n+2} = s] = \sum_{m,i} \mu_i^{E,m} P_{ij}^{(n-m)} P_{jk} P_{ks},$$

where $P^{(n-m)} = 0$ if $n < m$. A similar computation of $\Pr[p]$ is possible for any p such that p is false if E is never entered and p depends only on outcomes after E is entered.

As an application of this calculation, we can relate condition (4) to properties of P.

Proposition 10-6: Let $\{x_n\}$ be an extended stochastic process satisfying conditions (1), (2), and (3) for an extended chain. Then for $j \in E$,

$$\nu_j = (\mu^E N)_j.$$

Moreover, the process is an extended chain if and only if P is the transition matrix of a transient chain whose transient states are S and which may have b as an absorbing state.

PROOF: We have

$$\nu_j = \sum_n \Pr[x_n = j] = \sum_{n,m,i} \mu_i^{E,m} P_{ij}^{(n-m)} = \sum_{m,i} \mu_i^{E,m} N_{ij} = \sum_i \mu_i^E N_{ij}$$

or

$$\nu_j = (\mu^E N)_j.$$

Taking $E = \{j\}$, we find

$$\nu_j = \mu_j^E N_{jj}.$$

Now μ_j^E is assumed finite by (3) and it is strictly positive by (1). Hence ν_j is finite for all $j \in S$ if and only if all elements of S are transient for a Markov chain with transition matrix P. Furthermore, a cannot occur as a state, and if b occurs, it must be an absorbing state.

Corollary 10-7: Let $\{x_n\}$ be an extended stochastic process satisfying conditions (1), (2), and (3) for an extended chain. Then $\nu_j > 0$ for all j.

PROOF: We have $\nu_j = \mu_j^{\{j\}} N_{jj}$, and each factor on the right side is positive.

An important example, but by no means the most general example, of an extended chain is obtained as follows. Let P be a Markov chain with all states transient and let π be a starting distribution such that πN is strictly positive. (For instance, let π assign weight 2^{-n} to the nth state.) Let $\bar{P}$ be the enlarged chain obtained by adding the absorbing state b. Form an extended stochastic process by defining a measure on cylinder sets as follows: Every basic statement containing the assertion $x_n = i$ for $n < 0$ and $i \neq a$ or the assertion $x_m = a$ for $m \geq 0$ gets probability zero. The statement

$$x_{-m} = a \wedge \cdots \wedge x_{-1} = a \wedge x_0 = i \wedge x_1 = j \wedge x_2 = k$$
$$\wedge \cdots \wedge x_{n-1} = r \wedge x_n = s$$

gets probability $\pi_i \bar{P}_{ij} \bar{P}_{jk} \ldots \bar{P}_{rs}$. The probabilities of all other basic cylinder statements can be obtained from these by adding the probabilities of a suitable number of statements of the form just described. The claim is that this extended stochastic process $\{x_n\}$ is an extended chain with transition matrix $\bar{P}$. Property (1) follows from the fact that $\pi N > 0$. Property (2) is an immediate consequence of the definition of the process. Property (3) follows from Theorem 4-9; the starting distribution for the process watched starting in E is πB^E. Finally Property (4) comes from Proposition 10-6; alternatively we could compute directly that the mean number of times in state j is $(\pi N)_j$. We shall call this process the **extended chain associated with π and P.**

If $\{x_n\}$ is an extended chain and E is a finite set in S, we define $\nu^E = \mu^E N$. We know that the process watched starting in E is a Markov chain with starting distribution μ^E. Hence ν_j^E is the mean number of times in j in this Markov chain. That is, it is the mean number of times in j *after entering* E in the extended chain. From this interpretation we see that ν^E is monotone increasing in E and that $\nu_j^E = \nu_j$ for $j \in E$. In order to get an interpretation of ν^E and μ^E in terms of ν, we shall generalize the notion of balayage potential as defined in Chapter 8.

If P is a Markov chain with all states transient, if h is a non-negative finite-valued superregular function, and if E is a finite set of states, we define the **balayage potential** of h on E to be the function $B^E h$. By Lemma 8-22, $B^E h$ is a pure potential with support in E. Now $B^E h$ is the unique pure potential with support in E which agrees with h on E, since if $\bar{h}$ is another such potential, $\bar{h}$ must be the unique balayage potential of $B^E h$ on E (Theorem 8-46) and hence must equal $B^E h$. Moreover, if E_n is an increasing sequence of finite sets with union the set of all states, then the charges of $B^{E_n} h$, namely $(I - P)(B^{E_n} h)$, converge to $(I - P)h$, since

$$\lim_n PB^{E_n} h = Ph$$

by part (4) of Proposition 8-16 and by monotone convergence.

Let us dualize these results. Let γ be a non-negative finite-valued superregular measure, and let E be a finite set. Then there is a unique pure potential measure with support in E which agrees with γ on E. This potential is defined to be the balayage potential of γ on E. It has the property that if E_n is an increasing sequence of sets with union the set of all states, then the balayage charges of γ on E_n converge to $\gamma(I - P)$. By Proposition 6-16, the balayage potential of γ on E is

$$(\gamma_E \quad \gamma_E U({}^E N)_{\tilde{E}}),$$

where

$$P = \begin{pmatrix} T & U \\ R & Q \end{pmatrix}$$

and $({}^E N)_{\tilde{E}} = \sum Q^n$.

If we let γ^E be the balayage potential of γ on E and if β^E is the charge of γ^E, then $E \subset F$ implies $\beta^E = \beta^F B^E$. To see this equality, we note that β^E and $\beta^F B^E$ are both pure charges with support in E and that the potential of $\beta^F B^E$, when restricted to E, is

$$(\beta^F B^E N)_E = (\beta^F N - \beta^F {}^E N)_E = (\beta^F N)_E = \gamma_E = (\beta^E N)_E$$

by Lemma 8-17. Hence the potentials of β^E and $\beta^F B^E$ agree on E, and they must therefore agree everywhere. Thus

$$\beta^E = \beta^F B^E$$

by Theorem 8-4.

Our characterization of ν^E and μ^E in terms of balayage potentials is the content of the next proposition.

Proposition 10-8: For every extended chain with transition matrix P,

(1) ν is a superregular measure for P_S.
(2) $\nu(I - P_S) = \mu$, where $\mu = \lim_E \mu^E$ as E increases to S.
(3) ν^E and μ^E are the balayage potential and charge, respectively, for ν on E.

PROOF: We have

$$\nu^E P_S = \mu^E N P_S = \mu^E(N - I) = \nu^E - \mu^E.$$

Along any increasing sequence of sets E_n with union S, ν^E increases to ν. Hence by monotone convergence

$$\nu P_S = \nu - \lim_n \mu^{E_n}.$$

This equality implies that

$$\nu P_S = \nu - \lim_{E \uparrow S} \mu^E$$

and proves (1) and (2). To prove (3) we need only remark that ν^E is a pure potential with support in E which agrees with ν on E. Hence it must be the balayage potential.

Proposition 10-8 has as a converse the following theorem, which asserts roughly that any superregular measure can be represented as the vector of mean times in the states of some extended chain. This result will not be used until Section 11, and its proof, which is quite long, will not be given until after that section.

Theorem 10-9: Let P^* be the transition matrix of a transient chain with $P^*\mathbf{1} = \mathbf{1}$ and with at most one absorbing state b, and let ν be a non-negative finite-valued measure defined on the transient states and superregular for the restriction of P^* to the transient states. Let S be the support of ν, and suppose S has at least two elements. Then there is an extended chain $\{x_n\}$ having $P^*_{S\cup\{b\}}$ as transition matrix and having ν as its vector of mean times in the states of S. Furthermore, if

$$\nu_S = \mu N + \rho$$

is the unique decomposition of ν_S with ρ regular for P^*_S, then μN is contributed by the paths ω with $\mathbf{u}(\omega) > -\infty$ and ρ is contributed by the paths with $\mathbf{u}(\omega) = -\infty$.

To conclude this section we shall define what we mean by the reverse of an extended stochastic process, and we shall prove that the reverse of an extended chain is an extended chain. The transition matrix of the reverse, at least when restricted to S, will turn out to be the ν-dual of the transition matrix, restricted to S, of the original process. We need the following lemma, whose proof uses the calculation preceding Proposition 10-6.

Lemma 10-10: If $\{x_n\}$ is an extended chain with transition matrix P and if k is in a finite set E in S, then

$$\Pr[x_{v_E-2} = i \wedge x_{v_E-1} = j \wedge x_{v_E} = k] = \nu_i P_{ij} P_{jk} e^E_k,$$

where e^E is the escape vector for P.

PROOF:

$$\begin{aligned}
\Pr[x_{v_E-2} &= i \wedge x_{v_E-1} = j \wedge x_{v_E} = k] \\
&= \sum_n \Pr[x_n = i \wedge x_{n+1} = j \wedge x_{n+2} = k \\
&\qquad\qquad \wedge \{x_n\} \text{ not in } E \text{ after time } (n+2)] \\
&= \sum_{m,s,n} \mu^{E,m}_s P^{(n-m)}_{si} P_{ij} P_{jk} e^E_k \quad \text{by the calculation} \\
&= \sum_n \Pr[x_n = i] P_{ij} P_{jk} e^E_k \quad \text{by the calculation again} \\
&= \nu_i P_{ij} P_{jk} e^E_k.
\end{aligned}$$

For any point ω in Ω we define ω' to be the point in Ω with

$$x_n(\omega') = \begin{cases} x_{-n}(\omega) & \text{if } x_{-n}(\omega) \in S \\ a & \text{if } x_{-n}(\omega) = b \\ b & \text{if } x_{-n}(\omega) = a. \end{cases}$$

Definition 10-11: Let $\{x_n\}$ be an extended chain defined on Ω with measure μ. Set

$$\Pr'[\omega \in A] = \Pr[\omega' \in A]$$

for all sets A for which the right side is defined. The extended stochastic process on Ω defined by the measure $\Pr'$ is called the **reverse** of the extended chain.

Proposition 10-12: If $\{x_n\}$ is an extended chain with transition matrix P, then its reverse is also an extended chain and its transition matrix $\hat{P}$ satisfies

$$\hat{P}_{ji} = \nu_i P_{ij}/\nu_j$$

for all states i and j in S.

PROOF: From the definition of ω', we see that

$$\hat{\mathbf{u}}_E(\omega) = -\mathbf{v}_E(\omega').$$

Hence

$$\Pr'[\omega \in \text{domain } \hat{\mathbf{u}}_E] = \Pr[\omega \in \text{domain } \mathbf{v}_E] = \Pr[\omega \in \text{domain } \mathbf{u}_E] > 0$$

and (1) holds. Since the chain watched starting in E is in the finite set E infinitely often with probability 0 (second half of Proposition 10-6),

$$0 = \Pr[\mathbf{v}_E = +\infty] = \Pr'[\hat{\mathbf{u}}_E = -\infty].$$

Thus (2) holds.

Next, we show that the reverse process watched starting in E is a Markov chain with transition matrix $\hat{P}$. We shall compute only a typical conditional probability: Let $k \in E$ and first suppose $i \neq b$. Since

$$\begin{aligned}\Pr'[x_{\hat{\mathbf{u}}_E} = k \wedge x_{\hat{\mathbf{u}}_E+1} = j \wedge x_{\hat{\mathbf{u}}_E+2} = i] \\ = \Pr[x_{\mathbf{v}_E-2} = i \wedge x_{\mathbf{v}_E-1} = j \wedge x_{\mathbf{v}_E} = k] \\ = \nu_i P_{ij} P_{jk} e_k^E\end{aligned}$$

by Lemma 10-10, we have, provided the condition has positive $\Pr'$-measure,

$$\begin{aligned}\Pr'[x_{\hat{\mathbf{u}}_E+2} = i \mid x_{\hat{\mathbf{u}}_E} = k \wedge x_{\hat{\mathbf{u}}_E+1} = j] &= (\nu_i P_{ij} P_{jk} e_k^E)/(\nu_j P_{jk} e_k^E) \\ &= \nu_i P_{ij}/\nu_j \\ &= \hat{P}_{ji}.\end{aligned}$$

Next we compute the typical probability

$$\Pr'[x_{\hat{\mathbf{u}}_E} = k \wedge x_{\hat{\mathbf{u}}_E+1} = j \wedge x_{\hat{\mathbf{u}}_E+2} = b].$$

Since we know b is absorbing, we may assume $j \neq b$. Then this probability is

$$\Pr'[x_{\hat{u}_E} = k \wedge x_{\hat{u}_E+1} = j] - \Pr'[x_{\hat{u}_E} = k \wedge x_{\hat{u}_E+1} = j \wedge x_{\hat{u}_E+2} \in S]$$
$$= \nu_j P_{jk} e_k^E - \sum_{i \in S} \nu_i P_{ij} P_{jk} e_k^E.$$

Hence if $\Pr'[x_{\hat{u}_E} = k \wedge x_{\hat{u}_E+1} = j] > 0$, then

$$\Pr'[x_{\hat{u}_E+2} = b \mid x_{\hat{u}_E} = k \wedge x_{\hat{u}_E+1} = j] = (\nu_j - \sum_{i \in S} \nu_i P_{ij})/\nu_j.$$

(Notice this probability is non-negative because ν is P_S-superregular.) Therefore the reverse process watched starting in E is a Markov chain.

The total starting measure is finite for the reverse process watched starting in E because

$$\sum_i \Pr'[x_{\hat{u}_E} = i] = \sum_i \Pr[x_{v_E} = i]$$
$$= \sum_i \nu_i e_i^E \quad \text{by Lemma 10-10}$$
$$< \infty$$

by (4) for the given process and by the finiteness of the set E. Hence (3) holds.

Finally we prove (4) for the reverse process. The same argument as in Proposition 6-12 shows that

$$\hat{N}_{ji} = \nu_i N_{ij}/\nu_j.$$

Hence P is transient and (4) holds by Proposition 10-6.

3. Martin exit boundary

We now define the Martin exit boundary of a transient chain with respect to a given starting distribution. With this boundary we shall be able to describe the long-range behavior of the process and we shall obtain a Poisson integral-type representation for all finite-valued non-negative superregular functions which are integrable with respect to the starting distribution.

Let P be a Markov chain with all states transient and let π be a starting vector ($\pi \geq 0$ and $\pi\mathbf{1} = 1$). Throughout our discussion P and π will be fixed. The vector πN is non-negative, finite-valued, and superregular. Ordinarily we shall assume that π has been chosen so that πN is strictly positive; that is, so that there is positive probability of reaching any state eventually. But for technical reasons which will arise when we consider h-processes, it will be convenient to adopt conventions about what to do when πN has some zero entries. These conventions we shall discuss at the end of this section, and except when stated otherwise we shall assume that πN is everywhere positive. Let S be the state space of P.

Since $N_{\pi j} = (\pi N)_j$ has been assumed positive, we may define

$$K(i, j) = N_{ij}/N_{\pi j}.$$

The notation $K(\cdot, j)$ will mean $K(i, j)$ considered as a function of the first variable with j fixed. Then for each j, $K(\cdot, j)$ is a non-negative finite-valued superregular function with $\pi K(\cdot, j) = 1$. It is regular everywhere except at j, where it is strictly superregular.

For fixed i the function $K(i, \cdot)$ is bounded, since

$$K(i, j) = \frac{N_{ij}}{N_{\pi j}} = \frac{1}{N_{\pi i}}\left(\frac{N_{\pi i}N_{ij}}{N_{\pi j}}\right) = \frac{1}{N_{\pi i}}\hat{N}_{ji} \leq \frac{1}{N_{\pi i}}\hat{N}_{ii} = \frac{N_{ii}}{N_{\pi i}},$$

where carets denote duality with respect to πN.

The real-valued functions $d_i(j, j')$ defined on $S \times S$ by

$$d_i(j, j') = |K(i, j) - K(i, j')|$$

have the property that $\{K(i, j_n)\}$ is a Cauchy sequence for all i in S if and only if $\lim_{m,n\to\infty} d_i(j_m, j_n) = 0$ for all i. According to the bound we just computed for $K(i, j)$, we may lump the functions d_i into the single finite-valued function d defined by

$$d(j, j') = \sum_{i \in S} w_i N_{\pi i}|K(i, j) - K(i, j')|,$$

where the w_i are positive weights such that $\sum w_i N_{ii}$ is finite.

We shall show that d is a metric for S. Clearly d satisfies all the conditions of a metric except possibly that $d(j, j') = 0$ implies $j = j'$. But if $d(j, j') = 0$, then, since $w_i N_{\pi i} > 0$ for all i, we must have

$$K(\cdot, j) = K(\cdot, j').$$

Multiplying through by P and supposing that $j \neq j'$, we obtain

$$K(j', j) = \sum_i P_{j'i}K(i, j) = \sum_i P_{j'i}K(i, j') < K(j', j'),$$

the strict inequality holding, since $K(\cdot, j')$ is not regular at j'. We conclude that $j = j'$ and that d is a metric.

Proposition 10-13: A sequence $\{j_n\}$ in the metric space (S, d) is Cauchy if and only if the sequence of real numbers $\{K(i, j_n)\}$ is Cauchy for every i.

PROOF: If $\{j_n\}$ is Cauchy, then certainly $\{w_i N_{\pi i} K(i, j_n)\}$ is Cauchy. Since $w_i N_{\pi i} > 0$, $\{K(i, j_n)\}$ is Cauchy.

Conversely, let $\{K(i, j_n)\}$ be Cauchy for all i, and let $\epsilon > 0$ be given. Choose a finite set of states E such that

$$\sum_{k \in S - E} w_k N_{kk} < \epsilon/2,$$

and choose M sufficiently large that

$$|K(i, j_m) - K(i, j_n)| < \frac{\epsilon}{2 \sum w_k N_{kk}}$$

for $i \in E$ and for all $n, m \geq M$. Then $d(j_m, j_n) < \epsilon$.

We define S^* to be the Cauchy completion of the metric space (S, d), and we let $B = S^* - S$. The set B is the **Martin exit boundary** for the chain P started with distribution π.

The set B is not necessarily a boundary in the topological sense, since there are examples in which it is not a closed set in S^*, but the abuse of notation will not disturb us.

From Proposition 10-13 we see that $K(i, \cdot)$ is a uniformly continuous function on S. Hence it extends uniquely to a continuous function on S^*. We shall use the same notation $K(i, \cdot)$ for the function on S^*, but will normally denote points of $B = S^* - S$ by x or y.

The characterization of Cauchy sequences given in Proposition 10-13 shows that the nature of the space S^* does not depend upon the choice of the weights w_i. That is, the Cauchy completions of S corresponding to two different choices of weights are homeomorphic.

Since $K(i, \cdot)$ is continuous on S^*, it follows that the extension of d to S^* is simply

$$d(x, y) = \sum_{i \in S} w_i N_{\pi i} |K(i, x) - K(i, y)|.$$

A repetition of the argument in Proposition 10-13 then shows that $\{x_n\}$ is Cauchy in S^* if and only if $\{K(i, x_n)\}$ is Cauchy for each i. Applying this result to the sequence whose terms are alternately x and then y, we find that $x = y$ if and only if $K(i, x) = K(i, y)$ for all i. We state this conclusion as a proposition.

Proposition 10-14: $K(i, x) = K(i, y)$ for all i if and only if $x = y$.

Proposition 10-15: The space S^* is compact.

PROOF: Since S^* is a metric space, it is enough to prove that any sequence $\{x_n\}$ has a convergent subsequence. Now

$$K(i, x_n) \leq \sup_{j \in S} K(i, j) \leq N_{ii}/N_{\pi i} < \infty.$$

Thus by a diagonal process we may choose a subsequence $\{x_{n_k}\}$ such that $\{K(i, x_{n_k})\}$ is Cauchy for all i. Then $\{x_{n_k}\}$ is Cauchy in S^*. Since S^* is complete, $\{x_{n_k}\}$ is convergent.

The sets in the smallest Borel field containing the open sets of S^* are called **Borel sets**. The boundary B is a Borel set, since it is the complement of a countable set. Finite measures defined on the Borel sets are called **Borel measures**.

We conclude this section by agreeing on what conventions we shall adopt in case πN has some zero entries. If S is the set of all states, let

$$W = \{i \in S \mid (\pi N)_i > 0\}.$$

The special nature of W implies that $P_W = P^W$, and by Lemma 8-18 we see that the fundamental matrix for P_W is N_W. Thus if we agree that boundary theory for P and π is to be interpreted as boundary theory for P_W and π_W, we find that for i and j in W

$$K(i, j) = N_{ij}/N_{\pi j},$$

and $K(i, j)$ is not defined otherwise. Hence the metric is

$$d(j, j') = \sum_{i \in W} w_i N_{\pi i} |K(i, j) - K(i, j')|$$

for j and j' in W. Boundary theory is then done relative to the Cauchy completion of (W, d).

4. Convergence to the boundary

We continue to assume that P is a Markov chain with all states transient and that π is a starting distribution with $\pi N > 0$. Let $\bar{P}$ denote the enlarged chain obtained from P by adding the absorbing state b.

The main theorem of this section will be that with probability one every path ω has the property that either $x_n(\omega)$ converges in S^* or the process along ω disappears in finite time. From the results of the next sections we shall be able to sharpen the theorem by concluding that, when convergence takes place, it a.e. is to a nice subset of the boundary.

Lemma 10-16: Let g be a pure potential for P with charge f. If $\{x_n\}$ is an extended chain of total measure one with transition matrix $\bar{P}$ for which νf is finite, then the limit of $g(x_n)$ as n decreases to $\mathbf{u}$ exists a.e.

PROOF: Form the process watched starting in the finite set E. The claim is that for $n \geq 0$, $\{g(x_{u_E+n})\}$ is a non-negative supermartingale. It satisfies the supermartingale inequality because g is non-negative superregular. Thus, to show that the means are finite, it is sufficient to consider $\mathrm{M}[g(x_{u_E})]$. We have

$$\mathrm{M}[g(x_{u_E})] = \mu^E g = \mu^E N f = \nu^E f \leq \nu f < \infty,$$

since f is non-negative and $\nu^E \leq \nu$. Hence $\{g(x_{u_E+n})\}$ is a non-negative supermartingale.

If $0 \leq r < s$, then Proposition 3-11 applied to $-g$ (or Proposition 8-79) shows that the mean number of downcrossings of $[r, s]$ by $\{g(x_{u_E+j})\}$ up to time n is bounded by $s/(s-r)$ independently of n and of E. Let $n \to \infty$ and then let E increase so that $\mathbf{u}_E$ approaches $\mathbf{u}$. The mean number of downcrossings of $[r, s]$ remains bounded, and by monotone convergence the mean number of downcrossings after time $\mathbf{u}$ is finite. By the argument in the proof of Theorem 3-12, $g(x_n)$ converges a.e. as n decreases to $\mathbf{u}$. It can be shown that the limit is finite a.e., but this fact will not be needed.

Lemma 10-17: Let $\mathscr{G}$ be a Borel field of subsets of a set Ω, let S^* be a compact metric space, and for each n let $f_n\colon \Omega \to S^*$ be a function with the property that $f_n^{-1}(E)$ is in $\mathscr{G}$ for all Borel sets E. If $f_n(\omega) \to f(\omega)$ for all ω, then $f^{-1}(E)$ is in $\mathscr{G}$ for all Borel sets E.

PROOF: First consider the case of a compact set C. Let $N_\epsilon(C)$ be the open set of all points at a distance less than ϵ from C. Then

$$f^{-1}(C) = \bigcap_{k=1}^{\infty} \bigcup_{n=1}^{\infty} \bigcap_{m=n}^{\infty} f_m^{-1}(N_{1/k}(C)).$$

Let $\mathscr{C}$ be the class of all Borel sets E for which $f^{-1}(E) \in \mathscr{G}$. Then $\mathscr{C}$ is clearly closed under countable unions and complements. Since $\mathscr{C}$ contains all compact sets, it contains all Borel sets.

Theorem 10-18: Let the chain P with all states transient be started with a distribution π such that $\pi N > 0$. For each path ω let $\mathbf{v}(\omega)$ be the supremum of the n such that $x_n(\omega)$ is in S. Then a.e. either $\mathbf{v}(\omega) < +\infty$ and $x_{\mathbf{v}(\omega)}(\omega) \in S$ or $\mathbf{v}(\omega) = +\infty$ and $x_n(\omega)$ converges to a point $x_{\mathbf{v}(\omega)}(\omega) \in S^*$ as n tends to infinity. Furthermore, if $\mathscr{G}$ is the least Borel field containing the cylinder sets for P, then the set of ω where $x_{\mathbf{v}}$ is defined is in $\mathscr{G}$, and the inverse image under $x_{\mathbf{v}}$ of any Borel set in S^* is a set of $\mathscr{G}$.

PROOF: First we prove the measurability. The set where x_v is defined is the countable union of the sets $\{\mathbf{v} = 1\}, \{\mathbf{v} = 2\}, \ldots$ and the set $\{\mathbf{v} = \infty \wedge x_n(\omega) \text{ converges}\}$. All of these except possibly the last are certainly in $\mathscr{G}$. The last set is

$$A = \{\omega \mid \mathbf{v}(\omega) = \infty \wedge \limsup_n K(i, x_n(\omega)) = \liminf_n K(i, x_n(\omega)) \text{ for all } i\}$$

and is therefore in $\mathscr{G}$. Now the intersection of $\{\mathbf{v} = n\}$ with the inverse image under x_v of a Borel set E is certainly in $\mathscr{G}$. Therefore to complete the proof of the measurability part of the theorem it is sufficient to prove that the intersection of the set A defined above with $x_v^{-1}(E)$ is in $\mathscr{G}$ for every Borel set E. In Lemma 10-17 let Ω be the set A and let the field be the class of sets $A \cap G$, where $G \in \mathscr{G}$. Since $A \cap x_n^{-1}(E)$ is in the field for all E, the lemma applies and gives the result immediately.

Next we are to prove the almost-everywhere statement. Form the extended chain associated with π and P, as described in Section 2 after Corollary 10-7. All statements about this extended chain after time $n \geq 0$ have the same probabilities as the corresponding statements about P, and the vector ν of mean times in the various states is πN. It is therefore sufficient to show in the extended chain the convergence of $K(i, x_n)$ for all i.

Let $\{\hat{x}_n\}$ be the reverse of this extended chain. Since

$$K(i, j) = \hat{N}_{ji}/N_{\pi i},$$

it suffices to show for each i that in the reverse process $\hat{N}_{x_n, i}$ converges a.e. as n decreases to $\hat{\mathbf{u}}$. But $\hat{N}_{\cdot i}$ is the $\hat{P}$-potential of a unit charge at i. Since the charge has finite support, $\hat{\nu}$ times it must be finite. Therefore the theorem follows by applying Lemma 10-16 to the potential $\hat{N}_{\cdot i}$ for the reverse process $\{\hat{x}_n\}$.

By Theorem 10-18 the statement that x_v exists (or equivalently that $x_v \in S^*$) and the statement that $x_v \in E$, where E is a Borel set in S^*, are both measurable with respect to the least Borel field containing all cylinder sets. But $\Pr_i$ is defined on all such statements. Hence

$$\Pr_i[x_v \in E]$$

is defined if E is a Borel set in S^*.

From now on, we use the notation of Chapter 2 that $\mathscr{F}$ is the field of cylinder sets for P and $\mathscr{G}$ is the least Borel field containing $\mathscr{F}$.

Corollary 10-19: $\Pr_i[x_v \in S^*] = 1$.

PROOF: As in the proof of Theorem 10-18, form the extended chain associated with P and π. By Theorem 10-18 almost every path in the process P (started according to π) satisfies $x_v \in S^*$. Hence the same is true of the extended chain, and hence it is true of those paths in the extended chain which pass through i. On any path ω which passes through i, $x_v(\omega) \in S^*$ if and only if $x_v(\omega_{u_{(i)}}) \in S^*$. Therefore by Definition 10-4, the extended chain watched starting in $\{i\}$ satisfies

$$\Pr_{\mu^{\{i\}}}[x_v \in S^*] = \mu_i^{\{i\}}.$$

On the other hand,

$$\Pr_{\mu^{\{i\}}}[x_v \in S^*] = \mu_i^{\{i\}} \Pr_i[x_v \in S^*].$$

Since $\mu_i^{\{i\}} \neq 0$, we must have $\Pr_i[x_v \in S^*] = 1$.

It is to be emphasized that S^* has been constructed for the fixed starting distribution π and that Corollary 10-19 is not the same as Theorem 10-18 restated for the case where π assigns measure one to the state i: *The boundaries for different starting distributions may be different.*

5. Poisson–Martin Representation Theorem

The notation P, π, $K(i, x)$, S^*, $\mathscr{F}$, and $\mathscr{G}$ of Sections 3 and 4 is still in force. We shall use Pr to mean $\Pr_\pi$.

We recall that $\pi K(\cdot, j) = 1$ for all j in S. If $j_n \to x$ in S^*, then $K(\cdot, j_n) \to K(\cdot, x)$ and, for all n, $\pi K(\cdot, j_n) = 1$. Hence $\pi K(\cdot, x) \leq 1$ by Fatou's Theorem. Moreover, we know that $K(\cdot, j)$ is P-superregular for all j in S. If $j_n \to x$, then again by Fatou's Theorem,

$$\begin{aligned} PK(\cdot, x) &= P \lim K(\cdot, j_n) \leq \liminf PK(\cdot, j_n) \\ &\leq \liminf K(\cdot, j_n) = K(\cdot, x). \end{aligned}$$

That is, $K(\cdot, x)$ is P-superregular for all x in S^*. These remarks enable us to prove the following proposition.

Proposition 10-20: If ν is any Borel measure on S^* with $\nu(S^*) = 1$, then the function h, defined by

$$h_i = \int_{S^*} K(i, x) d\nu(x),$$

is finite-valued non-negative superregular and satisfies $\pi h \leq 1$.

PROOF: It is clearly non-negative and is finite-valued because $K(i, \cdot)$ is bounded. By Fubini's Theorem,

$$\pi h = \sum_i \int_{S^*} \pi_i K(i, x) d\nu(x) = \int_{S^*} \left[\sum_i \pi_i K(i, x)\right] d\nu(x) \leq \int_{S^*} d\nu(x) = 1$$

and

$$(Ph)_i = \sum_j \int_{S^*} P_{ij} K(j, x) d\nu(x) = \int_{S^*} \left[\sum_j P_{ij} K(j, x)\right] d\nu(x)$$
$$\leq \int_{S^*} K(i, x) d\nu(x) = h_i.$$

Thus Borel measures on S^* give rise to π-integrable non-negative superregular functions h. Our goal in this section will be to prove conversely that every non-negative (finite-valued) superregular function h arises as the integral over S^* of $K(i, x)$ with respect to some measure. We postpone the uniqueness question to Sections 6 and 7.

Throughout the remainder of this chapter we shall use "superregular" to mean "finite-valued superregular."

Harmonic measure μ is defined on the Borel sets E of S^* by

$$\mu(E) = \Pr[x_v \in E].$$

By Theorem 10-18 the definition of μ makes sense and $\mu(S^*) = 1$. The complete additivity of μ is a consequence of the complete additivity of Pr. Thus μ is a Borel measure. The proposition to follow gives a formula for $\Pr_i[x_v \in E]$ in terms of harmonic measure.

Proposition 10-21: For every Borel set E of S^*,

$$\Pr_i[x_v \in E] = \int_E K(i, x) d\mu(x).$$

PROOF: Let E_n be a fixed increasing sequence of finite sets of S with union S. Let $\mathbf{v}_n(\omega)$ be the last time (possible $+\infty$) that an outcome on the path ω is in E_n, and let $\mathbf{v}_n(\omega) = 0$ if no outcome on ω is in E_n. For any starting distribution γ, Proposition 4-28 implies that

$$\Pr[x_{v_n} = j] = \sum_{k=0}^{\infty} \Pr_\gamma[x_k = j \wedge x_m \notin E_n \text{ after time } k]$$
$$= \sum_{k=0}^{\infty} (\gamma P^k)_j e_j^{E_n}$$
$$= (\gamma N)_j e_j^{E_n}.$$

Hence

$$\Pr_i[x_{v_n} = j] = N_{ij} e_j^{E_n} = K(i, j) N_{\pi j} e_j^{E_n} = K(i, j) \Pr_\pi[x_{v_n} = j].$$

For Borel sets E of S^* define measures by

$$\mu_{in}(E) = \Pr_i[x_{v_n} \in E],$$
$$\mu_i(E) = \Pr_i[x_v \in E],$$
$$\mu_{\pi n}(E) = \Pr_\pi[x_{v_n} \in E],$$

and

$$\mu_\pi(E) = \Pr_\pi[x_v \in E] = \mu(E).$$

What we have just shown is that

$$\mu_{in}(E) = \int_E K(i, x) d\mu_{\pi n}(x).$$

Now if $f \geq 0$ is a Borel measurable function on S^*, the claim is that

$$\int_{S^*} f(x) d\mu_{in}(x) = \int_\Omega f(x_{v_n}(\omega))\, d\Pr_i(\omega).$$

The result for characteristic functions is just the definition of μ_{in}, and for general $f \geq 0$ it follows from the result for simple functions by monotone convergence. Then the result holds for continuous $f \geq 0$ and hence for all continuous f. Similarly for continuous f,

$$\int_{S^*} f(x) d\mu_i(x) = \int_\Omega f(x_v(\omega))\, d\Pr_i(\omega),$$

where we set $x_v(\omega) = 0$ when $\mathbf{v}$ is not defined.

As $n \to \infty$, $x_{v_n}(\omega) \to x_v(\omega)$ a.e. $[\Pr_i]$ by Corollary 10-19. When f is continuous, $f(x_{v_n}(\omega)) \to f(x_v(\omega))$ a.e. $[\Pr_i]$. Since continuous functions are bounded, we have

$$\lim_n \int_\Omega f(x_{v_n}(\omega))\, d\Pr_i(\omega) = \int_\Omega f(x_v(\omega))\, d\Pr_i(\omega)$$

by dominated convergence. Hence

$$\lim_n \int_{S^*} f(x) d\mu_{in}(x) = \int_{S^*} f(x) d\mu_i(x)$$

for all continuous f. Similarly

$$\lim_n \int_{S^*} f(x) d\mu_{\pi n}(x) = \int_{S^*} f(x) d\mu_\pi(x).$$

Since $K(i, \cdot)$ is continuous, so is $f(\cdot)K(i, \cdot)$. Therefore

$$\lim_n \int_{S^*} f(x) K(i, x) d\mu_{\pi n}(x) = \int_{S^*} f(x) K(i, x) d\mu_\pi(x)$$

for all continuous f. Since $\mu_{in}(E) = \int_E K(i, x) d\mu_{\pi n}(x)$, we obtain

$$\int_{S^*} f(x) d\mu_i(x) = \int_{S^*} f(x) K(i, x) d\mu_\pi(x)$$

for all continuous f. Therefore

$$\mu_i(E) = \int_E K(i, x) d\mu_\pi(x).$$

Corollary 10-22: $\int_{S^*} K(i, x) d\mu(x) = 1$ for all $i \in S$.

PROOF: Since $\Pr_i[x_v \in S^*] = 1$ by Corollary 10-19, the result follows from Proposition 10-21.

The corollary we have just proved is the representation theorem as it applies to the column vector which is identically one. We shall be able to get the general case by applying the corollary to a suitable modification of the h-process introduced in Chapter 8. We now re-define the h-process in such a way that we allow h to have some entries equal to zero.

Definition 10-23: If $h \geq 0$ is a finite-valued P-superregular function such that $\pi h = 1$, then h-process is defined to be the Markov chain with state space S and with the measure $\Pr^h$ defined by

$$\begin{aligned} \Pr{}^h[x_0 = c \wedge x_1 = d \wedge x_2 = e \wedge \cdots \wedge x_{n-1} = i \wedge x_n = j] \\ = \pi_c P_{cd} P_{de} \ldots P_{ij} h_j. \end{aligned}$$

We readily check that the h-process is indeed a Markov chain. If we define S^h by

$$S^h = \{i \in S \mid h_i > 0\},$$

then the transition matrix P^h of the h-process satisfies

$$P_{ij}^h = \begin{cases} \dfrac{P_{ij} h_j}{h_i} & \text{for } i \text{ and } j \text{ in } S^h, \\ 0 & \text{for } i \in S^h \text{ and } j \in S - S^h \end{cases}$$

and the starting vector π^h satisfies

$$\pi_i^h = \pi_i h_i \quad \text{for all } i.$$

If i is in $S - S^h$, then P_{ij}^h is not defined and we shall agree to take it to be zero. With this definition we compute directly that the fundamental matrix N^h satisfies

$$N_{ij}^h = \begin{cases} \dfrac{N_{ij} h_j}{h_i} & \text{if } i \text{ and } j \text{ are in } S^h \\ \delta_{ij} & \text{otherwise.} \end{cases}$$

Hence P^h has only transient states.

Lemma 10-24: If $h \geq 0$ is a P-superregular function and if $h_i = 0$ and $h_j > 0$, then $N_{ij} = 0$. If, in addition, $\pi h = 1$, then $(\pi^h N^h)_j = h_j(\pi N)_j$, and $(\pi^h N^h)_j > 0$ if and only if j is in S^h. For i and j in S^h,

$$K^h(i, j) = K(i, j)/h_i.$$

PROOF: If $h_i = 0$ and $h_j > 0$, then for every n

$$h_i \geq \sum_k P_{ik}^{(n)} h_k \geq P_{ij}^{(n)} h_j$$

and hence $P_{ij}^{(n)} = 0$. Therefore $N_{ij} = \sum_n P_{ij}^{(n)} = 0$. Consequently,

$$(\pi^h N^h)_j = \sum_{i \in S^h} \pi_i h_i \left(\frac{N_{ij} h_j}{h_i}\right) = h_j \sum_{i \in S^h} \pi_i N_{ij} = h_j \sum_{i \in S} \pi_i N_{ij} = h_j(\pi N)_j.$$

By assumption π is a vector such that πN is strictly positive. Therefore $(\pi^h N^h)_j = 0$ if and only if $h_j = 0$.

Finally, according to the convention at the end of Section 3 and the calculation just completed, $K^h(i, j)$ is defined if i and j are in S^h. We have

$$K^h(i, j) = N_{ij}^h/(\pi^h N^h)_j = \frac{N_{ij} h_j / h_i}{h_j(\pi N)_j} = K(i, j)/h_i.$$

Since the h-process has the property that $(\pi^h N^h)_j$ is positive exactly when j is in S^h, we can, as noted at the end of Section 3, define a metric d^h on $S^h \times S^h$ and we can form the Cauchy completion S^{h*} with boundary B^h. We shall agree to use the same weights in defining d^h that were used in defining d.

Lemma 10-25: The identity map from (S^h, d^h) into (S, d) is an isometry.

PROOF: Let j and j' be in S^h. Then

$$\begin{aligned}
d^h(j, j') &= \sum_{i \in S^h} w_i (\pi^h N^h)_i |K^h(i, j) - K^h(i, j')| \\
&= \sum_{i \in S^h} w_i N_{\pi i} h_i |K(i, j) - K(i, j')|/h_i \\
&= \sum_{i \in S^h} w_i N_{\pi i} |K(i, j) - K(i, j')| \\
&= \sum_{i \in S} w_i N_{\pi i} |K(i, j) - K(i, j')| \\
&= d(j, j').
\end{aligned}$$

It follows from Lemma 10-25 that S^{h*} can be canonically identified with a compact subset of S^*. Thus by continuity, $K^h(i, x) = K(i, x)/h_i$ for all i in S^h and x in S^{h*}. Harmonic measure for the h-process will be denoted μ^h. We can consider it to be defined on S^* (as well as on S^{h*}) if we set

$$\mu^h(E) = \mu^h(E \cap S^{h*})$$

for Borel sets E in S^*.

We are finally in a position to state and prove the existence half of the Markov chain analog of the Poisson–Martin Representation Theorem.

Theorem 10-26: If $h \geq 0$ is a finite-valued P-superregular function such that $\pi h = 1$, then

$$h_i = \int_{S^*} K(i, x) d\mu^h(x).$$

PROOF: Applying Corollary 10-22 to the h-process, we have

$$\int_{S^{h*}} K^h(i, x) d\mu^h(x) = 1$$

for i in S^h. That is, for i in S^h

$$h_i = \int_{S^{h*}} K(i, x) d\mu^h(x) = \int_{S^*} K(i, x) d\mu^h(x).$$

Now if $i \in S - S^h$, $N_{ij} = 0$ for all $j \in S^h$ by Lemma 10-24. Thus $K(i, j) = 0$ for such i and j. Since $K(i, x)$ is continuous on S^{h*}, $K(i, x) = 0$ for $i \in S - S^h$ and $x \in S^{h*}$. Therefore for such i,

$$h_i = 0 = \int_{S^{h*}} K(i, x) d\mu^h(x) = \int_{S^*} K(i, x) d\mu^h(x).$$

Of course, the representation theorem immediately extends to cover all P-superregular functions $h \geq 0$ for which πh is positive and finite. However, the probabilistic interpretation of the measure μ^h is lost. We shall return to this point in Theorem 10-41 of Section 7 after proving the uniqueness theorem.

6. Extreme points of the boundary

The measure μ^h is not necessarily the unique Borel measure which represents h in the sense of Theorem 10-26, and we consequently need another hypothesis to get uniqueness. What we shall do in this section is to define the set B_e of extreme points of the boundary and the subset $\bar{S} = S \cup B_e$ of S^*. In Section 7 we shall see that μ^h has all its mass on $\bar{S}$ and that μ^h is the unique measure with all its mass on $\bar{S}$ for which the representation in Theorem 10-26 holds.

There are three kinds of behavior of points of the boundary that we shall want to exclude:

(1) x has the property that $\pi K(\cdot, x) < 1$.
(2) x has the property that $K(\cdot, x)$ is not regular.
(3) x has the property that $K(\cdot, x)$ is regular but is a nontrivial convex combination of other non-negative regular functions.

The first two of these possibilities are the topic of the two lemmas to follow. The third possibility will require more of our attention, and we discuss it beginning with Definition 10-29 and Lemma 10-30.

We recall that $\pi K(\cdot, x) \le 1$ for all x in S^*.

Lemma 10-27: For almost every $x\,[\mu]$ in S^*, $\pi K(\cdot, x) = 1$. The set where the equality holds is a Borel set.

PROOF: For each i, the function $K(i, x)$ is continuous. Hence the countable sum $\pi K(\cdot, x)$ is Borel measurable. Therefore the set where it equals one is a Borel set.

By Corollary 10-22,

$$\int_{S^*} K(i, x) d\mu(x) = 1.$$

Thus by Fubini's Theorem,

$$1 = \pi 1 = \pi \int_{S^*} K(\cdot, x) d\mu(x) = \int_{S^*} \pi K(\cdot, x) d\mu(x).$$

But $\int_{S^*} 1 d\mu(x) = 1$ also, and since $1 - \pi K(\cdot, x) \ge 0$, we conclude that $\pi K(\cdot, x) = 1$ a.e. by Corollary 1-40.

We say that a function h is **normalized** if $\pi h = 1$. By Lemma 10-27, $K(\cdot, x)$ is normalized for a.e. $x\,[\mu]$.

We recall that $K(\cdot, x)$ is P-superregular for all $x \in S^*$.

Lemma 10-28: For almost every $x\,[\mu]$ in the boundary B of S^*, the function $K(\cdot, x)$ is regular. The set where it is regular is a Borel set.

PROOF: The set where $P_{i\cdot}K(\cdot, x) = K(i, x)$ is a Borel set since it is the set where a Borel measurable function takes on the value $K(i, x)$. The set where $K(\cdot, x)$ is regular is the countable intersection of these sets.

By Theorem 4-10 with the random time identically equal to one, we see that the column vector whose ith component is (see Proposition 10-21)

$$\int_B K(i, x)d\mu(x) = \Pr_i[x_v \in B]$$

is P-regular. By this observation and by Fubini's Theorem, we have

$$\int_B PK(\cdot, x)d\mu(x) = P\int_B K(\cdot, x)d\mu(x) = \int_B K(\cdot, x)d\mu(x).$$

Since $PK(\cdot, x) \leq K(\cdot, x)$, we must have $PK(\cdot, x) = K(\cdot, x)$ a.e. by Corollary 1-40.

Definition 10-29: A finite-valued function $h \geq 0$ is **minimal** if

(1) h is regular, and
(2) whenever $0 \leq h' \leq h$ with h' regular, then $h' = ch$.

Lemma 10-30: A normalized finite-valued regular function $h \geq 0$ is minimal if and only if it cannot be written as a nontrivial convex combination of two distinct normalized non-negative regular functions.

PROOF: If $h = c_1h_1 + c_2h_2$ is such a convex combination, then either h_1 or h_2, say h_1, is not equal to h. Since $h \geq c_1h_1$, we must have $c_1h_1 = ch$ if h is minimal. Multiplying through by π, we obtain $c_1 = c$. Since $c_1 \neq 0$, we conclude $h_1 = h$, contradiction.

Conversely, if $h \geq h' \geq 0$ with h' regular and h' not equal to 0 or h, then

$$h = (\pi h')\frac{h'}{(\pi h')} + [\pi(h - h')]\frac{h - h'}{\pi(h - h')}$$

exhibits h as a nontrivial convex combination of normalized regular functions, provided we can prove $0 < \pi h' < 1$. If so, then by hypothesis the two normalized functions must be equal to each other and hence equal to h. That is, $h' = (\pi h')h$. Thus we are to prove $0 < \pi h' < 1$. Let $h_j' > 0$. Since $(\pi N)_j > 0$, choose n so that $(\pi P^n)_j > 0$. Since h' is superregular, $h' \geq P^nh'$ and hence $\pi h' \geq \pi P^nh' \geq (\pi P^n)_jh_j > 0$. A similar argument applied to $h - h'$ shows that $\pi h' < 1$.

Definition 10-31: A point x in S^* is an **extreme point** of S^* if the function $K(\cdot, x)$ is minimal and normalized. The set of extreme points is denoted B_e. Let $\bar{S} = S \cup B_e$.

Since $K(\cdot, j)$ is not regular when $j \in S$, no point of S can be in B_e, and B_e must be entirely contained in the boundary B. The set $\bar{S}$ is the subset of S^* with respect to which the uniqueness theorem will be stated. We shall see eventually that $\bar{S}$ is a Borel set and that $\mu(\bar{S}) = 1$ (compare with Lemmas 10-27 and 10-28).

If we form an h-process, we know that $\overline{S^h} \subset S^{h*} \subset S^*$. The following lemma strengthens this conclusion and shows that actually $\overline{S^h} \subset \bar{S}$.

Lemma 10-32: Let $h \geq 0$ be a finite-valued normalized P-superregular function. If x is in S^{h*}, then

(1) $K^h(\cdot, x)$ is normalized if and only if $K(\cdot, x)$ is normalized.

(2) $K^h(\cdot, x)$ is regular for P^h restricted to S^h if and only if $K(\cdot, x)$ is P-regular.

(3) $K^h(\cdot, x)$ is minimal for P^h restricted to S^h if and only if $K(\cdot, x)$ is minimal for P.

Hence $B_e^h = B^h \cap B_e$ and $\overline{S^h} \subset \bar{S}$.

PROOF: Conclusions (1) and (2) follow from the identities

$$\sum_{i \in S^h} \pi_i^h K^h(i, x) = \pi K(\cdot, x)$$

$$\sum_{j \in S^h} P_{\cdot j}^h K^h(j, x) = PK(\cdot, x),$$

both of which use the fact that $K(i, x) = 0$ if i is not in S^h (see the proof of Theorem 10-26).

Thus in (3) we may assume that $K^h(\cdot, x)$ and $K(\cdot, x)$ are both regular. Multiplying both by the same constant, if necessary, we may assume for the purposes of this proof that they are normalized. We shall use Lemma 10-30 and show that a nontrivial decomposition exists for $K(\cdot, x)$ if and only if a nontrivial decomposition exists for $K^h(\cdot, x)$. In fact, if for $i \in S$

$$K(i, x) = c_1 h_i^{(1)} + c_2 h_i^{(2)}$$

nontrivially, then for $i \in S^h$,

$$K^h(i, x) = K(i, x)/h_i = c_1 \frac{h_i^{(1)}}{h_i} + c_2 \frac{h_i^{(2)}}{h_i}.$$

We have

$$\sum_{j \in S^h} P_{ij}^h \frac{h_j^{(1)}}{h_j} = \sum_{j \in S^h} \frac{1}{h_i} P_{ij} h_j^{(1)} = \sum_{j \in S} \frac{1}{h_i} P_{ij} h_j^{(1)} = \frac{h_i^{(1)}}{h_i}$$

and

$$\sum_{i \in S^h} \pi_i^h \frac{h_i^{(1)}}{h_i} = \sum_{i \in S^h} \pi_i h_i^{(1)} = \sum_{i \in S} \pi_i h_i^{(1)} = 1,$$

where the sums over S^h can be replaced by sums over S because $i \in S - S^h$ implies $0 \leq h_i^{(1)} \leq K(i, x)/c_1 = 0$ or $h_i^{(1)} = 0$. Consequently we may assume that

$$\frac{h_i^{(1)}}{h_i} = \frac{h_i^{(2)}}{h_i}$$

for $i \in S^h$. That is, $h_i^{(1)} = h_i^{(2)}$ for $i \in S^h$. But $h_i^{(1)} = h_i^{(2)} = 0$ for $i \in S - S^h$. Hence $h^{(1)} = h^{(2)}$.

Conversely, if for $i \in S^h$

$$K^h(i, x) = c_3 h_i^{(3)} + c_4 h_i^{(4)},$$

then

$$K(i, x) = c_3 h_i^{(3)} h_i + c_4 h_i^{(4)} h_i$$

for $i \in S^h$. Extend $\{h_i^{(3)} h_i\}$ and $\{h_i^{(4)} h_i\}$ to be defined for all $i \in S$ by setting them equal to zero for $i \in S - S^h$. The convex sum of them is still $K(i, x)$, and they are both regular normalized functions, since

$$\sum_j P_{ij} h_j^{(3)} h_j = \sum_{j \in S^h} P_{ij} h_j^{(3)} h_j = h_i^{(3)} h_i$$

and

$$\sum_i \pi_i h_i^{(3)} h_i = \sum_{i \in S^h} \pi_i h_i^{(3)} h_i = \pi^h h^{(3)} = 1.$$

Consequently we may assume that $h_i^{(3)} h_i = h_i^{(4)} h_i$ for all $i \in S$. That is, $h_i^{(3)} = h_i^{(4)}$ for all $i \in S^h$.

We now begin to derive properties of the set B_e of extreme points.

Lemma 10-33: If $h \geq 0$ is a normalized minimal function such that

$$h_i = \int_{S^*} K(i, x) d\nu(x)$$

for a Borel measure ν with $\nu(S^*) = 1$, then ν is concentrated at a single point and that point is extreme.

PROOF: Consider the functions

$$h^A = \frac{1}{\nu(A)} \int_A K(\cdot, x) d\nu(x)$$

as A ranges through the Borel sets with $0 < \nu(A) < 1$. For any such A, h^A and $h^{\tilde{A}}$ are superregular and satisfy $\pi h^A \leq 1$ and $\pi h^{\tilde{A}} \leq 1$ by Proposition 10-20. But

$$h = \nu(A) h^A + \nu(\tilde{A}) h^{\tilde{A}}$$

with h regular and normalized and with $\nu(A) + \nu(\tilde{A}) = 1$. Hence h^A

and $h^{\bar{A}}$ must both be regular and normalized. By Lemma 10-30, $h = h^A = h^{\bar{A}}$. Thus, if $0 < \nu(A) < 1$,

$$\nu(A)h_i = \int_A K(i, x)d\nu(x)$$

for each fixed i. The same is trivially true if $\nu(A) = 0$, and it is true by hypothesis if $\nu(A) = 1$. Hence it is true of all Borel sets. Therefore for fixed i,

$$K(i, x) = h_i \quad \text{a.e. } [\nu].$$

Thus $K(i, x) = h_i$ for *all* i almost everywhere $[\nu]$. Since $\nu(S^*) > 0$, there is at least one point x_0 where it is true. We have

$$K(i, x_0) = h_i$$

for all i. If there were another such point x', then we would have $K(\cdot, x_0) = K(\cdot, x')$, and hence $x_0 = x'$ by Proposition 10-14. Therefore the complement of $\{x_0\}$ has measure zero, or ν is concentrated at x_0. Now, we know that $K(\cdot, x_0) = h$ and h is normalized and minimal. Hence x_0 is extreme.

Lemma 10-34: If h, $h^{(1)}$, and $h^{(2)}$ are normalized non-negative superregular functions with

$$h = c_1 h^{(1)} + c_2 h^{(2)},$$

where $c_1 \geq 0$ and $c_2 \geq 0$, then

$$\mu^h = c_1 \mu^{h^{(1)}} + c_2 \mu^{h^{(2)}}.$$

PROOF: For a typical basic statement $x_0 = i \wedge x_1 = j \wedge x_2 = k$, we have, by Definition 10-23,

$$\Pr{}^h[x_0 = i \wedge x_1 = j \wedge x_2 = k] = \pi_i P_{ij} P_{jk} h_k.$$

Using the analogous identities for $h^{(1)}$ and $h^{(2)}$ and breaking up h_k as $c_1 h_k^{(1)} + c_2 h_k^{(2)}$, we obtain

$$\Pr{}^h[x_0 = i \wedge x_1 = j \wedge x_2 = k] = c_1 \Pr{}^{h^{(1)}}[x_0 = i \wedge x_1 = j \wedge x_2 = k] + c_2 \Pr{}^{h^{(2)}}[x_0 = i \wedge x_1 = j \wedge x_2 = k].$$

Hence the same is true of all statements in $\mathscr{F}$, and by the uniqueness half of Theorem 1-19 we find that

$$\Pr{}^h[p] = c_1 \Pr{}^{h^{(1)}}[p] + c_2 \Pr{}^{h^{(2)}}[p]$$

for all statements p measurable relative to $\mathscr{G}$.

Take p to be the statement $x_v \in E$, where E is a Borel set of S^*. The claim is that $x_v \in E$ if and only if $x_v^h \in E$, and similarly for $h^{(1)}$ and $h^{(2)}$. For each n we certainly have $x_n = x_n^h$. Hence $x_v = x_v^h$ when $\mathbf{v}$ is finite. When $\mathbf{v}$ is infinite, we have $x_v = x_v^h$ because $x_n = x_n^h$ for all n and because S^{h*} is isometric with a subset of S^*. Therefore

$$\Pr{}^h[x_v \in E] = c_1 \Pr{}^{h^{(1)}}[x_v \in E] + c_2 \Pr{}^{h^{(2)}}[x_v \in E]$$

or

$$\mu^h = c_1\mu^{h^{(1)}} + c_2\mu^{h^{(2)}}.$$

Proposition 10-35: Let $h = K(\cdot, x)$. Then $x \in \bar{S}$ if and only if h is normalized and $\mu^h(\{x\}) = 1$.

PROOF: First let $x \in S$. Then h is certainly normalized, and hence the h-process is defined. Suppose we can prove that the h-process disappears with probability one. Then by definition of μ^h, $\mu^h(S^h) = 1$. Hence by Theorem 10-26,

$$K(\cdot, x) = \int_{S^*} K(\cdot, y)d\mu^h(y) = \int_S K(\cdot, y)d\mu^h(y) = \sum_j N_{ij}\left(\frac{\mu^h(\{j\})}{N_{\pi j}}\right).$$

But $K(\cdot, x) = N_{ix}/N_{\pi x} = \sum_j N_{ij}(\delta_{jx}/N_{\pi j})$. By Theorem 8-4,

$$\frac{\delta_{jx}}{N_{\pi j}} = \frac{\mu^h(\{j\})}{N_{\pi j}} \quad \text{for all } j.$$

That is, $\mu^h(\{x\}) = 1$. Thus we are to prove that the h-process disappears with probability one. By the remarks following Definition 8-13, it suffices to show that $\sum_{j\in S^h} N_{ij}^h f_j = 1$ for some f. Take f to be a single mass $1/(N_{\pi x}h_x)$ at x, and the equality follows.

Next let $x \in B_e$. Then h is normalized by definition. By Theorem 10-26,

$$K(\cdot, x) = \int_{S^*} K(\cdot, y)d\mu^h(y).$$

In Lemma 10-33 take ν to be μ^h. Then $\mu^h(\{x_0\}) = 1$ for some x_0. But then $K(\cdot, x) = K(\cdot, x_0)$, and hence $x = x_0$ by Proposition 10-14.

Conversely, suppose h is normalized and $\mu^h(\{x\}) = 1$. If $x \notin S$, then $x \in B$, and by Lemmas 10-28 and 10-32 (conclusion 2), h must be regular. It remains to show that h is minimal. If $h = c_1h^{(1)} + c_2h^{(2)}$, then by Lemma 10-34

$$\mu^h = c_1\mu^{h^{(1)}} + c_2\mu^{h^{(2)}}.$$

Hence $\mu^{h^{(1)}}$ must put all its weight on x, and therefore $h = h^{(1)}$. By Lemma 10-30 we conclude that h is minimal.

7. Uniqueness of the representation

In this section we retain all of the notations of Sections 3 through 6. We are going to prove that μ^h is the unique Borel measure concentrated on $\bar{S}$ for which the representation in Theorem 10-26 holds. The first step will be to show that $\bar{S}$ is a Borel set of harmonic measure one.

We denote by S_N the set of points x in S^* for which $K(\cdot, x)$ is normalized. These are exactly the points for which the h-process with $h = K(\cdot, x)$ has been defined. By Lemma 10-27, S_N is a Borel set of harmonic measure one. Since S_N is a Borel subset of S^*, the notion of a Borel measurable function on S_N is well defined.

Lemma 10-36: If A is a fixed $\mathscr{G}$-measurable set in Ω, then $\Pr^{K(\cdot,x)}[A]$ is a Borel measurable function of x in S_N.

PROOF: By Definition 10-23,

$$\Pr^{K(\cdot,x)}[x_0 = i \wedge x_1 = j \wedge x_2 = k] = \pi_i P_{ij} P_{jk} K(k, x),$$

and the right side is continuous even for x in S^*. Since any cylinder set is the countable disjoint union of basic cylinder sets, the function $\Pr^{K(\cdot,x)}[\omega \in A]$ for $A \in \mathscr{F}$ is a denumerable sum of such functions and hence is Borel measurable.

Let $\mathscr{C}$ be the class of all sets in Ω for which $\Pr^{K(\cdot,x)}[\omega \in A]$ is Borel measurable. If $\{A_n\}$ and $\{B_n\}$ are, respectively, increasing and decreasing sequences of such sets, then

$$\Pr^{K(\cdot,x)}[\bigcup A_n] = \lim_n \Pr^{K(\cdot,x)}[A_n]$$

and

$$\Pr^{K(\cdot,x)}[\bigcap B_n] = \lim_n \Pr^{K(\cdot,x)}[B_n],$$

the latter equality holding since $\Pr^{K(\cdot,x)}$ is a finite measure. Hence $\bigcup A_n$ and $\bigcap B_n$ are both in $\mathscr{C}$. By the Monotone Class Lemma (see Halmos [1950], pp. 27–28), $\mathscr{C}$ contains $\mathscr{G}$.

Lemma 10-37: If A is in $\mathscr{G}$ and if C is a Borel set in S_N, then

$$\Pr[\omega \in A \wedge x_v \in C] = \int_C \Pr^{K(\cdot,x)}[A] d\mu(x).$$

PROOF: If p is the typical basic statement $x_0 = i \wedge x_1 = j \wedge x_2 = k$, then

$$\begin{aligned}\Pr[p \wedge x_v \in C] &= \Pr[p] \cdot \Pr[x_v \in C \mid p] \\ &= \pi_i P_{ij} P_{jk} \Pr_k[x_v \in C]\end{aligned}$$

by Theorem 4-9. By Proposition 10-21 and by Definition 10-23, this expression is equal to

$$= \pi_i P_{ij} P_{jk} \int_C K(k, x) d\mu(x)$$

$$= \int_C \pi_i P_{ij} P_{jk} K(k, x) d\mu(x)$$

$$= \int_C \Pr{}^{K(\cdot, x)}[p] d\mu(x).$$

By Lemma 10-36 the function $\Pr^{K(\cdot,x)}[A]$ is a Borel measurable function of x if A is in $\mathscr{G}$. Fixing C, we may therefore define a set function σ by

$$\sigma(A) = \int_C \Pr{}^{K(\cdot, x)}[A] d\mu(x).$$

Then σ is certainly non-negative, and it is completely additive by the Monotone Convergence Theorem. The calculation above shows that the two set functions $\sigma(A)$ and $\Pr[\omega \in A \wedge x_v \in C]$ agree on basic cylinder sets and hence on all of the field $\mathscr{F}$. By Theorem 1-19 they must agree on all of $\mathscr{G}$.

Proposition 10-38: The set $\bar{S}$ is a Borel set with $\mu(\bar{S}) = 1$.

PROOF: Applying Lemma 10-37 to the statement $x_v \in D$, where D is a Borel set of S^*, we have for any Borel subset C of S_N

$$\Pr[x_v \in D \wedge x_v \in C] = \int_C \Pr{}^{K(\cdot, x)}[x_v \in D] d\mu(x) = \int_C \mu^{K(\cdot, x)}(D) d\mu(x).$$

But by definition

$$\Pr[x_v \in D \wedge x_v \in C] = \mu(D \cap C) = \int_C \chi_D(x) d\mu(x).$$

The set on which two Borel measurable functions agree is a Borel set. Hence for fixed D the set of x's on which

$$\mu^{K(\cdot, x)}(D) = \chi_D(x)$$

holds is a Borel set in S_N. Since these two functions have the same μ-integral over all Borel sets C, they must be equal a.e. $[\mu]$.

We shall let D range over the intersection with S_N of all balls with centers in S and with rational radii. Let $\mathscr{T}$ be the collection of such balls and let

$$T = \{x \in S_N \mid \mu^{K(\cdot, x)}(D) = \chi_D(x) \quad \text{for all } D \text{ in } \mathscr{T}\}.$$

Remembering that S_N is a Borel set of harmonic measure one, we see that T is the denumerable intersection of Borel sets of measure one and is therefore a Borel set with $\mu(T) = 1$. We show $T = \bar{S}$.

First let $x \in T$. Choose s_n in S with $d(x, s_n) < 1/n$, and let D_n be the intersection with S_N of the ball with center s_n and radius $1/n$. By assumption

$$\mu^{K(\cdot,x)}(D_n) = \chi_{D_n}(x) = 1$$

for all n. Hence

$$\mu^{K(\cdot,x)}(\{x\}) = \mu^{K(\cdot,x)}(\cap D_n) = 1.$$

Therefore $x \in \bar{S}$ by Proposition 10-35.

Conversely, if $x \in \bar{S}$, then

$$\mu^{K(\cdot,x)}(\{x\}) = 1$$

by Proposition 10-35, and so

$$\mu^{K(\cdot,x)}(D) = \chi_D(x)$$

for all Borel sets D in S_N. Therefore $x \in T$, and $T = \bar{S}$.

Lemma 10-39: Let h be defined by

$$h_i = \int_{\bar{S}} K(i, x) d\nu(x),$$

where ν is a measure with $\nu(\bar{S}) = 1$. Then the h-process is well defined, and for any A in $\mathscr{G}$

$$\Pr{}^h[A] = \int_{\bar{S}} \Pr{}^{K(\cdot,x)}[A] d\nu(x).$$

PROOF: By Proposition 10-20, h is non-negative superregular. By Fubini's Theorem $\pi h = 1$, since $\pi K(\cdot, x) = 1$ for all x in $\bar{S}$. Hence the h-process is well defined. If p is a typical basic statement,

$$x_0 = i \wedge x_1 = j \wedge x_2 = k,$$

$$\begin{aligned}\Pr{}^h[p] &= \pi_i P_{ij} P_{jk} h_k \\ &= \int_{\bar{S}} \pi_i P_{ij} P_{jk} K(k, x) d\nu(x) \\ &= \int_{\bar{S}} \Pr{}^{K(\cdot,x)}[p] d\nu(x).\end{aligned}$$

Proceeding as in Lemma 10-37, we define σ by

$$\sigma(A) = \int_{\bar{S}} \Pr^{K(\cdot,x)}[A] d\nu(x).$$

Then σ is a finite measure on $\mathscr{G}$ which agrees with the measure $\Pr^h[A]$ on $\mathscr{F}$. By Theorem 1-19,

$$\Pr^h[A] = \sigma(A)$$

for all A in $\mathscr{G}$.

The theorem to follow is the uniqueness theorem mentioned at the beginning of this section.

Theorem 10-40: If $h \geq 0$ is a normalized superregular function such that

$$h_i = \int_{\bar{S}} K(i, x) d\nu(x)$$

for some measure ν on S^* with $\nu(\bar{S}) = \nu(S^*) = 1$, then $\nu = \mu^h$.

PROOF: By Lemma 10-39,

$$\mu^h(C \cap \bar{S}) = \Pr^h[x_v \in C \cap \bar{S}] = \int_{\bar{S}} \Pr^{K(\cdot,x)}[x_v \in C \cap \bar{S}] d\nu(x).$$

But by Proposition 10-35,

$$\Pr^{K(\cdot,x)}[x_v \in C \cap \bar{S}] = \mu^{K(\cdot,x)}(C \cap \bar{S}) = \chi_{C \cap \bar{S}}(x).$$

Hence for all Borel sets $C \cap \bar{S}$

$$\mu^h(C \cap \bar{S}) = \int_{\bar{S}} \chi_{C \cap \bar{S}}(x) d\nu(x) = \nu(C \cap \bar{S}).$$

Since $\nu(\bar{S}) = 1$, we must have $\mu^h(\bar{S}) = 1$ and hence $\mu^h(C) = \nu(C)$ for all Borel sets C in S^*.

We have not yet proved that the representation in Theorem 10-40 does hold for the measure μ^h, but this fact is a consequence of the theorem to follow, which will summarize the results of the past three sections.

Theorem 10-41: The π-integrable non-negative P-superregular functions h stand in one-one correspondence with the non-negative finite measures ν on the Borel sets of $\bar{S}$, the correspondence of h to ν being

$$h_i = \int_{\bar{S}} K(i, x) d\nu(x)$$

and satisfying $\pi h = \nu(\bar{S})$. The unique representation $h = Nf + r$ with

r regular arises by decomposing the integral over $\bar{S}$ into a part over S and a part over B_e. If h is normalized, then the measure that corresponds to h is μ^h.

PROOF: By Fubini's Theorem any function of the form

$$h_i = \int_{\bar{S}} K(i, x) d\nu(x)$$

with ν finite is non-negative superregular and has $\pi h = \nu(\bar{S})$.

Conversely, let h be given. If $\pi h = 0$, then the superregularity of h implies that

$$0 = \pi h \geq \pi P^n h \geq 0$$

for all n and hence $\pi P^n h = 0$ for all n. Thus $(\pi N)h = 0$. Since πN is everywhere positive, $h = 0$. Thus existence and uniqueness of ν follow if $\pi h = 0$. Next, let πh be positive. Since h and ν must be related by $\pi h = \nu(\bar{S})$, we may, for both existence and uniqueness, divide h by an appropriate positive constant to obtain $\pi h = 1$. Uniqueness of ν and the fact that $\nu = \mu^h$ then follow from Theorem 10-40. Existence of ν follows from Theorem 10-26 provided we can show that $\mu^h(S^* - \bar{S}) = 0$. By Lemma 10-32,

$$\overline{S^h} \subset S^{h*} \cap \bar{S}.$$

Hence

$$\mu^h(\bar{S}) = \mu^h(S^{h*} \cap \bar{S}) \geq \mu^h(\overline{S^h}).$$

But the right side equals one by Proposition 10-38. Thus $\mu^h(\bar{S}) = 1$ and $\mu^h(S^* - \bar{S}) = 0$.

Finally we have $\bar{S} = S \cup B_e$ disjointly, and an application of Fubini's Theorem shows that

$$\int_{B_e} K(i, x) d\nu$$

is regular. Since

$$\int_S K(i, x) d\nu = \sum_{j \in S} N_{ij} \left(\frac{\nu(\{j\})}{N_{\pi j}} \right),$$

the representation $h = Nf + r$ is as asserted.

8. Analog of Fatou's Theorem

In the classical case of the disk, which was discussed in Section 1, normalized Lebesgue measure m on the circle has the distinguishing property that it corresponds to the function 1. If h is the non-negative harmonic function corresponding to a measure ν and if $\nu = fm + \mu_s$ is the Lebesgue decomposition of ν with respect to m, then Fatou's Theorem asserts that for almost every x $[m]$ on the circle, $h(re^{i\theta}) \to f(x)$ whenever $re^{i\theta}$ converges to x nontangentially. In this section we shall prove a Markov chain analog of this theorem in terms of the Martin boundary and the measures given by Theorem 10-41. As expected, harmonic measure μ will play the role of Lebesgue measure.

Our procedure will be first to derive an almost everywhere statement in terms of the measure on the probability space and then to translate this statement into a result in terms of harmonic measure. As a preliminary to the first step, we consider a special case in the lemma below.

We shall be dealing with expressions of the form $\lim_{n\to\infty} h(x_n(\omega))$ in this section, where h is non-negative and P-regular, and we shall adopt the convention that $h(x_n(\omega)) = 0$ if $n > \mathbf{v}$. This definition is motivated by the following consideration: If $\bar{P}$ is the enlarged chain for P, then a P-regular function h extends to be regular for $\bar{P}$ if h is defined to be zero at the absorbing state; consequently if πh is finite, $\{h(x_n)\}$ is a martingale with $\mathrm{M}[h(x_0)] = \pi h$.

Lemma 10-42: If $h \geq 0$ is a normalized bounded regular function, then μ^h is absolutely continuous with respect to μ and

$$h_i = \int_{\bar{S}} K(i, x) f(x) d\mu(x) = \int_{B_e} K(i, x) f(x) d\mu(x),$$

where f is the Radon–Nikodym derivative of μ^h with respect to μ. The function f may be taken to be zero on S, and if it is, then

$$\Pr[\lim_{n\to\infty} h(x_n) = f(x_v)] = 1.$$

PROOF: Since h is bounded, $h \leq c\mathbf{1}$ for some constant $c > 1$. Noting that

$$\mathbf{1} = \frac{c-1}{c}\left[\frac{1}{c-1}(c\mathbf{1} - h)\right] + \frac{1}{c} h,$$

set $g = (c-1)^{-1}(c\mathbf{1} - h)$. Then g and h are non-negative normalized superregular functions, and Lemma 10-34 shows that

$$\mu = \frac{c-1}{c}\mu^g + \frac{1}{c}\mu^h.$$

Thus $\mu^h \leq c\mu$ and μ^h is absolutely continuous with respect to μ. By the Radon–Nikodym Theorem there is a Borel function f such that

$$\mu^h(C) = \int_C f(x) d\mu(x)$$

for all Borel sets C. Since h is regular, $\mu^h(S) = 0$ and we may take f to vanish on S.

By Theorem 10-41 we have

$$h_i = \int_{\bar{S}} K(i, x) d\mu^h(x)$$

$$= \int_{\bar{S}} K(i, x) f(x) d\mu(x)$$

$$= \int_{S^*} K(i, x) f(x) d\mu(x).$$

Now the argument in the proof of Proposition 10-21 shows that

$$\int_{S^*} K(i, x) f(x) d\mu(x) = \int_{\Omega} f(x_v(\omega))\, d\mathrm{Pr}_i(\omega).$$

Thus if $\Pr_\pi[x_0 = j \wedge \cdots \wedge x_n = i] > 0$,

$$\mathrm{M}_\pi[f(x_v) \mid x_0 = j \wedge \cdots \wedge x_n = i] = \mathrm{M}_i[f(x_v)] = h_i.$$

Similarly, if b denotes the absorbing state in the enlarged chain, then

$$\mathrm{M}_\pi[f(x_v) \mid x_0 = j \wedge \cdots \wedge x_n = b] = 0 = h_b.$$

That is, if $\mathscr{R}_n$ is the partition generated by $\{x_0, \ldots, x_n\}$, then

$$\mathrm{M}_\pi[f(x_v) \mid \mathscr{R}_n] = h(x_n).$$

On one hand, the Borel field generated by the $\mathscr{R}_n$ is $\mathscr{G}$, and on the other hand $f(x_v)$ is measurable over $\mathscr{G}$, since it is the composition of a Borel function and a function for which the inverse image of every Borel set is in $\mathscr{G}$. By Proposition 3-18,

$$\Pr[\lim_{n \to \infty} h(x_n) = f(x_v)] = 1.$$

The general case of the almost everywhere statement in terms of the measure on the probability space is covered by the following theorem.

Theorem 10-43: Let $h \geq 0$ be a normalized regular function and let $\mu^h = f\mu + \mu_s$ be the Lebesgue decomposition of μ^h with respect to μ (where f is taken to be zero on the state space S); then

$$\Pr[\lim_{n \to \infty} h(x_n) = f(x_v)] = 1.$$

PROOF: If we let $k = \frac{1}{2}1 + \frac{1}{2}h$, then k is a non-negative normalized superregular function. Since k is strictly positive, we have $S^k = S$

and $S^{k*} = S^*$. By Lemma 10-32, $B_e^k = B_e$. The function h/k is a bounded regular function for the k-process with

$$\sum_i \pi_i^k \left(\frac{h_i}{k_i}\right) = \sum_i \pi_i h_i = 1,$$

and hence Lemma 10-42 yields a function g with

$$\left(\frac{h}{k}\right)_i = \int_{\bar{S}} K^k(i, x) g(x) d\mu^k(x)$$

and

$$\Pr{}^k\left[\lim \frac{h(x_n)}{k(x_n)} = g(x_v^k)\right] = 1.$$

As was pointed out in the proof of Lemma 10-34, x_v^k is identical with x_v and also

$$\Pr{}^k[p] = \tfrac{1}{2}\Pr{}^1[p] + \tfrac{1}{2}\Pr{}^h[p]$$

for all p. Thus $\Pr^k[p]$ cannot be one unless $\Pr^1(p)$ and $\Pr^h[p]$ both equal one, and we conclude that

$$\Pr\left[\lim \frac{h(x_n)}{k(x_n)} = g(x_v)\right] = 1$$

or

$$\Pr\left[\lim \frac{h(x_n)}{\frac{1}{2} + \frac{1}{2}h(x_n)} = g(x_v)\right] = 1.$$

Since $\{h(x_n)\}$ is a non-negative martingale, $\lim h(x_n)$ exists a.e. [Pr] and is finite. Therefore the above identity implies that

$$\Pr\left[\lim h(x_n) = \frac{g(x_v)}{2 - g(x_v)}\right] = 1.$$

Thus to complete the proof, it suffices to prove that

$$f(x) = \frac{g(x)}{2 - g(x)} \quad \text{a.e. } [\mu].$$

First we identify g as the Radon–Nikodym derivative of μ^h with respect to μ^k. On one hand, we have

$$\begin{aligned} h_i &= k_i \int_{\bar{S}} K^k(i, x) g(x) d\mu^k(x) \\ &= \int_{\bar{S}} K(i, x) g(x) d\mu^k(x). \end{aligned}$$

On the other hand, Theorem 10-41 gives

$$h_i = \int_{\bar{S}} K(i, x) d\mu^h(x),$$

and thus the uniqueness part of Theorem 10-41 gives $\mu^h = g\mu^k$.

Now, by Lemma 10-34, we have $\mu^k = \frac{1}{2}\mu + \frac{1}{2}\mu^h$. Hence

$$f\mu + \mu_s = \mu^h = g\mu^k = \tfrac{1}{2}g\mu + \tfrac{1}{2}g\mu^h = \tfrac{1}{2}g\mu + \tfrac{1}{2}gf\mu + \tfrac{1}{2}g\mu_s$$

or

$$(2f - g - fg)\mu = (g - 2)\mu_s.$$

Since μ and μ_s are singular with respect to each other, each side is the zero measure. For the left side, this statement means that

$$2f - g - fg = 0 \quad \text{a.e. } [\mu]$$

or

$$f = \frac{g}{2 - g} \quad \text{a.e. } [\mu].$$

The corollary to this theorem is the analog of Fatou's Theorem; it is a translation of the theorem into a result in terms of harmonic measure. The statement of the corollary needs a way of singling out for attention a single point x of S^*. One way of proceeding is to use the $K(\cdot, x)$ process, at least if x is in $\bar{S}$; for in that case Proposition 10-35 shows that

$$\Pr{}^{K(\cdot,x)}[x_v(\omega) = x] = 1.$$

Corollary 10-44: Let $h \geq 0$ be a normalized regular function, and let $\mu^h = f\mu + \mu_s$ (with f equal to zero on S) be the Lebesgue decomposition of μ^h with respect to μ. Then for almost every x $[\mu]$ for which $K(\cdot, x)$ is normalized,

$$\Pr{}^{K(\cdot,x)}[\lim_{n\to\infty} h(x_n) = f(x)] = 1.$$

PROOF: By Theorem 10-43,

$$\Pr[\lim h(x_n) = f(x_v)] = 1.$$

Then by Lemma 10-39,

$$1 = \Pr[\lim h(x_n) = f(x_v)] = \int_{\bar{S}} \Pr{}^{K(\cdot,x)}[\lim h(x_n) = f(x_v)]d\mu(x).$$

Since

$$\Pr{}^{K(\cdot,x)}[\lim h(x_n) = f(x_v)] \leq 1$$

for every x, equality must hold for almost every x $[\mu]$. But

$$\Pr{}^{K(\cdot,x)}[x_v = x] = 1$$

or

$$\Pr{}^{K(\cdot,x)}[f(x_v) = f(x)] = 1$$

for almost all x in $\bar{S}$. Since $\mu(\bar{S}) = 1$, the corollary follows.

The results above clearly extend to all π-integrable regular $h \geq 0$ provided we replace μ^h everywhere by the unique measure ν associated to h by Theorem 10-41.

Since the function f in all three of the above results is equal to zero on S, we may think of f as a function on B_e. If f is so restricted, it is called the **fine boundary function** of h.

9. Fine boundary functions

The results of Section 8 may be extended to non-negative π-integrable superregular functions with the help of the proposition below. Our convention that $h(x_n(\omega)) = 0$ if $n > \mathbf{v}(\omega)$ is still in force.

Proposition 10-45: If g is a π-integrable function of the form Nf with $f \geq 0$, then

$$\Pr[\lim_{n \to \infty} g(x_n) = 0] = 1.$$

For almost every x $[\mu]$ for which $K(\cdot, x)$ is normalized,

$$\Pr^{K(\cdot,x)}[\lim_{n \to \infty} g(x_n) = 0] = 1.$$

PROOF: In the first statement, g is non-negative superregular and πg is finite; hence $\{g(x_n)\}$ is a non-negative supermartingale and $g(x_n) \to z \geq 0$ a.e. [Pr]. Now $g \geq P^n g$ for all n, and $P^n g \to 0$. By dominated convergence $\pi P^n g \to 0$. Thus

$$\mathrm{M}_\pi[z] \leq \lim \mathrm{M}_\pi[g(x_n)] = \lim (\pi P^n g) = 0$$

and $z = 0$ a.e. [Pr]. The proof of the second statement in the proposition is the same as the proof of Corollary 10-44.

Thus if $h \geq 0$ is π-integrable and superregular, we may write, according to Theorem 5-10, $h = Nf + r$, with r regular and Nf and r both π-integrable and non-negative. Corollary 10-44 and Proposition 10-45 may therefore be combined into a single result whenever necessary.

The fine boundary function f of a normalized minimal regular function h takes on an especially simple form. By Lemma 10-33, we must have $h = K(\cdot, y)$ for some y in B_e, and, by Proposition 10-35,

$$\mu^h(\{y\}) = 1.$$

There are two cases. First, if $\mu(\{y\}) = 0$, then μ^h is singular with respect to μ and the fine boundary function of h is zero. By Lemma

10-42, h is unbounded. Second, if $\mu(\{y\}) = a > 0$, then the fine boundary function may be taken as $1/a$ at y and 0 elsewhere, since

$$\mu^h(C) = \chi_C(y) = \int_C f(x)d\mu(x)$$

for all Borel sets C. Moreover, h is bounded by $1/a$ since

$$1 \geq \Pr_i[x_v(\omega) = y] = \int_{\{y\}} K(i, x)d\mu(x) = aK(i, y) = ah_i.$$

The class of π-integrable regular functions $h \geq 0$ with a given μ-integrable non-negative function f as fine boundary function is exactly the class of functions h for which

$$\nu = f\mu + \mu_s$$

is the Lebesgue decomposition with respect to μ of the measure ν associated to h. Thus the class of such functions h is exactly the class of functions of the form

$$h_i = \int_{B_e} K(i, x)f(x)d\mu(x) + \int_{B_e} K(i, x)d\mu_s(x),$$

where μ_s is any Borel measure singular with respect to μ. In this class there is a unique smallest such function

$$\bar{h}_i = \int_{B_e} K(i, x)f(x)d\mu(x).$$

On one hand, Theorem 10-43 gives $\lim h(x_n) = f(x_v)$ a.e. [Pr], and hence

$$\mathrm{M}_\pi[\lim h(x_n)] = \mathrm{M}_\pi[f(x_v)] = \int_{B_e} f(x)d\mu(x) = \pi\bar{h}.$$

On the other hand,

$$\lim \mathrm{M}_\pi[h(x_n)] = \lim \pi P^n h = \pi h.$$

If $h_j > \bar{h}_j$ for some j, choose n so that $(\pi P^n)_j > 0$. Then

$$\pi h = \pi P^n h > \pi P^n \bar{h} = \pi\bar{h}.$$

Thus by Proposition 1-52, $\{h(x_n)\}$ is uniformly integrable if and only if $h = \bar{h}$.

There is a different topology for S^* which is occasionally referred to in connection with fine boundary functions. The **fine topology** for S^* is defined in terms of its neighborhood system as follows: For any point x in $S^* - B_e$ every set in S^* containing x is to be a neighborhood of x. For x in B_e, the neighborhoods of x are the sets A in S^* such that x is in A and

$$\Pr{}^{K(\cdot, x)}[x_n \in A \text{ from some time on}] = 1.$$

Evidently x is in each of its neighborhoods, the intersection of two neighborhoods is a neighborhood, and a superset of a neighborhood is a neighborhood. We must check that S^* is a neighborhood of x; that is, that the $K(\cdot, x)$-process disappears with probability zero. But this fact is a consequence of Proposition 10-35.

If x is in B_e and if A is an open set containing x in the metric topology of S^*, then

$$\begin{aligned} 1 &\geq \Pr^{K(\cdot,x)}[x_n \in A \text{ from some time on}] \\ &\geq \Pr^{K(\cdot,x)}[x_v \in A] = \mu^{K(\cdot,x)}(A) \geq \mu^{K(\cdot,x)}(\{x\}) = 1. \end{aligned}$$

Therefore A is open in the fine topology, and the fine topology is stronger than the metric topology.

The next lemma and proposition show that a zero-one law holds for the probabilities which define the fine topology. The lemma by itself is of value in checking whether a non-negative regular function is minimal.

Lemma 10-46: A non-negative normalized regular function h is minimal for P if and only if the only bounded regular functions for P^h (restricted to S^h) are constants.

PROOF: If h is minimal, then the regularity of h implies that $\mathbf{1}$ is regular for P^h restricted to S^h. By adding a suitable multiple of $\mathbf{1}$ to a given bounded regular function for P^h, we may assume that it is a non-negative bounded regular function for P^h. Thus let $\bar{h}$ with $0 \leq \bar{h} \leq c\mathbf{1}$ be a regular function for P^h restricted to S^h. Set

$$k_i = \begin{cases} \bar{h}_i h_i & \text{if } i \in S^h \\ 0 & \text{otherwise.} \end{cases}$$

Then k is P-regular and satisfies $0 \leq k_i \leq ch_i$. Since h is minimal, $k_i = c'h_i$ for all i and hence $\bar{h}_i = c'$ for i in S^h. That is, $\bar{h} = c'\mathbf{1}$.

Conversely, if h is not minimal, find a regular function k with $0 \leq k \leq h$ and $k \neq ch$. Set

$$\bar{h}_i = k_i/h_i \quad \text{for } i \in S^h.$$

Then $\bar{h}$ is a non-constant regular function for P^h restricted to S^h.

Proposition 10-47: If x is in B_e, then, for any subset A of S,

$$\Pr^{K(\cdot,x)}[x_n \in A \text{ from some time on}] = 0 \text{ or } 1.$$

PROOF: The h-process for $h = K(\cdot, x)$ disappears with probability zero, and thus

$$\Pr^{K(\cdot,x)}[x_n \in S] = 1.$$

Therefore

$$\begin{aligned}\Pr^{K(\cdot,x)}[x_n \in A \text{ from some time on}] &= \Pr^{K(\cdot,x)}[x_n \in S - A \text{ only finitely often}] \\ &= 1 - \Pr^{K(\cdot,x)}[x_n \in S - A \text{ infinitely often}].\end{aligned}$$

By Lemma 10-46, the only bounded regular functions for this process are the constants, and thus, by Proposition 5-19, this last expression must equal zero or one.

Proposition 10-47 shows that fine neighborhoods of x in B_e are those sets A in S^* for which x is in A and

$$\Pr^{K(\cdot,x)}[x_n \in A \text{ from some time on}] > 0.$$

The complement of a fine neighborhood of x is called **thin** at x. Such sets A are characterized by the property

$$\Pr^{K(\cdot,x)}[x_n \in A \text{ infinitely often}] = 0.$$

By Proposition 10-47 this probability must again be zero or one.

Let $h \geq 0$ be a π-integrable regular function with fine boundary function f. The fine topology for S^* has the property that the function $h \cup f$ obtained by extending h to S^* by f is continuous at almost every $[\mu]$ point x in S^*. In fact, the statement is trivial for x not in B_e, and it therefore suffices by Corollary 10-44 to prove the result for every x in B_e for which

$$\Pr^{K(\cdot,x)}[\lim h(x_n) = f(x)] = 1.$$

Let $a < f(x) < b$. We are to produce a fine neighborhood of x such that

$$a < (h \cup f)(y) < b$$

for all y in that neighborhood. Let A be the set of points of S for which $a < h < b$ and form the set $A \cup \{x\}$. We shall show this is a neighborhood of x. The convergence of $h(x_n)$ to $f(x)$ implies that

$$\Pr^{K(\cdot,x)}[a < h(x_n) < b \text{ from some time on}] = 1$$

or

$$\Pr^{K(\cdot,x)}[x_n \in A \text{ from some time on}] = 1.$$

Therefore

$$\Pr^{K(\cdot,x)}[x_n \in A \cup \{x\} \text{ from some time on}] = 1$$

and $A \cup \{x\}$ is a fine neighborhood of x.

10. Martin entrance boundary

The Martin exit boundary was defined in terms of P and a vector $\pi \geq 0$ such that πN is strictly positive. The completion S^* of S in the metric d served for a representation of all π-integrable superregular functions $h \geq 0$ in terms of finite measures on this space.

In this section we shall introduce a different completion $^*S$ of S which will serve for the representation of P-superregular measures. As the analog of π, we fix once and for all a function $f \geq 0$ such that Nf is everywhere positive and finite-valued. The representation theorem will be for superregular measures $\sigma \geq 0$ for which σf is finite.

The formalism is as follows: For i and j in S we define

$$J(i, j) = \frac{N_{ij}}{(Nf)_i}.$$

Then the measure $J(i, \cdot)$ is P-regular everywhere except at i, where it is strictly superregular, and it satisfies $J(i, \cdot)f = 1$ for all i. For each j the function is bounded by $N_{jj}/(Nf)_j$ because, if g is defined as Nf, then

$$J(i, j) = \frac{N_{ij}}{g_i} = \frac{N^g_{ij}}{g_j} \leq \frac{N^g_{jj}}{g_j} = \frac{N_{jj}}{g_j}.$$

We define

$$d'(j, j') = \sum_i w_i(Nf)_i|J(j, i) - J(j', i)|,$$

where the w_i are positive weights such that $\sum w_i N_{ii}$ is finite. The bound we have just computed shows that d' is finite-valued, and d' is a metric if we can show that $d'(j, j') = 0$ implies $j = j'$. But if $d'(j, j') = 0$, then

$$J(j, \cdot) = J(j', \cdot).$$

Multiplying through by P and supposing that $j \neq j'$, we obtain

$$J(j, j') = \sum_i J(j, i)P_{ij'} = \sum_i J(j', i)P_{ij'} < J(j', j'),$$

the strict inequality holding since $J(j', \cdot)$ is not regular at j'. Thus $j = j'$ and d' is a metric.

We define $^*S$ to be the Cauchy completion of S in the metric d'; the set $^*S - S$ is the **Martin entrance boundary**. A sequence $\{j_n\}$ is Cauchy in S if and only if the sequence $\{J(j_n, i)\}$ is Cauchy for every i. Consequently $J(\cdot, i)$ extends to a continuous function on $^*S$. Then $\{x_n\}$ is Cauchy in $^*S$ if and only if $\{J(x_n, i)\}$ is Cauchy for every i. Two points x and y are equal if and only if $J(x, \cdot) = J(y, \cdot)$, and the space $^*S$ is compact.

A P-superregular measure σ is **normalized** if $\sigma f = 1$. It is **minimal** if it is regular and if $0 \leq \bar{\sigma} \leq \sigma$ with $\bar{\sigma}$ regular implies $\bar{\sigma} = c\sigma$. Application of Fatou's Theorem shows that $J(x, \cdot)$ is a P-superregular measure for each x and that it satisfies $J(x, \cdot)f \leq 1$. A point x of $^*S$ is **extreme** if $J(x, \cdot)$ is minimal and normalized, and the set of extreme points is denoted B^e. Then $B^e \cap S$ is empty.

The next theorem is the Markov chain analog for the Poisson–Martin Representation Theorem for P-superregular measures.

Theorem 10-48: The sets S and B^e are Borel subsets of $^*S$. The non-negative P-superregular measures σ with σf finite stand in one-one correspondence with the non-negative finite measures ν on the Borel sets of $S \cup B^e$, the correspondence of σ to ν being

$$\sigma_i = \int_{S \cup B^e} J(x, i) d\nu(x)$$

and satisfying $\sigma f = \nu(S \cup B^e)$. The unique representation $\sigma = \gamma N + \rho$ with ρ regular arises by decomposing the integral over $S \cup B^e$ into a part over S and a part over B^e.

The proof will be accomplished by using duality, but we shall isolate several of the steps beforehand. Let $\alpha > 0$ be a superregular measure such that $\alpha f = 1$. Duality in the remainder of this section will mean α-duality. Let $\hat{P} = \text{dual } P$ and $\hat{\pi} = \text{dual} f$. Since

$$0 < \text{dual } g = \text{dual } (Nf) = \hat{\pi}\hat{N}$$

and

$$1 = \alpha f = (\text{dual} f)(\text{dual } \alpha) = \hat{\pi}\mathbf{1},$$

the exit boundary of $\hat{P}$ relative to $\hat{\pi}$ is defined. Let $\hat{d}$ be the defining metric of the exit boundary, and let $\hat{S}^*$ be the completion of S under $\hat{d}$. We have

$$\text{dual } J(j, \cdot) = \frac{J(j, \cdot)}{\alpha.} = \frac{N_{j\cdot}}{(Nf)_j \alpha.} = \frac{\hat{N}_{\cdot j}}{(\hat{\pi}\hat{N})_j} = \hat{K}(\cdot, j).$$

Lemma 10-49: If the same weights w_i are used in defining $\hat{d}$ and d', then the identity map on S extends to an isometry of $(^*S, d')$ onto $(\hat{S}^*, \hat{d})$.

Proof: We have

$$\begin{aligned} d'(j, j') &= \sum_i w_i (Nf)_i |J(j, i) - J(j', i)| \\ &= \sum_i w_i \alpha_i (Nf)_i \left| \frac{J(j, i)}{\alpha_i} - \frac{J(j', i)}{\alpha_i} \right| \\ &= \sum_i w_i (\hat{\pi}\hat{N})_i |\hat{K}(i, j) - \hat{K}(i, j')| \\ &= \hat{d}(j, j'). \end{aligned}$$

Under the identification of $*S$ and $\hat{S}*$,

$$J(x, \cdot) = \text{dual } \hat{K}(\cdot, x)$$

by the continuity of the functions $J(\cdot, i)$ and $\hat{K}(i, \cdot)$. Under duality a $\hat{P}$-superregular function h corresponds to a P-superregular row vector $\sigma = \text{dual } h$. Since $\hat{P} = \text{dual } P$ and $\hat{\pi} = \text{dual} f$, regularity and normalization are both preserved. But minimality is also preserved since duality preserves inequalities. Therefore $B^e = \hat{B}_e$.

PROOF OF THEOREM 10-48: Let $h = \text{dual } \sigma$, and apply Theorem 10-41. Then

$$h_i = \int_{S \cup \hat{B}_e} \hat{K}(i, x) d\nu(x)$$

with $\hat{\pi} h = \nu(S \cup \hat{B}_e)$. Therefore

$$\frac{\sigma_i}{\alpha_i} = \int_{S \cup \hat{B}_e} \frac{1}{\alpha_i} J(x, i) d\nu(x)$$

and $\sigma f = \nu(S \cup \hat{B}_e)$. If we identify B^e and $\hat{B}_e$ and call ν by the same name on B^e as on $\hat{B}_e$, then we have

$$\sigma_i = \int_{S \cup B^e} J(x, i) d\nu(x)$$

as required. Conversely, if σ is given by

$$\sigma_i = \int_{S \cup B^e} J(x, i) d\nu(x),$$

then

$$(\text{dual } \sigma)_i = \int_{S \cup \hat{B}_e} \hat{K}(i, x) d\nu(x).$$

Therefore dual σ is $\hat{P}$-superregular and σ must be superregular. Since there is only one measure which represents dual σ, there can be only one measure which represents σ. The last statement of the theorem follows from the fact that the unique decomposition $\sigma = \gamma N + \rho$ is transformed under duality into the unique decomposition of dual σ in the $\hat{P}$-process.

If $\sigma > 0$, we can take $\alpha = \sigma$ and then the proof of Theorem 10-48 gives us the following interpretation for $\nu : \nu$ is harmonic measure for the chain with transition matrix σ-dual P and starting distribution σ-dual f.

11. Application to extended chains

In order to assign a probabilistic interpretation to the entrance boundary, we must pass to extended chains. Except where noted, we therefore assume throughout this section that $\{z_n\}$ is an extended chain with state space $S \cup \{a\} \cup \{b\}$, with (enlarged) transition matrix $\bar{P}$, and with vector ν of mean times in the various states of S. Let π and f be row and column vectors, respectively, such that $\pi \geq 0$ $f \geq 0$, $\pi 1 = 1$, $\nu f = 1$, $\pi N > 0$, and $Nf > 0$. By Theorem 8-3, Nf is finite-valued. We continue to use the notations of Sections 3 through 10 for boundary theory of P relative to π and f, and we shall use the vectors μ^E and ν^E defined for extended chains in Section 2.

Our first pair of propositions will give the "final" and "starting" distributions for the extended chain. If x is in $^*S$ and if E ranges over finite sets in S, we define

$$p(x) = \lim_{E \uparrow S} J(x, \cdot)e^E.$$

Similarly, if y is in S^*, we define

$$q^{\nu}(y) = \lim_{E \uparrow S} \mu^E K(\cdot, y).$$

We first show that these limits always exist.

Lemma 10-50: For every x in $^*S$ and y in S^*, both $p(x)$ and $q^{\nu}(y)$ exist, possibly being infinite. Their defining limits are increasing limits as $E \uparrow S$. Each is a Borel measurable function on its domain, and the functions satisfy

$$p(x) = \lim_{E \uparrow S} \lim_{j \to x} \frac{h_j^E}{(Nf)_j}$$

and

$$q^{\nu}(y) = \lim_{E \uparrow S} \lim_{j \to y} \frac{\nu_j^E}{(\pi N)_j}.$$

PROOF: For any finite set E,

$$h_j^E = \sum_{k \in E} N_{jk} e_k^E$$

and

$$\frac{h_j^E}{(Nf)_j} = \sum_{k \in E} J(j, k) e_k^E.$$

Since E is finite, we may pass to the limit as $j \to x$ to get

$$\lim_{j \to x} \frac{h_j^E}{(Nf)_j} = J(x, \cdot)e^E.$$

For each fixed j, the expression $h_j^E/(Nf)_j$ increases with E, and hence so does the limit as $j \to x$. Therefore $p(x)$ exists, is defined by an increasing limit, and has the asserted value. If we choose an increasing sequence of finite sets E_n with union S, then $p(x)$ is exhibited as the limit of a *sequence* of continuous functions and is therefore Borel measurable.

Similarly we have

$$\nu_j^E = \sum_{k \in E} \mu_k^E N_{kj}$$

and hence

$$\lim_{j \to y} \frac{\nu_j^E}{(\pi N)_j} = \mu^E K(\cdot, y).$$

Thus the same argument applies to $q^{\mathbf{v}}(y)$, since ν^E increases with E.

Proposition 10-51: On almost every path,

$$z_{\mathbf{v}}(\omega) = \lim_{n \uparrow \mathbf{v}} z_n(\omega)$$

exists in S^*. On almost every path for which the limit exists, either $\mathbf{v} < +\infty$ and $z_{\mathbf{v}} \in S$ or else $\mathbf{v} = +\infty$ and $z_{\mathbf{v}} \in B_e$. Moreover, for any Borel set C in S^*,

$$\Pr[z_{\mathbf{v}} \in C] = \int_C q^{\mathbf{v}}(y) d\mu(y).$$

PROOF: The process watched starting in E is a Markov chain with transition matrix P and starting distribution μ^E if E is a finite set in S. By Proposition 10-21, $z_{\mathbf{v}}$ exists and has the required values for almost every path in this chain, and in this chain

$$\begin{aligned}
\Pr_{\mu^E}[z_{\mathbf{v}} \in C] &= \sum_{i \in E} \mu_i^E \Pr_i[z_{\mathbf{v}} \in C] \\
&= \sum_{i \in E} \mu_i^E \int_C K(i, x) d\mu(x) \\
&= \int_C \mu^E K(\cdot, x) d\mu(x).
\end{aligned}$$

As E increases to S, we obtain almost all paths of $\{z_n\}$ in this way. Hence $z_{\mathbf{v}}$ exists and is in $S \cup B_e$ a.e., and, by Definition 10-4,

$$\Pr[z_{\mathbf{v}} \in C] = \lim_{E \uparrow S} \int_C \mu^E K(\cdot, y) d\mu(y).$$

By Lemma 10-50, the integrand increases with E. Replacing the sets E by an increasing sequence E_n and applying monotone convergence, we obtain

$$\begin{aligned}
\Pr[z_{\mathbf{v}} \in C] &= \int_C \lim_{E \uparrow S} \mu^E K(\cdot, y) d\mu(y) \\
&= \int_C q^{\mathbf{v}}(y) d\mu(y).
\end{aligned}$$

Proposition 10-52: On almost every path,

$$z_u(\omega) = \lim_{n \downarrow u} z_n(\omega)$$

exists in $^*S$. On almost every path for which the limit exists, either $\mathbf{u} > -\infty$ and $z_n \in S$ or else $\mathbf{u} = -\infty$ and $z_u \in B^e$. Moreover, for any Borel set C in $^*S$,

$$\Pr[z_u \in C] = \int_C p(x) d\mu^\nu(x),$$

where μ^ν is the unique measure on $S \cup B^e$ representing ν.

PROOF: Form the reverse process and call its measure Pr′. Its transition matrix restricted to the states of S is the ν-dual of P, denoted $\hat{P}$. Form the exit and entrance boundaries of $\hat{P}$ relative to $\hat{\pi} = \nu$-dual f and $\hat{f} = \nu$-dual π. By Proposition 10-51, the limit

$$z_v = \lim_{n \uparrow v} z_n$$

exists in $\hat{S}^*$ a.e. [Pr′], and either $\mathbf{v}$ is finite and $z_v \in S$ or else $\mathbf{v}$ is infinite and a.e. $z_v \in \hat{B}_e$. Therefore

$$z_u = \lim_{n \downarrow u} z_n$$

exists in $\hat{S}^*$ a.e. [Pr], and either $\mathbf{u}$ is finite and $z_u \in S$ or else $\mathbf{u}$ is infinite and a.e. $z_u \in \hat{B}_e$. Since $\hat{S}^*$ and $^*S$ are canonically identified according to Lemma 10-49, the first part of the proposition follows. Moreover, we have

$$\Pr[z_u \in C] = \Pr'[z_v \in C] = \int_C \lim_{E \uparrow S} [\mu^E \hat{K}(\cdot, x)] d\mu^\nu(x),$$

where μ^ν is the measure on $\hat{S}^*$ which represents $\mathbf{1}$, that is, the measure on $^*S$ which represents ν. The above expression is

$$= \int_C \lim_{E \uparrow S} [J(x, \cdot)(\text{dual } \mu^E)] d\mu^\nu(x).$$

But μ^E, by Proposition 10-8, is the balayage charge of ν on E and its dual is thus the balayage charge of 1 on E; that is, the charge e^E of h^E.

We thus obtain $\int p(x) d\mu^\nu(x)$ and $\int q^\nu(y) d\mu(y)$ as starting and final distributions for the extended chain. The fact that ν is normalized ($\nu f = 1$) implies that μ^ν is a probability measure. Note that $p(x)$ depends only on P, π, and f and not on the extended chain or even the vector ν, but that $q^\nu(y)$ depends on P, π, f, and ν. We shall return to this point shortly.

Conversely, if we select any probability measure μ^σ on $S \cup B^e$, then Theorem 10-48 yields a unique normalized P-superregular measure σ represented by μ^σ. If σ is positive, Theorem 10-9 assures the existence of at least one extended chain with σ as its vector of mean times. Any such chain starts in C with probability

$$\int_C p(x) d\mu^\sigma(x).$$

By varying μ^σ we may change the total measure on $\{z_n\}$, even as to whether it is finite or infinite.

We consider two special cases. First, if 0 is in S, let

$$\sigma_j = J(0, j) = N_{0j}/(Nf)_0.$$

By the uniqueness of the representation we must have $\mu^\sigma(\{0\}) = 1$, and hence we may represent σ (provided it is positive) by an extended chain which has a.e. path starting at 0. The total measure of the process must be

$$\int_{*S} p(x) d\mu^\sigma(x) = p(0)\mu^\sigma(\{0\}) = p(0),$$

and so the interpretation of σ_j as the mean number of times in state j shows that

$$\sigma_j = p(0)N_{0j}.$$

As a check, we can compute $p(0)$ directly. We have

$$p(0) = \lim_{E \uparrow S} \frac{h_0^E}{(Nf)_0} = \frac{1}{(Nf)_0},$$

as required.

Second, choose $\sigma = J(x, \cdot)$, where x is in B^e. Then $\mu^\sigma(\{x\}) = 1$, and (if σ is positive) an extended chain representing σ in the sense of Theorem 10-9 must start a.e. at x with total measure $p(x)$. Therefore, $J(x, \cdot)$ may be interpreted as the vector of mean times for any extended chain which is started almost surely at x, and, if $p(x)$ is finite, then

$$\frac{J(x, \cdot)}{p(x)}$$

is the conditional mean of the number of times in the various states given that the process starts at x.

Returning to the case of a general ν and supposing that $p(x)$ is finite a.e. $[\mu^\nu]$, we see that the identity

$$\nu_j = \int_{*S} \frac{J(x, j)}{p(x)} \cdot p(x) d\mu^\nu(x)$$

may be interpreted as follows: ν_j may be computed by choosing a starting state x according to the starting distribution and then weighting it by the conditional mean of the number of times in j given that the process starts at x.

As we noted earlier, there is an asymmetry between $p(x)$ and $q^{\nu}(y)$ in that one of these quantities depends on ν and the other does not. This asymmetry is cleared up by a discussion of h-processes for P; we shall see that $p(x)$ depends on h and that $q^{\nu}(y)$ does not. The special case we have been considering so far will be seen to be the case $h = 1$.

In discussing the entrance boundary for P^h, where $h > 0$ is superregular and normalized, we choose

$$f_i^h = f_i/h_i$$

in analogy with the choice of π^h. Define σ by

$$\sigma_i = \nu_i h_i.$$

Then σ is positive, is P^h-superregular, and satisfies $\sigma f^h = \nu f = 1$. Direct calculation shows that

$$\sigma\text{-dual}\ (P^h) = \nu\text{-dual}\ P = \hat{P}$$

and that

$$\sigma\text{-dual}\ (f^h) = \nu\text{-dual}\, f = \hat{\pi}.$$

Therefore the entrance boundary for P^h and f^h is the same as that for P and f.

Now let $\{z_n'\}$ be an extended chain representing P^h and σ (existence by Theorem 10-9), and take π^h and f^h as the functions relative to which the exit and entrance boundaries are formed. The process starts in $^*S$ and goes to $S^{h*} = S^*$. To obtain the starting and final distributions, we need the functions p and q. For p we have

$$\begin{aligned} p^h(x) &= \lim_{E \uparrow S} \lim_{j \to x} \frac{(h^E)_j^h}{(N^h f^h)_j} \\ &= \lim_{E \uparrow S} \lim_{j \to x} \frac{\sum_{k \in E} (B^E)_{jk}^h}{(Nf)_j/h_j} \\ &= \lim_{E \uparrow S} \lim_{j \to x} \frac{(B^E h)_j}{g_j}. \end{aligned}$$

For q we have

$$\begin{aligned} q^{\sigma,h}(y) &= \lim_{E \uparrow S} (\mu^E)^h K^h(\cdot, y) \\ &= \lim_{E \uparrow S} \sum_{i \in E} \mu_i^E h_i K(i, y)/h_i \\ &= \lim_{E \uparrow S} \mu^E K(\cdot, y) \\ &= q^{\nu}(y). \end{aligned}$$

Therefore, in the chain $\{z'_n\}$ we have as starting and final distributions

$$\Pr{}^h[z'_u \in C] = \int_C p^h(x)d\mu^\nu(x)$$

and

$$\Pr{}^h[z'_v \in C] = \int_C q^\nu(y)d\mu^h(y).$$

These relations show the symmetric roles of p and q and give us symmetric interpretations for μ^ν and μ^h.

In the special case of positive minimal ν and h, we must have

$$\nu = J(x_0, \cdot)$$

and

$$h = K(\cdot, y_0),$$

where x_0 and y_0 are in B^e and B_e, respectively. In this case

$$\mu^\nu(\{x_0\}) = \mu^h(\{y_0\}) = 1.$$

Thus the process $\{z'_n\}$ has almost every path starting at x_0 and going to y_0. The paths which start at x_0 have measure $p^h(x_0)$, and the paths which go to b have measure $q^\nu(y_0)$. Therefore we must have

$$p^h(x_0) = q^\nu(y_0).$$

12. Proof of Theorem 10-9

(1) *Construction of the extended stochastic process.* Let

$$\bar{P} = P^*_{S\cup\{b\}} \quad \text{and} \quad P = \bar{P}_S.$$

Redefine ν as being restricted to the domain S. Let ν^E be the balayage potential of ν on a finite set E with μ^E the balayage charge. For each E we shall consider a Markov chain with transition matrix P and starting distribution μ^E, and we shall combine these into a single extended stochastic process by choosing a common time scale. For the sake of convenience we assume that $S = \{0, 1, \ldots\}$. Our measure space will, as usual, be the double sequence space obtained from S, a, and b. To define the process $\{x_n\}$ it is sufficient to assign probabilities consistently to basic statements.

We shall use $\{y_n\}$ to denote the outcome functions for the Markov chain P with various μ^E as starting distributions. These vectors are finite measures, but not necessarily probability measures. Since S is the set of non-negative integers, it has a natural ordering on it. For each path ω in the ordinary sequence space of $\bar{P}$, let $\mathbf{s}(\omega)$ be the smallest numbered state on path ω and let $\mathbf{t}(\omega)$ be the time that $\mathbf{s}(\omega)$ is first entered. (Then $\mathbf{s}$ and $\mathbf{t}$ are defined everywhere except on the one path $(b, b, b, b, \ldots)$.)

There are two kinds of probability assignments to be made on the basic cylinder sets. First, if $i \in S$, define

$$\Pr[x_n = i \wedge x_{n+1} = j \wedge \cdots \wedge x_{n+m} = k]$$
$$= \Pr_{\mu^E}[y_{\mathbf{t}+n} = i \wedge y_{\mathbf{t}+n+1} = j \wedge \cdots \wedge y_{\mathbf{t}+n+m} = k],$$

where $E = \{i\}$ and where the right side is taken as 0 for the set of ω on which $\mathbf{t}(\omega) + n < 0$. Second, define

$$\Pr[x_{-n-r} = a \wedge \cdots \wedge x_{-n-2} = a \wedge x_{-n-1} = a$$
$$\wedge\, x_{-n} = i \wedge x_{-n+1} = j \wedge \cdots \wedge x_{-n+m} = k]$$
$$= \mu_i \Pr_i[y_1 = j \wedge \cdots \wedge y_m = k \wedge \mathbf{t} = n].$$

The effect of these definitions is to fix a time scale with this property: For each path there is a designated state such that the first entry into that state occurs at time 0. Then for $i \in S$, $\Pr[x_n = i]$ is finite for all n. To show that $\{x_n\}$ is an extended stochastic process, we must check that the above definitions on cylinder sets are consistent. That is, we must show typically that

$$\sum_k \Pr[x_n = i \wedge x_{n+1} = j \wedge x_{n+2} = k] = \Pr[x_n = i \wedge x_{n+1} = j],$$

when i and j do not both equal a, and that

$$(*) \quad \sum_i \Pr[x_{-n-1} = i \wedge x_{-n} = j \wedge x_{-n+1} = k]$$
$$= \Pr[x_{-n} = j \wedge x_{-n+1} = k],$$

when j and k do not both equal b. The first identity is immediate from the above definitions. But to prove the second identity, we need some alternate expressions for the left side.

Let $i \in S$, let $E = \{i\}$, let F be a finite set with $E \subset F$, and let q be the statement $y_{\mathbf{t}+n} = i \wedge y_{\mathbf{t}+n+1} = j \wedge y_{\mathbf{t}+n+2} = k$. By the first definition above,

$$\Pr[x_n = i \wedge x_{n+1} = j \wedge x_{n+2} = k] = \Pr_{\mu^E}[q].$$

We shall show that

$$\Pr_{\mu^F}[q] = \Pr_{\mu^E}[q].$$

Since the truth of q depends only on events after the time $\mathbf{t}_E$ when E is reached, Theorem 4-10 gives

$$\Pr_{\mu^F}[q] = \sum_{m \in E} \Pr_{\mu^F}[y_{\mathbf{t}_E} = m] \Pr_m[q]$$
$$= \sum_{m \in E} (\mu^F B^E)_m \Pr_m[q].$$

Now $\mu^F B^E = \mu^E$, since μ^F and μ^E are balayage charges (see Section 2). Therefore

$$\Pr_{\mu^F}[q] = \sum_{m \in E} (\mu^E)_m \Pr_m[q] = \Pr_{\mu^E}[q].$$

It follows that

$$\Pr_{\mu^E}[q] = \lim_{F \uparrow S} \Pr_{\mu^F}[q],$$

which is the first of the two alternate identities we need.

Next, we note that $\nu(I - P) = \mu$ by Theorem 5-10. Hence $\mu_i = \lim_{F \uparrow S} \mu_i^F$ by the property of balayage charges that they converge to the product of the given superregular measure and $I - P$. Thus

$$\begin{aligned}
\Pr[x_{-n-2} = a \wedge x_{-n-1} = a \wedge x_{-n} = i \wedge x_{-n+1} = j \wedge x_{-n+2} = k] \\
= \mu_i \Pr_i[y_1 = j \wedge y_2 = k \wedge \mathbf{t} = n] \\
= \lim_F \mu_i^F \Pr_i[y_1 = j \wedge y_2 = k \wedge \mathbf{t} = n] \\
= \lim_F \Pr_{\mu^F}[y_0 = i \wedge y_1 = j \wedge y_2 = k \wedge \mathbf{t} = n],
\end{aligned}$$

which is our second alternate identity.

Now we are in a position to prove (*). In the proof we shall use the abbreviation

$$Y = \lim_{F \uparrow S} \sum_{m=1}^{\infty} \sum_{i \in S - F} \Pr_{\mu^F}[y_{m-1} = i \wedge y_m = j \wedge y_{m+1} = k \wedge \mathbf{t} = m + n].$$

We begin by using our two alternate identities and by performing a direct calculation. Recall that i cannot equal b, since $j \neq b$.

$$\begin{aligned}
\sum_{\substack{i \in S \\ \text{or } i = a}} \Pr[x_{-n-1} = i \wedge x_{-n} = j \wedge x_{-n+1} = k] \\
= \lim_{F \uparrow S} \sum_{i \in F} \Pr_{\mu^F}[y_{t-n-1} = i \wedge y_{t-n} = j \wedge y_{t-n+1} = k] \\
+ \lim_{F \uparrow S} \Pr_{\mu^F}[y_0 = j \wedge y_1 = k \wedge \mathbf{t} = n] \\
= \lim_F \Big\{ \sum_{m=1}^{\infty} \sum_{i \in F} \Pr_{\mu^F}[y_{m-1} = i \wedge y_m = j \wedge y_{m+1} = k \wedge \mathbf{t} = m + n] \\
+ \Pr_{\mu^F}[y_0 = j \wedge y_1 = k \wedge \mathbf{t} = n] \Big\} \\
= \lim_F \sum_{m=0}^{\infty} \Pr_{\mu^F}[y_m = j \wedge y_{m+1} = k \wedge \mathbf{t} = m + n] - Y \\
= \lim_F \Pr_{\mu^F}[y_{t-n} = j \wedge y_{t-n+1} = k] - Y \\
= \Pr[x_{-n} = j \wedge x_{-n+1} = k] - Y.
\end{aligned}$$

Thus we are to prove $Y = 0$. Now

$$\begin{aligned}
0 \leq Y &\leq \lim_F \sum_{m=1}^{\infty} \sum_{i \in S-F} \Pr_{\mu^F}[y_{m-1} = i \wedge y_m = j] \\
&= \lim_F \sum_{m=1}^{\infty} \sum_{i \in S-F} (\mu^F P^{m-1})_i P_{ij} \\
&= \lim_F \sum_{i \in \tilde{F}} (\mu^F N)_i P_{ij} \\
&= \lim_F \sum_{i \in \tilde{F}} \nu_i^F P_{ij} \\
&\leq \lim_F \sum_{i \in \tilde{F}} \nu_i P_{ij}.
\end{aligned}$$

Since $\sum_i \nu_i P_{ij}$ is convergent, the right side is zero. Hence $Y = 0$. Therefore $\{x_n\}$ is an extended stochastic process.

(2) *Verification of some of the properties of an extended chain.* We must now check that $\{x_n\}$ satisfies the four properties of an extended chain. First we check (2). Consider the special case in which E is the finite set $\{0, 1, 2, \ldots, i\}$. This set has the property that on any path beginning in a state of E the statement $\mathbf{t} \geq n$ implies the statement that there is an $m \geq n$ with $y_m \in E$. Therefore

$$\begin{aligned}
&\lim_n \Pr[x_{-l} \in E \text{ for some } l \geq n] \\
&\qquad = \lim_n \sum_{l=n}^{\infty} \Pr[x_{-l} \in E \wedge x_{-l+1} \notin E \wedge \cdots \wedge x_{-n} \notin E] \\
&\qquad = \lim_n \sum_{l=n}^{\infty} \Pr_{\mu^E}[y_{t-l} \in E \wedge \cdots \wedge y_{t-n} \notin E] \\
&\qquad = \lim_n \Pr_{\mu^E}[y_{t-l} \in E \text{ for some } l \geq n] \\
&\qquad \leq \lim_n \Pr_{\mu^E}[\mathbf{t} \geq n] \\
&\qquad \leq \lim_n \Pr_{\mu^E}[(\exists m) \text{ with } m \geq n \text{ and } y_m \in E].
\end{aligned}$$

Since P is transient and μ^E is finite, the right side is zero. Hence so is the left side. By Corollary 1-17 we conclude that

$$\Pr[\mathbf{u}_E = -\infty] = 0.$$

For the general case of a finite set F, choose a finite set E of the above form which contains it. Then

$$\Pr[\mathbf{u}_F = -\infty] \leq \Pr[\mathbf{u}_E = -\infty] = 0.$$

Hence (2) holds.

Now suppose for the moment that we have proved (3) in the form that for any finite set E the process watched starting in E is a Markov chain with starting distribution μ^E and transition matrix P, and suppose we have identified the given measure ν as the vector of mean times in the various states. Then (4) follows, since ν was assumed finite-valued. To prove (1) that the domain of $\mathbf{u}_E$ has positive measure, let $j \in E$. Then

$$0 < \nu_j = \nu_j^E = \sum_{i \in E} \mu_i^E N_{ij}.$$

Hence $\mu_i^E > 0$ for some $i \in E$. But $\mu^E 1$ is the measure of the set on which E is ever entered for the first time. That is, it is the measure of the set where $\mathbf{u}_E > -\infty$. Hence (1) holds.

Thus we are to prove that the process watched starting in the finite set E is a Markov chain with starting distribution μ^E and transition matrix $\bar{P}$, and we are to identify ν. Suppose we have proved this assertion for all finite sets of the form $E = \{0, 1, 2, \ldots, e\}$, and suppose F is an arbitrary finite set. We show first how the assertion follows for F. Choose a set E of the special form containing F. Then watching the process beginning in F is the same as watching the process watched starting in E beginning in F. Thus since the E-process is a Markov chain, so is the F-process by Theorem 4-9. The starting distribution for the F-process is

$$\Pr_{\mu^E}[y_{t_F} = j] = \sum_{i \in S} \mu_i^E \Pr_i[y_{t_F} = j] = \sum_{i \in S} \mu_i^E B_{ij}^F = \mu_j^F.$$

Hence we may assume that E is a set of the form $E = \{0, 1, \ldots, e\}$. Let $\{z_n\}$ be the outcome functions for the process watched starting in E. We shall compute the probability of the typical basic statement p defined as $z_0 = i \wedge z_1 = j \wedge z_2 = k$, where $i \in E$, by considering separately the contributions from paths of $\{x_n\}$ with $\mathbf{u} > -\infty$ and paths with $\mathbf{u} = -\infty$. The notation $\tilde{E}$ will mean $S - E$.

(3) *Contribution to* Pr[p] *from paths with* $\mathbf{u} > -\infty$. The contribution to Pr[p] of paths with $\mathbf{u} > -\infty$ is

$$\begin{aligned}
\sum_{n=-\infty}^{\infty} \sum_{m=0}^{\infty} \Pr[x_{-m-n-1} = a \wedge x_{-m-n} \in \tilde{E} \wedge \cdots \wedge x_{-n-1} \in \tilde{E} \wedge x_{-n} = i \\
\wedge x_{-n+1} = j \wedge x_{-n+2} = k] \\
= \sum_{n=-\infty}^{\infty} \sum_{m=0}^{\infty} \Pr_{\mu}[y_0 \in \tilde{E} \wedge \cdots \wedge y_{m-1} \in \tilde{E} \wedge y_m = i \wedge y_{m+1} = j \\
\wedge y_{m+2} = k \wedge \mathbf{t} = m + n].
\end{aligned}$$

The fact that E is of the special form $\{0, 1, \ldots, e\}$ means that if

$$y_0(\omega) \in \tilde{E} \wedge \cdots \wedge y_{m-1}(\omega) \in \tilde{E} \wedge y_m(\omega) = i,$$

then since $i \in E$ we must have $\mathbf{t}(\omega) \geq m$. Hence we need sum on n only over $n \geq 0$. If $n \geq 0$, then for any ω for which $y_0, \ldots, y_{m-1}$ are not in E, we have that $\mathbf{t}(\omega) = m + n$ if and only if $\mathbf{t}(\omega_m) = n$. Therefore we may apply Theorem 4-9 to Pr[—]. In symbols the argument is that the contribution is

$$= \sum_{n \geq 0} \sum_{m \geq 0} \Pr_\mu[y_0 \in \tilde{E} \wedge \cdots \wedge \mathbf{t} = m + n]$$

$$= \sum_{n \geq 0} \sum_{m \geq 0} \Pr_\mu[y_0 \in \tilde{E} \wedge \cdots \wedge y_{m-1} \in \tilde{E} \wedge y_m = i] \times \Pr_i[y_1 = j \wedge y_2 = k \wedge \mathbf{t} = n].$$

Let $B^{E,m}$ denote the matrix of probabilities of entry to E at time m. The above expression is

$$= \sum_{n \geq 0} \sum_{m \geq 0} (\mu B^{E,m})_i \Pr_i[y_1 = j \wedge y_2 = k \wedge \mathbf{t} = n]$$

$$= \sum_{n \geq 0} (\mu B^E)_i \Pr_i[y_1 = j \wedge y_2 = k \wedge \mathbf{t} = n]$$

$$= (\mu B^E)_i \Pr_i[y_1 = j \wedge y_2 = k]$$

$$= (\mu B^E)_i \bar{P}_{ij} \bar{P}_{jk}.$$

Thus the contribution is $(\mu B^E)_i \bar{P}_{ij} \bar{P}_{jk}$.

(4) *Contribution to* $\Pr[p]$ *from paths with* $\mathbf{u} = -\infty$. The contribution to $\Pr[p]$ of paths with $\mathbf{u} = -\infty$ is

$$\sum_{n=-\infty}^{\infty} \lim_{m \to \infty} \Pr[x_{-n-m} \in \tilde{E} \wedge \cdots \wedge x_{-n-1} \in \tilde{E} \wedge x_{-n} = i \wedge x_{-n+1} = j \wedge x_{-n+2} = k]$$

$$= \sum_n \lim_m \sum_{s \in \tilde{E}} \Pr[x_{-n-m} = s \wedge x_{-n-m+1} \in \tilde{E} \wedge \cdots \wedge x_{-n+2} = k]$$

$$= \sum_n \lim_m \sum_{s \in \tilde{E}} \lim_{F \uparrow S} \Pr_{\mu^F}[y_{\mathbf{t}-m-n} = s \wedge y_{\mathbf{t}-m-n+1} \in \tilde{E} \wedge \cdots \wedge y_{\mathbf{t}-n+2} = k]$$

$$= \sum_n \lim_m \sum_{s \in \tilde{E}} \lim_F \sum_{l=0}^{\infty} \Pr_{\mu^F}[y_l = s \wedge y_{l+1} \in \tilde{E} \wedge \cdots \wedge y_{l+m+2} = k \wedge \mathbf{t} = l + m + n].$$

In the expression $\Pr_{\mu^F}$[—], for fixed n we may assume that m is large, say $m \geq |n|$. Then $\mathbf{t} \geq l$. Now since E is of the special form $\{0, 1, \ldots, e\}$ and since $i \in E$, state i is lower in the ordering on S than any of the states of $\tilde{E}$. Hence

$$y_l(\omega) = s \wedge y_{l+1}(\omega) \in \tilde{E} \wedge \cdots \wedge \mathbf{t}(\omega) = l + m + n$$

for $m \geq n$ if and only if

$$y_0(\omega) \in \tilde{G} \wedge \cdots \wedge y_{l-1}(\omega) \in \tilde{G} \wedge y_l(\omega) = s \\ \wedge y_{l+1}(\omega) \in \tilde{E} \wedge \cdots \wedge \mathbf{t}(\omega) = l + m + n,$$

where $G = \{0, 1, \ldots, i\}$. In the presence of the information

$$y_0(\omega) \in \tilde{G} \wedge \cdots \wedge y_{l-1}(\omega) \in \tilde{G} \wedge y_l(\omega) = s,$$

it is true that $\mathbf{t}(\omega) = l + m + n$ if and only if $\mathbf{t}(\omega_l) = m + n$. Theorem 4-9 now applies. The contribution to $\Pr[p]$ therefore is

$$= \sum_n \lim_m \sum_{s \in \tilde{E}} \lim_F \sum_{l \geq 0} \Pr_{\mu^F}[y_0 \in \tilde{G} \wedge \cdots \wedge y_{l-1} \in \tilde{G} \wedge y_l = s] \\ \times \Pr_s[y_1 \in \tilde{E} \wedge \cdots \wedge y_{m+2} = k \wedge \mathbf{t} = m + n].$$

Next, we operate on the second factor. In the presence of the information $y_0(\omega) = s \wedge y_1(\omega) \in \tilde{E} \wedge \cdots \wedge y_{m-1}(\omega) \in \tilde{E} \wedge y_m(\omega) = i$, we know by the special form of E that $\mathbf{t}(\omega) = m + n$ if and only if $\mathbf{t}(\omega_m) = n$. Hence, by Theorem 4-9, we have

$$\Pr_s[y_1 \in \tilde{E} \wedge \cdots \wedge y_{m+2} = k \wedge \mathbf{t} = m + n] \\ = \Pr_s[y_1 \in \tilde{E} \wedge \cdots \wedge y_m = i] \cdot \Pr_i[y_1 = j \wedge y_2 = k \wedge \mathbf{t} = n].$$

The second factor on the right side of this last equation does not depend on m and therefore factors out. We may sum it on n, and we get $\bar{P}_{ij}\bar{P}_{jk}$. Hence the contribution is

$$= \lim_m \sum_{s \in \tilde{E}} \lim_F \sum_{l \geq 0} \Pr_{\mu^F}[y_0 \in \tilde{G} \wedge \cdots \wedge y_{l-1} \in \tilde{G} \wedge y_l = s] \\ \times \Pr_s[y_1 \in \tilde{E} \wedge \cdots \wedge y_m = i]\bar{P}_{ij}\bar{P}_{jk}.$$

The first factor here is

$$\sum_{l \geq 0} \Pr_{\mu^F}[y_0 \in \tilde{G} \wedge \cdots \wedge y_{l-1} \in \tilde{G} \wedge y_l = s] = \mu_s^F + \sum_{l > 0} \Pr_{\mu^F}[—] \\ = \mu_s^F + \sum_{l > 0} (\mu^F\,{}^GP^{l-1}P)_s \\ = \mu_s^F + (\mu^F\,{}^GNP)_s.$$

Applying the identity ${}^GN = N - B^GN$ of Lemma 8-17, we have

$$\begin{aligned} \mu^F\,{}^GNP &= \mu^F(N - B^GN)P \\ &= \mu^FNP - \mu^FB^GNP \\ &= \mu^FNP - \mu^GNP \quad \text{if } F \supset G \\ &= \nu^FP - \nu^GP. \end{aligned}$$

Since $\nu^F \to \nu$ monotonically and since $\mu_s^F \to \mu_s$, we have

$$\lim_F \sum_{l \ge 0} \Pr_{\mu^F}[y_0 \in \tilde{G} \wedge \cdots \wedge y_{l-1} \in \tilde{G} \wedge y_l = s] = \mu_s + (\nu P)_s - (\nu^G P)_s$$
$$= \nu_s - (\nu^G P)_s,$$

since $\nu(I - P) = \mu$. Thus the contribution is

$$= \lim_m \sum_{s \in \tilde{E}} (\nu - \nu^G P)_s \Pr_s[y_1 \in \tilde{E} \wedge \cdots \wedge y_{m-1} \in \tilde{E} \wedge y_m = i]\bar{P}_{ij}\bar{P}_{jk}.$$

But

$$\Pr_s[y_1 \in \tilde{E} \wedge \cdots \wedge y_{m-1} \in \tilde{E} \wedge y_m = i] = ({}^E P^{m-1} P)_{si}.$$

If

$$P = \begin{pmatrix} T & U \\ R & Q \end{pmatrix}, \quad \text{then} \quad {}^E P = \begin{pmatrix} 0 & 0 \\ 0 & Q \end{pmatrix}$$

and

$${}^E P^{m-1} P = \begin{pmatrix} 0 & 0 \\ Q^{m-1}R & Q^m \end{pmatrix}.$$

From this representation of ${}^E P^{m-1} P$, we see that the sum $\sum_{s \in \tilde{E}}$ may now be extended over all of S, and the contribution is

$$= \lim_m [(\nu - \nu^G P)({}^E P^{m-1} P)]_i \bar{P}_{ij}\bar{P}_{jk}.$$

Now

$$\nu^G P({}^E P^{m-1})P \le \nu^G P^{m+1} \to 0$$

since ν^G is a potential. Therefore the contribution is

$$= \lim_m (\nu\, {}^E P^{m-1} P)_i (\bar{P}_{ij}\bar{P}_{jk}).$$

Since i is a state of E, this expression is

$$= \lim_m (\nu_{\tilde{E}} Q^{m-1} R)_i (\bar{P}_{ij}\bar{P}_{jk}).$$

From the identity $\nu = \nu P + \mu$, we have

$$\nu_{\tilde{E}} = \nu_E U + \nu_{\tilde{E}} Q + \mu_{\tilde{E}}$$

or

$$\nu_{\tilde{E}} Q = \nu_{\tilde{E}} - (\nu_E U + \mu_{\tilde{E}}).$$

Iterating, we find that

$$\nu_{\tilde{E}} Q^{m-1} = \nu_{\tilde{E}} - (\nu_E U + \mu_{\tilde{E}})(I + Q + \cdots + Q^{m-2})$$

and hence

$$\nu_{\tilde{E}} Q^{m-1} R = \nu_{\tilde{E}} R - (\nu_E U + \mu_{\tilde{E}})(I + Q + \cdots + Q^{m-2})R.$$

By monotone convergence,

$$\lim_m \nu_{\tilde{E}} Q^{m-1} R = \nu_{\tilde{E}} R - (\nu_E U + \mu_{\tilde{E}})^E N_{\tilde{E}} R.$$

As we noted in Section 2, the balayage potential ν^E satisfies

$$\nu^E = (\nu_E \qquad \nu_E U\, {}^E N_{\tilde{E}}).$$

Hence

$$\lim_m \nu_{\tilde{E}} Q^{m-1} R = (\nu - \nu^E)_{\tilde{E}} R - \mu_{\tilde{E}}\, {}^E N_{\tilde{E}} R.$$

If we twice use the fact that ν and ν^E agree on E, we obtain

$$(\nu - \nu^E)_{\tilde{E}} R = [(\nu - \nu^E) P]_E = [(\nu - \mu) - (\nu^E - \mu^E)]_E = \mu^E_{\tilde{E}} - \mu_E.$$

Thus

$$\lim_m \nu_{\tilde{E}} Q^{m-1} R = \mu^E_{\tilde{E}} - (\mu_E + \mu_{\tilde{E}}\, {}^E N_{\tilde{E}} R) = (\mu^E - \mu B^E)_E.$$

We conclude that the contribution to $\Pr[p]$ is

$$= (\mu^E - \mu B^E)_i \bar{P}_{ij} \bar{P}_{jk}.$$

(5) *Completion of proof.* From steps 3 and 4, we find that

$$\Pr[z_0 = i \wedge z_1 = j \wedge z_2 = k] = \mu^E_i \bar{P}_{ij} \bar{P}_{jk} \quad \text{for } i \in E.$$

Hence

$$\Pr[z_2 = k \mid z_0 = i \wedge z_1 = j] = \bar{P}_{jk}$$

if $\Pr[z_0 = i \wedge z_1 = j] > 0$. That is, $\{z_n\}$ is a Markov chain with matrix $\bar{P}$ and with starting vector μ^E, provided $E = \{0, 1, \ldots, e\}$. We have seen how the proof for a general finite set follows from this special case.

We must compute the mean number of times in state j in $\{x_n\}$. The mean number starting in E is $(\mu^E N)_j = \nu^E_j$. Hence the mean total number is $\lim_E \nu^E_j = \nu_j$ by monotone convergence.

Finally we show that the contribution to ν of the paths with $\mathbf{u} > -\infty$ is μN. On these paths the process behaves as if it were started with distribution μB^E (see step 3 if $E = \{0, 1, \ldots, e\}$ and use the identity $B^E B^F = B^F$, where $E \supset F$, in the general case). Therefore the contribution is

$$\lim_E (\mu B^E) N = \mu N - \lim_E \mu\, {}^E N$$

by Lemma 8-17. Since ${}^E N \leq N$ and $\lim {}^E N = 0$, $\mu\, {}^E N \to 0$ by dominated convergence. Thus the contribution is

$$\lim_E (\mu B^E) N = \mu N.$$

13. Examples

EXAMPLE 1: Basic example.

For the basic example, we recall from Chapter 5 that

$$N_{ij} = \begin{cases} \dfrac{\beta_j}{\beta_\infty} & \text{if } i \le j \\ \dfrac{\beta_j}{\beta_\infty} - \dfrac{\beta_j}{\beta_i} & \text{if } i > j. \end{cases}$$

In a process of this kind in which all pairs of states communicate, there is no gain in choosing a complicated starting vector π. Indeed, if we choose π to be a unit mass at 0, then *all* finite-valued P-regular functions are certainly π-integrable, and therefore no other choice of π will make the representation theorem yield additional regular functions.

With π chosen as a unit mass at 0, $K(i, j)$ becomes

$$K(i, j) = \frac{N_{ij}}{N_{0j}} = \begin{cases} 1 & \text{if } i \le j \\ \dfrac{\beta_i - \beta_\infty}{\beta_i} & \text{if } i > j. \end{cases}$$

Since K is of a particularly simple form, it is possible to compute the metric for the exit boundary. If $j < j'$, then

$$\begin{aligned} d(j, j') &= \sum_i w_i N_{0i} |K(i, j) - K(i, j')| \\ &= \sum_{j < i \le j'} w_i \frac{\beta_i}{\beta_\infty} \left| \left(1 - \frac{\beta_\infty}{\beta_i}\right) - 1 \right| \\ &= \sum_{j < i \le j'} w_i. \end{aligned}$$

The metric space (S, d) thus is isometric to the subset of the real line consisting of the points $0, w_1, w_1 + w_2, w_1 + w_2 + w_3, \ldots$. Its Cauchy completion contains the one extra point corresponding to $\sum_{i \ge 1} w_i$.

Alternatively we can find S^* directly without using the metric. In fact, if $\{j_n\}$ is a sequence of points in S, then $\{K(i, j_n)\}$ is clearly Cauchy if and only if either $\{j_n\}$ is eventually constant or $\{j_n\}$ tends to infinity.

Either way, there is exactly one boundary point, which we may call $+\infty$, and the relative topology for S is discrete. Moreover,

$$K(i, +\infty) = \lim K(i, j_n) = 1.$$

Since $\mathbf{1}$ is regular and since there is only one boundary point, the boundary point must be extreme. Moreover, every regular function $h \geq 0$ must be constant. Since the process does not disappear, harmonic measure μ must assign unit mass to $+\infty$.

Next, we consider the entrance boundary. If f is a unit mass at 0, then J is given by

$$J(j, i) = \frac{N_{ji}}{N_{j0}} = \begin{cases} \beta_i/\beta_0 & \text{if } j = 0 \\ \beta_i/\beta_0 & \text{if } j > 0,\ j > i \\ \dfrac{\beta_i}{\beta_\infty}\Big/\left(\dfrac{\beta_0}{\beta_\infty} - \dfrac{\beta_0}{\beta_i}\right) & \text{if } j > 0,\ j \leq i. \end{cases}$$

The sequence 1, 2, 3, ... again is Cauchy, so that there is necessarily exactly one limit point in $^*S$. But

$$\lim_{n \to \infty} J(n, i) = J(0, i),$$

and consequently $^*S = S$ and the entrance boundary is empty. The state 0 is a limit point and thus S does not have the discrete topology. Since the entrance boundary is empty, there are no non-zero regular measures $\sigma \geq 0$. (We had arrived at this conclusion in Chapter 5, too.)

EXAMPLE 2: The p-q random walk.

The process P is the Markov chain with state space the integers which moves one step to the right with probability p or one step to the left with probability $q = 1 - p$. We shall assume $0 < q < p < 1$. A calculation like that in Section 5-8 shows that

$$H_{ij} = \begin{cases} 1 & \text{if } i \leq j \\ (q/p)^{i-j} & \text{if } i \geq j \end{cases}$$

and then

$$N_{ij} = \begin{cases} (p - q)^{-1} & \text{if } i \leq j \\ (q/p)^{i-j}(p - q)^{-1} & \text{if } i \geq j. \end{cases}$$

Take π and f to be unit masses at state 0; as in Example 1, we may make these choices since all pairs of states communicate. We obtain

$$K(i, j) = \frac{N_{ij}}{N_{0j}} = \begin{cases} 1 & \text{for } j \geq i \text{ and } j \geq 0 \\ (q/p)^i & \text{for } j \leq i \text{ and } j \leq 0. \end{cases}$$

There are two distinct infinite Cauchy sequences, one corresponding to $+\infty$ and the other corresponding to $-\infty$. For these points,

$$K(i, +\infty) = 1$$
$$K(i, -\infty) = (q/p)^i.$$

Both functions are regular and minimal, and thus the boundary contains two points, both extreme. No point of S is a limit point. Since the process goes to $+\infty$ with probability one, we have $\mu(\{+\infty\}) = 1$. (The concentration of μ can also be deduced analytically from the representation of $\mathbf{1}$.)

The entrance boundary is treated most quickly by reversing the process with respect to the regular measure $\alpha = \mathbf{1}^T$. Then $\hat{P}$ is a process of the same type but with the roles of p and q interchanged. Consequently

$$J(-\infty, j) = 1$$
$$J(+\infty, j) = (p/q)^j,$$

and $\mu^\alpha(\{-\infty\}) = \hat{\mu}(\{-\infty\}) = 1$.

The function $p(x)$, which does not depend upon α, satisfies

$$p(-\infty) = \lim_{E \uparrow S} \lim_{j \to -\infty} \frac{h_j^E}{N_{j0}} = \lim_{E \uparrow S} \frac{1}{N_{00}} = p - q.$$

For $p(+\infty)$, we note that if j is large, then $h_j^E = H_{jm}$, where m is the last state in E. Also

$$\frac{H_{jm}}{N_{j0}} = \frac{(q/p)^{j-m}}{(p - q)^{-1}(q/p)^j} = \frac{(q/p)^{-m}}{(p - q)^{-1}}.$$

As $E \uparrow S$, $m \to +\infty$ and therefore

$$p(+\infty) = 0.$$

Any extended chain representing $\mathbf{1}^T$ starts at $-\infty$ with probability $p - q$ and goes to $+\infty$.

The p-q random walk can be used to show what happens if π is chosen in such a way that there is a P-regular function which is not π-integrable. In fact, set

$$\pi_i = \begin{cases} (p/q)^i\left(\dfrac{p - q}{p}\right) & \text{for } i \le 0 \\ 0 & \text{for } i > 0. \end{cases}$$

Then $\pi\mathbf{1} = 1$ and the regular function $(q/p)^i$ will not be integrable. We must recompute $K(i, j)$. First we note that for $j > 0$,

$$N_{\pi j} = \sum_{i \le 0} \pi_i N_{ij} = (p - q)^{-1} \sum \pi_i = (p - q)^{-1},$$

and that for $j \le 0$,

$$\begin{aligned} N_{\pi j} &= (p - q)^{-1} \sum_{i \le 0} \pi_i H_{ij} = p^{-1} \sum_{i \le 0} (p/q)^i H_{ij} \\ &= p^{-1} \sum_{i \le j} (p/q)^i + p^{-1} \sum_{j < i \le 0} (p/q)^i (q/p)^{i-j} \\ &= (p - q)^{-1}(p/q)^j + p^{-1} j (p/q)^j. \end{aligned}$$

Then

$$K(i,j) = \frac{N_{ij}}{N_{\pi j}} = \begin{cases} 1 & \text{for } j \geq i \text{ and } j \geq 0 \\ \dfrac{(p - q)^{-1}(q/p)^i}{(p - q)^{-1} + p^{-1}j} & \text{for } j \leq i \text{ and } j \leq 0. \end{cases}$$

Again there are two boundary points $+\infty$ and $-\infty$, but this time

$$K(i, +\infty) = 1$$

and

$$K(i, -\infty) = 0.$$

Thus $-\infty$ is not an extreme boundary point. The nonintegrability of $(q/p)^i$ introduced a boundary point whose associated function was identically zero. This example is typical of the general situation, but we shall not pursue the details.

There is still a second way a zero boundary point can arise, and the degenerate case where $p = 1$ and $q = 0$ in the example above illustrates the point. If h is regular for this process, then

$$h_i = P_{i,i+1}h_{i+1} = h_{i+1},$$

and hence h must be constant. On the other hand, it is readily seen that, for any choice of π for which $\pi N > 0$, there are two boundary points $+\infty$ and $-\infty$ and that

$$K(i, +\infty) = 1$$

and

$$K(i, -\infty) = 0.$$

Thus again there is a zero boundary point, but this time no function is missing.

EXAMPLE 3: Symmetric random walk in three dimensions.

The symmetric random walk in three dimensions is the sums of independent random variables process on the integer lattice in three-dimensional space with transition probability $\frac{1}{6}$ from any point to each of its six neighbors. We recall from Proposition 5-20 that the only bounded regular functions for this process are the constants. We now prove, using in part the methods of transient boundary theory, the deeper result that the only non-negative regular functions are the constants.

In fact, if we choose π to be a unit mass at 0, then every regular function is π-integrable, and the representation theorem assures us that it is enough to prove that $K(\cdot, x) = 1$ for every x in the boundary. That is, it suffices to prove that for every i

$$\lim_{j \to \infty} K(i, j) = 1.$$

We shall prove in a moment the following estimate for N_{0j}: If $j \neq 0$, then

$$N_{0j} = \frac{3}{2\pi|j|} + O(|j|^{-2}).$$

Once we have this estimate, we also have, for $i \neq j$,

$$K(i,j) = \frac{N_{ij}}{N_{0j}} = \frac{\dfrac{3}{2\pi|j-i|} + O(|j-i|^{-2})}{\dfrac{3}{2\pi|j|} + O(|j|^{-2})}$$

$$= \frac{|j-i|^{-1} + O(|j|^{-2})}{|j|^{-1} + O(|j|^{-2})}$$

$$= \frac{\dfrac{|j|}{|j-i|} + O(|j|^{-1})}{1 + O(|j|^{-1})}$$

$$= \frac{|j|}{|j-i|} + O(|j|^{-1}).$$

Hence, $\lim_j K(i, j) = 1$, as asserted.

Thus we are to prove the estimate for N_{0j}. We first show that

$$N_{0j} = 3 \int_{-1/2}^{1/2} \int_{-1/2}^{1/2} \int_{-1/2}^{1/2} \frac{e^{-2\pi i (j \cdot x)} dx_1 dx_2 dx_3}{3 - \cos 2\pi x_1 - \cos 2\pi x_2 - \cos 2\pi x_3},$$

where $x = (x_1, x_2, x_3)$. Call the right side of the equation h_j. The singularity in the denominator of the integrand is of the order of

$$\frac{1}{3 - (1 - 2\pi^2 x_1^{\,2}) - (1 - 2\pi^2 x_2^{\,2}) - (1 - 2\pi^2 x_3^{\,2})} = \frac{1}{2\pi^2 |x|^2}$$

near $x = 0$, and $|x|^{-2}$ is integrable in a neighborhood of $x = 0$. Thus h_j is certainly well defined, and moreover $h_j \to 0$ as $j \to \infty$ by the Riemann–Lebesgue Lemma. We claim that

$$(I - P)_{m\cdot} h_{\cdot} = \delta_{m0}.$$

Call the six neighbors of 0 by the names a_s, and call the region

$$-\tfrac{1}{2} < x_1, x_2, x_3 \leq \tfrac{1}{2}$$

by the name E. Then

$$\begin{aligned}(I - P)_{m\cdot}h_{\cdot} &= h_m - \tfrac{1}{6}\sum_s h_{m+a_s} \\ &= 3\int_E (3 - \cos 2\pi x_1 - \cos 2\pi x_2 - \cos 2\pi x_3)^{-1} \\ &\qquad \times \left(e^{-2\pi i(m\cdot x)} - \tfrac{1}{6}\sum_s e^{-2\pi i(m+a_s)\cdot x}\right)dx \\ &= 3\int_E (3 - \cos 2\pi x_1 - \cos 2\pi x_2 - \cos 2\pi x_3)^{-1} \\ &\qquad \times e^{-2\pi i m\cdot x}\left(1 - \tfrac{1}{6}\sum_s e^{-2\pi i a_s\cdot x}\right)dx \\ &= \int_E e^{-2\pi i m\cdot x}dx \\ &= \delta_{m0}.\end{aligned}$$

Now $N_{0\cdot} = N_{\cdot 0}$ is also a function tending to zero (see Proposition 7-10) and satisfying

$$(I - P)_{m\cdot}N_{\cdot 0} = \delta_{m0},$$

and thus $N_{\cdot 0} - h_{\cdot}$ is a bounded regular function. By Proposition 5-20 it must be constant, and that constant must be zero, since both functions vanish at infinity. Thus the formula for N_{0j} is established.

We now have N_{0j} exhibited as the jth Fourier coefficient of a certain periodic function of three variables. The remainder of the proof will presuppose some knowledge of Fourier transforms and tempered distributions.

Let $\varphi(x)$ be an infinitely differentiable function defined on R^3 such that

(1) $0 \leq \varphi \leq 1$.
(2) $\varphi(x) = 1$ for x in a small neighborhood of 0.
(3) $\varphi(x) = 0$ outside of a slightly larger neighborhood of 0.

For simplicity, let $\|x\|^4 = x_1^4 + x_2^4 + x_3^4$. Denote by $\bar{f}(x)$ the periodic continuation of the function

$$f(x) = \varphi(x)\left[\frac{3}{2\pi^2|x|^2} + \frac{\|x\|^4}{2|x|^4}\right].$$

The difference

$$\bar{f}(x) - \frac{3}{3 - \cos 2\pi x_1 - \cos 2\pi x_2 - \cos 2\pi x_3}$$

is a bounded periodic function which is infinitely differentiable away from 0 and is $O(|x|^2)$ near 0. Moreover, its Laplacian is integrable in $E - \{0\}$. These facts imply that the Fourier coefficients of the difference are $O(|j|^{-2})$ as $j \to \infty$. Hence it is enough to show that the Fourier coefficients of $\tilde{f}$ are equal to

$$\frac{3}{2\pi|j|} + O(|j|^{-2}).$$

It is clear from the definitions that the Fourier coefficients of $\tilde{f}$ satisfy

$$\int_E \tilde{f}(x)e^{-2\pi i j\cdot x}dx = \int_{R^3} f(x)e^{-2\pi i j\cdot x}dx.$$

Here the right side is the Fourier transform of f evaluated at j. We are thus to estimate the Fourier transform of f. Write

$$f(x) = [\varphi(x) - 1]\left[\frac{3}{2\pi^2|x|^2} + \frac{\|x\|^4}{2|x|^4}\right] + \frac{3}{2\pi^2|x|^2} + \frac{\|x\|^4 - \frac{3}{5}|x|^4}{2|x|^4} + \frac{3}{10},$$

and take the Fourier transform of both sides in the sense of distributions. We shall consider the four terms on the right separately.

The transform of the constant $\frac{3}{10}$ is the distribution defined by the measure which assigns weight $\frac{3}{10}$ to the origin. It has the property that it is supported, say, entirely in the unit ball.

In the third term, the numerator is a homogeneous harmonic polynomial, and the transform of the term is known from Fourier analysis to be

$$\frac{15}{8\pi}\,\mathrm{PV}\left[\frac{\|y\|^4 - \frac{3}{5}|y|^4}{2|y|^7}\right],$$

where

$$(\mathrm{PV}[g], \psi) = \lim_{\epsilon\to 0}\int_{|x|\geq\epsilon} g(x)\psi(x)dx.$$

This distribution has the property that it is the sum of a function and a distribution which is supported in the unit ball. The function is

$$\frac{15}{8\pi}\frac{\|y\|^4 - \frac{3}{5}|y|^4}{2|y|^7} \quad \text{for } |y| \geq 1,$$

and it is $O(|y|^{-3})$ as $y \to \infty$.

The transform of the second term is known from Fourier analysis to be the distribution arising from the function

$$\frac{3}{2\pi|y|}.$$

We come to the first term. Since the transform of the left side is a function, the transform of the first term must be the sum of a function and a distribution supported in the unit ball. On the other hand, the first term is an infinitely differentiable function. If we iterate taking the Laplacian of it enough times, the resulting function is integrable and has a bounded function as Fourier transform. Since the operation of taking the Laplacian goes into multiplication by $-4\pi^2|y|^2$ under the Fourier transform, we obtain the inequality of functions

$$(4\pi^2|y|^2)^k \text{ FT(first term)} \leq A_k, \qquad k \text{ sufficiently large.}$$

This result implies that the function part of the distribution FT(first term) is $O(|y|^{-2k})$ as $y \to \infty$.

Since the sum of the transforms of the four terms is a function, we must have, outside the unit ball,

$$\text{FT}(f)(y) = O(|y|^{-2k}) + \frac{3}{2\pi|y|} + O(|y|^{-3}).$$

We conclude therefore that

$$\text{FT}(f)(j) = \frac{3}{2\pi|j|} + O(|j|^{-3})$$

and hence that

$$N_{0j} = \frac{3}{2\pi|j|} + O(|j|^{-2}),$$

as required.

Example 4:

This example will be a process with an exit boundary point x such that $K(\cdot, x)$ is regular and normalized but is not minimal.

The state space will be

$$S = \{a_i, a_i', b_i \mid i = 0, 1, 2, \ldots\}$$

and the non-zero transition probabilities are

$$P_{a_{i-1}a_i} = p_i, \qquad P_{a_{i-1}b_{i-1}} = q_i = 1 - p_i,$$

$$P_{b_{i-1}b_i} = r_i,$$

$$P_{a'_{i-1}a'_i} = p_i, \qquad P_{a'_{i-1}b_{i-1}} = q_i = 1 - p_i.$$

The picture for this process is as follows.

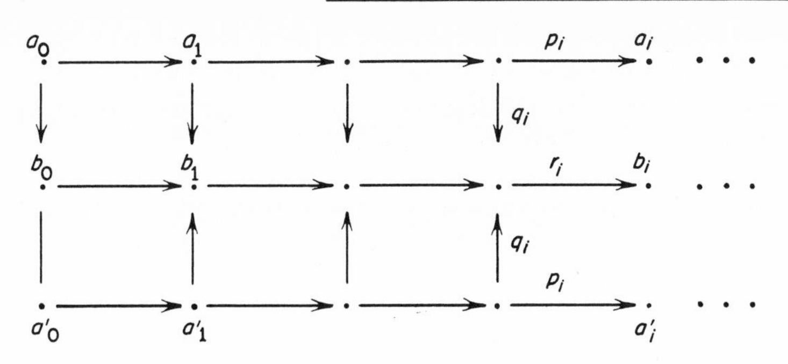

Let

$$\beta_0 = 1, \qquad \beta_i = \prod_{k=1}^{i} p_k, \qquad \beta_\infty = \lim_{i\to\infty} \beta_i,$$

$$\gamma_0 = 1, \qquad \gamma_i = \prod_{k=1}^{i} r_k,$$

$$\sigma_{-1} = 0, \qquad \sigma_i = \sum_{k=0}^{i} \beta_k q_{k+1}/\gamma_k, \qquad \sigma_\infty = \lim_{i\to\infty} \sigma_i.$$

We shall assume for this example that $\sigma_\infty = +\infty$. At the end of the discussion of the example we shall show that this condition is possible.

Any state can be reached either from a_0 or from a'_0. Thus if we set

$$\pi_{a_0} = \pi_{a'_0} = \tfrac{1}{2},$$

then πN is strictly positive. Since the process is in any given state at most once, we must have $H = N$. It is clear that

$$H_{a_i a_j} = H_{a'_i a'_j} = \begin{cases} 0 & \text{if } j < i \\ \beta_j/\beta_i & \text{if } j \geq i, \end{cases}$$

$$H_{b_i b_j} = \begin{cases} 0 & \text{if } j < i \\ \gamma_j/\gamma_i & \text{if } j \geq i, \end{cases}$$

and $H_{a'_i a_j} = H_{a_i a'_j} = 0 = H_{b_i a_j} = H_{b_i a'_j}$. We must compute $H_{a_i b_j}$ (and $H_{a'_i b_j}$). If $j < i$, then $H_{a_i b_j} = 0$. Otherwise we obtain a set of alternatives by considering the first time the process switches from the a-row to the b-row. Then

$$H_{a_i b_j} = \sum_{k=i}^{j} H_{a_i a_k} q_{k+1} H_{b_k b_j}.$$

Substituting, we find

$$H_{a_i b_j} = H_{a'_i b_j} = \begin{cases} 0 & \text{if } j < i \\ \dfrac{\gamma_j}{\beta_i}(\sigma_j - \sigma_{i-1}) & \text{if } j \geq i. \end{cases}$$

In determining the exit boundary we use the observation that whenever finitely many Cauchy sequences exhaust S, then those Cauchy sequences define all of the boundary points. In the present example the claim is that the three sequences $\{a_i\}$, $\{a_i'\}$, and $\{b_i\}$ are each Cauchy sequences.

First consider $K(\cdot, a_j)$ for large j. We have

$$K(a_i', a_j) = K(b_i, a_j) = 0$$

and, for $i \leq j$,

$$K(a_i, a_j) = \frac{H_{a_i a_j}}{H_{\pi a_j}} = \frac{H_{a_i a_j}}{\frac{1}{2}H_{a_0 a_j}} = \frac{2}{\beta_i}.$$

Hence $a_\infty = \lim a_j$ is a boundary point, and it satisfies

$$K(a_i, a_\infty) = \frac{2}{\beta_i} \quad \text{and} \quad K(a_i', a_\infty) = K(b_i, a_\infty) = 0.$$

We can check directly that $K(\cdot, a_\infty)$ is a regular function; moreover, it is normalized because

$$\pi K(\cdot, a_\infty) = \tfrac{1}{2}K(a_0, a_\infty) = \tfrac{1}{2}\cdot 2 = 1.$$

Similarly we find that $\{a_j'\}$ is a Cauchy sequence. If we call its limit a_∞', then

$$K(a_i', a_\infty') = \frac{2}{\beta_i} \quad \text{and} \quad K(a_i, a_\infty') = K(b_i, a_\infty') = 0.$$

Since $K(\cdot, a_\infty) \neq K(\cdot, a_\infty')$, a_∞ and a_∞' are distinct boundary points. We see also that $K(\cdot, a_\infty')$ is regular and normalized.

Finally we consider the sequence $\{b_j\}$ for large j. For $i \leq j$, we have

$$K(a_i, b_j) = \frac{H_{a_i b_j}}{H_{\pi b_j}} = \frac{H_{a_i b_j}}{H_{a_0 b_j}} = \frac{1}{\beta_i}\left(\frac{\sigma_j - \sigma_{i-1}}{\sigma_j}\right)$$

and

$$K(b_i, b_j) = \frac{H_{b_i b_j}}{H_{\pi b_j}} = \frac{H_{b_i b_j}}{H_{a_0 b_j}} = \frac{1}{\gamma_i \sigma_j}.$$

Since, by hypothesis, $\sigma_\infty = +\infty$, we have

$$\lim_j K(a_i, b_j) = \lim_j K(a_i', b_j) = \frac{1}{\beta_i}$$

and

$$\lim_j K(b_i, b_j) = 0.$$

Therefore $\{b_j\}$ is Cauchy. Denoting its limit in S^* by b_∞, we conclude

$$K(a_i, b_\infty) = K(a_i', b_\infty) = 1/\beta_i \quad \text{and} \quad K(b_i, b_\infty) = 0.$$

We can check that $K(\cdot, b_\infty)$ is regular and normalized.

Thus there are exactly three boundary points, and each is associated with a regular normalized function. But

$$K(\cdot, b_\infty) = \tfrac{1}{2}K(\cdot, a_\infty) + \tfrac{1}{2}K(\cdot, a'_\infty)$$

and b_∞ therefore cannot be an extreme point. Then by the representation theorem all non-negative regular functions are generated by $K(\cdot, a_\infty)$ and $K(\cdot, a_{-\infty})$. Hence their linear independence as functions on S implies that they are both minimal. That is, $B_e = \{a_\infty, a'_\infty\}$.

In this example, β_∞ is positive if and only if harmonic measure assigns positive weight to the boundary. In fact,

$$\mu(\{a_\infty\}) = \tfrac{1}{2}\beta_\infty = \mu(\{a'_\infty\}).$$

The remainder of the weight is put on the states from which the process can disappear. We obtain

$$\mu(\{b_j\}) = \Pr_\pi[\text{process disappears from } b_j] = (\gamma_j - \gamma_{j+1})\sigma_j.$$

Note that if $\beta_\infty = 0$, then there exist non-negative regular functions, but none of them is bounded.

We still must show that the condition $\sigma_\infty = +\infty$ puts no restriction on whether β_∞ must be positive. To get $\beta_\infty = 0$, take

$$p_j = q_j = r_j = \tfrac{1}{2}.$$

Then $\beta_j = \gamma_j = 2^{-j}$ and $\sigma_j = \tfrac{1}{2}(j + 1)$. Hence $\sigma_\infty = +\infty$. To get $\beta_\infty > 0$, take

$$p_j = \frac{j(j+2)}{(j+1)^2}, \qquad q_j = 1 - p_j = \frac{1}{(j+1)^2}, \qquad r_j = \frac{j+1}{j+2}.$$

Then

$$\beta_j = \frac{j+2}{2(j+1)}, \qquad \beta_\infty = \frac{1}{2}, \qquad \gamma_j = \frac{2}{j+2},$$

and

$$\sigma_\infty = \sum_{j=0}^{\infty} \frac{j+2}{2(j+1)} \frac{1}{(j+2)^2} \frac{j+2}{2} = \sum_{j=0}^{\infty} \frac{1}{4(j+1)} = +\infty.$$

EXAMPLE 5:

This example will be a process in which S^* has every point of S as a limit point.

Let S be the set of all finite strings of positive integers of the form

$$(1, k_1, k_2, \ldots, k_n)$$

as n and the k's vary. The transition probability from $(1, k_1, \ldots, k_n)$

to $(1, k_1, \ldots, k_n, m)$ is to be 2^{-m} and all other transition probabilities are zero.

Since each state is entered at most once, we have $N = H$. The hitting probability from

$$I = (1, j_1, \ldots, j_n)$$

to a distinct state

$$J = (1, k_1, \ldots, k_{n'}, m)$$

is

$$H_{IJ} = \begin{cases} 2^{-(k_{n+1} + \cdots + k_{n'} + m)} & \text{if } n \leq n' \text{ and } j_1 = k_1, \ldots, j_n = k_n \\ 0 & \text{otherwise.} \end{cases}$$

Take π to be a unit mass at (1). If

$$0 = (1), \qquad J = (1, k_1, \ldots, k_n), \qquad J_m = (1, k_1, \ldots, k_n, m),$$

the claim is that

$$\lim_{m \to \infty} K(I, J_m) = K(I, J)$$

for all I. It would therefore follow that every point J in S is a limit point of S^*.

For the proof we observe first that $K(I, J) = K(I, J_m) = 0$ unless I is an initial segment of some J_m. Also if $I = J_{m_0}$, then

$$K(I, J) = \lim K(I, J_m) = 0.$$

Thus we may suppose that I is an initial segment of J. In this case, if $I = (1, k_1, \ldots, k_s)$,

$$K(I, J_m) = \frac{H_{IJ_m}}{H_{0J_m}} = 2^{k_1 + \cdots + k_s}$$

and

$$K(I, J) = \frac{H_{IJ}}{H_{0J}} = 2^{k_1 + \cdots + k_s}.$$

Hence $K(I, J) = \lim K(I, J_m)$ always.

Example 6: Space-time coin tossing.

Let S be the set of lattice points (n, i) in the plane with $0 \leq i \leq n$. A point of S is to be identified with i heads in n tosses of a fair coin. Thus we take

$$P_{(n,i),(n+1,i)} = P_{(n,i),(n+1,i+1)} = \tfrac{1}{2}$$

with all other transition probabilities equal to zero. In this example, $N = H$ and also

$$H_{(m,i),(n,j)} = \begin{cases} 2^{m-n}\dbinom{n-m}{j-i} & \text{if } n \geq m,\ j \geq i \\ 0 & \text{otherwise.} \end{cases}$$

Let π be a unit mass at (0, 0). Then $\pi N > 0$ since every state can be reached from (0, 0). We obtain

$$K((m, i), (n, j)) = \begin{cases} 2^m \binom{n - m}{j - i} \Big/ \binom{n}{j} & \text{if } n \geq m,\ j \geq i \\ 0 & \text{otherwise.} \end{cases}$$

As a first step in obtaining the exit boundary, we shall prove that if (n_k, j_k) is Cauchy, then $\lim_k (j_k/n_k)$ exists. In fact, set $i = 0$ and $m = 1$ and consider, for $n_k \geq 1$, the expression

$$K((1, 0), (n_k, j_k)) = 2\binom{n_k - 1}{j_k} \Big/ \binom{n_k}{j_k} = \frac{2(n_k - j_k)}{n_k} = 2 - 2\frac{j_k}{n_k}.$$

As $k \to \infty$, the left side converges; hence so does the right side, and $\lim j_k/n_k$ must exist.

Conversely suppose that (n_k, j_k) is an infinite sequence such that $t = \lim j_k/n_k$ exists. We claim that (n_k, j_k) is Cauchy. Fix (m, i) and denote by (n, j) a term of the sequence (n_k, j_k). Then

$$\begin{aligned} K((m, i), (n, j)) &= 2^m \binom{n - m}{j - i} \Big/ \binom{n}{j} \\ &= 2^m \frac{[(j)\ldots(j - i + 1)][(n - j)\ldots(n - j - m + i + 1)]}{(n)\ldots(n - m + 1)}. \end{aligned}$$

Noting that both numerator and denominator have exactly m factors and that m is fixed, divide the numerator and the denominator by n^m and pass to the limit in each factor separately. In each factor we have $j/n \to t$ and, for instance, $(i - 1)/n \to 0$. We get

$$\lim K((m, i), (n, j)) = 2^m t^i (1 - t)^{m-i}.$$

Therefore the classes of infinite Cauchy sequences are in one-to-one correspondence with the rays to infinity, that is, in one-to-one correspondence with points t in [0, 1]. The functions associated to these sequences are

$$K((m, i), t) = 2^m t^i (1 - t)^{m-i}.$$

It is easy to check that all of these functions are regular and normalized, and hence the points in question form the entire boundary B (and nothing more).

Next, we check that the topology of B is the usual topology for the unit interval. The map of the unit interval onto B is continuous since each function $K((m, i), \cdot)$ is continuous in the unit interval topology. Since the map is one-one from a compact space onto a Hausdorff space, it is a homeomorphism.

To see that every point of B is extreme, we assume, on the contrary, that $2^m t_0^i(1 - t_0)^{m-i}$ is not minimal. By Theorem 10-41 there is a measure ν on the Borel sets of $B_e \subset [0, 1]$ with

$$2^m t_0^i(1 - t_0)^{m-i} = \int_{B_e} 2^m t^i(1 - t)^{m-i} d\nu(t).$$

Specializing to the case $m = i$ and extending ν to be defined on $[0, 1]$, we find that

$$t_0^i = \int_0^1 t^i d\nu(t)$$

for all i. But by the Weierstrass Approximation Theorem, ν is completely determined by its integral against all polynomials t^i. Hence ν is a point mass at t_0, and t_0 is extreme.

We can summarize our results so far by saying that no point of S is a limit point, that the boundary B is homeomorphic to the unit interval under the correspondence

$$K((m, i), t) = 2^m t^i(1 - t)^{m-i},$$

and that every point of the boundary is extreme.

The Strong Law of Large Numbers, when applied to coin tossing, is equivalent with the statement that harmonic measure μ concentrates all its mass at $t = \frac{1}{2}$. We can verify this result directly: We do have

$$\int_0^1 2^m(t)^i(1 - t)^{m-i} d\delta_{1/2}(t) = 2^m(\tfrac{1}{2})^i(1 - \tfrac{1}{2})^{m-i} = 1,$$

and, by the uniqueness half of Theorem 10-41, $\delta_{1/2}$ can be the only measure yielding 1. Hence $\delta_{1/2}$ is harmonic measure; that is, harmonic measure concentrates all its mass at $t = \frac{1}{2}$.

By Theorem 10-41 every measure μ^h on the unit interval gives rise to a non-negative regular function h for the process and μ^h is harmonic measure for the h-process. We shall discuss only two special cases.

If μ^h is the unit mass at a point t_0 other than 0 or 1, then

$$h_{(m,i)} = 2^m t_0^i(1 - t_0)^{m-i}$$

and

$$P^h_{(m,i),(m+1,i)} = 1 - t_0,$$

$$P^h_{(m,i),(m+1,i+1)} = t_0.$$

Thus the h-process is space-time coin tossing in which there is probability t_0 of heads and probability $1 - t_0$ of tails. The fact that

$\mu^h(\{t_0\}) = 1$ means that with probability one the ratio of heads to total tosses tends to t_0, again in agreement with the Strong Law of Large Numbers.

If, instead, μ^h is Lebesgue measure, then

$$h_{(m,i)} = 2^m \int_0^1 t^i(1-t)^{m-i}dt$$
$$= 2^m \frac{\Gamma(i+1)\Gamma(m-i+1)}{\Gamma((m+1)+1)}$$
$$= 2^m \frac{i!(m-i)!}{(m+1)!}.$$

Consequently, the transition probabilities in the h-process are again computable, and we find

$$P^h_{(m,i),(m+1,i)} = \frac{m-i+1}{m+2}$$

and

$$P^h_{(m,i),(m+1,i+1)} = \frac{i+1}{m+2}.$$

The significance of this process is discussed in Problems 28 and 29.

14. Problems

1. If π assigns positive weight only to finitely many states, show that $\pi K(\cdot, x) = 1$ for every boundary point x. What is the corresponding condition so that $K(\cdot, x)$ is regular for every boundary point?
2. Consider the identity

$$\int_{S^*} PK(\cdot, x)d\mu(x) = P\int_{S^*} K(\cdot, x)d\mu(x) = P1.$$

What can we conclude about μ if $P1 = 1$? What if $(P1)_i < 1$?

Problems 3 to 12 deal with sums of independent random variables on the integers with $p_{-1} = \frac{2}{9}$, $p_1 = \frac{3}{9}$, $p_2 = \frac{4}{9}$. This example was discussed in Section 5-8, and the results there obtained will be useful.

3. Find N.
4. Let $\pi_0 = 1$, and compute $K(i, j)$.
5. Show that the two boundary points are $\pm\infty$, and find the two minimal functions.
6. From the form of these two functions determine what weight μ must put at each point. Does this agree with your understanding of the 'ong-range behavior of the chain?
7. Let $h_i = (\frac{1}{4})^i$. Find P^h and μ^h. Check the representation of h.

8. Show that the h of Problem 7 is minimal by proving that P^h has only constants for bounded regular functions.
9. Let $f_i = \delta_{i0}$ and $g = Nf$. Find $J(i, j)$.
10. Show that the entrance boundary is "the same" as the exit boundary, and find the two minimal measures.
11. Find $p(-\infty)$ and $p(+\infty)$.
12. Construct extended chains representing the two minimal measures of Problem 10 (in the sense of Theorem 10-9). For each chain, verify that

$$\Pr[z_u \in C] = \int_C p(x)d\mu^\nu(x).$$

Problems 13 to 20 refer to a "double basic example." Let

$$S = \{0, 1, 2, \ldots, 0', 1', 2', \ldots\}.$$

The non-zero entries of P are

$$P_{i-1,i} = p_i; \quad P_{i-1,0'} = 1 - p_i; \quad P_{(i-1)',i'} = p_{i'}; \quad P_{(i-1)',0} = 1 - p_{i'}.$$

Let

$$\beta_0 = \beta_{0'} = 1, \quad \beta_i = \prod_{k=1}^{i} p_i, \quad \beta_{i'} = \prod_{k=1}^{i} p_{i'},$$

$$\beta_\infty = \lim \beta_i, \qquad \beta_{\infty'} = \lim \beta_{i'}.$$

Assume that $\beta_\infty > 0$ and $\beta_{\infty'} > 0$.

13. Find H. [*Hint:* Use Propositions 4-14 and 4-16.]
14. Let $\pi_0 = 1$, and compute K.
15. Show that there are two boundary points which correspond to ∞ and ∞', and find the minimal functions.
16. From the forms of the two minimal functions, find μ. Give an intuitive interpretation for the two weights.
17. Find the most general non-negative normalized regular function h. Show that h is a convex combination of the two minimal functions.
18. Show that any such h tends to μ^h/μ at each boundary point.
19. Let $h = \frac{1}{5}K(\cdot, \infty) + \frac{4}{5}K(\cdot, -\infty)$. Show that P^h goes to ∞ with probability $\frac{1}{5}$ by computing $\mu^h(\infty)$.
20. For the example of Problem 19, compute P^h explicitly if

$$p_i = \frac{i(i+2)}{(i+1)^2} \quad \text{and} \quad p_{i'} = \frac{(i+1)(i+3)}{(i+2)^2}.$$

Verify the conclusion of Problem 19 by computing h_0^E for

$$E = \{i, i+1, i+2, \ldots\}$$

and by letting i tend to ∞.

Problems 21 to 27 deal with a certain tree-process. The states are all finite sequences of H and T. The empty sequence is the starting state and is denoted state 0. From state $a = (a_1, a_2, \ldots, a_n)$ the process is equally likely to go to $(a_1, a_2, \ldots, a_n, H)$ or to $(a_1, a_2, \ldots, a_n, T)$.

21. Find N.

22. Let $\pi_0 = 1$, and compute K.

23. Show that the boundary is the set of all infinite sequences of H and T.

24. What is the topology of the boundary? [It is *not* that of the unit interval.]

25. Use the identity
$$\Pr_a[x_v \in C] = \int_C K(a, x)d\mu(x)$$
to find the measure μ.

26. Let h be defined by
$$h_a = \begin{cases} 1 & \text{if } a = 0 \\ 2 & \text{if } a = (H) \\ 4 & \text{if } a \text{ begins with } a_1 = a_2 = H \\ 0 & \text{otherwise.} \end{cases}$$
Prove that h is a normalized regular function. Find P^h and μ^h. Show that h is continuous and that $h(x) = d\mu^h/d\mu$ for a.e. x.

27. Let $f_a = 1/m$ if a consists of exactly m H's and m T's, and let f_a equal 0 otherwise. Show that f is a charge, and verify that
$$\Pr[\lim g(x_n) = 0] = 1.$$

Problems 28 and 29 deal with an instance of Polya's urn scheme. An urn contains some white balls and some black balls. A drawing is made with each ball equally likely to be drawn; the ball drawn is then replaced and another of the same color is added to the urn. This scheme is repeated over and over.

28. Let the pair (m, i) stand for $m + 2$ balls in the urn with $i + 1$ of them white. Show that if the outcomes of the Polya urn scheme are taken as such pairs (m, i), then the resulting process is a Markov chain. Note that the transition matrix for this chain is identical with the one for the h-process considered at the very end of Section 13.

29. Let the scheme be started with 1 white ball and 1 black ball, and let r_m be the fraction of white balls at time m. Use the observation in Problem 28 to compute
$$\limsup_m \Pr[|r_m - \tfrac{1}{2}| < \epsilon].$$

Problems 30 to 34 establish the necessity of a necessary and sufficient condition that a transient chain P with all pairs of states communicating have a non-zero non-negative regular measure α. Number the states $0, 1, 2, \ldots$. Let ${}^kL_{ij}$ be the probability that the process started at i reaches j and that the first visit is immediately preceded by a visit to a state $\geq k$. For instance, ${}^0L_{ij} = \bar{H}_{ij}$. The condition on P if α exists is as follows: There must exist an infinite set E of states such that
$$\lim_{k \to \infty} \lim_{\substack{i \to \infty \\ i \in E}} \frac{{}^kL_{ij}}{H_{ij}} = 0 \quad \text{for all } j.$$

30. Prove that

$$\frac{{}^kL_{ij}}{H_{ij}} \leq \frac{{}^kL_{ij}}{N_{i0}} \frac{N_{00}}{P_{0j}^{(n)}}$$

for any n with $P_{0j}^{(n)} > 0$.

31. Let

$${}^jP_{ik}^{(n)} = \Pr_i[x_n = k \wedge x_m \neq j \quad \text{if} \quad 0 \leq m < n].$$

Show that

$${}^kL_{ij} = \sum_{r=k}^{\infty} \sum_{n=0}^{\infty} {}^jP_{ir}^{(n)} P_{rj},$$

and derive as a consequence that

$${}^kL_{ij} \leq N_{ij} - \sum_{r=0}^{k-1} N_{ir} P_{rj}.$$

32. Define f by $f_j = \delta_{j0}$. If an $\alpha > 0$ exists with $\alpha P = \alpha$, show that there is a point x_0 of $*S$ for which $J(x_0, \cdot)$ is a minimal measure. Choose E to be the set of states in some Cauchy sequence converging to x_0.

33. Prove that

$$\lim_{k\to\infty} \lim_{\substack{i\to\infty \\ i\in E}} \left[\frac{N_{ij}}{N_{i0}} - \sum_{r=0}^{k-1} \frac{N_{ir}}{N_{i0}} P_{rj}\right] = 0.$$

34. Put together the preceding results to obtain a proof of the necessity of the stated condition.

Problems 35 to 38 refer to a transient chain with absorbing states. As usual, we write the transition matrix in the form

$$P = \begin{pmatrix} I & 0 \\ R & Q \end{pmatrix}.$$

35. Put

$$P' = \begin{pmatrix} 0 & 0 \\ R & Q \end{pmatrix}.$$

Show that P' has only transient states. Let π be any starting distribution for P' with $\pi N' > 0$. Prove that if i is an absorbing state of P, then the harmonic measure $\mu(i)$ in P' is equal to the probability in P of absorption in i.

36. If P is the infinite drunkard's walk with $p = \frac{2}{3}$, find harmonic measure for P' when $\pi_1 = 1$.

37. Suppose $P\mathbf{1} = \mathbf{1}$. Find a necessary and sufficient condition on harmonic measure for P' that P should have been an absorbing chain.

38. Let Q be any chain with all states transient and with the enlarged chain $\bar{Q}$ absorbing. Use the condition of Problem 37 to give a new proof that Q has no non-zero bounded regular functions $h \geq 0$.

CHAPTER 11

RECURRENT BOUNDARY THEORY

1. Entrance boundary for recurrent chains

Boundary theory for recurrent chains proceeds along altogether different lines from the approach in Chapter 10. A clue to the difficulty is that every non-negative superregular function is constant, and hence the representation of such functions degenerates. Moreover, since a recurrent chain is in every state infinitely often with probability one, an almost-everywhere convergence theorem is out of the question.

But intuitively, at least for some recurrent chains, there is some limiting behavior going on. For instance, with the one-dimensional symmetric random walk the Central Limit Theorem implies that the probability that the process is on either half-axis after time n tends to $\frac{1}{2}$ as n increases.

If P is a normal chain and E is a finite set, then $(P^nB^E)_{ij}$ is the probability that the process started in i enters E after time n at state j, and its limit λ_j^E is the "long run" probability of entering E at j. As we let E swell to the whole state space S, it is not clear from this interpretation just what happens. Consider therefore the following alternate interpretation: $(P^nB^E)_{ij}$ is the probability that the process started in i at time $-n$ enters E after time 0 at state j, and the limit λ_j^E is the probability that the process started at time $-\infty$ enters E after time 0 at state j. In this interpretation, if we let E swell to S and pass to the limit in the appropriate sense, we can expect λ^E to converge to an entrance distribution for the chain.

Suppose we compute instead the limit of P^nB^E on E followed by the limit on n. By dominated convergence, we have $\lim_{E\to S} P^nB^E = P^n$. Therefore, if we can justify the interchange of limits on E and n, we see that each row of P^n can be expected to converge to an entrance distribution for the chain. This conjecture is in contrast with the situation for transient chains, where P^n converges to an exit distribution.

In other words, the limiting behavior of P^n has something to do with the past history of the process. Our procedure will therefore be as follows: We start with a recurrent chain P, make it disappear after it reaches 0, and apply transient entrance boundary theory to the resulting process. The boundary obtained will be the entrance boundary for P, and it will be suitable in a wide class of chains for describing the limiting behavior of both λ^E and P^n. We shall see that this procedure is canonical in that it does not depend upon the choice of the state 0.

For the remainder of this chapter, let P be a recurrent chain, let 0 be a distinguished state, and let α be a finite-valued positive regular measure; α is unique up to a constant factor.

We define a transient chain Q associated to P and 0 by

$$Q_{ij} = \begin{cases} P_{ij} & \text{if } i \neq 0 \\ 0 & \text{if } i = 0. \end{cases}$$

For this transient chain $N_{ij} = {}^0N_{ij} + \delta_{j0}$. Choose f to be a column vector which places a unit weight at 0. Then $Nf = 1$.

Form the Martin entrance boundary of Q with respect to the reference function f. We have

$$J(i, j) = \frac{N_{ij}}{(Nf)_i} = {}^0N_{ij} + \delta_{j0}.$$

The compact metric space $*S$ is the completion of S in the metric described in Chapter 10, and we shall see in Proposition 11-1 that $*S$ does not depend upon the choice of the state 0. The spaces $*S$ and $*S - S$ can unambiguously be called the **completed space** and the **recurrent entrance boundary**, respectively, of the chain P. The set B^e of extreme points of $*S - S$ will also be independent of state 0, and we are therefore free to speak of extreme points of the recurrent boundary.

Since $J(x, j)$ is well defined for x in $*S$, the above expression for J shows that ${}^0N(x, j)$ is also well defined if we put

$${}^0N(x, j) = \begin{cases} {}^0N_{ij} & \text{if } x = i \in S \\ \lim_n {}^0N_{i_n j} & \text{if } x \in *S - S \text{ and } x = \lim i_n. \end{cases}$$

Proposition 11-1: If $*S_0$ and $*S_1$ are the completed spaces of P formed with respect to two distinct states 0 and 1, then the identity map on S extends to a homeomorphism of $*S_0$ onto $*S_1$. Under the homeomorphism the extreme parts of the boundary correspond.

PROOF: We first establish the identity

$$^1N_{ij} = {}^0N_{ij} + {}^1N_{0j} - \frac{\alpha_j}{\alpha_1}\,{}^0N_{i1}. \tag{*}$$

In fact, the mean number of times the process started at i visits j from the time of the first visit to 1 until before the first visit to 0 is ${}^0H_{i1}\,{}^0N_{1j}$. Thus if we compute in two ways the mean number of times the process started at i visits j before reaching both 0 and 1, we obtain

$$^1N_{ij} + {}^0H_{i1}\,{}^0N_{1j} = {}^0N_{ij} + {}^1H_{i0}\,{}^1N_{0j}.$$

Substitute ${}^1H_{i0} = 1 - {}^0H_{i1}$ and get

$$^1N_{ij} = {}^0N_{ij} + {}^1N_{0j} - {}^0H_{i1}({}^1N_{0j} + {}^0N_{1j}).$$

Applying Lemma 9-9 to $\hat{P}$ under the identification $i \to 1$, $j \to 0$, and $k \to j$, we find that the term in parentheses equals $(\alpha_j/\alpha_1)\,{}^0N_{11}$. From this relation, (*) follows.

Equation (*) and the expression for the kernel J show immediately that Cauchy sequences of S in the chain relative to state 0 are Cauchy relative to state 1, and by symmetry the converse is also true. Thus the statement about $^*S_0$ and $^*S_1$ follows.

For the assertion about the extreme points, let J_0 and J_1 be the kernels for the two transient chains, and suppose, for instance, that $J_1(y, \cdot)$ is normalized minimal and that $J_0(y, \cdot)$ is not. In any case, $J_0(y, \cdot)$ is normalized since $J_0(y, 0) = 1$. Thus

$$J_0(y, \cdot) = \int_{^*S} J_0(x, \cdot)\,d\nu(x)$$

by Theorem 10-48, where $\nu(^*S) = 1$ and ν does not concentrate all its mass at one point. Extending equation (*) to $^*S$ and using the connection between the kernels and the N's, we have also

$$J_1(x, j) = J_0(x, j) + J_1(0, j) - \frac{\alpha_j}{\alpha_1}\,J_0(x, 1) - \delta_{j0}.$$

Integration of this equation against ν gives

$$J_1(y, j) = \int_{^*S} J_1(x, j)\,d\nu(x).$$

Since $J_1(y, \cdot)$ is normalized minimal and ν concentrates its weight at more than one point, this last equation contradicts the dual of Lemma 10-33.

2. Measures on the entrance boundary

Let $\delta_{0\cdot}$ and $\delta_{\cdot 0}$ denote the 0th row and column, respectively, of the identity matrix. In this section we consider finite-valued row vectors ν for the Markov chain P with the following three properties:

(1) $\nu \geq 0$.
(2) $\nu_0 = 0$.
(3) $\nu P \leq \nu + \delta_{0\cdot}$.

We shall associate to each such row vector a probability measure β^ν on the entrance boundary and show how β^ν is related to ν probabilistically.

Direct calculation shows that the row vector whose jth entry is $\bar{\nu}_j = \nu_j + \delta_{0j}$ is non-negative and Q-superregular. We introduce a process which, when $\bar{\nu} > 0$, is the $\bar{\nu}$-dual of Q. Let $S^\nu = \{i \mid \nu_i > 0\}$ and define

$$Q^\nu_{ij} = \frac{\nu_j P_{ji}}{\nu_i + \delta_{0i}} \quad \text{for } i, j \in S^\nu \cup \{0\}.$$

All other entries of Q^ν are taken to be 0. Note that

$$N^\nu_{0j} = \frac{\bar{\nu}_j N_{j0}}{\bar{\nu}_0} = \bar{\nu}_j = \nu_j + \delta_{0j}.$$

Proposition 11-2: If ν satisfies (1), (2), and (3), then there is a unique probability measure β^ν on the Borel sets of $S \cup B^e$ such that

$$\nu = \int_{S \cup B^e} {}^0N(x, \cdot) d\beta^\nu(x).$$

PROOF: By Theorem 10-48

$$\bar{\nu} = \int_{S \cup B^e} J(x, \cdot) d\beta^\nu(x)$$

for a unique measure β^ν. Since $\bar{\nu}_0 = 1$, $\bar{\nu}$ is normalized and $\beta^\nu(S \cup B^e) = 1$. Hence

$$\nu = \int_{S \cup B^e} {}^0N(x, \cdot) d\beta^\nu(x).$$

If another such probability measure is given, we can reverse the argument and conclude the measure is β^ν by the uniqueness half of Theorem 10-48.

We define θ^E_j to be the probability in the Q^ν-process started at 0 that there is a last time the process is in E and that this occurrence is at state j. Note that θ^E depends on P, 0, ν, and E and that $\theta^E_j = 0$ for j not in E.

Proposition 11-3: If E is any subset of S containing 0 and if ν satisfies (1), (2), and (3), then

$$\nu_E(I - P^E) = \theta^E_E - \delta_{0\cdot}.$$

PROOF: Using as a set of alternatives the time when the Q^ν-process is last in E, we have for, $j \in E$,

$$\begin{aligned}\theta^E_j &= \sum_{n=0}^{\infty} (Q^\nu)^{(n)}_{0j} \Pr^\nu_j[\text{process leaves } E \text{ immediately and never returns}] \\ &= N^\nu_{0j}\left[1 - \sum_{k \in E}\left(Q^\nu_{jk} + \sum_{c,d \in \bar{E}} Q^\nu_{jc}\, {}^E N^\nu_{cd} Q^\nu_{dk}\right)\right].\end{aligned}$$

Now $N^\nu_{0j} = \nu_j + \delta_{0j}$, and therefore

$$N^\nu_{0j} Q^\nu_{jk} = \nu_k P_{kj}$$

and

$$N^\nu_{0j} Q^\nu_{jc}\, {}^E N^\nu_{cd} Q^\nu_{dk} = \nu_k P_{kd}\, {}^E N_{dc} P_{cj}.$$

for *all* states, not just those in $S^\nu \cup \{0\}$. Substitution and use of Lemma 6-6 give

$$\theta^E_j = \nu_j + \delta_{0j} - \sum_{k \in E} \nu_k P^E_{kj}.$$

In general, θ^E has total measure less than one, since the Q^ν process either may fail to reach E or may return to it infinitely often. If E is finite and $0 \in E$, neither of these alternatives has positive probability and thus $\theta^E \mathbf{1} = 1$. This conclusion also follows from Proposition 11-3, since finite matrices associate.

The special case $E = S$ yields the following corollary. Let $\bar{\beta}^\nu$ stand for the restriction to S of the measure β^ν defined in Proposition 11-2.

Corollary 11-4: If ν satisfies (1), (2), and (3), then

$$\nu(I - P) = \bar{\beta}^\nu - \delta_{0\cdot}.$$

PROOF: Let $E = S$ in Proposition 11-3. The only way the process can leave S is to disappear, and thus θ^S_j is the probability that the Q^ν-process disappears from state j. But this is just the definition of $\beta^\nu(j)$, since β^ν was defined as harmonic measure for the Q^ν-process started at 0.

We conclude this section by noting the connection between θ^E and β^ν. The proof is contained in the proof of Proposition 10-21.

Proposition 11-5: If ν satisfies (1), (2), and (3) and if $\{E_k\}$ is an increasing sequence of finite sets of states with union S, then the measures θ^{E_k} converge to β^ν weak-star on $^*S$.

3. Harmonic measure for normal chains

We come now to the first convergence theorem. We shall prove that if P is a normal chain, then the measures λ^E converge weak-star to a measure β which will play the role of an entrance distribution for P. This result agrees with the statement in Section 1.

Lemma 11-6: If P is normal, then the row vector $\nu = G_{0\cdot}$ satisfies conditions (1), (2), and (3). Also, for any finite set E containing 0,

$$\nu_E(I - P^E) = \lambda_E^E - \delta_{0\cdot}.$$

PROOF: We know that $G_{0j} \geq 0$ and $G_{00} = 0$. Hence (1) and (2) hold. Condition (3) follows by multiplying the definition of $G_{0\cdot}$ through on the right by P and applying Fatou's Theorem.

Form the K-matrix of Definition 9-80 with respect to the distinguished state 0. By Lemma 9-81, $K_{0j} = G_{0j}$. Hence the formula for $\nu_E(I - P^E)$ is the 0th row of the formula of Corollary 9-86.

Harmonic measure β for a normal chain P is defined to be the measure $\beta = \beta^\nu$ of Proposition 11-2 for $\nu = G_{0\cdot}$. The justification for a name independent of 0 is contained in the following theorem.

Theorem 11-7: If $\{E_k\}$ is an increasing sequence of finite sets with union S, then the measures λ^{E_k} converge weak-star to β on $^*S$. The measure β is independent of the distinguished state 0. Also

$$G_{ij} = \int_{S \cup B^e} {}^iN(x, j)\, d\beta(x)$$

and

$$G(I - P) = \mathbf{1}\beta - I.$$

PROOF: Ultimately the sets E_k contain 0, and Proposition 11-3 and Lemma 11-6 apply. From these results we obtain $\lambda^{E_k} = \theta^{E_k}$, and from Proposition 11-5 we conclude that λ^{E_k} converges to β. Thus β is given a characterization independent of the state 0 (since λ^E does not depend on 0). Since β does not depend on 0, we can use any state i as distinguished state in Proposition 11-2 and Corollary 11-4. The two formulas of the theorem then follow.

4. Continuous and T-continuous functions

The first convergence theorem, which was proved in Section 3, resulted from examining a particular row vector ν satisfying conditions (1), (2), and (3) of Section 2, namely $\nu = G_{0\cdot}$. We now begin to work toward the second convergence theorem, and we do so by examining the vector $\nu = {}^0N(x, \cdot)$. For the present we do not assume that P is normal.

We start by checking that $\nu = {}^0N(x, \cdot)$ does satisfy the three conditions and by identifying θ^E for ν. In fact, $\nu \geq 0$ and $\nu_0 = 0$ because ${}^0N(i, \cdot) \geq 0$ and ${}^0N(i, 0) = 0$. Thus (1) and (2) hold. Moreover we have, for every $i \neq 0$,

$$ {}^0N(i, \cdot)P \leq {}^0N(i, \cdot) + \delta_0. $$

Thus we can let $i \to x$, using Fatou's Theorem, and obtain (3).

The procedure for calculating θ^E is to do so for the approximating vectors ${}^0N(i, \cdot)$ first.

Lemma 11-8: If E is a finite set containing 0 and j, then

$$ \sum_{k \in E} {}^0N_{ik}(I - P^E)_{kj} = B^E_{ij} - \delta_{0j}. $$

PROOF: Apply Lemma 9-37 and conclusion (1) of Theorem 9-15.

REMARK: The lemma remains valid for infinite sets E containing 0 and j, and the proof consists in computing the dual of the left side by means of Propositions 6-16 and 6-17 and Lemma 9-14.

We shall say that a column vector h is **continuous** if it has a (necessarily unique) extension to a continuous function $h(x)$ defined on all of $^*S$.

In this notation the right side of the identity in Lemma 11-8 is continuous for fixed $j \in E$, and hence $B^E_{\cdot j}$ is continuous if $j \in E$. But $B^E_{\cdot j}$ is identically zero for $j \notin E$ and any state can be taken as the reference state 0. Thus we are justified in writing $B^E(x, j)$ for the continuous extension of the jth column of E whenever E is a nonempty finite set. By an **elementary continuous function** is meant a finite linear combination of such functions.

Passing to the limit $i \to x$ in Lemma 11-8, we have

$$ \sum_{k \in E} {}^0N(x, k)(I - P^E)_{kj} = B^E(x, j) - \delta_{0j} $$

whenever $0 \in E$, $j \in E$, and E is finite. Consequently $\theta^E = B^E(x, \cdot)$

by Proposition 11-3, provided E is a finite set containing 0. (Note both sides are zero for states j not in E.)

Let us denote the measure β^ν obtained from Proposition 11-2 for $\nu = {}^0N(x, \cdot)$ by β^x. If $\{E_k\}$ is an increasing sequence of finite sets with union S, then Proposition 11-5 asserts that $B^{E_k}(x, \cdot)$ converges weak-star to β^x. We have thus proved the following proposition.

Proposition 11-9: If h is a continuous column vector and if $\{E_k\}$ is an increasing sequence of finite sets with union S, then

$$\lim_k B^{E_k}(x, \cdot)h = \int_{*S} h(y)d\beta^x(y)$$

pointwise for x in $*S$.

If $\nu = {}^0N(x, \cdot)$, then the associated Q-superregular measure is $\bar{\nu} = J(x, \cdot)$. Since β^x is the measure for $J(x, \cdot)$, β^x concentrates its mass at x if x is in S or in B^e. Thus the right side of the identity of Proposition 11-9 equals $h(x)$ for all such x.

Define a linear transformation T of continuous functions to bounded functions on $*S$ by

$$Th(x) = \int_{*S} h(y)d\beta^x(y).$$

A continuous function h such that $Th = h$ is said to be **T-continuous.** (The motivation for this name appears as Problem 2 at the end of the chapter.)

We conclude this section by characterizing the T-continuous functions. Notice that if every boundary point is extreme, then every continuous function is T-continuous.

Lemma 11-10: If h is an elementary continuous function, then $Th = h$.

PROOF: It suffices to consider the function $B^E_{\cdot j}$ where E is a finite set. If $E \subset E_k$, then $B^{E_k}B^E = B^E$ by Proposition 5-8. Passing to the limit as $i \to x$ with k fixed, we have

$$B^{E_k}(x, \cdot)B^E_{\cdot j} = B^E(x, j).$$

Hence the left side of the identity in Proposition 11-9 is $B^E(x, j)$. But the right side is $TB^E(x, j)$.

Lemma 11-11: If h is T-continuous and $\{E_k\}$ is an increasing sequence of finite sets with union S, then for any $\epsilon > 0$ some convex combination of the functions $B^{E_k}(x, \cdot)h$ is within ϵ of h uniformly for x in $*S$.

PROOF: By Proposition 11-9, the functions $B^{E_k}(x, \cdot)h$ converge pointwise to $h(x)$. Thus by dominated convergence their integrals against any Borel measure on $^*S$ converge to the integral of h. That is, the functions converge to h weakly. Thus h is in the weak closure of the set $\{B^{E_k}(x, \cdot)h\}$, is certainly in the weak closure of the convex hull of the set, and must therefore be in the strong closure of the convex hull of the set. (See Dunford and Schwartz [1958], p. 422, Corollary 14.)

Proposition 11-12: The set of T-continuous functions is exactly the uniform closure of the set of elementary continuous functions.

PROOF: Every T-continuous function is contained in the uniform closure according to Lemma 11-11, since $B^{E_k}(x, j)$ vanishes unless j is in E_k. Conversely, every elementary continuous function is T-continuous by Lemma 11-10, and the uniform limit of T-continuous functions is T-continuous, since T has norm no greater than one.

5. Normal chains and convergence to the boundary

As an application of the machinery of Section 4, we can prove the second convergence theorem—that each row of P^n converges weak-star to the harmonic measure β in a suitable class of normal chains. This result was suggested in the discussion in Section 1, and it was pointed out that the key to the proof should be a certain interchange of limits. In fact, this interchange has already taken place and is concealed in the proof of Lemma 11-11.

We begin with a particularly sharp form of the convergence theorem.

Theorem 11-13: If P is normal, if i is any state in S, and if h is T-continuous, then

$$\lim_n \, (P^n h)_i = \int_{^*S} h(x) d\beta(x).$$

Conversely, if this equation holds for all states i and all T-continuous h, then P is normal.

PROOF: Let $\epsilon > 0$ be given and let $\{E_k\}$ be an increasing sequence of finite sets with union S. Since h is continuous, we can choose k_0 large enough so that

$$\left| \int h(x) d\beta(x) - \lambda^{E_k} h \right| < \epsilon$$

for all $k \geq k_0$ by Theorem 11-7. Truncate the sequence of sets so that it contains only those sets $\{E_k\}$ with $k \geq k_0$. Since h is T-continuous, we can apply Lemma 11-11 to the truncated sequence to obtain a convex combination of the functions $B^{E_k}(x, \cdot)h$ which is uniformly within ϵ of h. If, say,

$$\left\| \sum c_j B^{E_j} h - h \right\| < \epsilon,$$

then also

$$\left| P^n_{i\cdot}\left(\sum c_j B^{E_j} h - h\right) \right| < \epsilon$$

since $P^n 1 = 1$ and $P^n \geq 0$. Consequently,

$$\begin{aligned} \left| \int h(x) d\beta(x) - (P^n h)_i \right| &\leq \left| \int h(x) d\beta(x) - \sum c_j \lambda^{E_j} h \right| \\ &\quad + \left| \sum c_j (\lambda^{E_j} - P^n_{i\cdot} B^{E_j}) h \right| \\ &\quad + \left| P^n_{i\cdot}\left(\sum c_j B^{E_j} h - h\right) \right| \\ &< 2\epsilon + \left| \sum c_j (\lambda^{E_j} - P^n_{i\cdot} B^{E_j}) h \right|. \end{aligned}$$

The sum on the right is a finite sum and in each summand only finitely many entries of $\lambda^{E_j} - P^n_{i\cdot} B^{E_j}$ can be non-zero. Since $P^n_{i\cdot} B^{E_j} \to \lambda^{E_j}$ pointwise, we conclude that the sum on the right side is less than ϵ for n sufficiently large.

The converse follows by applying the assumption to columns of B^E for two-point sets E.

The convergence theorem is as follows.

Theorem 11-14: If P is normal and if $B = B^e$, then each row of P^n converges weak-star to the harmonic measure β.

PROOF: Apply Theorem 11-13. If $B = B^e$, then every continuous function is T-continuous.

Thus for normal chains with $B = B^e$, the measure β indicates what the chain is "near" in the long run. In the case of null chains with an additional property, we can show that the chain is near the boundary in the long run.

Proposition 11-15: If P is a normal null chain with $B = B^e$ in which every one-point set in S is open, then $\beta(S) = 0$. That is, the measure is entirely on the boundary.

PROOF: If i is given, then the characteristic function of i is continuous. By Theorem 11-14, $\lim P^{(n)}_{0i} = \beta(i)$. But for a null chain, P^n tends to zero.

Under the hypotheses of Theorem 11-15, the number $\beta(X)$ for any Borel subset X of B may be interpreted as the probability that the process is near this part of the boundary in the long run. For example, if x is at a positive distance from other boundary points, then any sufficiently small neighborhood E of x will have a continuous characteristic function. By Theorem 11-14,

$$\lim_n \Pr_i[x_n \in E] = \sum_{j \in E} P_{ij}^{(n)} \to \beta(x).$$

Let us see what our results say for a noncyclic ergodic chain P. Such a chain is necessarily normal, and α may be chosen to have total measure one. If this choice is made, then $G(I - P) = 1\alpha - I$ by Corollary 9-51. Comparison with Theorem 11-7 shows that $\alpha = \bar{\beta}$. Thus the harmonic measure is concentrated entirely on S, in contrast with Proposition 11-15. The measure β is a generalization to all normal chains of the measure α for noncyclic ergodic chains. Thus our results generalize to all normal chains results known for ergodic chains. For example, the representation

$$G_{ij} = \int_{S \cup B^e} {}^iN(x, j) d\beta(x)$$

is a generalization of the identity $G_{ij} = \sum_k \alpha_k \, {}^iN_{kj}$, which holds for noncyclic ergodic chains. (Theorem 9-26 gives

$$G_{ij} = {}^i\nu_j = \lim \sum P_{mk}^{(n)} \, {}^iN_{kj},$$

and Proposition 1-57 yields $G_{ij} = \sum \alpha_k \, {}^iN_{kj}$.) As a second example, $(P^n h)_i$ converges to αh for any bounded function h if P is noncyclic ergodic (Lemma 9-52). This result is generalized in Theorem 11-13, but in this theorem we had to make a stronger assumption about h. The difference arises because in a noncyclic ergodic chain each row of P^n actually converges to α in the norm topology of measures, not just in the weak-star topology (see Theorem 6-38).

6. Representation Theorem

Beginning in this section, we connect the results we have obtained so far with the results of Chapter 9 on potential theory. We start by proving a representation theorem. For the moment we do not assume that P is normal.

If μ and ν are row vectors with $\mu = \nu(I - P)$, then μ will be called the **deviation** (from regularity) of ν. If $\mu 1$ is finite, then we say that ν

is of **totally finite deviation.** Dually $f = (I - P)h$ is the deviation of h, and h is of totally finite deviation if αf is finite.

Theorem 11-16: If ν is a non-negative row vector whose deviation μ is totally finite, then $\mu 1 \leq 0$ and there is a probability measure π on B^e such that

$$\nu_j = \nu_0(\alpha_j/\alpha_0) + (\mu\, {}^0N)_j - (\mu 1)\int_{B^e} {}^0N(x, j)\,d\pi(x).$$

If $\mu 1 \neq 0$, then the probability measure π is uniquely determined.

PROOF: We know that

$$N_{ij} = {}^0N_{ij} + \delta_{0j}.$$

If we put $\bar{\mu} = \nu(I - Q)$, we have $\bar{\mu}_i = \mu_i + \nu_0 P_{0i}$ by direct calculation. We therefore get

$$\begin{aligned}(\bar{\mu}N)_j &= (\mu\, {}^0N)_j + (\mu 1)\delta_{0j} + \nu_0\, {}^0\bar{N}_{0j} \\ &= (\mu\, {}^0N)_j + (\mu 1)\delta_{0j} + \nu_0(\alpha_j/\alpha_0).\end{aligned}$$

From Proposition 8-7 applied to the chain Q, we see that $\nu = \bar{\mu}N + \rho$, where ρ is regular and non-negative. The calculation of $\bar{\mu}N$ yields

$$\rho_j = \nu_j - (\bar{\mu}N)_j = \begin{cases} -(\mu 1) & \text{if } j = 0 \\ \nu_j - \nu_0(\alpha_j/\alpha_0) - (\mu\, {}^0N)_j & \text{if } j \neq 0. \end{cases}$$

Since $\rho_0 \geq 0$, $\mu 1 \leq 0$. If $\mu 1 = 0$, then ρ is a Q-regular measure ≥ 0 with $\rho_0 = 0$. Since it is possible to get from any state to 0 eventually in the Q-process, ρ vanishes identically. The representation then follows immediately from the expression for ρ.

Thus suppose $\mu 1 \neq 0$. Define σ by

$$\sigma_j = -(\mu 1)^{-1}[\nu_j - \nu_0(\alpha_j/\alpha_0) - (\mu\, {}^0N)_j].$$

We claim that σ satisfies conditions (1), (2), and (3). It is clear that $\sigma_0 = 0$, and the fact that $\sigma \geq 0$ follows from what we have shown above. For (3), we have

$$(\sigma P)_j = -(\mu 1)^{-1}(\rho Q)_j = -(\mu 1)^{-1}\rho_j = \sigma_j + \delta_{0j},$$

and thus equality holds. Thus, except for the assertion that π is concentrated on B^e, the rest of the theorem is immediate from Proposition 11-2. So simply note from the proof of that proposition that if we have equality in (3), then π is concentrated on B^e.

We define the **exit boundary** of P to be the entrance boundary of $\hat{P}$. That is, $S^* = {}^*\hat{S}$ and $B_e = \hat{B}^e$. We obtain a dual for Theorem 11-16.

Theorem 11-17: If h is a non-negative column vector whose deviation f is totally finite, then $\alpha f \leq 0$ and there is a probability measure π on B_e such that

$$h_i = h_0 + ({}^0Nf)_i - \frac{\alpha f}{\alpha_i} \int_{B_e} {}^0\hat{N}(x, i) d\pi(x).$$

If $\alpha f \neq 0$, then the probability measure π is uniquely determined.

We turn to results connected with potential theory. We first apply Theorem 11-17 to give elementary continuous functions a characterization which is valid for all recurrrent chains, normal or not.

Proposition 11-18: If h is an elementary continuous function, then

(1) $(I - P)h = f$ has finite support.
(2) $h = c\mathbf{1} + {}^0Nf$ for some c.
(3) $\alpha f = 0$.
(4) $B^E h = h$ for any set E containing the support of f.

Conversely, if (1) and (2) hold, then h is an elementary continuous function.

PROOF: $(I - P)B^E$ is equal to $I - P^E$ for states in E and is 0 otherwise. Hence $\alpha[(I - P)B^E] = 0$ and (1) and (3) hold for columns of B^E. Since an elementary continuous function is a finite linear combination of such functions for finite sets E, (1) and (3) follow. In particular, any column of B^E is of totally finite deviation and Theorem 11-17 applies. The representation in that theorem establishes (2) for columns of B^E, and the general result follows by linearity.

We shall complete the proof by showing that (2) implies (4) and that (1) and (4) imply that h is an elementary continuous function. Suppose (2) holds. Let $0 \in E$. Then ${}^0N = B^E\,{}^0N + {}^EN$ by Lemma 9-14. If E contains also the support of f, then ${}^ENf = 0$, and we see directly that $B^E h = h$. So (4) holds. If (1) and (4) hold, then (4) holds for some finite set E, and h is exhibited as a finite linear combination of the columns of B^E.

We now assume that P is normal and we shall prove statements for T-continuous functions that look like applications of Proposition 11-18 followed by a passage to the limit.

Proposition 11-19: If P is normal, then a function h of finite deviation is T-continuous if and only if $h = c\mathbf{1} + g$ for a T-continuous potential g. If the representation holds, then

$$c = \int_{*S} h(x)d\beta(x).$$

PROOF: If $(I - P)h = f$ and if h is T-continuous, then

$$(I + P + \cdots + P^n)f = (I - P^{n+1})h \to h - c\mathbf{1}$$

by Theorem 11-13. If h is of finite deviation, then f is a charge and the limit $h - c\mathbf{1}$ is a potential. This potential is T-continuous, since $\mathbf{1}$ is. (Notice that $\mathbf{1}$ is the 0th column of $B^{\{0\}}$.) The converse is clear.

Corollary 11-20: If P is normal and if g is a T-continuous potential, then

$$\int_{*S} g(x)d\beta(x) = 0.$$

PROOF: Potentials are of finite deviation by definition. Applying Proposition 11-19 and using the fact that $\mathbf{1}$ is not a potential (since its charge would have to be zero), we see that $c = 0$.

Corollary 11-20 is a recurrent analog of the statement for transient boundary theory that potentials tend to zero along almost all paths of the process.

Corollary 11-21: If P is normal and if h is a T-continuous function whose deviation f is totally finite, then $\alpha f = 0$ and $h = c\mathbf{1} + {}^0Nf$.

PROOF: By Proposition 11-19, h differs from the potential g of f by a constant. But $g = g_0\mathbf{1} + {}^0Nf$ by Theorem 9-15.

Our final proposition enables us to give an interpretation to θ^E for certain infinite sets provided we have an interpretation for finite sets.

Proposition 11-22: If E_k is an increasing sequence of finite sets with union E and if $\theta^E\mathbf{1} = 1$, then $\lim \theta^{E_k} = \theta^E$ pointwise.

PROOF: Since $\theta_j^{E_k}$ is an exit probability, it is decreasing in k from some point on. Hence it tends to a limit, say θ_j. Since $\theta^{E_k}\mathbf{1} = 1$, we have $\theta\mathbf{1} \leq 1$ by Fatou's Theorem. For large k, we have also $\theta_j^E \leq \theta_j^{E_k}$ and hence $\theta_j^E \leq \theta_j$. Thus $1 = \theta^E\mathbf{1} \leq \theta\mathbf{1} \leq 1$. This statement is consistent with the inequality $\theta^E \leq \theta$ only if $\theta^E = \theta$.

As an application of Proposition 11-22, suppose P is normal and we choose $\nu_j = G_{0j}$. Then $\theta^E = \lambda^E$ for finite sets containing 0, and θ^E is thus a limit of λ^E-measures for any set with $\theta^E 1 = 1$, whether or not λ^E exists for the infinite set E. The condition that $\theta^E 1 = 1$ means that the transient chain Q^ν leaves E with probability one; this condition is exactly the statement that E be an equilibrium set for Q^ν. For such a set E and for any reference state $i \in E$, we have

$$[G_E(I - P^E)]_{i\cdot} = \theta^E_E - \delta_{i\cdot}$$

by Proposition 11-3. Since, for two different reference states i in E, we have θ^E as the limit of the same sequence λ^{E_k} (by Proposition 11-22), we may write $G_E(I - P^E) = 1\theta^E_E - I$. Then

$$(\theta^E_E G_E)(I - P^E) = 0$$

and hence $\theta^E_E G_E = k\alpha$ for some constant k. If we form the K-matrix relative to state 0, then Lemma 9-81 gives

$$G_{ij} = K_{ij} + (G_{i0} - C_{i0})(\alpha_j/\alpha_0).$$

Consequently, $\theta^E_E K_E = k'\alpha$, provided

$$\sum_{i\in E} \theta_i G_{i0} < \infty \quad \text{and} \quad \sum_{i\in E} \theta_i C_{i0} < \infty.$$

The conclusion that $\theta^E_E K_E = k'\alpha$ is exactly the statement that k' be the generalized capacity of E and θ^E be the (generalized) recurrent equilibrium charge for E in the sense of Definitions 9-88 and 9-113. However, in the part of Chapter 9 where these points were discussed, we restricted ourselves to finite sets. By means of boundary theory we see we can extend these results to certain infinite sets.

Finally, let us consider the special case where the space $*S$ for P has only one limit point ∞. Then a neighborhood of ∞ is simply the complement of a finite set not containing ∞. Hence the probability of being within this neighborhood tends to one if P is null. Therefore $P^n h$ tends to $h(\infty)$ for any such null chain and any continuous function h.

If P is null and if P and $\hat{P}$ both have only a single limit point, then P must be normal. For such a chain,

$$G_{ij} = {}^iN(\infty, j) \quad \text{and} \quad C_{ij} = (\alpha_j/\alpha_i)\,{}^j\hat{N}(\hat{\infty}, i),$$

and the representation theorem takes on the simpler form

$$h_i = h_0 + ({}^0Nf)_i - \frac{\alpha f}{\alpha_0} C_{i0}.$$

From Proposition 9-45 we know that

$${}^0N_{ij} - (\alpha_j/\alpha_0)C_{i0} = C_{0j} - C_{ij}.$$

Hence

$$h = h_0 \mathbf{1} + (C_{0\cdot} - C_{i\cdot})f.$$

Thus if Cf is finite-valued, then h differs from Cf by a constant.

7. Sums of independent random variables

The results of this chapter take on an especially neat-looking form where P is a recurrent sums of independent random variables process with state space the lattice in N-dimensional space. Such a chain is always null, and Spitzer [1962] showed it is always normal, has only minimal boundary points, and has no points of S as limit points. In two dimensions there is always a unique boundary point, whereas in one dimension there are one or two, depending on whether the distribution has infinite or finite variance.

In the case of a single boundary point, a continuous function h is one having a limit at infinity, and we have already noted that the convergence of $P^n h$ is trivial. In one dimension with finite variance, the two distinct boundary points correspond to $-\infty$ and $+\infty$, and h, to be continuous, must have limits in both directions. For such a function Theorem 11-14 states that $P^n h$ converges to the average of these two limits. This result also follows from the Central Limit Theorem.

There are such chains in one dimension for which $P^n h$ fails to converge for as nice a function as the characteristic function of the positive integers. However, this behavior can occur only if the variance is infinite; then there is only one boundary point, and h is not continuous.

If $h \geq 0$ is a function whose deviation f is totally finite and if Cf is finite-valued, then the representation theorem takes the form

$$h = -Cf + \text{const.}$$

for the case of one boundary point and

$$h_i = -(Cf)_i + ai + b$$

for the two-boundary-point one-dimensional case. We have already seen that the former identity holds for any chain with a single limit point. The latter follows from the representation theorem together with special knowledge of the nature of C obtained by Spitzer for sums of independent random variables.

8. Examples

EXAMPLE 1: Sums of independent random variables, $p_{-1} = \frac{2}{3}$, $p_2 = \frac{1}{3}$.

This example shows the best possible boundary behavior for a null chain and illustrates the points discussed in Section 7. To determine the recurrent entrance boundary we can work either with the kernel $J(i, j)$ associated to the chain Q or with the matrix ${}^0N_{ij}$ associated to P, since

$$J(i, j) = {}^0N_{ij} + \delta_{j0}.$$

We use the latter.

To compute ${}^0N$, we can proceed in the familiar way—first finding ${}^0H$ and then computing ${}^0N$ from the identities ${}^0\bar{H} = P\,{}^0H$, ${}^0N_{jj} = 1/(1 - {}^0\bar{H}_{jj})$, and ${}^0N_{ij} = {}^0H_{ij}\,{}^0N_{jj}$. As usual, the calculation of ${}^0H$ involves a difference equation valid for certain intervals of i's and j's. In this case, the equation is

$${}^0H_{ij} = \tfrac{2}{3}\,{}^0H_{i-1,j} + \tfrac{1}{3}\,{}^0H_{i+2,j}.$$

The general solution as a function of i is

$${}^0H_{ij} = A + Bi + C(-2)^i$$

and is valid always for a slightly larger interval of i's. For instance, if $0 < j$ and if we are considering i's satisfying $0 \le i \le j$, then the difference equation is valid for $0 < i < j$ and the solution applies for $0 \le i \le j + 1$. The initial conditions in this instance are ${}^0H_{0j} = 0$, ${}^0H_{jj} = 1$ and ${}^0H_{j+1,j} = 1$, and they determine A, B, and C. Similar remarks apply for the other intervals of i's and j's, and the result is:

for $0 < j$

$${}^0N_{ij} = \begin{cases} \tfrac{1}{3}[1 - (-2)^{-j}][1 - (-2)^i] & i \le 0 \\ i - \tfrac{1}{3}(-2)^{-j}[1 - (-2)^i] & 0 \le i \le j \\ j + \tfrac{1}{3}[1 - (-2)^{-j}] & j \le i, \end{cases}$$

for $j < 0$

$${}^0N_{ij} = \begin{cases} \tfrac{1}{3}(-2)^i[(-2)^{-j} - 1] - j & i \le j \\ \tfrac{1}{3}[1 - (-2)^i] - i & j \le i \le 0 \\ 0 & 0 \le i. \end{cases}$$

Therefore there are two boundary points, $+\infty$ and $-\infty$, no point of S is a limit point, and

$${}^0N(+\infty, j) = \begin{cases} j + \tfrac{1}{3}[1 - (-2)^{-j}] & j \ge 0 \\ 0 & j \le 0 \end{cases}$$

$${}^0N(-\infty, j) = \begin{cases} \tfrac{1}{3}[1 - (-2)^{-j}] & j \ge 0 \\ |j| & j \le 0. \end{cases}$$

From the Central Limit Theorem we know that for large n the process is very likely to be far from 0 and equally likely to be on the right or on the left. Hence $\beta(+\infty) = \beta(-\infty) = \frac{1}{2}$. Since β assigns positive mass to each boundary point, both boundary points are extreme and consequently every continuous function is T-continuous. From Theorem 11-7 we have

$$G_{0j} = \tfrac{1}{2}\,{}^0N(+\infty, j) + \tfrac{1}{2}\,{}^0N(-\infty, j),$$

and for sums of independent random variables we have also

$$C_{ij} = G_{ij} = G_{0,j-i}.$$

Therefore

$$C_{ij} = G_{ij} = \begin{cases} \frac{1}{2}|j - i| + \frac{1}{3}[1 - (-2)^{i-j}] & j \geq i \\ \frac{1}{2}|j - i| & j \leq i. \end{cases}$$

Let us see what the representation theorem says for this process. We first need to know ${}^0\hat{N}(x, i)$. We can take $\alpha = \mathbf{1}^T$, and we see that $\hat{P}_{ij} = P_{ji}$ and ${}^0\hat{N}_{ij} = {}^0N_{ji}$. For $\hat{P}$ we again have two boundary points, and

$${}^0\hat{N}(+\infty, i) = \begin{cases} \frac{1}{3}[1 - (-2)^i] & i \leq 0 \\ |i| & 0 \leq i, \end{cases}$$

$${}^0\hat{N}(-\infty, i) = \begin{cases} |i| + \frac{1}{3}[1 - (-2)^i] & i \leq 0 \\ 0 & 0 \leq i. \end{cases}$$

Now let $h \geq 0$ be a function whose deviation f is totally finite, and suppose Cf is finite-valued. The representation theorem gives

$$h_i = h_0 + ({}^0Nf)_i - (\alpha f)[\pi(+\infty)\,{}^0\hat{N}(+\infty, i) + \pi(-\infty)\,{}^0\hat{N}(-\infty, i)].$$

If we use the identity

$${}^0N_{ij} - (\alpha_j/\alpha_0)C_{i0} = C_{0j} - C_{ij},$$

we obtain

$$h_i = -(Cf)_i + (\alpha f)[\pi(-\infty) - \tfrac{1}{2}]i + (h_0 + (Cf)_0).$$

This last equation is an example of the formula

$$h_i = -(Cf)_i + ai + b$$

discussed in Section 7.

EXAMPLE 2: Basic example, null case.
For the basic example we have

$${}^0N_{ij} = \begin{cases} {}^i\bar{N}_{ij} = \alpha_j/\alpha_i & \text{if } j \geq i > 0 \\ 0 & \text{otherwise.} \end{cases}$$

Hence, for fixed j, ${}^0N_{ij} = 0$ if i is sufficiently large. Consequently there is a single limit point p in $*S$, and ${}^0N(p, i) = 0$ for all i. Since ${}^0N(p, i) = {}^0N(0, i)$, we have $p = 0$; the limit point is the state 0. Thus the boundary is empty.

We know from Section 9-6 that in a null basic example $\lambda_0^E = 1$ if E is a finite set containing zero. Therefore the harmonic measure β, which is the weak-star limit of such measures, assigns unit mass to state 0. Thus by Theorem 11-7, $G_{0j} = {}^0N(0, 0) = 0$, in agreement with the result obtained in Chapter 9. This example shows that the condition on one-point sets in Proposition 11-15 cannot be omitted.

We can also check directly, using the results of Section 9-6, that $G_{ij} = {}^iN(0, j)$ and $G(I - P) = \mathbf{1}\delta_{0\cdot} - I$, again in agreement with Theorem 11-7.

The reverse $\hat{P}$ of the basic example has

$$
{}^0\hat{N}_{ij} = \begin{cases} 1 & \text{if } i \geq j > 0 \\ 0 & \text{otherwise.} \end{cases}
$$

Again there is one limit point p', but this time we have ${}^0\hat{N}(p', j) = 1 - \delta_{0j}$. The measure $\hat{J}(p', \cdot)$ associated with the transient chain

$$
\hat{Q}_{ij} = \begin{cases} \hat{P}_{ij} & \text{if } i \neq 0 \\ 0 & \text{if } i = 0 \end{cases}
$$

satisfies $\hat{J}(p', j) = {}^0\hat{N}(p', j) + \delta_{j0} = 1$ and is easily seen to be $\hat{Q}$-regular, and it follows that p' is a boundary point and is extreme. Put $+\infty = p'$.

It is clear that $\lambda_m^E = 1$ if m is the largest element of the finite set E and if $\hat{P}$ is null. Hence $\beta(+\infty) = 1$, in agreement with Proposition 11-15. Therefore $\hat{G}_{ij} = {}^i\hat{N}(+\infty, j)$, and we find that $\hat{G}(I - \hat{P}) = -I$.

EXAMPLE 3: Three-line example.

This example is designed to show that P and $\hat{P}$ can have different limiting behavior and to show that in a normal null chain some additional assumption is needed (such as $B = B_e$ in Theorem 11-14) to ensure that each row of P^n converges weak-star to a limiting measure.

Let S consist of three copies of the non-negative integers with typical elements denoted by i, i', and i'', respectively. The process P moves deterministically to the left on the first and third lines and moves from 0 or $0''$ to $0'$ also with probability one. On the middle line it moves one step to the right or moves to one of the other lines, as shown in the accompanying figure. The quantities on the arrows are transition probabilities, and p and q are positive numbers with sum one. The

p_i and q_i are chosen as in the basic example, except that $q_1 = 0$, β_i is defined as the product of the p_j's up through p_i as in the basic example, and we require that $\beta_i \to 0$ and $\sum \beta_i = +\infty$.

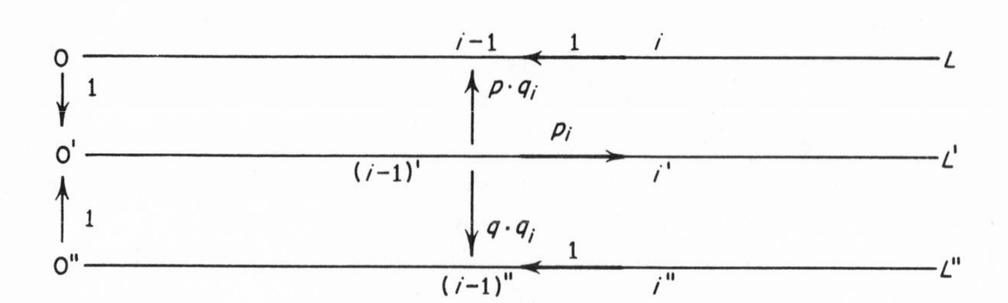

Clearly $\bar{H}_{0'0'} = 1 - \lim \beta_i = 1$, and thus P is recurrent. If α is defined by

$$\alpha_i = p\beta_i, \qquad \alpha_{i'} = \beta_i, \qquad \alpha_{i''} = q\beta_i,$$

then α is a regular measure. Since $\alpha 1 = 2 \sum \beta_i = +\infty$, the chain is null.

To see that P is normal, we compute ${}^{0'}\lambda$ and ${}^{0'}\hat{\lambda}$. Since P is null, the process after a long time is likely to be far to the right of any state we consider. Thus

$$ {}^{0'}\lambda_a = \begin{cases} p & \text{if } a \in L \\ q & \text{if } a \in L'' \\ 0 & \text{if } a \in L' \text{ and } a \neq 0'. \end{cases}$$

The reverse process $\hat{P}$ moves a step to the right on L (or L'') or switches to L'. On L' it moves deterministically to the left until it reaches $0'$, and then it goes to 0 with probability p and $0''$ with probability q. Thus

$$ {}^{0'}\hat{\lambda}_a = \begin{cases} 0 & \text{if } a \in L \text{ or } a \in L'' \\ 1 & \text{if } a \in L' \text{ and } a \neq 0'. \end{cases}$$

By Theorem 9-26, P is normal.

We now determine the boundaries of P and $\hat{P}$. If we choose $0'$ as the distinguished state, we see that if a and b are any two states different from $0'$, not necessarily on the same line, then

$$ {}^{0'}N_{ab} = \begin{cases} 1 & \text{if } a, b \in L \text{ or } a, b \in L'', a \text{ to the right of } b \\ p & \text{if } a \in L', b \in L, a \text{ to the right of } b \\ q & \text{if } a \in L', b \in L'', a \text{ to the right of } b \\ \alpha_b/\alpha_a & \text{if } a \in L', a \text{ to the left of } b \\ 0 & \text{otherwise.} \end{cases}$$

Thus for each b, ${}^{0'}N_{ab}$ tends to a limit along each line, but, as functions of b, the limits are different for each line. The result is three boundary points ∞, ∞', and ∞'', and we have

$$
{}^{0'}N(\infty, b) = \begin{cases} 1 & \text{if } b \in L \\ 0 & \text{otherwise} \end{cases}
$$

$$
{}^{0'}N(\infty', b) = \begin{cases} p & \text{if } b \in L \\ q & \text{if } b \in L'' \\ 0 & \text{otherwise} \end{cases}
$$

$$
{}^{0'}N(\infty'', b) = \begin{cases} 1 & \text{if } b \in L'' \\ 0 & \text{otherwise.} \end{cases}
$$

The measures $J(x, \cdot)$ defined for the associated transient chain satisfy

$$
J(x, b) = {}^{0'}N(x, b) + \delta_{b0'},
$$

and direct calculation shows that

$$
J(\infty', \cdot) = pJ(\infty, \cdot) + qJ(\infty'', \cdot).
$$

We conclude that ∞ and ∞'' are extreme boundary points, whereas ∞' is not.

If E is a large finite set of states containing 0 and $0''$, then the chain P is most likely to be to the right of E. Hence $\lambda_k^E = p$ and $\lambda_{m''}^E = q$, where k and m'' are the last elements of E on the first and third lines, respectively. From the form of λ^E, we deduce that

$$
\beta(\infty) = p \quad \text{and} \quad \beta(\infty'') = q.
$$

We may interpret this result as follows: L, L', and L'' are neighborhoods of ∞, ∞', and ∞'', respectively. In the long run, the chain is typically far to the right in one of these sets. If it is in L or L'', it must remain there for a long time; but if it is in L', it can leave in one step by switching to L or L''. This behavior is what makes ∞' nonminimal. In other words, far out in L is near ∞, far out in L'' is near ∞'', but far out in L' means near ∞ with probability p and near ∞'' with probability q.

Now let us consider the boundary of $\hat{P}$. We are to look at the limiting behavior of ${}^{0'}\hat{N}_{ab} = \alpha_b\, {}^{0'}N_{ba}/\alpha_a$ along sequences of a's. But for fixed b, this quantity tends to the same limit along all three lines. We thus have just one limit point $\hat{\infty}$, and the corresponding measure is

$$
\hat{J}(\hat{\infty}, b) = {}^{0'}\hat{N}(\hat{\infty}, b) + \delta_{b0'} = \begin{cases} 1 & \text{if } b \in L' \\ 0 & \text{otherwise.} \end{cases}
$$

This measure is regular for $\hat{Q}$, and consequently $\hat{\infty}$ is an extreme boundary point. By Proposition 11-15, $\hat{\beta}(\hat{\infty}) = 1$. For $\hat{P}$ the chain either is far out in L' (and hence has to stay there for a long time) or else is in a set from which it can move in one step at any moment to a position far out in L'. The three boundary points for P collapse into one for $\hat{P}$ because a position far out on L or L'' is only one step away from being far out on L'.

We conclude by sketching a proof that the 0'th row of P^n does not converge weak-star as n tends to infinity if the p_i's are chosen appropriately.

Thus some condition such as $B = B_e$ is needed in Theorem 11-14, and some condition such as T-continuity is needed in Theorem 11-13. Let h be the characteristic function of L'. Then h is continuous, since $h(\infty) = h(\infty'') = 0$ and $h(\infty') = 1$; but $(Th)(\infty') = 0$ and thus h is not T-continuous. Let

$$a_n = (P^n h)_{0'} = \Pr_{0'}[x_n \in L'].$$

We shall show that $\{a_n\}$ does not converge if the p_i's are chosen suitably, and hence $P^n_{0'\cdot}$ does not converge weak-star. In fact, we shall indicate that $\{a_n\}$ can fail to be even Abel summable. We define

$$A(t) = \sum a_n t^n, \qquad P(t) = \sum P^{(n)}_{0'0'} t^n$$
$$B(t) = \sum \beta_n t^n, \qquad F(t) = \sum \bar{F}^{(n)}_{0'0'} t^n.$$

If we let k be the last time before n that the process is in $0'$, we see that

$$a_n = \sum_{k=0}^{n} P^{(k)}_{0'0'} \beta_{n-k}$$

and hence

$$A(t) = P(t)B(t).$$

For any chain, we have

$$P(t) = \frac{1}{1 - F(t)}.$$

For this chain,

$$\bar{F}^{(2n+2)}_{0'0'} = \beta_n q_{n+1} = \beta_n - \beta_{n+1},$$

whereas $\bar{F}^{(m)}_{0'0'} = 0$ if m is not of the form $2n + 2$. Thus

$$F(t) = 1 - (1 - t^2)B(t^2).$$

Combining our results, we have

$$A(t) = \frac{B(t)}{(1 - t^2)B(t^2)}.$$

The Abel limit of the sequence a_n is

$$\lim_{t \uparrow 1} (1 - t)A(t) = \frac{1}{2} \lim_{t \uparrow 1} \frac{B(t)}{B(t^2)}.$$

Now the sequence $\{\beta_n\}$ which defines $B(t)$ is an arbitrary decreasing sequence of positive numbers subject only to the conditions that $\lim \beta_n = 0$, $\sum \beta_n = +\infty$, and $\beta_0 = \beta_1 = 1$. Such a sequence can be found so that the expression $B(t)/B(t^2)$ oscillates as $t \uparrow 1$.

9. Problems

1. For sums of independent random variables on the integers with $p_{-1} = \frac{2}{3}$ and $p_2 = \frac{1}{3}$, show that if h is continuous and is of finite deviation, then $\alpha f = 0$.
2. Prove that if h and Th are both continuous, then h is T-continuous.
3. Show that for any normal chain $(I - P)C = b\alpha - I$. Identify the vector b.
4. What identities hold for $(I - P)K$ and $K(I - P)$ in a normal chain?
5. Let P be normal and let h be continuous. The balayage potential of h on a small ergodic set E is $B^E h - (\lambda^E h)\mathbf{1}$ (see Proposition 9-43). Let $\{E_k\}$ be an increasing sequence of finite sets with union S. Prove that the balayage potentials of h on E_k converge to $h - (\int h d\beta)\mathbf{1}$ on S.
6. Show that if P is a normal null chain with $B = B_e$ and if x is a point of the boundary with a neighborhood in S^* containing no other limit points, then $\lim_n \Pr[x_n \in E]$ exists for all sufficiently small neighborhoods E and is the same for all such sets E.
7. Prove that in a normal chain the elementary continuous functions are exactly the functions that can be written as the sum of a constant and a potential of finite support. [*Hint:* Use Theorems 9-15 and 11-18.]
8. Prove that if P is a normal null chain and if $P_{\cdot j}$ has only finitely many non-zero entries, then $\hat{\beta}_j = 0$. [*Hint:* Consider columns of $I - P$ as charges.]
9. Show that every T-continuous potential in a normal chain is the uniform limit of potentials of finite support.

Problems 10 to 15 refer to the null chain of Chapter 9, Problems 11 to 22. We shall use the notation and results there developed.

10. Show that the recurrent entrance boundary is empty and that $\beta_0 = 1$.
11. Show that $\hat{B}$ is the unit interval when parametrized by $t = \lim (x/n)$ and that
$$\hat{J}(t, z) = \binom{n-2}{x-1} t^{x-1}(1 - t)^{y-1}.$$
12. What is the form of the most general non-negative function regular for P except at 0?

13. In Problem 12, let $h_0 = 0$ and use Lebesgue measure on $\hat{B}$. Verify that h is regular except at 0.
14. Show that $\hat{\beta}$ is Lebesgue measure on $\hat{B}$.
15. Verify the interpretations of β and $\hat{\beta}$ as limits of rows of P^m and $\hat{P}^m$ by showing this: For either chain n is most probably large in the long run. For $\hat{P}$, the ratio x/n cannot change much in a few steps (if n is large), but for P a transition to 0 is always possible.

CHAPTER 12

INTRODUCTION TO RANDOM FIELDS

DAVID GRIFFEATH

1. Markov fields

One means of generalizing denumerable stochastic processes $\{x_n\}$ with time parameter set $\mathbb{N} = \{0, 1, \ldots\}$ is to consider **random fields** $\{x_t\}$, where t takes on values in an arbitrary countable parameter set T. Roughly, a random field with denumerable state space S is described by a probability measure μ on the space $\Omega = S^T$ of all configurations of values from S on the generalized time set T. In this chapter we discuss certain extensions of Markov chains, called **Markov fields**, which have been important objects of study in the recent development of probability theory. Only some of the highlights of this rich theory will be covered; we concentrate especially on the case $T = \mathbb{Z}$ = the integers, where the connections with classical Markov chain theory are deepest.

Proceeding to the formal definitions, assume as usual that the **state space** S is a countable set of integers including 0, but let the **time parameter set** T be any countable set. The **configuration space** $\Omega = S^T$ is the space of all functions ω from T to S. An element $\omega = \{\omega_t; t \in T\}$ of Ω is called a **configuration**, and is to be thought of as an assignment of values from S to the **sites** t of T. The **outcome function** x_t from Ω to S takes the configuration ω to its value ω_t at site t. Let $\mathscr{B}$ be the minimal complete σ-algebra with respect to which all the outcome functions x_t, $t \in T$, are measurable. In this context, we introduce the following definition.

Definition 12-1: A **random field** is given by $(\Omega, \mathscr{B}, \mu, \{x_t\})$, where μ is a probability measure on $(\Omega, \mathscr{B})$ such that

$$\Pr[x_t = i_t; t \in A] > 0$$

for all finite (non-empty) $A \subset T$ and arbitrary $i_t \in S$. (As always, $\Pr[p] = \mu(\{\omega | p\})$.)

As in the case of stochastic processes, we often identify a random field with its outcome functions $\{x_t\}$ (the remaining structure being understood). At other times it will be more convenient to think of μ as the random field. Our positivity assumption on cylinder probabilities ensures that all conditional probability statements are well-defined. The role of transition probabilities at this new level of generality is played by **characteristics**, which we define next.

Definition 12-2: Given a random field $\{x_t\}$, and finite (non-empty) sets A and Λ such that $A \subset \Lambda$, the (A, Λ)-**characteristic** is the real-valued function on Ω given by

$$\mu_A^\Lambda(\iota) = \Pr[x_t = i_t \text{ for all } t \in A \mid x_t = i_t \text{ for all } t \in \Lambda - A]$$

when evaluated at the configuration $\iota = \{i_t; t \in T\}$. For $a \in T$, we abbreviate $\mu_a^\Lambda = \mu_{\{a\}}^\Lambda$; the collection $\{\mu_a^\Lambda; a \in \Lambda \subset T\}$ is called the **local characteristics** of the random field.

Throughout this chapter A and Λ will *always* be *finite* subsets of T, even when not explicitly identified as such.

Our immediate objective is to formulate the notion of a **Markov field**. As motivation, we return briefly to the setting of Chapter 4.

Definition 12-3: A denumerable stochastic process $\{x_n\}$ satisfies the **two-sided Markov property** if

$$\Pr[x_n = i_n \mid x_k = i_k; k \in \{m, m+1, \ldots, M\} - \{n\}] = \begin{cases} \Pr[x_0 = i_0 \mid x_1 = i_1] & \text{if } 0 = m = n < M \\ \Pr[x_n = i_n \mid x_{n-1} = i_{n-1} \wedge x_{n+1} = i_{n+1}] & \text{if } 0 \le m < n < M \end{cases}$$

whenever $\Pr[x_k = i_k; m \le k \le M] > 0$.

Proposition 12-4: Any Markov chain $\{x_n\}$ satisfies the two-sided Markov property.

PROOF: Let π be the starting distribution, P the transition matrix for $\{x_n\}$. When $\Pr[x_k = i_k; m \le k \le M] > 0$, we consider the quantity $\Pr[x_n = i_n \mid x_k = i_k; k \in \{m, m+1, \ldots, M\} - \{n\}]$. If $0 = m = n < M$ this is simply $\Pr[x_0 = i_0 \mid x_1 = i_1]$ by reversibility of the Markov property (Proposition 6-44). Otherwise the above conditional probability may be evaluated as

$$\frac{(\pi P^m)_{i_m} \prod_{k=m}^{M-1} P_{i_k i_{k+1}}}{(\pi P^m)_{i_m} \left(\prod_{k=m}^{n-2} P_{i_k i_{k+1}}\right) P^{(2)}_{i_{n-1} i_{n+1}} \left(\prod_{k=n+1}^{M-1} P_{i_k i_{k+1}}\right)}$$

$$= P_{i_{n-1} i_n} P_{i_n i_{n+1}} / P^{(2)}_{i_{n-1} i_{n+1}}$$

$$= (\pi P^{n-1})_{i_{n-1}} P_{i_{n-1} i_n} P_{i_n i_{n+1}} / (\pi P^{n-1})_{i_{n-1}} P^{(2)}_{i_{n-1} i_{n+1}}$$

$$= \Pr[x_n = i_n \mid x_{n-1} = i_{n-1} \wedge x_{n+1} = i_{n+1}].$$

The two-sided Markov property, unlike the ordinary Markov property, generalizes to any parameter set T which has a **neighbor system**, i.e., a collection $\partial = \{\partial a; a \in T\}$ of *finite* subsets of T such that (i) $a \notin \partial a$, and (ii) $a \in \partial b$ if and only if $b \in \partial a$, $a, b \in T$. The sites $b \in \partial a$ are called the **neighbors** of a. We write $\bar{a} = \{a\} \cup \partial a$. Also, for $A \subset V$ let

$$\partial A = \{b \in T - A : b \in \partial a \quad \text{for some } a \in A\}; \quad \bar{A} = A \cup \partial A.$$

Definition 12-5: Let T have neighbor system ∂. The random field $\{x_t; t \in T\}$ is a **Markov field** (with respect to ∂) if

$$\mu_a^A = \mu_a^{\bar{a}} \quad \text{whenever } \bar{a} \subset A \subset T,\ A \text{ finite}.$$

We shall usually assume an underlying neighbor system for T, and simply refer to the Markov field $\{x_t\}$. Note that any Markov chain with strictly positive cylinder probabilities is a random field, where $T = \mathbb{N}$. The natural neighbor structure on $\mathbb{N}$ is $\partial 0 = 1$, and for $n \geq 1$, $\partial n = \{n - 1, n + 1\}$. In this case the Markov random field condition is precisely the two-sided Markov property. Proposition 12-4 shows that any Markov chain with positive cylinders may be considered as a Markov field on $\Omega = S^{\mathbb{N}}$. Later we will see that the classes of Markov processes with positive cylinders and Markov fields on $S^{\mathbb{N}}$ actually coincide.

A random field is called **finite** when T is a finite set. Such fields have an elementary theory, which will be developed in the next two sections. First, though, we note that the Markov field property simplifies somewhat when T is finite.

Proposition 12-6: Let $\{x_t\}$ be a finite random field. Then the following three conditions are equivalent:

(1) $\{x_t\}$ is a Markov field.
(2) $\mu_a^T = \mu_a^{\bar{a}}$ for all $a \in T$.
(3) $\mu_a^T(\iota) = \mu_a^T(\iota')$ whenever $a \in T$ and $i_t = i'_t$ for all $t \in \bar{a}$.

PROOF: When T is finite, (2) is simply the Markov field property with $\Lambda = T$, while the fact that $\mu_a^{\bar{a}}$ depends only on $t \in \bar{a}$ together with (2) implies $\mu_a^T(\iota) = \mu_a^{\bar{a}}(\iota) = \mu_a^{\bar{a}}(\iota') = \mu_a^T(\iota')$ whenever ι and ι' agree on $\bar{a}$. To see that (3) implies (2), fix $a \in T$ and let $\kappa = \{k_t; t \in T - \bar{a}\}$ be any prescription of values from S on $T - \bar{a}$. Denote by p, q, and r the statements

$$x_a = i_a, \quad x_t = i_t \text{ for all } t \in \partial a, \quad x_t = k_t \text{ for all } t \in T - \bar{a},$$

respectively. Then (3) asserts that

$$\Pr[p \mid q \wedge r_\kappa] = c \quad \text{for all } \kappa,$$

or equivalently,

$$\Pr[p \wedge q \wedge r_\kappa] = c \Pr[q \wedge r_\kappa].$$

Summing over all possible κ, we obtain $\Pr[p|q] = c$. Thus $\Pr[p \mid q \wedge r_\kappa] = \Pr[p|q]$, which is precisely (2) when $k_t = i_t$ for all $t \in T - \bar{a}$. Finally, to show that (2) implies (1) we choose $\bar{a} \subset \Lambda \subset T$ and $\iota = \{i_t\} \in \Omega$. Since $\mu_a^{\bar{a}}$ depends only on sites in $\bar{a}$, (2) yields

$$\begin{aligned}\Pr[x_t = i_t \quad &\text{for all } t \in \Lambda \wedge x_t = k_t \quad \text{for all } t \in T - \Lambda] \\ &= \Pr[x_t = i_t \quad \text{for all } t \in \Lambda - \{a\} \wedge x_t = k_t \quad \text{for all } t \in T - \Lambda]\, \mu_a^{\bar{a}}(\iota),\end{aligned}$$

where $\kappa = \{k_t; t \in T - \Lambda\}$ is any prescription of values from S on $T - \Lambda$. Summing over all possible κ, we conclude that $\mu_a^\Lambda(\iota) = \mu_a^{\bar{a}}(\iota)$.

2. Finite Gibbs fields

In this section we introduce an extremely useful representation for the measure μ of an arbitrary *finite* random field. The inspiration behind this approach (and hence most of its terminology) is derived from statistical mechanics, where random fields may be considered as equilibrium distributions for a variety of physical systems.

Definition 12-7: A **potential** U on a finite set T is a family $\{U_A(\iota); A \subset T\}$ of functions from Ω to the real line $\mathbb{R}$ with the property that $U_A(\iota) = U_A(\iota')$ whenever $i_t = i_t'$ for all $t \in A$, and such that $U_\varnothing = 0$. The **energy** H_U of the potential U is given by

$$H_U = \sum_{A \subset T} U_A.$$

U is said to be **normalized** if $U_A(\iota) = 0$ whenever $i_a = 0$ for some $a \in A$. When T has a neighbor system ∂, a set $C \subset T$ is called a **clique** if $b \in \partial a$ whenever $a, b \in C$, $a \neq b$, i.e. if every two distinct sites in C are neighbors. Let $\mathscr{C}$ be the class of all cliques in T. U is called a **neighbor potential** if $U_A = 0$ whenever $A \notin \mathscr{C}$.

Definition 12-8: A finite random field $\{x_t\}$ is a **Gibbs field with potential** U if

$$\mu(\{\iota\}) = z^{-1}e^{H_U(\iota)} \quad \iota \in \Omega,$$

where $z = \sum_{\iota\in\Omega} \exp\{H_U(\iota)\}$ is often called the **partition function**. Implicit is the assumption $z \neq +\infty$, which imposes a condition on U. If U is a neighbor potential, then $\{x_t\}$ is called a **neighbor Gibbs field**, and $H_U = \sum_{C\in\mathscr{C}} U_C$.

We remark that the potential and energy of random field theory should not be confused with those of Markov chain theory presented in earlier chapters. These terms have common origins in classical physics.

Example 12-9: T is sometimes called a **cubic lattice** if no clique in the neighbor system for T has more than two elements. The most important examples are subsets of the d-dimensional integer lattices $\mathbb{Z}^d$, where $\partial a = \{b \in T: |a - b| = 1\}$. When T is a finite cubic lattice and U is a neighbor potential, the energy function becomes

$$H_U = \sum_{a\in T} U_{\{a\}} + \sum_{\{a,b\}\in\mathscr{N}} U_{\{a,b\}} \quad \text{where } \mathscr{N} = \{\{a, b\}: b \in \partial a\}.$$

In this case U is called a **neighbor pair potential.**

Two lemmas prepare the way for the representation theorem for finite random fields. Given $\iota = \{i_t\}$ and $A \subset T$, the modification $\iota^A = \{i_t^A\}$ of ι has values

$$i_t^A = \begin{cases} i_t & \text{for } t \in A \\ 0 & \text{otherwise.} \end{cases}$$

We abbreviate $\iota^{A+a} = \iota^{A\cup\{a\}}$ when $a \notin A$ and $\iota^{A-a} = \iota^{A-\{a\}}$ when $a \in A$.

Lemma 12-10: If $\{x_t\}$ is a finite random field, $\iota = \{i_t\} \in \Omega$, $A \subset T$ and $a \notin A$, then

$$\frac{\mu_a^T(\iota^A)}{\mu_a^T(\iota^{A+a})} = \frac{\mu(\{\iota^A\})}{\mu(\{\iota^{A+a}\})}.$$

PROOF: $\mu_a^T(\iota^A) = \Pr[x_t = i_t^A \text{ for all } t \in T]/\Pr[x_t = i_t^A \text{ for all } t \in T - \{a\}]$. When we replace ι^A by ι^{A+a} the denominator is unchanged, since $i_t^A = i_t^{A+a}$ for $t \in T - \{a\}$.

Lemma 12-11 (Möbius inversion formula): Let Λ be finite, and let Φ and Ψ be real-valued set functions defined on all subsets of Λ. Then

$$\Phi(A) = \sum_{B\subset A} (-1)^{|A-B|}\Psi(B) \quad \text{for all } A \subset \Lambda$$

if and only if

$$\Psi(A) = \sum_{B \subset A} \Phi(B) \qquad \text{for all } A \subset \Lambda.$$

(Here $|A|$ denotes the cardinality of the set A.)

PROOF: Assume that the first condition holds. Then

$$\begin{aligned}\sum_{B \subset A} \Phi(B) &= \sum_{B \subset A} \sum_{D \subset B} (-1)^{|B-D|} \Psi(D) \\ &= \sum_{D \subset A} \Psi(D) \left[\sum_{E \subset A-D} (-1)^{|E|} \right] \qquad (E = B - D) \\ &= \Psi(A),\end{aligned}$$

since the bracketed sum above is 1 if $D = A$ and 0 otherwise. The opposite implication is verified by an analogous computation.

Theorem 12-12: Let $\{x_t\}$ be a finite random field with local characteristics $\{\mu_a^A\}$. Then $\{x_t\}$ is a Gibbs field with **canonical potential** V defined by $V_\varnothing = 0$, and for $A \neq \varnothing$,

$$\begin{aligned}V_A(\iota) &= \sum_{B \subset A} (-1)^{|A-B|} \ln \mu(\{\iota^B\}), \\ &= \sum_{B \subset A} (-1)^{|A-B|} \ln \mu_a^T(\{\iota^B\})\end{aligned}$$

for any fixed $a \in A$. Moreover, V is the unique normalized potential for $\{x_t\}$.

PROOF: Let $\mathbf{0}$ denote the configuration with a 0 at every site of T. Fix $\iota \in \Omega$. For $A \subset T$, define $\Psi(A) = \ln[\mu(\{\iota^A\})/\mu(\{\mathbf{0}\})]$ and $\Phi(A) = V_A(\iota)$, where V is the potential given by the first sum in the Theorem. When $A \neq \varnothing$ we have $\sum_{B \subset A} (-1)^{|A-B|} = 0$, and hence

$$\begin{aligned}\Phi(A) = V_A(\iota) &= \left[\sum_{B \subset A} (-1)^{|A-B|} \ln \mu(\{\iota^B\}) \right] - \ln \mu(\{\mathbf{0}\}) \left[\sum_{B \subset A} (-1)^{|A-B|} \right] \\ &= \sum_{B \subset A} (-1)^{|A-B|} \Psi(B).\end{aligned}$$

When $A = \varnothing$, $\Phi(\varnothing) = V_\varnothing(\iota) = 0 = \ln[\mu(\{\iota^\varnothing\})/\mu(\{\mathbf{0}\})] = \Psi(\varnothing)$, since $\iota^\varnothing = \mathbf{0}$ for any ι. Applying Lemma 12-11 with $\Lambda = T$ we conclude that

$$\ln \frac{\mu(\{\iota\})}{\mu(\{\mathbf{0}\})} = \ln \frac{\mu(\{\iota^T\})}{\mu(\{\mathbf{0}\})} = \sum_{B \subset T} V_B(\iota) = H_V(\iota),$$

and hence

$$\mu(\{\iota\}) = \mu(\{\mathbf{0}\}) e^{H_V(\iota)} = z^{-1} e^{H_V(\iota)}.$$

Thus $\{x_t\}$ is a Gibbs field with potential V. For any $a \in A \subset T$ we can write

$$V_A(\iota) = \sum_{B \subset A - a} (-1)^{|A-B|}[\ln \mu(\{\iota^B\}) - \ln \mu(\{\iota^{B+a}\})].$$

This shows that V is normalized, since $\iota^B = \iota^{B+a}$ whenever $i_a = 0$. By Lemma 12-10 the right hand side of the last equation may be rewritten as

$$\sum_{B \subset A - a} (-1)^{|A-B|}[\ln \mu_a^T(\iota^B) - \ln \mu_a^T(\iota^{B+a})]$$
$$= \sum_{B \subset A} (-1)^{|A-B|} \ln \mu_a^T(\iota^B),$$

which establishes the second expression for V in the statement of the Theorem. Finally, suppose that U is any normalized potential for $\{x_t\}$. Then $H_U(\mathbf{0}) = 0$ and $U_D(\iota^B) = 0$ unless $D \subset B$. Therefore

$$\ln \frac{\mu(\{\iota^B\})}{\mu(\{\mathbf{0}\})} = H_U(\iota^B) = \sum_{D \subset B} U_D(\iota^B) = \sum_{D \subset B} U_D(\iota^A)$$

whenever $B \subset A \subset T$. If we apply Lemma 12-11 with $\Lambda = A$, $\Phi(D) = U_D(\iota^A)$ and $\Psi(B) = \ln[\mu(\{\iota^B\})/\mu(\{\mathbf{0}\})]$, the conclusion is

$$U_A(\iota) = \sum_{B \subset A} (-1)^{|A-B|} \ln \frac{\mu(\{\iota^B\})}{\mu(\{\mathbf{0}\})} = V_A(\iota),$$

since the last sum is 0 when $A = \varnothing$, and otherwise

$$\ln \mu(\{\mathbf{0}\}) \textstyle\sum_{B \subset A} (-1)^{|A-B|} = 0.$$

Corollary 12-13: A finite random field is completely determined by its local characteristics $\{\mu_a^A\}$.

PROOF: The second equation in Theorem 12-12 shows that the canonical potential V for μ is determined by the local characteristics, and V determines μ.

Proposition 12-14: Let $\{x_t\}$ be a finite Gibbs field with potential U. Then the canonical potential V for $\{x_t\}$ is related to U by

$$V_A(\iota) = \sum_{B \subset A \subset D \subset T} (-1)^{|A-B|} U_D(\iota^B), \quad A \neq \varnothing.$$

PROOF: Since $\{x_t\}$ is Gibbs,

$$\ln \frac{\mu(\{\iota^B\})}{\mu(\{\iota^{B+a}\})} = \sum_{D \subset T} [U_D(\iota^B) - U_D(\iota^{B+a})]$$

for any $a \in T$ and $B \subset T - \{a\}$. Using the first equation for V in Theorem 12-12, it now follows that whenever $a \in A \subset T$,

$$\begin{aligned} V_A(\iota) &= \sum_{B \subset A - a} (-1)^{|A-B|} \ln \frac{\mu(\{\iota^B\})}{\mu(\{\iota^{B+a}\})} \\ &= \sum_{D \subset T} \sum_{B \subset A} (-1)^{|A-B|} U_D(\iota^{B \cap D}) \\ &= \sum_{D \subset T} \sum_{B_1 \subset D \cap A} \left[(-1)^{|A-B_1|} U_D(\iota^{B_1}) \left\{ \sum_{B_2 \subset \tilde{D} \cap A} (-1)^{|(\tilde{D} \cap A) - B_2|} . \right\} \right] \end{aligned}$$

The inner sum in the last expression is 0 unless $\tilde{D} \cap A = \varnothing$, i.e., unless $A \subset D$.

Corollary 12-15: Given two potentials U' and U'', let $\Delta_A = U'_A - U''_A$. U' and U'' determine the same finite Gibbs field if and only if

$$\sum_{B \subset A \subset D \subset T} (-1)^{|A-B|} \Delta_D(\iota^B) = 0 \quad \text{for every } A \neq \varnothing .$$

PROOF: Letting V' and V'' be the canonical potentials corresponding to U' and U'' respectively, Proposition 12-14 shows that the given equation is equivalent to $V'_A = V''_A$.

3. Equivalence of finite Markov and neighbor Gibbs fields

We now prove an important equivalence theorem which states that the finite Markov fields are precisely those for which the canonical V is a neighbor potential.

Theorem 12-16: Let $\{x_t\}$ be a finite random field with canonical potential V. Then $\{x_t\}$ is a Markov field if and only if V is a neighbor potential.

PROOF: Fix $a \in T$ and $\iota, \iota' \in \Omega$ such that $i_t = i'_t$ whenever $t \in \bar{a}$. Let ${}^s\iota$ and ${}^s\iota'$ be the modifications of ι and ι' respectively obtained by replacing the value at site a with $s \in S$. If V is a neighbor potential, then

$$\begin{aligned} H_V({}^s\iota) &= \sum_{A \subset T - \{a\}} V_A({}^s\iota) + \sum_{a \in A \subset \bar{a}} V_A({}^s\iota) + \sum_{a \in A \not\subset \bar{a}} V_A({}^s\iota) \\ &= \qquad \textstyle\sum_1 \qquad + \qquad \textstyle\sum_2(s) \qquad + \qquad 0, \end{aligned}$$

where $\sum_1$ is **independent** of s. Let $\sum'_1$ and $\sum'_2(s)$ be the corresponding sums when ${}^{s}\iota$ is replaced by ${}^{s}\iota'$. Since ${}^{s}\iota$ and ${}^{s}\iota'$ agree on $\bar{a}$, we have $\sum_2(s) = \sum_2'(s)$ for all s. Thus

$$\mu_a^T({}^{s_0}\iota) = \frac{e^{\Sigma_1 + \Sigma_2(s_0)}}{\sum_{s \in S} e^{\Sigma_1 + \Sigma_2(s)}} = \frac{e^{\Sigma_2(s_0)}}{\sum_{s \in S} e^{\Sigma_2(s)}} = \frac{e^{\Sigma_1' + \Sigma_2'(s_0)}}{\sum_{s \in S} e^{\Sigma_1' + \Sigma_2'(s)}} = \mu_a^T({}^{s_0}\iota')$$

for any $s_0 \in S$. Taking $s_0 = i_a = i_a'$ we have verified (3) of Proposition 12-6; this shows that $\{x_t\}$ is Markov. Conversely, if $\{x_t\}$ is Markov, we claim that the canonical potential V is a neighbor potential. To see this, choose $a, b \in A \subset T$ such that $b \notin \bar{a}$. Expand V as

$$V_A(\iota) = \sum_{B \subset A - \{a,b\}} (-1)^{|A-B|} \ln\left[\frac{\mu_a^T(\iota^B)/\mu_a^T(\iota^{B+a})}{\mu_a^T(\iota^{B+b})/\mu_a^T(\iota^{B+a+b})}\right].$$

Since $b \notin \bar{a}$, Proposition 12-6 shows that $\mu_a^T(\iota^{B+b}) = \mu_a^T(\iota^B)$ and $\mu_a^T(\iota^{B+a+b}) = \mu_a^T(\iota^{B+a})$, yielding the desired result.

Corollary 12-17: There is a one-to-one correspondence between the local characteristics $\{\mu_a^T; a \in T\}$ for finite Markov fields and normalized (canonical) potentials V for finite neighbor Gibbs fields, given by

$$V_A(\iota) = \sum_{B \subset A} (-1)^{|A-B|} \ln \mu_a^T(\iota^B) \quad a \in A \in \mathscr{C}$$
$$(\equiv 0 \qquad\qquad A \notin \mathscr{C}),$$

and

$$\mu_a^T(\iota) = z^{-1} \exp\left\{\sum_{a \in B \subset \bar{a}} V_B(\iota)\right\},$$

where z is the appropriate normalizing constant.

Proof: If $\{x_t\}$ is Markov, then $V_A(\iota) \equiv 0$ for $A \notin \mathscr{C}$, by the last theorem. The rest of the first equation above was proved in Theorem 12-12, and the second equation was derived in Theorem 12-16.

Another consequence of Theorem 12-16 is an alternative formulation for finite Markov fields.

Corollary 12-18: A finite random field $\{x_t\}$ is Markov if and only if

$$\mu_A^A(\iota) = \mu_A^{\bar{A}}(\iota) \quad \text{whenever } \bar{A} \subset \Lambda \subset T.$$

Proof: Let $\{x_t\}$ be Markov, with canonical neighbor potential V.

A computation similar to the one in the proof of Theorem 12-16 shows that

$$\mu_A^A(\iota) = z^{-1} \exp\left\{ \sum_{B \subset \bar{A}: B \cap A \neq \varnothing} V_B(\iota) \right\} = \mu_A^A(\iota')$$

whenever $i_t = i_t'$ for all $t \in \bar{A}$. The claim now follows from the straightforward generalization of Proposition 12-6 obtained by replacing $\{a\}$ with A.

Example 12-19: The Markov process case. Let $\{x_n; 0 \leq n \leq N\}$ be a denumerable Markov process viewed from time 0 to time N. Suppose that $\{x_n\}$ has starting distribution π and one-step transition matrix P_n at time n, where

$$P_{n,ij} = \Pr[x_{n+1} = j \mid x_n = i].$$

If π and all the P_n are strictly positive, then $\{x_n\}$ is a Markov field with neighbor system ∂, where $\partial 0 = \{1\}$, $\partial N = \{N - 1\}$, and $\partial n = \{n - 1, n + 1\}$ for $1 \leq n < N$. The local characteristics for $\{x_t\}$ are given by

$$\mu_0^T(\iota) = \pi(i_0)P_{0,i_0 i_1} \Big/ \sum_{i \in S} \pi(i)P_{0,i i_1}$$

$$\mu_n^T(\iota) = \frac{P_{n-1,i_{n-1}i_n}P_{n,i_n i_{n+1}}}{\sum_{i \in S} P_{n-1,i_{n-1}i}P_{n,i i_{n+1}}} \qquad 1 \leq n < N,$$

$$\mu_N^T(\iota) = P_{N-1,i_{N-1}i_N}$$

The canonical potential for the process is then given by

$$V_{\{n\}}(\iota) = \ln \mu_n^T(\iota)/\mu_n^T(\mathbf{0}),$$

$$V_A(\iota) = \ln \frac{\mu_n^T(\iota^A)\mu_n^T(\mathbf{0})}{\mu_n^T(\iota^{\{n-1\}})\mu_n^T(\iota^{\{n\}})} \qquad A = \{n - 1, n\},$$

$$V_A(\iota) = 0 \qquad \text{otherwise},$$

$0 \leq n \leq N$, and $\{x_n\}$ is a neighbor Gibbs field with normalized potential V. Conversely, suppose that $\{x_n; 0 \leq n \leq N\}$ is a Markov field with the above neighbor system ∂. A routine calculation using the explicit representation for μ in terms of V shows that

$$\Pr[x_{n+1} = j \mid x_n = i_n, \ldots, x_0 = i_0] = \Pr[x_{n+1} = j \mid x_n = i_n]$$

whenever $0 \leq n < N$, so $\{x_n\}$ is a Markov process. On a finite time parameter set we see that the one and two sided Markov properties are equivalent, so that Markov fields are precisely the Markov processes with positive cylinder probabilities.

4. Markov fields and neighbor Gibbs fields: the infinite case

For the remainder of the chapter, T will be a countably infinite set with some neighbor system ∂. A **neighbor potential** U on T is defined just as in the previous sections. But we can no longer define the probability measure μ for a neighbor Gibbs field with potential U explicitly; instead we must make use of the local characteristics.

Definition 12-20: An infinite random field $\{x_t\}$ is a **neighbor Gibbs field with neighbor potential** U if

$$\mu_a^\Lambda(\iota) = z^{-1} \exp\left\{ \sum_{a \in B \subset \bar{a}} U_B(\iota) \right\} \quad \text{for all finite } \Lambda \supset \bar{a},$$

where z is the appropriate normalizing constant. If $s \in S$ and ${}^s\iota$ is the modification obtained by replacing i_a with s, then

$$z = \sum_{s \in S} \exp\left\{ \sum_{a \in B \subset \bar{a}} U_B({}^s\iota) \right\}.$$

Theorem 12-21: Let $\{x_t\}$ be an infinite random field. Then $\{x_t\}$ is a neighbor Gibbs field if and only if it is a Markov field.

Proof: Suppose $\{x_t\}$ is neighbor Gibbs. From the definition we see that $\mu_a^\Lambda(\iota) = \mu_a^\Lambda(\iota')$ whenever $\Lambda \supset \bar{a}$ and ι' agrees with ι on $\bar{a}$. Just as in Proposition 12-6, this implies that $\mu_a^\Lambda(\iota) = \mu_a^{\bar{a}}(\iota)$ for all finite $\Lambda \supset \bar{a}$, so $\{x_t\}$ is Markov. Conversely, if $\{x_t\}$ is Markov we define a potential V by $V_\varnothing = 0$ and

$$V_\Lambda(\iota) = \sum_{B \subset \Lambda} (-1)^{|\Lambda - B|} \ln \mu_a^\Lambda(\iota^B) \quad a \in \Lambda.$$

The argument of Theorem 12-16 shows that V is a neighbor potential, and the only normalized potential determined by the $\{\mu_a^\Lambda\}$. An application of the Möbius inversion formula shows that V satisfies the conditions in Definition 12-20, so $\{x_t\}$ is a neighbor Gibbs field. We remark here that Corollary 12–18 also holds for infinite fields; the proof is routine.

Let $\mathscr{G}_V = \mathscr{G}_V(T)$ be the class of neighbor Gibbs fields on the infinite set T with canonical potential V. The bijection of Corollary 12-17, which carries over to the *finite* subsets of T, shows that we may consider equivalently the class of Markov fields with local characteristics $\{\mu_a^\Lambda\}$ corresponding to V. Since we will be considering many fields simultaneously, the elements of $\mathscr{G}_V$ will be thought of as measures μ

governing $\{x_t\}$. In the finite case we have seen that $\mathscr{G}_V$ always contains a single field. When T is infinite this need not be so; indeed, there are exactly three possibilities:

$$\text{(i) } \mathscr{G}_V = \varnothing \qquad \text{(ii) } |\mathscr{G}_V| = 1 \qquad \text{(iii) } |\mathscr{G}_V| = \infty.$$

This follows from the fact that $\mathscr{G}_V$ is **convex**, i.e., if $\mu_1, \mu_2 \in \mathscr{G}_V$ and $0 \le \alpha \le 1$, then $\alpha\mu_1 + (1 - \alpha)\mu_2 \in \mathscr{G}_V$. Examples of (i)–(iii) will be presented in Section 5.

Definition 12-22: When $|\mathscr{G}_V| = \infty$ we say that there is **phase multiplicity** (or **phase transition**) for V. A measure $\mu \in \mathscr{G}_V$ is **extreme** if whenever $\mu = \alpha\mu_1 + (1 - \alpha)\mu_2$, $\mu_1, \mu_2 \in \mathscr{G}_V$, $0 < \alpha < 1$, then $\mu_1 = \mu_2 = \mu$. The class of extreme elements of $\mathscr{G}_V$ is denoted by $\mathscr{E}_V$.

Since $\mathscr{G}_V$ is convex, in the case of phase multiplicity one would hope for an integral representation in terms of $\mathscr{E}_V$. We will obtain such a representation, along with a number of other results on the structure of $\mathscr{G}_V$, by connecting neighbor Gibbs fields with Martin boundaries for certain Markov chains. The remainder of this section is devoted to the study of general structural properties of $\mathscr{G}_V$ with the aid of the boundary theory developed in Chapter 10.

To begin, we fix a neighbor potential V, assume $\mathscr{G}_V \neq \varnothing$, and choose a **reference measure** $\nu \in \mathscr{G}_V$. Also we fix an increasing sequence $\{\Lambda(n), n = 0, 1, \ldots\}$ of finite subsets of T, such that $\overline{\Lambda(n)} \subset \Lambda(n + 1)$ and $\Lambda(n) \uparrow T$ as $n \to \infty$. Write $K(0) = \Lambda(0), K(n) = \Lambda(n) - \Lambda(n - 1)$ for $n \ge 1$. Then any configuration $\iota \in \Omega$ may be thought of as a sequence of subconfigurations $(\kappa^0, \kappa^1, \ldots)$, where $\kappa^n = \{k_t^n; t \in K(n))$ satisfies $k_t^n = i_t$ when $t \in K(n)$. For brevity's sake we denote $\{\omega \mid x_t(\omega) = k_t^n \text{ for all } t \in K(n)\}$ simply as $[\kappa^n]$. Similarly, $[\kappa^0, \kappa^1, \ldots, \kappa^n]$ means $\{\omega \mid x_t(\omega) = k_t^r \text{ for all } t \in K(r), 0 \le r \le n\}$, and so forth. Also, we write $\nu(A|B) = \nu(A \cap B)/\nu(B)$ when convenient (and, of course $\nu(B) > 0$). With these notations in effect, observe the following key property of neighbor Gibbs fields.

Proposition 12-23: If $\iota = (\kappa^0, \kappa^1, \ldots) \in \Omega$, then

$$\nu([\kappa^{m+1}, \ldots, \kappa^n] \mid [\kappa^0, \ldots, \kappa^m]) = \nu([\kappa^{m+1}, \ldots, \kappa^n] \mid [\kappa^m]) \qquad 1 \le m < n < \infty.$$

PROOF: If $\{\nu_A^A\}$ are the characteristics of ν, then the left hand side divided by the right hand side is equal to $\nu_{\Lambda(m-1)}^{\Lambda(n)}(\iota)/\nu_{\Lambda(m-1)}^{\Lambda(m)}(\iota) = 1$, since the numerator and denominator of this last quotient both equal $\nu_{\Lambda(m-1)}^{\overline{\Lambda(m-1)}}(\iota)$ by the Markov field property.

The above result reveals a Markovian structure for ν which can be exhibited explicitly in terms of a Markov chain. The states for the chain are all possible pairs (n, κ^n), $n \geq 0$, $\kappa^n \in S^{K(n)}$. The transition matrix is

$$P_{(n,\kappa^n)(n+1,\kappa^{n+1})} = \nu([\kappa^{n+1}] \mid [\kappa^n]).$$

Proposition 12-23 and a simple induction show that the n-step transition matrix is given by

$$P^{(n)}_{(m,\kappa^m)(m+n,\kappa^{m+n})} = \nu([\kappa^{m+n}] \mid [\kappa^n]).$$

If the initial distribution is $\pi_{(0,\kappa^0)} = \nu([\kappa^0])$, and $\{y_n\}$ denotes the resulting chain, then we obtain the simple relationship

$$\Pr_\pi[y_n = (n, \kappa^n)] = \nu([\kappa^n]).$$

Next, we connect P-regular functions with fields in $\mathscr{G}_V$. A lemma will be useful for this purpose.

Lemma 12-24: If $\mu \in \mathscr{G}_V$, $\iota = (\kappa^0, \kappa^1, \ldots) \in \Omega$, then

$$\frac{\mu([\kappa^0, \kappa^1, \ldots, \kappa^n])}{\nu([\kappa^0, \kappa^1, \ldots, \kappa^n])} = \frac{\mu([\kappa^m, \ldots, \kappa^n])}{\nu([\kappa^m, \ldots, \kappa^n])} \qquad 1 \leq m \leq n < \infty.$$

PROOF: The left hand expression may be rewritten as

$$\frac{\mu([\kappa^m, \ldots, \kappa^n])}{\nu([\kappa^m, \ldots, \kappa^n])} \frac{\mu^{A(n)}_{A(m-1)}(\iota)}{\nu^{A(n)}_{A(m-1)}(\iota)},$$

and the two characteristics agree, being identical functionals of the potential V.

As in Section 10-6, we call a non-negative P-regular function h **normalized** if

$$\pi h = \sum_{\kappa^0 \in S^{K(0)}} \pi_{(0,\kappa^0)} h_{(0,\kappa^0)} = 1,$$

and **minimal normalized** if it cannot be written as a non-trivial convex combination of two distinct non-negative regular functions.

Theorem 12-25: There is a one-to-one correspondence between non-negative normalized P-regular functions h and neighbor Gibbs fields $\mu \in \mathscr{G}_V$ given by

$$h_{(n,\kappa^n)} = \mu([\kappa^n])/\nu([\kappa^n])$$

and

$$\mu([\kappa^0, \ldots, \kappa^n]) = \nu([\kappa^0, \ldots, \kappa^n]) h_{(n,\kappa^n)},$$

$n \geq 0$, $\kappa^r \in S^{K(r)}$. Moreover, h is minimal normalized if and only if μ is in $\mathscr{E}_V$.

PROOF: Given $\mu \in \mathscr{G}_V$, define h according to the first equation in the theorem. By Lemma 12-24,

$$\sum_{\kappa^{n+1}} P_{(n,\kappa^n)(n+1,\kappa^{n+1})} h_{(n+1,\kappa^{n+1})} = \sum_{\kappa^{n+1}} \frac{\nu([\kappa^n, \kappa^{n+1}])}{\nu([\kappa^n])} \frac{\mu([\kappa^n, \kappa^{n+1}])}{\nu([\kappa^n, \kappa^{n+1}])}$$
$$= \sum_{\kappa^{n+1}} \frac{\mu([\kappa^n, \kappa^{n+1}])}{\nu([\kappa^n])}$$
$$= \frac{\mu([\kappa^n])}{\nu([\kappa^n])} = h_{(n,\kappa^n)},$$

so h is P-regular. Clearly, h is also non-negative and normalized. Conversely, let h be any non-negative normalized P-regular function. We claim that the second equation of the theorem prescribes cylinder probabilities for a unique measure $\mu \in \mathscr{G}_V$ determined by h. Note first that h must be strictly positive. To see this, write

$$\sum_{\kappa^{n+1}} \nu([\kappa^{n+1}]) h_{(n+1,\kappa^{n+1})} = \pi P^{n+1} h = 1,$$

and

$$h_{(n,\kappa^n)} = (Ph)_{(n,\kappa^n)} = \sum_{\kappa^{n+1}} \frac{\nu([\kappa^n, \kappa^{n+1}])}{\nu([\kappa^n])} h_{(n+1,\kappa^{n+1})}.$$

The first equation above implies that $h_{(n+1,\kappa^{n+1})} > 0$ for some κ^{n+1}, and hence the second shows that $h_{(n,\kappa^n)} > 0$ for all (n, κ^n). Thus all cylinder measures for μ are evidently positive. Next, we use Proposition 12-23 to compute

$$\sum_{\kappa^{n+1}} \mu([\kappa^0, \ldots, \kappa^{n+1}]) = \nu([\kappa^0, \ldots, \kappa^n]) \sum_{\kappa^{n+1}} \frac{\nu([\kappa^0, \ldots, \kappa^{n+1}])}{\nu([\kappa^0, \ldots, \kappa^n])} h_{(n+1,\kappa^{n+1})}$$
$$= \nu([\kappa^0, \ldots, \kappa^n])(Ph)_{(n,\kappa^n)}$$
$$= \nu([\kappa^0, \ldots, \kappa^n]) h_{(n,\kappa^n)} = \mu([\kappa^0, \ldots, \kappa^n]).$$

This shows that the measures $\mu([\kappa^0, \ldots, \kappa^n])$ on $S^{\Lambda(n)}$ are consistent for $n \geq 0$, and also by induction that

$$\sum_{\kappa^0,\ldots,\kappa^n} \sum_{\kappa^{n+1}} \mu([\kappa^0, \ldots, \kappa^{n+1}]) = \sum_{\kappa^0,\ldots,\kappa^n} \mu([\kappa^0, \ldots, \kappa^n]) = 1$$

for all $n \geq 0$, since $\sum_{\kappa^0} \mu([\kappa^0]) = \pi h = 1$. By the usual extension theorems we obtain a unique μ with the desired cylinder probabilities. To see that $\mu \in \mathscr{G}_V$, fix finite A, Λ, with $\bar{A} \subset \Lambda$, and choose n large enough that $\Lambda \subset \Lambda(n)$. Let $\iota = \{i_t; t \in T\} = (\kappa^0, \kappa^1, \ldots) \in \Omega$. Introduce the sets

$$[\iota_A] = \{\omega \mid \omega_t = i_t \text{ for all } t \in A\},$$
$$[\iota_{\Lambda - A}] = \{\omega \mid \omega_t = i_t \text{ for all } t \in \Lambda - A\},$$
$$[\psi] = \{\omega \mid \omega_t = \psi_t \text{ for all } t \in \Lambda(n-1) - \Lambda\} \quad \text{where } \psi \in S^{\Lambda(n-1)-\Lambda}.$$

Then by construction,

$$\mu([\iota_A]) = \sum_{\psi} \sum_{\kappa^n} \nu([\iota_A] \cap [\psi] \cap [\kappa^n]) h_{(n,\kappa^n)}$$

and

$$\mu([\iota_{\Lambda - A}]) = \sum_{\psi} \sum_{\kappa^n} \nu([\iota_{\Lambda - A}] \cap [\psi] \cap [\kappa^n]) h_{(n,\kappa^n)},$$

where ψ is summed over $S^{\Lambda(n-1)-\Lambda}$, κ^n over $S^{K(n)}$. If ι^{ψ,κ^n} denotes the modification of ι obtained by replacing its values on $\Lambda(n-1) - \Lambda$ with ψ, and its values on $K(n)$ with κ^n, then $\nu_A^{\Lambda(n)}(\iota^{\psi,\kappa^n}) = \nu_A^{\bar{A}}(\iota)$ for all ψ and κ^n, since $\nu \in \mathscr{G}_V$. Thus

$$\begin{aligned}\mu([\iota_A]) &= \sum_{\psi} \sum_{\kappa^n} \nu_A^{\Lambda(n)}(\iota^{\psi,\kappa^n}) \nu([\iota_{\Lambda - A}] \cap [\psi] \cap [\kappa^n]) h_{(n,\kappa^n)} \\ &= \nu_A^{\bar{A}}(\iota) \mu([\iota_{\Lambda - A}]),\end{aligned}$$

and so $\mu_A^{\Lambda}(\iota) = \nu_A^{\bar{A}}(\iota) = \nu_A^{\Lambda}(\iota)$. Hence $\mu \in \mathscr{G}_V$. To check the one-to-one correspondence, let $\mathscr{H}$ denote the normalized non-negative P-regular functions. We have defined mappings $\rho\colon \mathscr{G}_V \to \mathscr{H}$ and $\sigma\colon \mathscr{H} \to \mathscr{G}_V$. One easily verifies that $\sigma(\rho(\mu)) = \mu$ and $\rho(\sigma(h)) = h$, as desired. Finally, ρ is obviously a cone homomorphism in the sense that

$$\rho(\alpha\mu_1 + (1 - \alpha)\mu_2) = \alpha\rho(\mu_1) + (1 - \alpha)\rho(\mu_2),$$

$0 \le \alpha \le 1$, $\mu_1, \mu_2 \in \mathscr{G}_V$. This implies the last assertion in the theorem, and the proof is complete,

Using the Martin boundary theory for the chain P with starting distribution π, as constructed from the reference measure $\nu \in \mathscr{G}_V$, we will now derive several structure properties of $\mathscr{G}_V$.

Definition 12-26: When $\mu \in \mathscr{G}_V$ and $n \ge 1$, set

$$\mu^{\kappa^n}([\kappa^0, \ldots, \kappa^{n-1}]) = \mu_{\Lambda(n-1)}^{\Lambda(n)}(\iota), \quad \iota = (\kappa^0, \kappa^1, \ldots) \in \Omega.$$

For each fixed κ^n on $K(n)$, $\mu^{\kappa^n}(\cdot)$ defines a finite random field on $S^{\Lambda(n-1)}$. The fields μ^{κ^n} are said to have **thermodynamic limit** μ_∞, $t\text{-}\lim_{n\to\infty} \mu^{\kappa^n} = \mu_\infty$ in notation, if there is a measure μ_∞ on Ω such that

$$\lim_{n\to\infty} \mu^{\kappa^n}([\kappa^0, \ldots, \kappa^m]) = \mu_\infty([\kappa^0, \ldots, \kappa^m])$$

for all $m \geq 0$ and all configurations $(\kappa^0, \ldots, \kappa^m)$ on $S^{A(m)}$.

Theorem 12-27:

$$\nu\left(\left\{\omega \mid t\text{-}\lim_{n\to\infty} \nu^{\kappa^n(\omega)} \in \mathscr{E}_V\right\}\right) = 1,$$

where $\kappa^n(\omega)$ is the configuration of ω restricted to $K(n)$. In particular, $\mathscr{E}_V \neq \varnothing$ whenever $\mathscr{G}_V \neq \varnothing$.

PROOF: Let B_e comprise the extreme points of the Martin boundary for P started with π, and let λ denote harmonic measure. Theorem 10-41 applied to the constant regular function $h = 1$ shows that $\Pr_\pi[y_v \in B_e] = \lambda(B_e) = 1$. Since $\{y_n\}$ visits each state (m, κ^m) at most once, the Martin kernel is given by

$$K((m, \kappa^m), (n, \kappa^n)) = P^{(n-m)}_{(m,\kappa^m)(n,\kappa^n)} \Big/ \sum_{\kappa^0} \pi(\kappa^0) P_{(0,\kappa^0)(n,\kappa^n)}, \quad n \geq m,$$

$(=0$ otherwise). Thus $y_v(\omega) \in B_e$ means that

$$K((m, \kappa^m), x) = \lim_{n\to\infty} \frac{\nu([\kappa^m, \kappa^n(\omega)])}{\nu([\kappa^m])\nu([\kappa^n(\omega)])}$$

exists for every (m, κ^m), and is a minimal regular function of (m, κ^m). By the last theorem, $K(\cdot, x)$ is minimal regular if and only if $\nu([\kappa^0, \ldots, \kappa^m])K((m, \kappa^m), x) = \mu_\infty([\kappa^0, \ldots, \kappa^m])$ for a (unique) $\mu_\infty \in \mathscr{E}_V$. In this case, we deduce from Proposition 12-23 that

$$\frac{\mu_\infty([\kappa^0, \ldots, \kappa^m])}{\nu([\kappa^0, \ldots, \kappa^m])} = \frac{\lim_{n\to\infty} \nu([\kappa^0, \ldots, \kappa^m] \mid [\kappa^n(\omega)])}{\nu([\kappa^0, \ldots, \kappa^m])}$$

$$= \frac{\lim_{n\to\infty} \nu^{\kappa^n(\omega)}([\kappa^0, \ldots, \kappa^m])}{\nu([\kappa^0, \ldots, \kappa^m])}$$

for all m, κ^r. This shows that $t\text{-}\lim_{n\to\infty} \nu^{\kappa^n(\omega)} \in \mathscr{E}_V$ on $\{\omega : y_v(\omega) \in B_e\}$. Using the reference measure $\nu \in \mathscr{G}_V$, we have produced a set of measures in $\mathscr{E}_V$ which has ν-measure 1.

Theorem 12-28: The elements of $\mathscr{E}_V$ are in one-to-one correspondence with those of B_e. If $\mu^x \in \mathscr{E}_V$ corresponds to $x \in B_e$, then there is a bijection between the probability measures on B_e and the neighbor Gibbs fields in $\mathscr{G}_V$, given by the equations

$$\mu = \int_{x\in B_e} \mu^x d\lambda^\mu(x)$$

and

$$\lambda^{\mu}(E) = \mu\left(\left\{\omega \mid t\text{-}\lim_{n\to\infty} \mu^{\kappa^n(\omega)} = \mu^x \quad \text{for some } x \in E\right\}\right),$$

$$E \quad \text{Borel in } B_e.$$

PROOF: We know from Chapter 10 that B_e is in one-to-one correspondence with the class $\mathscr{H}_e$ of minimal normalized regular functions, while Theorem 12-25 gives a bijection between $\mathscr{H}_e$ and $\mathscr{E}_V$. Hence there is a one-to-one correspondence $x \leftrightarrow \mu^x$ between B_e and $\mathscr{E}_V$. For $\mu \in \mathscr{G}_V$, apply Theorem 10-41 to $\rho(\mu) \in \mathscr{H}$, and use Lemma 12-24, to get

$$\frac{\mu([\kappa^0, \ldots, \kappa^m])}{\nu([\kappa^0, \ldots, \kappa^m])} = \int_{x \in B_e} \frac{\mu^x([\kappa^0, \ldots, \kappa^m])}{\nu([\kappa^0, \ldots, \kappa^m])}\, d\lambda^{\rho(\mu)}(x),$$

where λ^h is harmonic measure for the h-process. A routine computation using the explicit form of the Martin kernel, derived in the proof of the previous theorem, shows that

$$\begin{aligned}\lambda^{\rho(\mu)}(E) &= \mu\left(\left\{\omega \mid t\text{-}\lim_{n\to\infty} \nu^{\kappa^n(\omega)} = \mu^x \quad \text{for some } x \in E\right\}\right) \\ &= \mu\left(\left\{\omega \mid t\text{-}\lim_{n\to\infty} \mu^{\kappa^n(\omega)} = \mu^x \quad \text{for some } x \in E\right\}\right),\end{aligned}$$

(The $\rho(\mu)$ process changes the reference measure to μ, and $\mu^{\kappa^n} = \nu^{\kappa^n}$ since both random fields are defined in terms of the same characteristics.) Setting $\lambda^{\mu} = \lambda^{\rho(\mu)}$, the theorem follows.

Corollary 12-29: If $\mu \in \mathscr{E}_V$, then there is an $\iota = (\kappa^0, \kappa^1, \ldots) \in \Omega$ such that

$$\mu = t\text{-}\lim_{n\to\infty} \mu^{\kappa^n}.$$

PROOF: The uniqueness of the integral representation implies that if $\mu = \mu^x \in \mathscr{E}_V$, then $\lambda^{\mu}(\{x\}) = \mu(\{\omega: t\text{-}\lim_{n\to\infty} \mu^{\kappa^n(\omega)} = \mu\}) = 1$. The desired ι may therefore be any configuration from a full ω-set with respect to μ.

The entire development of this section has proceeded on the assumption that $\mathscr{G}_V$ is not empty. The problem of determining from the potential V just *when* $\mathscr{G}_V$ is not empty turns out to be a difficult one if the state space S is countably infinite. In the case where T is the integers, we shall have more to say about this later. When S is *finite* on the other hand, it will now be proved that $\mathscr{G}_V$ is *always* non-empty.

Theorem 12-30: If S is finite, then $\mathscr{G}_V \neq \varnothing$. Moreover, the limits

$$\mu^+([\kappa^0, \ldots, \kappa^m]) = \lim_{n \to \infty} \max_{\kappa^n \in S^{K(n)}} \mu^{\kappa^n}([\kappa^0, \ldots, \kappa^m])$$

and

$$\mu^-([\kappa^0, \ldots, \kappa^m]) = \lim_{n \to \infty} \min_{\kappa^n \in S^{K(n)}} \mu^{\kappa^n}([\kappa^0, \ldots, \kappa^m])$$

exist for all $m \geq 0$, $(\kappa^0, \ldots, \kappa^m) \in S^{\Lambda(m)}$, and there is phase multiplicity for V if and only if

$$\mu^+([\kappa^0, \ldots, \kappa^m]) \neq \mu^-([\kappa^0, \ldots, \kappa^m]) \quad \text{for some } m \text{ and } (\kappa^0, \ldots, \kappa^m).$$

(Recall that $\mu^{\kappa^n}([\kappa^0, \ldots, \kappa^m])$ is a certain characteristic which is completely and uniquely determined by V.)

To prove the theorem, we first need the following lemma.

Lemma 12-31: For given $\mu \in \mathscr{G}_V$, $m > 0$ and fixed $(\kappa^0, \ldots, \kappa^m) \in S^{\Lambda(m)}$, abbreviate

$$\mu_n^+ = \max_{\kappa^n \in S^{K(n)}} \mu^{\kappa^n}([\kappa^0, \ldots, \kappa^m]),$$

$$\mu_n^- = \min_{\kappa^n \in S^{K(n)}} \mu^{\kappa^n}([\kappa^0, \ldots, \kappa^m]).$$

Then

(1) $0 < \mu_n^- \leq \mu([\kappa^0, \ldots, \kappa^m]) \leq \mu_n^+$ for each $n > m$, and
(2) μ_n^- is increasing and μ_n^+ is decreasing as $n \to \infty$.

Proof: (1) μ_n^- is a minimum over a finite set of strictly positive probabilities, hence strictly positive. When $n > m$,

$$\begin{aligned} \mu([\kappa^0, \ldots, \kappa^m]) &= \sum_{\kappa^n} \mu^{\kappa^n}([\kappa^0, \ldots, \kappa^m])\mu([\kappa^n]) \\ &\geq \mu_n^- \sum_{\kappa^n} \mu([\kappa^n]) = \mu_n^-, \end{aligned}$$

and an analogous estimate establishes the remaining inequality.
(2) For any $n > m$ and $\kappa^{n+1} \in K(n+1)$,

$$\begin{aligned} \mu^{\kappa^{n+1}}([\kappa^0, \ldots, \kappa^m]) &= \sum_{\kappa^n} \mu([\kappa^0, \ldots, \kappa^m] \mid [\kappa^n, \kappa^{n+1}])\mu([\kappa^n] \mid [\kappa^{n+1}]) \\ &= \sum_{\kappa^n} \mu^{\kappa^n}([\kappa^0, \ldots, \kappa^m])\mu([\kappa^n] \mid [\kappa^{n+1}]) \\ &\geq \mu_n^- \sum_{\kappa^n} \mu([\kappa^n] \mid [\kappa^{n+1}]) = \mu_n^- \end{aligned}$$

This shows that μ_n^- is increasing with n; a similar estimate proves that μ_n^+ is decreasing.

PROOF OF THEOREM 12-30: The limits μ^+ and μ^- are well-defined by the monotonicity established in the lemma. We now show that for any given configuration $(\kappa^0, \ldots, \kappa^m)$ on $\Lambda(m)$, there is a random field $\underline{\mu} \in \mathscr{G}_V$ such that $\underline{\mu}([\kappa^0, \ldots, \kappa^m]) = \mu^-([\kappa^0, \ldots, \kappa^m])$. Let $\underline{\kappa}^n$ denote the configuration on $K(n)$ for which the minimum value μ_n^- is attained, and define measures $\underline{\mu}_n$, $n \geq 1$, on Ω by

$$\underline{\mu}_n\colon \begin{cases} \underline{\mu}_n([\iota]) = \mu^{\kappa^n}([\iota_\Lambda]) & \text{whenever } \Lambda \subset \Lambda(n-1) \\ \underline{\mu}_n([\kappa^n]) = 1 & \\ \underline{\mu}_n(\{\omega\colon x_t = 0\}) = 1 & \text{whenever } t \notin \Lambda(n) \end{cases}$$

(The notation $[\iota_\Lambda]$ was introduced in the proof of Theorem 12-25.) These specifications are clearly consistent, so each $\underline{\mu}_n$ is well-defined. Now by the finiteness of S, we can use a diagonal argument (like the one in Proposition 1-63) to choose a subsequence $\{\underline{\mu}_{n'}\}$ such that $\underline{\mu}_{n'}([\iota_\Lambda]) \to \underline{\mu}([\iota_\Lambda])$ for all possible configurations on every finite $\Lambda \subset T$. By the extension theorem, these cylinder limits give rise to a unique μ on Ω. Now observe that

$$\underline{\mu}([\kappa^0, \ldots, \kappa^m]) = \lim_{n' \to \infty} \underline{\mu}_{n'}([\kappa^0, \ldots, \kappa^m]) = \lim_{n' \to \infty} \mu^{\underline{\kappa}^{n'}}([\kappa^0, \ldots, \kappa^m]),$$

the last limit being equal to $\mu^-([\kappa^0, \ldots, \kappa^m])$ by definition. To verify that $\underline{\mu} \in \mathscr{G}_V$, we first note that the $\underline{\mu}_{n'}$ measures of $[\iota_\Lambda]$ are bounded away from 0 for n' sufficiently large, since $\underline{\mu}_{n'}([\iota_\Lambda]) = \mu^{\underline{\kappa}^{n'}}([\iota_\Lambda])$ is strictly positive as soon as $\Lambda \subset \Lambda(n' - 1)$ and increases with n' by Lemma 12-31. It follows that $\underline{\mu}$ is strictly positive on finite cylinders, and that the neighbor Gibbs property is inherited from the $\underline{\mu}_{n'}$ (i.e., the limit may be interchanged with the operations defining the characteristics.) This completes the proof that $\mathscr{G}_V$ is always non-empty. By an analogous construction we find a neighbor Gibbs field $\bar{\mu} \in \mathscr{G}_V$ with $\bar{\mu}([\kappa^0, \ldots, \kappa^m]) = \mu^+([\kappa^0, \ldots, \kappa^m])$. If $\mu^+ \neq \mu^-$ (for some $(\kappa^0, \ldots, \kappa^m)$), then evidently $|\mathscr{G}_V| > 1$. If $\mu^+ = \mu^-$, then Lemma 12-31 shows that any $\mu \in \mathscr{G}_V$ is uniquely determined on all events $[\kappa^0, \ldots, \kappa^m]$, $m \geq 0$, hence on all $[\iota_\Lambda]$, finite $\Lambda \subset T$. This shows that $|\mathscr{G}_V| = 1$, completing the proof.

Unfortunately, the general criterion for phase multiplicity just given is often difficult to apply, since μ^+ and μ^- may not be readily computable. A more detailed theory is available for certain "attractive potentials," examples of which will be mentioned in Section 6.

5. Homogeneous Markov fields on the integers

Throughout this section the time parameter set T will be $\mathbb{Z}$ = the integers, $\mathbb{N} = \{0, 1, \ldots\}$ or $-\mathbb{N} = \{\ldots, -1, 0\}$, with the standard neighbor structure ∂ (i.e. $\partial n = \{n' \in T: |n' - n| = 1\}$). Our objective is to treat, in some detail, the classes of Markov fields on these infinite linear lattices.

First let us consider the "one-sided" cases. We show here the previously mentioned fact that the Markov random fields on $\mathbb{N}$ are simply the Markov processes with strictly positive cylinders.

Proposition 12-32: If $T = \mathbb{N}$ and $\mu \in \mathscr{G}_V$, then $\{x_n\}$ is a Markov process. In particular, there are probability measures π_n and transition matrices P_n, $n \geq 0$, such that

$$\pi_{n,i} = \Pr[x_n = i], \quad P_{n,ij} = \Pr[x_{n+1} = j \mid x_n = i],$$

and $\pi_n P_n = \pi_{n+1}$. Similarly, if $T = -\mathbb{N}$ and $\mu \in \mathscr{G}_V$, then $\{x_n\}$ is a Markov process on $-\mathbb{N}$, and there are probability measures π_n and transition matrices P_n, $n < 0$, satisfying the above relations.

PROOF: Without loss of generality, assume that V is normalized. It suffices to assume $T = \mathbb{N}$ and check the Markov property (the proof for $T = -\mathbb{N}$ being analogous). Fix $n > 0$, and set $A = \{1, 2, \ldots, n-1\}$. For any $\iota \in \Omega$, define

$$z_n(i_0, i_n) = \sum_{\kappa \in S^A} \exp\left\{\sum_{B \subset \bar{A}: B \cap A \neq \varnothing} V_B(\iota^\kappa)\right\},$$

where ι^κ is the modification obtained by replacing the values of ι on A with those of κ. Now choose a particular $\iota \in \Omega$, and let ι' be the configuration obtained by replacing the value i_0 at site 0 with a 0. Then, using the Markov field property at 0, we have

$$\frac{\Pr[x_0 = i_0 \wedge x_n = i_n]}{\Pr[x_0 = 0 \wedge x_n = i_n]} = \frac{\mu_0^{\bar{A}}(\iota)\,\mu_A^{\bar{A}}(\iota')}{\mu_0^{\bar{A}}(\iota')\,\mu_A^{\bar{A}}(\iota)} = \frac{\mu_0^{\bar{0}}(\iota)\,\mu_A^{\bar{A}}(\iota')}{\mu_0^{\bar{0}}(\iota')\,\mu_A^{\bar{A}}(\iota)}.$$

Writing the characteristics according to Definition 12-20, the last term is

$$\exp\{V_{\{0\}}(\iota) + V_{\{0,1\}}(\iota)\}\frac{z_n(0, i_n)^{-1}\exp\left\{\sum_{B \subset \bar{A}: B \cap A \neq \varnothing} V_B(\iota')\right\}}{z_n(i_0, i_n)^{-1}\exp\left\{\sum_{B \subset \bar{A}: B \cap A \neq \varnothing} V_B(\iota)\right\}}$$

$$= \frac{z_n(i_0, i_n)}{z_n(0, i_n)}\, e^{V_{\{0\}}(\iota)}$$

after cancellation. Hence

$$
\begin{aligned}
&\Pr[x_0 = i_0 \wedge \cdots \wedge x_n = i_n] \\
&\quad = \mu_A^{\bar{A}}(\iota) \Pr[x_0 = i_0 \wedge x_n = i_n] \\
&\quad = z_n(i_0, i_n)^{-1} \exp\left\{ \sum_{B \subset \bar{A}: B \cap A \neq \varnothing} V_B(\iota) \right\} \Pr[x_0 = i_0 \wedge x_n = i_n] \\
&\quad = z_n(0, i_n)^{-1} \exp\left\{ V_{\{0\}}(\iota) + \sum_{B \subset \bar{A}: B \cap A \neq \varnothing} V_B(\iota) \right\} \Pr[x_0 = 0 \wedge x_n = i_n].
\end{aligned}
$$

Finally, after further cancellation, we obtain

$$
\begin{aligned}
&\Pr[x_n = i_n \mid x_0 = i_0 \wedge \cdots \wedge x_{n-1} = i_{n-1}] \\
&\quad = \frac{z_{n-1}(0, i_{n-1})}{z_n(0, i_n)} [e^{V_{\{n-1\}}(\iota) + V_{\{n-1,n\}}(\iota)}] \frac{\Pr[x_0 = 0 \wedge x_n = i_n]}{\Pr[x_0 = 0 \wedge x_{n-1} = i_{n-1}]},
\end{aligned}
$$

a conditional probability depending only on $n - 1$, i_{n-1} and i_n, which we may set equal to $P_{n-1, i_{n-1} i_n}$.

Proposition 12-33: Let $T = \mathbb{N}$, and suppose that $\nu \in \mathscr{G}_V$ has initial measure π_0 and transition matrices P_n. Then any $\mu \in \mathscr{G}_V$ is Markovian with initial measure π_0' and transition matrices P_n' which satisfy $\pi'_{0,i_0} = \pi_{0,i_0} h_{(0,i_0)}$ and $P'_{n,i_n i_{n+1}} = P_{n,i_n i_{n+1}} h_{(n+1,i_{n+1})} / h_{(n,i_n)}$ for some solution h of the equations

$$
h_{(n,i_n)} = \sum_{i_{n+1} \in S} P_{n,i_n i_{n+1}} h_{(n+1,i_{n+1})}.
$$

Conversely, any μ arising in this way is in $\mathscr{G}_V$.

PROOF: Apply the construction of the previous section with $T = \mathbb{N}$ and $\Lambda(n) = \{0, 1, \ldots, n\}$. Then the correspondence of Theorem 12-25 is clearly equivalent to the one asserted here, because ν and μ are both Markovian by Proposition 12-32.

Of course, an obvious analogue of the last result holds in case $T = -\mathbb{N}$. A concrete example of phase multiplicity on a half-line will be given later in this section.

We turn our attention now to the "two-sided" case, $T = \mathbb{Z}$. In contrast to the one-sided setting, we shall see shortly that there are Markov fields on $\mathbb{Z}$ which are *not* Markov processes. Since the neighbor structure ∂ on $\mathbb{Z}$ commutes with translation, it is possible to define a class of homogeneous Markov fields.

Definition 12-34: A Markov field $\{x_t\}_{t \in \mathbb{Z}}$ is **homogeneous** if

$$\mu_m^{\overline{m}}(\iota) = \mu_{m+n}^{\overline{m+n}}(\theta^{-n}\iota) \quad \text{for all } m, n \in \mathbb{Z}, \iota \in \Omega,$$

where the value of $\theta^{-n}\iota$ at site t is i_{t-n}. A neighbor potential U is **homogeneous** if

$$U_{\{m\}}(\iota) = u_j \quad \text{and} \quad U_{\{m,m+1\}}(\iota) = u_{jk} \quad \text{whenever } i_m = j, i_{m+1} = k.$$

Two facts follow immediately from the definitions. First, if $\{x_t\}$ is homogeneous Markov, then the canonical V of Theorem 12-21 is a homogeneous neighbor potential. Second, any neighbor Gibbs field with homogeneous neighbor potential U is homogeneous Markov. For such a U, if we set $Q_{jk} = \exp\{\frac{1}{2}u_j + u_{jk} + \frac{1}{2}u_k\}$, then μ is a Gibbs field with potential U if and only if

$$\mu_m^{\overline{m}}(\iota) = \frac{e^{u_j + u_{ij} + u_{jk}}}{\sum_{s \in S} e^{u_s + u_{is} + u_{sk}}} = \frac{Q_{ij}Q_{jk}e^{-(u_i + u_k)/2}}{\sum_{s \in S} Q_{is}Q_{sk}e^{-(u_i + u_k)/2}} = \frac{Q_{ij}Q_{jk}}{(Q^2)_{ik}}$$

whenever $i_{m-1} = i$, $i_m = j$ and $i_{m+1} = k$. On the other hand, any strictly positive Q defines consistent local characteristics by means of the above equation, and these in turn give rise to a homogeneous neighbor Gibbs potential for μ, according to Theorem 12-21. Thus we obtain a multiplicative representation for the characteristics in terms of Q which is more convenient than the one involving the canonical potential V. Let $\mathscr{G}_Q$ denote the class of Markov fields determined by Q in this manner. An immediate requirement for $\mathscr{G}_Q$ to be non empty is $Q^2 < \infty$, and since the local characteristics determine all the characteristics, it follows from this assumption that $Q^n < \infty$ for all n, so that

$$\mu_{[m,n]}^{\Lambda}(\iota) = \mu_{[m,n]}^{[m-1,n+1]}(\iota) = \frac{Q_{i_{m-1}i_m}Q_{i_m i_{m+1}} \cdot \cdots \cdot Q_{i_{n-1}i_n}Q_{i_n i_{n+1}}}{(Q^{n-m+2})_{i_{m-1}i_{n+1}}}$$

whenever $m \leq n$ and $[m-1, n+1] \subset \Lambda$. (Here $[m, n]$ denotes $\{m, m+1, \ldots, n\}$.)

We have seen that any homogeneous neighbor Gibbs field is in $\mathscr{G}_Q$ for some Q. But many matrices Q' give rise to the same characteristics, just as in the potential representation. By definition, either $\mathscr{G}_Q = \mathscr{G}_{Q'}$ or $\mathscr{G}_Q \cap \mathscr{G}_{Q'} = \varnothing$; say that Q is **equivalent** to Q' in the former case ($Q \approx Q'$ in notation).

Proposition 12-35: Strictly positive matrices Q' and Q'' are equivalent if and only if

$$\frac{Q''_{ij}}{Q'_{ij}} = c\frac{h_j}{h_i} \quad i, j \in S,$$

for some $c > 0$ and strictly positive $h = (h_k)_{k \in S}$.

PROOF: The statements and proofs of Proposition 12-14 and Corollary 12-15 are easily modified to apply to neighbor potentials on an infinite parameter set by replacing T with $\bar{A}$. In our case, if U' and U'' are the homogeneous neighbor potentials such that $u'_k = u''_k = 0$ for all k and $u'_{ij} = \ln Q'_{ij}$, $u''_{ij} = \ln Q''_{ij}$, then U' and U'' determine the same Gibbs fields if and only if

$$\sum_{B \subset A \subset D \subset \bar{A}} (-1)^{|A-B|} \Delta_D(\iota^B) = 0 \quad \text{for every } A = \{m\}, \{m, m+1\},$$

where $\Delta_A = U'_A - U''_A$. Setting $\delta_{ij} = u'_{ij} - u''_{ij}$, these equations become

$$\begin{aligned} &(1) \qquad \delta_{k0} + \delta_{0k} - 2\delta_{00} = 0 \\ &(2)\ \delta_{ij} - \delta_{i0} - \delta_{0j} + \delta_{00} = 0 \end{aligned} \qquad i, k, j \in S.$$

The combination (2) + $\frac{1}{2}$(1 with $k = i$) + $\frac{1}{2}$(1 with $k = j$) yields

$$(3)\ (u'_{ij} - u''_{ij}) + w_i - w_j - z = 0,$$
$$\text{where } w_k = \tfrac{1}{2}(\delta_{0k} - \delta_{k0}), \quad z = \delta_{00}.$$

Defining $h_k = e^{w_k}$ and $c = e^z$, the desired equation in terms of Q' and Q'' follows. Conversely, if the hypothesis holds then

$$\frac{Q''_{ij}Q''_{jk}}{(Q''^2)_{ik}} = \frac{c^2 Q'_{ij}(h_j/h_i)Q'_{jk}(h_k/h_j)}{c^2 \sum_{s \in S} Q'_{is}(h_s/h_i)Q'_{sk}(h_k/h_s)} = \frac{Q'_{ij}Q'_{jk}}{(Q'^2)_{ik}},$$

so that Q' and Q'' determine the same local characteristics, i.e., $Q' \approx Q''$.

The remainder of this section will be devoted to an analysis of $\mathscr{G}_Q$, the class of homogeneous Markov fields on $\mathbb{Z}$ determined by a strictly positive matrix Q. We let $\mathscr{E}_Q$ comprise the extreme measures in $\mathscr{G}_Q$. It will now be proved, using the Martin boundary arguments of the previous section, that these *extreme* Markov fields on $\mathbb{Z}$ are always Markov processes.

Theorem 12-36: If $\mu \in \mathscr{E}_Q$, then $\{x_n\}_{n \in \mathbb{Z}}$ is a Markov process. Thus μ is determined by measures π_n and transition matrices P_n, $n \in \mathbb{Z}$, where

$$\pi_{n,i} = \Pr[x_n = i], \quad P_{n,ij} = \Pr[x_{n+1} = j \mid x_n = i],$$

and $\pi_n P_n = \pi_{n+1}$.

PROOF: It suffices to check the Markov property. For $n \geq 0$, let $\Lambda(n) = \{-n, -n+1, \ldots, n\}$. By Corollary 12-29, if $\mu \in \mathscr{E}_Q$, then $\mu = t\text{-}\lim_{n\to\infty} \mu^{\kappa^n}$ for some $(\kappa^0, \kappa^1, \ldots) \in \Omega$. This implies that there are states $k_n \in S$ for $n \in \mathbb{Z}$ such that

$$\begin{aligned} \Pr[x_l = i_l \wedge x_{l+1} = i_{l+1} \wedge \cdots \wedge x_m = i_m] \\ = \lim_{n\to\infty} \Pr[x_l = i_l \wedge x_{l+1} = i_{l+1} \wedge \cdots \wedge x_m = i_m \mid \\ x_{-n} = k_{-n} \wedge x_n = k_n] \end{aligned}$$

whenever $l \leq m$. Therefore

$$\begin{aligned}\Pr[x_m = i_m \mid x_l = i_l \wedge x_{l+1} = i_{l+1} \wedge \cdots \wedge x_{m-1} = i_{m-1}] \\ = \lim_{n\to\infty} \Pr[x_m = i_m \mid x_{-n} = k_{-n} \wedge x_l = i_l \wedge x_{l+1} = i_{l+1} \wedge \\ \cdots \wedge x_{m-1} = i_{m-1} \wedge x_n = k_n].\end{aligned}$$

By the Markov field property, for all n sufficiently large the right hand side becomes

$$\begin{aligned}&\lim_{n\to\infty} \Pr[x_m = i_m \mid x_{-n} = k_{-n} \wedge x_{m-1} = i_{m-1} \wedge x_n = k_n] \\ &= \lim_{n\to\infty} \frac{\Pr[x_{m-1} = i_{m-1} \wedge x_m = i_m \mid x_{-n} = k_{-n} \wedge x_n = k_n]}{\Pr[x_{m-1} = i_{m-1} \mid x_{-n} = k_{-n} \wedge x_n = k_n]} \\ &= \Pr[x_m = i_m \mid x_{m-1} = i_{m-1}] = P_{m-1,i_{m-1}i_m}.\end{aligned}$$

Next we present a useful representation theorem for the extreme homogeneous Gibbs states with matrix Q.

Theorem 12-37: If $\mu \in \mathscr{E}_Q$, then there are strictly positive functions $g_{(n,i)}$ and $h_{(n,i)}$ $(n \in \mathbb{Z}, i \in S)$ such that

(1) $g_{(n+1,i_{n+1})} = \sum_{i_n \in S} g_{(n,i_n)} Q_{i_n i_{n+1}},$

(2) $h(n, i_n) = \sum_{i_{n+1} \in S} Q_{i_n i_{n+1}} h_{(n+1,i_{n+1})},$

and

(3) $\sum_{i_n \in S} g_{(n,i_n)} h_{(n,i_n)} = 1,$

and such that the measures π_n and transition matrices P_n for $\{x_n\}$ are given by

$$\pi_{n,j} = g_{(n,j)} h_{(n,j)} \quad \text{and} \quad P_{n,jk} = Q_{jk} h_{(n+1,k)} / h_{(n,j)}.$$

Moreover, there are constants $c', c'' > 0$ and $k_n \in S$ $(n \in \mathbb{Z})$ such that

$$g_{(m,i_m)} = c' \lim_{n\to-\infty} (Q^{m-n})_{k_n i_m} / (Q^{-n})_{k_n 0} \qquad m < 0,\ i_m \in S$$

$$h_{(m,i_m)} = c'' \lim_{n\to\infty} (Q^{n-m})_{i_m k_n} / (Q^n)_{0 k_n} \qquad m > 0,\ i_m \in S$$

Proof: Since $\{x_n\}$ is a Markov process,

$$\mu_{n+1}^{\overline{n+1}}(\iota) = \frac{Q_{jk} Q_{k0}}{(Q^2)_{j0}} = \frac{P_{n,jk} P_{n+1,k0}}{(P_n P_{n+1})_{j0}}$$

and

$$\mu_{n+2}^{\overline{n+2}}(\iota) = \frac{Q_{k0}Q_{00}}{(Q^2)_{k0}} = \frac{P_{n+1,k0}P_{n+2,00}}{(P_{n+1}P_{n+2})_{k0}}$$

whenever $i_n = j$, $i_{n+1} = k$, $i_{n+2} = i_{n+3} = 0$. Let $\bar{h}_{(n,j)} = (Q^2)_{j0}/(P_nP_{n+1})_{j0}$, $c_n = Q_{00}/P_{n+1,00}$. Then dividing the first equation by the second and rearranging terms,

$$\frac{P_{n,jk}}{Q_{jk}} = \frac{1}{c_{n+1}}\frac{\bar{h}_{(n+1,k)}}{\bar{h}_{(n,j)}}.$$

Choose $\bar{c}_n$, $n \in \mathbb{Z}$, so that $\bar{c}_0 = 1$ and $\bar{c}_{n-1}/\bar{c}_n = c_n$. Now define $h_{(n,j)} = \bar{c}_n\bar{h}_{(n,j)}$ to get

$$P_{n,jk} = Q_{jk}h_{(n+1,k)}/h_{(n,j)}$$

as desired. Equation (2) follows immediately from the fact that P_n is a transition matrix. If we define $g_{(n,j)} = \pi_{n,j}/h_{(n,j)}$, then (3) holds because π_n is a probability measure, while the equation $\pi_nP_n = \pi_{n+1}$ implies (1). It therefore remains only to derive the representation of g and h as ratio limits of powers of Q. To this end, choose $\Lambda(n)$ and k_n as in the proof of the previous theorem. Then for $m > 0$,

$$\begin{aligned}\Pr[x_m &= i_m \mid x_0 = 0]\\ &= \lim_{n\to\infty} \Pr[x_m = i_m \mid x_{-n} = k_{-n} \wedge x_0 = 0 \wedge x_n = k_n]\\ &= \lim_{n\to\infty} \Pr[x_m = i_m \mid x_0 = 0 \wedge x_n = k_n],\end{aligned}$$

the last since $\{x_n\}$ is a Markov field. In terms of g, h and Q we have

$$\frac{g_{(0,0)}(Q^m)_{0i_m}h_{(m,i_m)}}{g_{(0,0)}h_{(0,0)}} = \lim_{n\to\infty}\frac{g_{(0,0)}(Q^m)_{0i_m}(Q^{n-m})_{i_mk_n}h_{(n,k_n)}}{g_{(0,0)}(Q^n)_{0k_m}h_{(n,k_n)}}$$

so that

$$h_{(m,i_m)} = h_{(0,0)}\lim_{n\to\infty}(Q^{n-m})_{i_mk_n}/(Q^n)_{0k_n}.$$

An analogous computation yields the result for g when $m < 0$.

With the aid of the above representation, cases where $|\mathscr{G}_Q| = 0$, 1 and ∞ will now be discussed.

Example 12-38: Let $S = \mathbb{Z}$, and consider any matrix Q of the form

$$Q_{ij} = q_{j-i} > 0,\ i, j \in \mathbb{Z},\quad \text{where} \sum_{i\in\mathbb{Z}} q_i = 1.$$

Suppose that $\mu \in \mathscr{E}_Q$, with g and h the functions of Theorem 12-37. Equations (1) and (2) of that theorem say that $g_{(-m,i)}$ and $h_{(m,i)}$, $m \geq 0$, are space-time harmonic for sums of independent random variables. By analogy to Example 6 of Section 10-13, one can show that the extreme such functions are of the form $c^m t^i$ for some $c > 0$, $t > 0$ ($q_i > 0$ excludes degenerate solutions). Thus,

$$g_{(0,i)}h_{(0,i)} = \int_0^\infty \int_0^\infty [s^i t^i] d\nu(s) d\mu(t)$$

for some measures ν and μ on $(0, \infty)$. It follows that

$$\sum_{i \in \mathbb{Z}} g_{(0,i)}h_{(0,i)} = \int_0^\infty \int_0^\infty \left[1 + \sum_{i=1}^\infty (st)^i + \sum_{i=1}^\infty \frac{1}{(st)^i}\right] d\nu(s) d\mu(t)$$
$$= \infty,$$

since one of the two infinite sums must diverge for any given s and t. This contradicts equation (3) of Theorem 12-37, so $\mathscr{E}_Q$ is empty. $\mathscr{G}_Q$ is therefore empty by Theorem 12-28.

If $Q > 0$ is an ergodic transition matrix, then there is a unique probability measure $\alpha > 0$ such that $\alpha Q = \alpha$. In this case $\mathscr{G}_Q$ clearly contains the stationary process with

$$\pi_{n,i} = \alpha_i \quad \text{and} \quad P_{n,ij} = Q_{ij} \quad \text{for all } n \in \mathbb{Z}.$$

Thus, whenever $Q' \approx Q$ for some ergodic Q, then $|\mathscr{G}_{Q'}| \neq 0$. Our next goal is to show that when S is *finite* $\mathscr{G}_{Q'}$ contains exactly one Markov field for *any* $Q' > 0$, and that this field is a stationary Markov chain on $\mathbb{Z}$. The first step is contained in a lemma.

Lemma 12-39: If S is finite, then any $Q' > 0$ is equivalent to a strictly positive transition matrix Q (with $Q1 = 1$).

PROOF: We show that there is a vector $\bar{h} > 0$ such that $Q'\bar{h} = \bar{c}\bar{h}$ for some constant $\bar{c} > 0$. Then, defining

$$Q_{ij} = \frac{Q'_{ij}\bar{h}_j}{\bar{c}\bar{h}_i},$$

it follows that Q is a transition matrix, while $Q' \approx Q$ by Proposition 12-35. To get $\bar{h}$, let $\mathscr{S} = \{h = (h_i)_{i \in S} : h \geq 0 \text{ and } \sum_{i \in S} h_i = |S|\}$, and define $\bar{c} = \sup\{c : Q'h \geq ch \text{ for some } h \in \mathscr{S}\}$. Easy estimates prove that $0 < \min_{i,j} Q'_{ij} \leq \bar{c} \leq |S| \max_{i,j} Q'_{ij} < \infty$. By definition of $\bar{c}$ and Proposition 1-63, there are constants $c^{(n)}$ and elements $h^{(n)}$ of $\mathscr{S}$, $n = 0, 1, \ldots$, such that $c^{(n)} \to \bar{c}$, $Q'h^{(n)} \geq c^{(n)}h^{(n)}$, and $\lim_{n\to\infty} h^{(n)} = \bar{h}$ for some $\bar{h} \in \mathscr{S}$. This implies that $Q'\bar{h} \geq \bar{c}\bar{h}$. Now if $(Q'\bar{h})_j > \bar{c}\bar{h}_j$ for some j, then $Q'(Q'\bar{h}) > \bar{c}(Q'\bar{h})$ so $Q'(Q'\bar{h}) > (\bar{c} + \epsilon)(Q'\bar{h})$ for small ϵ.

But this contradicts the definition of $\bar{c}$ once we normalize $Q'\bar{h}$. Hence $\bar{c}\bar{h} = Q'\bar{h} > 0$, which shows that $\bar{h} > 0$.

Theorem 12-40: If S is finite, then $|\mathscr{G}_{Q'}| = 1$. $\mathscr{G}_{Q'}$ consists of the Markov chain with transition matrix Q and stationary measure α, where Q is defined as in Lemma 12-39 and α is the regular probability measure for Q.

PROOF: Suppose $\mu \in \mathscr{E}_Q$, and let k_n, $n \in \mathbb{Z}$, be the states in the ratio limit representation of Theorem 12-37 for the functions $g_{(m,i)}$ and $h_{(m,i)}$. Since S is finite, there is some state $j'' \in S$ and an infinite sequence $n'' \to \infty$ such that $k_{n''} \equiv j''$. Hence

$$h_{(m,i_m)} = c'' \frac{\lim\limits_{n''\to\infty} (Q^{n''-m})_{i_m j''}}{\lim\limits_{n''\to\infty} (Q^{n''})_{0j''}} = c'' \frac{\alpha_{j''}}{\alpha_{j''}} = c'',$$

by the convergence theorem for noncyclic ergodic chains. Thus h is constant. Similarly, there is some j' and sequence $n' \to -\infty$ with $k_{n'} \equiv j'$, whereby

$$g_{(m,i_m)} = c' \frac{\lim\limits_{n'\to-\infty} (Q^{m-n'})_{j' i_m}}{\lim\limits_{n'\to-\infty} (Q^{-n'})_{j'0}} = c' \frac{\alpha_{i_m}}{\alpha_0} = c_0 \alpha_{i_m}.$$

It follows that $\pi_{n,j} \equiv \alpha_j$ and $P_{n,jk} \equiv Q_{jk}$. In other words, μ is uniquely determined as the process described in the statement of the theorem. For any strictly positive finite matrix Q' with equivalent transition matrix Q, we therefore have $|\mathscr{E}_{Q'}| = |\mathscr{E}_Q| = 1$. Theorem 12-28 now implies $|\mathscr{G}_Q| = 1$.

Next we present a concrete example of a matrix Q with phase multiplicity, for which all the elements of $\mathscr{E}_Q$ can be exhibited explicitly.

Example 12-41: Let $S = \{0, 1, \ldots\}$, and consider the strictly positive matrix Q given by

$$Q_{ij} = \sum_{k=0}^{i\wedge j} \left[\binom{i}{k}(\tfrac{1}{2})^i\right][e^{-1/2}(\tfrac{1}{2})^{j-k}/(j-k)!] \qquad (i \wedge j = \min\{i, j\}).$$

Q may be thought of as describing transition from i to j particles in a population. First particles disappear independently with probability 1/2, and then an independent Poisson distributed population of mean 1/2 is added to those which remain. This interpretation makes it clear that $Q1 = 1$. We remark next that Q is ergodic, with regular measure $\alpha_i = e^{-1}/i!$, $i \in S$. In fact, Q is α-reversible;

$$\alpha_i Q_{ij} = e^{-3/2} \sum_{k=0}^{i \wedge j} \left[\frac{\binom{i}{k}}{i!(j-k)!} \right] (\tfrac{1}{2})^{i+j-k}$$

$$= e^{-3/2} \sum_{k=0}^{i \wedge j} \left[\frac{\binom{j}{k}}{j!(i-k)!} \right] (\tfrac{1}{2})^{i+j-k} = \alpha_j Q_{ji},$$

since the terms in square brackets agree for every i, j and k. Thus $\mathscr{G}_Q$ contains the stationary Markov chain with $\pi_{n,i} = \alpha_i$ and $P_{n,ij} = Q_{ij}$ for every $n \in \mathbb{Z}$. In order to determine the other elements of $\mathscr{G}_Q$, we first compute the powers of Q. For this purpose it is convenient to introduce the generating functions $\gamma_i^n(s) = \sum_{j=0}^{\infty} (Q^n)_{ij} s^j$ $(|s| \leq 1)$, $i \in \mathbb{Z}$. We claim:

$$\gamma_i^n(s) = [1 - (\tfrac{1}{2})^n(1 - s)]^i \exp\{-[1 - (\tfrac{1}{2})^n](1 - s)\}.$$

When $n = 0$ both sides are s^i; the result follows by induction from the following considerations. If we start with i particles at time 0, and S_n denotes the number at time $n + 1$, then $S_{n+1} = X + \sum_{i=1}^{S_n} Y_i$, where X and the Y_i are independent, X Poisson with mean 1/2, the Y_i taking on values 0 and 1 each with probability 1/2. Since X has generating function $e^{-(1-s)/2}$ and each Y_i has generating function $(1 + s)/2$, it follows from the formula for the generating function of a random sum that $\gamma_i^{n+1}(s)$, the generating function of S_{n+1}, equals $e^{-(1-s)/2}\gamma_i^n((1 + s)/2)$ for $n \geq 0$. The desired formula for $\gamma_i^n(s)$ satisfies this recursion relation. Hence $(Q^n)_{ij}$, the coefficient of s^j in the power series expansion of $\gamma_i^n(s)$, is given by

$$(Q^n)_{ij} = \sum_{k=0}^{i \wedge j} \left[\binom{i}{k} (\tfrac{1}{2})^{nk} (1 - (\tfrac{1}{2})^n)^{i-k} \right] [e^{-(1-(1/2)^n)}(1 - (\tfrac{1}{2})^n)^{j-k}/(j - k)!].$$

If $\mu \in \mathscr{E}_Q$, then according to Theorem 12-37, the limits

$$g_{(m,i)} = c' \lim_{n \to \infty} (Q^{n+m})_{k_{-n}i}/(Q^n)_{k_{-n}0}, \ m \in \mathbb{Z}, \ i \in S,$$

exist and are strictly positive, for some $c' > 0$ and fixed sequence k_n, $n \in \mathbb{Z}$. We show next that these limits exist if and only if $\lim_{n \to \infty} k_{-n}/2^n = \theta$ for some θ (≥ 0). Under this assumption

$$(1 - (\tfrac{1}{2})^{n+m})^{k_{-n}} \to e^{-\theta(1/2)^m}$$

and

$$\binom{k_{-n}}{l}((\tfrac{1}{2})^{n+m})^l \to \frac{\theta^l(1/2)^{ml}}{l!}, \; l \geq 0.$$

Hence

$$\lim_{n\to\infty} (Q^{n+m})_{k_{-n}i} = \sum_{l=0}^{i} \left[\frac{\theta^l(1/2)^{lm}}{l!} e^{-\theta(1/2)^m}\right][e^{-1}/(i-l)!]$$

$$= e^{-(\theta(1/2)^m+1)} \sum_{l=0}^{i} \binom{i}{l} [\theta(1/2)^m]^l/i!$$

$$= e^{-(\theta(1/2)^m+1)}(\theta(1/2)^m + 1)^i/i!.$$

We can take this last quantity to be $g_{(m,i)}$ by choosing $c' = e^{-(\theta+1)}$. On the other hand, if $\lim_{n\to\infty} k_{-n}/2^n$ does not exist then we must have $k_{-n}/2^n \to \infty$ as $n \to \infty$, for otherwise the defining ratios would converge to distinct limits along different subsequences. But the ratio $(Q^{n+1})_{k_{-n}i}/(Q^n)_{k_{-n}0}$, for example, is a sum of positive terms including

$$\left[\frac{e^{-(1-(1/2)^{n+1})}}{e^{-(1-(1/2)^n)}}\right]\left[\frac{1-(\frac{1}{2})^{n+1}}{1-(\frac{1}{2})^n}\right]^{k_{-n}} \frac{[1-(\frac{1}{2})^{n+1}]^i}{i!} \geq c_i\left(1+\frac{1}{2^{n+1}}\right)^{k_{-n}}, c_i > 0,$$

and this last expression tends to ∞ as $n \to \infty$ if $k_{-n}/2^n \to \infty$. We have therefore shown that the limits defining $g_{(m,i)}$ exist if and only if $k_{-n}/2^n \to \theta \in [0, \infty)$ as $n \to \infty$, in which case $g_{(m,i)}$ as a function of i may be taken to be Poisson with mean $\theta(\frac{1}{2})^m + 1$. Since Q is α-reversible, $\alpha_i(Q^n)_{ij} = \alpha_j(Q^n)_{ji}$, and hence the function $h_{(m,i)}$ of Theorem 12-37 can be computed as

$$h_{(m,i)} = c'' \lim_{n\to\infty} \frac{(Q^{n-m})_{ik_n}}{(Q^n)_{0k_n}} = c'' \lim_{n\to\infty} \frac{\alpha_0(Q^{n-m})_{k_n i}}{\alpha_i(Q^n)_{k_n 0}}$$
$$= c''' e^{-(\eta 2^m+1)}(\eta 2^m + 1)^i$$

if $k_n(\frac{1}{2})^n \to \eta \in [0, \infty)$; otherwise the limit does not exist. Condition (3) of Theorem 12-37 now dictates $c''' = e^{1-\theta\eta}$. In summary, we have proved that if $\mu \in \mathscr{E}_Q$, then μ is one of the Markov process measures $\mu_{\theta\eta}$, $\theta, \eta \in [0, \infty)$. Here θ determines g, η determines h, and then g and h determine $\mu_{\theta\eta}$ according to Theorem 12-37. Finally, it remains to check that all the $\mu_{\theta\eta}$ are extreme, so that in fact $\mathscr{E}_Q = \{\mu_{\theta\eta};\, \theta, \eta \in [0, \infty)\}$. One first verifies, as in Example 6 of Section 10-13, that the topology of the Martin boundary B associated with $\mathscr{G}_Q$ is the usual topology on $\mathbb{R}^2_+ = \{(\theta, \eta)\colon \theta \geq 0, \eta \geq 0\}$. Then by Theorem 12-28,

$$\mu_{\bar\theta\bar\eta} = \int_0^\infty \int_0^\infty \mu_{\theta\eta} d\lambda(\theta, \eta)$$

for a unique probability measure λ on $\mathbb{R}_2^+$ such that $\lambda(\{(\theta, \eta): \mu_{\theta\eta} \in \mathscr{E}_Q\}) = 1$. Evaluating both sides on $\{\omega \mid \omega_n = i, \omega_{n+1} = j\}$, we derive the equation

$$e^{-\bar{\theta}\bar{\eta}}\{e^{-(\bar{\theta}(1/2)^n+1)}(\bar{\theta}(\tfrac{1}{2})^n + 1)^i/i!\}\{e^{-(\bar{\eta}2^{n+1}+1)}(\bar{\eta}2^{n+1} + 1)^j/j!\}$$

$$= \int_0^\infty \int_0^\infty e^{-\theta\eta}\{e^{-\theta(1/2)^n+1}(\theta(\tfrac{1}{2})^n + 1)^i/i!\}$$

$$\times \{e^{-(\eta 2^{n+1}+1)}(\eta 2^{n+1} + 1)^j/j!\}d\lambda(\theta, \eta).$$

Multiply both sides by $u^i v^j$ ($u, v \in \mathbb{R}$), sum over $i, j \in S$, and interchange summation and integration, to obtain

$$e^{-\bar{\theta}\bar{\eta}}\{e^{-\bar{\theta}(1/2)^n(1-u)}\}\{e^{-\bar{\eta}2^{n+1}(1-v)}\}$$

$$= \int_0^\infty \int_0^\infty e^{-\theta\eta}\{e^{-\theta(1/2)^n(1-u)}\}\{e^{-\eta 2^{n+1}(1-v)}\}d\lambda(\theta, \eta).$$

If we make the change of variables: $x = (1 - u)/2^n$, $y = 2^{n+1}(1 - v)$, then for $x \geq 0$ and $y \geq 0$ the right hand side is the double Laplace transform of the measure $e^{-\theta\eta}d\lambda(\theta, \eta)$. Since the equation is satisfied when λ concentrates at $(\bar{\theta}, \bar{\eta})$, the uniqueness theorem for Laplace transforms implies that λ must be this measure. We conclude that $\mu_{\bar{\theta}\bar{\eta}} \in \mathscr{E}_Q$, as desired.

Whenever Q is a transition matrix and $\mu \in \mathscr{E}_Q$ admits a representation with $h_{(n,i)} \equiv 1$, then $\pi_{n,i} = g_{(n,i)}$ and $P_n \equiv Q$. A family of probability distributions π_n, $n \in \mathbb{Z}$, such that $\pi_n Q = \pi_{n+1}$ is called an **entrance law** for Q. For example, the Poisson distributions π_n with mean $\theta(\frac{1}{2})^n + 1$ for fixed $\theta \in [0, \infty)$ constitute an entrance law for the matrix Q of the last example. The case $\theta = 0$ yields the stationary Markov process with regular measure α; when $\theta > 0$ the process $\{x_n\}$ with measure $\mu_{\theta\eta}$ "comes down from infinity" in the sense that $\lim_{n\to-\infty} \Pr[x_n = i] = 0$ for any $i \in S$. A more surprising type of entrance law is described in Problem 8 at the end of the chapter.

The matrix Q of Example 12-41 may also be used to illustrate phase multiplicity on the non-negative half-line. Namely, for $\eta \in [0, \infty)$, let $h^\eta_{(n,i)} = e^{-\eta 2^n}(\eta 2^n + 1)^i$, and define $\pi^\eta_{0,i} = \alpha_i h^\eta_{(0,i)}$, $P^\eta_{n,jk} = Q_{jk} h^\eta_{(n+1,k)}/h^\eta_{(n,j)}$. If we set $\pi_0 = \alpha$, $P_n \equiv Q$, $\pi'_0 = \pi^\eta_0$ and $P'_n = P^\eta_n$ in Proposition 12-33, then the hypotheses there are clearly satisfied. Thus we have constructed a large family of one-sided Markov fields with the same characteristics as the Markov chain with transition matrix Q and initial distribution α.

As a final application of Example 12-41, we exhibit an element of $\mathscr{G}_Q$ which is not a Markov process. Consider the field $\{x_n\}$ given by the convex combination $\mu = \frac{1}{2}\mu_{00} + \frac{1}{2}\mu_{11}$. Clearly $\mu \in \mathscr{G}_Q$, and to see that μ is not the measure for a Markov process it suffices to check that

$$\Pr[x_2 = 0 \mid x_0 = 0 \wedge x_1 = 0] \neq \Pr[x_2 = 0 \mid x_1 = 0].$$

The π_n and P_n for μ_{11} satisfy $\pi_{0,0} = e^{-4}$, $\pi_{1,0} = e^{-9/2}$, $P_{0,00} = e^{-1}Q_{00}$ and $P_{1,00} = e^{-2}Q_{00}$. Thus

$$\begin{aligned}\Pr[x_2 = 0 \mid x_0 = 0 \wedge x_1 = 0] &= \frac{\frac{1}{2}(e^{-1}Q_{00}Q_{00}) + \frac{1}{2}(e^{-4}e^{-1}Q_{00}e^{-2}Q_{00})}{\frac{1}{2}(e^{-1}Q_{00}) + \frac{1}{2}(e^{-4}e^{-1}Q_{00})}\\ &= \frac{1 + e^{-6}}{1 + e^{-4}}Q_{00},\end{aligned}$$

while

$$\begin{aligned}\Pr[x_2 = 0 \mid x_1 = 0] &= \frac{\frac{1}{2}(e^{-1}Q_{00}) + \frac{1}{2}(e^{-9/2}e^{-2}Q_{00})}{\frac{1}{2}e^{-1} + \frac{1}{2}e^{-9/2}}\\ &= \frac{1 + e^{-11/2}}{1 + e^{-7/2}}Q_{00}.\end{aligned}$$

Hence $\{x_n\}$ has the two-sided Markov property, but *not* the one-sided Markov property.

We conclude this section by mentioning without proof the deepest result to date in the theory of denumerable Markov fields on $\mathbb{Z}$. When T has a group structure it is natural to consider the class $\mathscr{G}_V^0$ consisting of all those $\mu \in \mathscr{G}_V$ which are invariant under translations. The following theorem completely determines the possibilities for $\mathscr{G}_{Q'}^0$, the translation invariant Markov fields on $T = \mathbb{Z}$ with characteristics determined by the strictly positive matrix Q'.

Theorem 12-42: If $\mu \in \mathscr{G}_{Q'}$, then $Q' \approx Q$ for some strictly positive Q such that $Q1 \leq 1$. If $Q1 = 1$ and Q is ergodic, then $|\mathscr{G}_{Q'}^0| = 1$. In this case the unique member of $\mathscr{G}_{Q'}^0$ is the stationary Markov chain on $\mathbb{Z}$ with transition matrix Q and $\pi_n \equiv$ the Q-regular measure α. In all other cases $\mathscr{G}_{Q'}^0 = \varnothing$.

6. Examples of phase multiplicity in higher dimensions

When $T = \mathbb{Z}^d$ for $d \geq 2$, the conclusion of Theorem 12-40 ceases to hold. In other words, there are instances of phase multiplicity for homogeneous potentials U with S finite. This phenomenon is undoubtedly the most important in the theory of random fields, but an adequate treatment is far beyond the scope and purpose of this chapter. Instead, we briefly discuss two examples. Suggestions for further reading are included in the Additional Notes.

Example 12-43: Tree processes. $S = \{0, 1, \ldots, N\}$ and T is a countable **tree** (i.e., T is endowed with a neighbor structure ∂ which defines a connected graph with no loops). Let P be a strictly positive N-state transition matrix with regular measure α, and assume that P is α-reversible. The random field $\{x_t\}$ on S^T is called a **tree process** for (α, P) if μ is defined by the following three properties:

(i) $\Pr[x_t = i] = \alpha_i$ for all $i \in S, t \in T$;

(ii) $\Pr[x_{t_0} = i_0 \wedge x_{t_1} = i_1 \wedge \cdots \wedge x_{t_l} = i_l] = \alpha_{i_0} P_{i_0 i_1} \cdot \cdots \cdot P_{i_{l-1} i_l}$ whenever $\{t_0, t_1, \ldots, t_l\}$ is a finite path in the tree T and $i_0, i_1, \ldots, i_l \in S$;

(iii) $\Pr[x_{t_0} = i_0 \wedge \cdots \wedge x_{t_l} = i_l \mid x_{t_r} = i_r \text{ for all } r \in \Lambda]$
$= \Pr[x_{t_0} = i_0 \wedge \cdots \wedge x_{t_l} = i_l \mid x_{t_0} = i_0]$
for any finite $\Lambda \subset T$ and path $\{t_0, \ldots, t_l\} \subset T$ which intersect only at site t_0, and any $i_0, \ldots, i_l \in S$.

For given α and Q, conditions (i)–(iii) determine a well-defined and unique Markov field. According to (i) and (ii) the process behaves like a reversible Markov chain along paths (reversibility ensures that cylinder probabilities are independent of the direction we travel along a path in (ii)), while paths "patch together" because of condition (iii). As a special case, suppose that $S = \{0, 1\}$ and T is the tree with three neighbors for every site (sometimes called the 3-Bethe lattice). Consider the tree processes with measures μ' and μ'' induced by (α', P') and (α'', P'') respectively, where

$$\alpha' = (\tfrac{1}{3}, \tfrac{2}{3}) \qquad P' = \begin{pmatrix} \frac{2}{3} & \frac{1}{3} \\ \frac{1}{6} & \frac{5}{6} \end{pmatrix},$$

$$\alpha'' = (\tfrac{4}{5}, \tfrac{1}{5}) \qquad P'' = \begin{pmatrix} \frac{8}{9} & \frac{1}{9} \\ \frac{4}{9} & \frac{5}{9} \end{pmatrix}.$$

It is not hard to verify that both fields have the same local characteristics, so there is phase transition for the potential V corresponding to these characteristics. For instance,

$$\mu'(\{x_a = 0\} \mid \{x_t = 0 \text{ for all } t \in \partial a\}) = \frac{(\frac{1}{3})(\frac{2}{3})^3}{(\frac{1}{3})(\frac{2}{3})^3 + (\frac{2}{3})(\frac{1}{6})^3} = \tfrac{32}{33},$$

$$\mu''(\{x_a = 0\} \mid \{x_t = 0 \text{ for all } t \in \partial a\}) = \frac{(\frac{4}{5})(\frac{8}{9})^3}{(\frac{4}{5})(\frac{8}{9})^3 + (\frac{1}{5})(\frac{4}{9})^3} = \tfrac{32}{33}.$$

In this setting, if $P = \begin{pmatrix} 1-p & p \\ q & 1-q \end{pmatrix}$ then $\alpha = \left(\dfrac{q}{p+q}, \dfrac{p}{p+q}\right)$ and P is always α-reversible. The (α, P)-process is called **attractive** when $p + q \leq 1$. Roughly, attractiveness means that a 1 at site t increases the likelihood of 1's near t (and similarly for 0's). More precisely this implies that in Theorem 12-30 $\mu^{\kappa^n}(\{x_t = 1\})$ is maximized when κ^n consists of "all 1's" on $K(n)$. Using this fact it is possible to show that in the attractive case there is phase multiplicity for the potential corresponding to (α, P) if and only if $(p - q)^2 - 2(p + q) + 1 \geq 0$ and $(p, q) \neq (\frac{1}{4}, \frac{1}{4})$. The (α, P)-process is **repulsive** when $p + q \geq 1$, and in this case one can prove by entirely different methods that phase multiplicity occurs if and only if $p + q > \frac{3}{2}$.

Random fields on countable trees T have the advantage that most physically meaningful quantities can be computed explicitly; the fact that T has no loops enables one to use inductive methods. The natural setting for statistical mechanics is $T = \mathbb{Z}^d$, however, and in this case the theory is immensely more difficult. We summarize some of the leading results for the simplest Markov fields on the two-dimensional integer lattice in our last example.

Example 12-44: Two-dimensional Ising model. $S = \{0, 1\}$, $T = \mathbb{Z}^2$. V is a normalized potential of the form:

$$V_{\{a\}}(\iota) = v_0 \quad V_{\{a,b\}}(\iota) = v_1$$

whenever $|a - b| = 1$ and $i_a = i_b = 1$, with $V_A(\iota) = 0$ in all other cases. V is **attractive** if $v_1 \geq 0$, **repulsive** otherwise; the intuitive interpretation is the same as in the previous example. When V is attractive there is phase multiplicity if and only if $v_0 + 2v_1 = 0$ and $v_1 > 2 \ln(\sqrt{2} + 1)$. For repulsive V there is phase multiplicity in an open neighborhood of the line segment $\{(v_0, v_1): v_0 + 2v_1 = 0$ and $v_1 < K\}$, with K sufficiently negative. Similar results hold in higher dimensions, though less is known.

7. Problems

1. Show that if $S = \{0, 1\}$, T is a finite subset of $\mathbb{Z}^d$, and V is a normalized neighbor potential, then the energy H_V of V may be expressed as

$$H_V(\iota) = \tfrac{1}{2} \sum_{a \in T} \sum_{b \in \bar{a}} v_{ab} i_a i_b \quad \iota = \{i_t\} \in \Omega,$$

for some $v = \{v_{ab} \in \mathbb{R}, |a - b| \neq 1\}$ satisfying $v_{ab} = v_{ba}$.

2. Give an example of a finite Gibbs field which cannot be represented in terms of a pair potential.
3. Let ∂ be a given neighbor system. The K-**neighbor set** $\partial^K a$ $(K = 1, 2, \ldots)$ of $a \in T$ is defined recursively by $\partial^1 a = \partial a$, $\bar{\partial}^1 a = \bar{a}$, and for $K > 1$, $\partial^{K+1} a = \partial(\bar{\partial}^K a)$, $\bar{\partial}^{K+1} a = (\partial^{K+1} a) \cup (\bar{\partial}^K a)$. Thus a K-**neighbor** t of a is a site which can be reached from a in K steps to neighboring sites, and no fewer. A random field $\{x_t\}$ is called a K-**Markov field** (with respect to ∂) if $\mu_a^A = \mu_a^{\bar{\partial}^K a}$ whenever $\bar{\partial}^K a \subset A \subset T$, A finite. A potential U is called

a K-**neighbor potential** if $U_A = 0$ whenever A contains two sites which are not K-neighbors. Show that $\{x_t\}$ is K-Markov if and only if the canonical potential V for $\{x_t\}$ is K-neighbor. What does this say when $K = 0$? [*Hint:* Define a new neighbor system.]

4. Suppose that there is a metric d defined on T, and say that $\{x_t\}$ is L-**Markov** ($L \geq 0$) if $\mu_a^A = \mu_a^{B(a,L)}$ whenever $B(a, L) = \{t \in T: d(a, t) \leq L\} \subset A \subset T$. What property for the canonical potential V is equivalent to the L-Markov property for $\{x_t\}$? State and prove a theorem to justify your assertion. Describe the $\sqrt{2}$-Markov fields on $\mathbb{Z}^2$.
5. Show that if $\{x_t\}$ has *any* neighbor potential U, then $\{x_t\}$ is a Markov field. Give an example of a Markov field with potential U, such that U is *not* a neighbor potential. [*Hint:* For the first part, look carefully at the proof of Theorem 12-16.]
6. Let $T = \mathbb{Z}$. Prove that $\mathscr{G}_Q = \varnothing$ if

$$\lim_{n\to\infty} \sup_{i,k\in S} \frac{(Q^n)_{ij}(Q^n)_{jk}}{(Q^{2n})_{ik}} = 0 \quad \text{for some } j \in S.$$

[*Hint:* Show that this condition forces $\Pr[x_0 = j] = 0$.]

7. Give an example of a Markov process $\{x_n\}_{n\in\mathbb{Z}}$ which is *not* extreme in its class of Markov fields. [*Hint:* Use the matrix Q of Example 12-41.]
8. Let $S = \{0, 1, \ldots\}$, $T = \mathbb{Z}$. Define α_i inductively by $\alpha_0 = \frac{1}{2}$, and $\alpha_i = (\alpha_{i-1}//3) + \frac{1}{2}(\frac{1}{4})^i$ for $i \geq 1$. Let

$$Q_{ij} = \begin{cases} \alpha_j & i = 0, j \in S \\ [\frac{1}{2}(\frac{1}{4})^i + \frac{1}{3}\delta_{i-1,j}](\alpha_j/\alpha_i) & i \geq 1, j \in S \end{cases}.$$

Finally, put

$$\pi_{n,i} = \begin{cases} \alpha_i + (\delta_{(-n-1)i} - \alpha_i) \displaystyle\prod_{k=-n-1}^{\infty} (\alpha_k/3\alpha_{k+1}) & n < 0 \\ \alpha_i & n \geq 0 \end{cases}.$$

Show that α and all of the π_n are strictly positive probability vectors on S, and that Q is a strictly positive transition matrix with $Q1 = 1$. Finally, prove that $\alpha Q = \alpha$, $\pi_n Q = \pi_{n+1}$, and $\pi_n \neq \alpha$ for $n < 0$. Thus Q has an entrance law which agrees with the stationary one from time 0 on, but *not* before time 0.

9. Let V be any neighbor potential on $\mathbb{Z}$. Show that if $\mu \in \mathscr{E}_V$, then $\{x_n\}$ is a Markov process.
10. Suppose g and h satisfy (1)–(3) of Theorem 12-37 for some $Q > 0$. Show that $\pi_{n,j}$ and $P_{n,jk}$ as prescribed in that theorem give rise to a well-defined field $\{x_n\}$. Is $\{x_n\}$ in $\mathscr{G}_Q$? Is it in $\mathscr{E}_Q$?

NOTES

Chapter 3:

Stochastic processes with the martingale property were first studied by Lévy [1937]. Lévy considered, in Sections 67 to 70, partial sums of sequences $\{f_n\}$ such that

$$M[f_{n+1} \mid f_0 \wedge \cdots \wedge f_n] = 0.$$

These are a natural generalization of sums of independent random variables with mean 0. He proved theorems such as a central limit theorem suggested by comparison with sums of independent random variables. Ville [1939] recognized the importance of studying processes representing a fair game and for which system theorems should hold. He called these processes martingales. Although he did not prove any convergence theorems, he did prove the inequality given in Problem 7. From this he was able to conclude that non-negative martingales had finite lim sup with probability one. He made application of this to the study of sample paths of coin tossing. In particular, he proved one half of the law of the iterated logarithm. The basic convergence theorem, Theorem 3-12, for martingales was proved by Doob [1940]. In his book on stochastic processes, Doob [1953] introduced submartingales (called semi-martingales in that book) and made a systematic study of the system theorems and convergence theorems for these processes. The proof of Proposition 3-11 for martingales is due to Doob [1940]. The proof given here and the extension to submartingales is due to Snell [1952]. Additional applications of martingale theory to Markov chains may be found in Lamperti [1960a] and [1963a].

Chapter 4:

Markov chains with a finite number of states were introduced by Markov [1907]. Kolmogorov [1936] considered the case of a denumerable number of states. Important contributions in the foundations of Markov chains were made by Doeblin [1938]. There are a number of books devoted to the study of finite Markov chains. Among these are Fréchet [1938], Romanovskiĭ [1949], Kemeny and Snell [1960], Lahres [1964], and Gorden [1965]. The theory of denumerable Markov chains is the subject of a book by Chung [1960].

Finite random walks have been analyzed in some detail in Kemeny and Snell [1960], Chapter 7. See also Kac [1947a]. The books by Spitzer [1964] and Kemperman [1961] give detailed studies of Markov chain problems applied to sums of independent random variables. The class of random walks discussed in Example 8 was introduced by Karlin and McGregor [1959] who made an extensive study of these processes. There is a large literature on branching processes. References to this literature as well as an account of the theory of these processes may be found in a book of Harris [1963]. The process called the basic example in this book is often referred to as a "renewal process."

The recognition of the need for and the importance of system theorems is due to Doob. See Doob [1953], Chapter VII. The strong Markov property which holds for any denumerable Markov chain does not hold for general Markov processes where time is allowed to be continuous and the state space is the real line. For a discussion of this problem in the more general setting, see Blumenthal [1957].

A discussion of system theorems and a rather systematic use of these theorems in Markov chain theory may also be found in Chung [1960].

Chapter 5:

Kemeny and Snell [1960], Chapter III, showed that the fundamental matrix N could be used to obtain moments of many descriptive quantities for finite absorbing chains. The extension of this use of N to denumerable chains was made in Kemeny and Snell [1961b]. Theorem 5-10 is the analog of the Riesz Decomposition Theorem for superregular functions. A systematic discussion of results of this type which exploit the analogy between superregular functions for a Markov chain and classical superharmonic functions may be found in Feller [1956] and Doob [1959]. The proof of Proposition 5-20 is due to Dynkin and Malyutov [1961].

Proposition 5-22 is true even if we drop the hypothesis of finitely many k-values. This theorem is due to Chung and Erdös [1951], and a simplified proof of this result was given by Chung and Ornstein [1962].

Chapter 6:

Theorem 6-9 is due to Derman [1954]. He proved existence by showing that $\alpha_j = {}^i\bar{N}_{ij}$ is a regular measure. His proof of uniqueness is less elementary than ours and uses the Doeblin ratio theorem applied to the chain reversed by α. This ratio theorem proved by Doeblin [1938] states for recurrent chains that $\lim_{n\to\infty} (N_{ij}^{(n)}/N_{kl}^{(n)})$ exists and is independent of i and k. Derman also used the identification of this limit as α_j/α_l.

Proposition 6-24 is due to Kac [1947b].

Doeblin [1938] proved Proposition 6-32 and then applied limit theorems for sums of independent random variables to obtain limit theorems for Markov chains. For details of this technique and resulting limit theorems, see Chung [1960], pp. 75–106. The converse to Proposition 6-32 is due to Yosida and Kakutani [1940].

The fact that $\lim_{n\to\infty} P^n$ exists for a noncyclic recurrent chain is due to Kolmogorov [1936], [1937]. The extension of this result given in Theorem 6-38 is due to Orey [1962], and the first proof (including Lemmas 6-36 and 6-37) is a somewhat simplified version of his proof. The proof using the Renewal Theorem may be found in Feller [1957]. Theorem 6-43 is new.

Chapter 7:

The details of the connection between Brownian motion and classical potential theory are discussed by Knapp [1965]. The recognition of the importance of identifying these two theories started with Kakutuni [1944]. Doob [1954] made significant extensions of the results of Kakutani by further identifying martingale and submartingale theory with the theory of harmonic and superharmonic functions. Important contributions again exploiting connections between Brownian motion and classical potential theory were made by Kac [1951]. The next major contribution was made by Hunt [1957], [1958]. Hunt showed that one could develop a potential theory for essentially the most general Markov process. He considers continuous time and abstract state space. He showed conversely that under rather minimal requirements for a potential theory one can construct a Markov process associated with this theory. His work related to the potential theory that goes with transient processes.

Although many of Hunt's results go over easily to the Markov chain case, even for transient chains new problems arise, and a whole new theory must be developed for the recurrent case. These extensions were made by Kemeny and Snell [1961b] for general Markov chains and by Spitzer [1962] for the important class of Markov chains which arise from sums of lattice-valued independent random variables.

The fact that the symmetric random walk in one and two dimensions is recurrent, whereas in dimension three or greater it is transient was first proved by Polya [1921]. The proof of Proposition 7-10 was supplied by Lamperti.

Chapter 8:

The notion of h-regular function was introduced into the study of potential theory by Brelot [1956] and the corresponding idea of a function regular in the h-process for chains was discussed in Feller [1956] and Doob [1959].

A discussion of equilibrium potential, equilibrium charge, and capacity as they arise in electrostatics and Newtonian potential theory may be found in the book of Kellogg [1929]. A somewhat more modern approach may be found in Brelot [1959]. In the classical theory the Green's function which plays the role of the matrix N is always symmetric. The fact that there is an interesting potential theory even for nonsymmetric operators in probability was first shown by Hunt [1957], [1958].

The results of Sections 1 and 2 of this chapter were for the most part in Doob [1959]. Those of Sections 3 and 4 are specializations to the Markov chain case of results obtained by Hunt [1957], [1958] for more general Markov processes.

Choquet and Deny [1956–57] investigated the problem of the relation between the various potential principles for the case of potentials of the form $g = Gf$, where G is an arbitrary non-negative finite matrix. If G has an inverse they proved that the Principles of Balayage and Domination are equivalent and that each implies the Principle of Lower Envelope. Here every non-negative function f is a charge. They showed further that if G satisfies the Principle of Lower Envelope, then there is a unique permutation of the columns of G such that the resulting matrix satisfies the Principle of Balayage. Also they showed that G satisfies the Principle of Balayage if and only if it is of the form $G = A \sum_{p=0}^{\infty} S^p$, where A is a diagonal matrix with strictly positive diagonal entries and S is a non-negative matrix. Thus the most general operator here is only slightly more general than the class of all matrices of the form $N = (I - Q)^{-1}$, where Q is a finite transient chain. Some investigation of this problem for denumerable matrices was made by Kemeny and Snell [1961b].

The notion of energy seems to be significant only in the case of reversible chains and even here it does not have the nice probabilistic interpretations that the other potential theory concepts have.

The results of Section 8 are taken from Kemeny and Snell [1961b]. The idea of developing a potential theory for supermartingales was suggested by Doob [1961] where he also indicated the proof of Proposition 8-79.

Chapter 9:

The potential theory for recurrent chains discussed in this chapter was introduced by Kemeny and Snell [1961b].

The existence of the limit in Theorem 9-4 was first proved by Doeblin [1938]. The identification of the limit was made by Chung [1950]. The present proof is from Kemeny [1962].

Theorem 9-7 was proved under a mild assumption in Kemeny and Snell [1961b]. The case $i = j$ was proved in general by a method due to Chung in Chung [1961] and Kemeny and Snell [1961c]. The general case was proved by Kemeny [1963].

The fact that all ergodic chains are normal follows from Theorem 4, Chapter 1, Section 11 of Chung [1961]. The remaining results in the first three sections are taken primarily from Kemeny and Snell [1961b].

Proposition 9-65 was first proved by Lamperti [1960b]. Results of the form of Propositions 9-67 and 9-68 may be found in Chung [1960] Chapter 1, Section 11.

The notion of strong ergodic chains introduced here is new. The matrix Z was introduced in Kemeny and Snell [1960]. It was shown in this book that the matrix Z for finite ergodic chains could be used to express the moments of many interesting descriptive quantities and hence played for recurrent chains a role similar to the matrix N for finite absorbing chains.

All sums of independent random variables processes which form aperiodic recurrent Markov chains are normal. This was proved by Kemeny and Snell [1961a] for the case of finite variance and in general by Spitzer [1962].

The operator K was introduced by Kemeny and Snell [1963b]. Most of the results of Sections 8 and 9 are taken from this paper.

The method of associating denumerable chains with electric circuits discussed in Section 10 was carried out by Nash–Williams [1959] under slightly more restrictions on the chain than we impose. He proved also Lemma 9-129.

Chapter 10:

The Martin boundary for Markov chains was introduced independently by Doob [1959] and Watanabe [1960a]. Doob and Watanabe used the methods which were developed in the study of the classical Martin boundary relevant to Newtonian potentials. Details of this approach may be found in Brelot [1956] and Doob [1957] or Watanabe [1960a].

Hunt [1960] gave a new and more probabilistic treatment of Martin boundary theory for Markov chains and completed the work of Doob and Watanabe in several ways. In particular, he introduced a new class of processes called approximate P-chains. These are slightly more general than the processes we have called extended chains. Our treatment of the Martin boundary is for the most part a rewriting of Hunt's paper with more detail supplied, except that we have used a slightly different definition of boundary than that listed by Hunt. The difference is as follows. Doob introduced a metric on the state space, the one we have used if π assigns all its weight to one point, and completed the state space in terms of this metric. A point was called a boundary point by Doob in the completed space if it was a limit point of the original states. It is possible for one of the original states to be such a limit point. To avoid this peculiarity, Hunt modified Doob's metric slightly to make such a point into a new point. Since these new points are always nonminimal points and appear to play no essential role in the theory, we have followed Doob's metric. However, we have chosen to call the boundary simply the new points added by the completion. The observation that $\pi N > 0$ is the only condition on π that is needed appears in Orey [1964].

One can also use G. Choquet's theory of convex cones to develop Martin boundary theory. This approach has been carried out by Neveu [1964]. See also Hennequin and Tortrat [1965].

Brelot [1956] showed that the Martin boundary was ideally suited to the study of the first boundary problem, or the Dirichlet problem. His approach was to generalize the method developed by Perron and Wiener (see Kellogg [1929]) for regions in Euclidean space.

The probabilistic approach to the first boundary problem was first suggested by Kakutani [1945] and done more generally using the Martin boundary by Doob [1958]. The method presented in this book is the probabilistic approach of Kakutani and Doob.

The discussion of fine boundary limits follows that of Doob [1957], who considered these problems for superharmonic functions using Brownian motion theory.

The Martin boundary has now been worked out for several important classes of Markov chains. In particular, Doob, Snell and Williamson [1960] have worked out the boundary for general sums of independent random variables. Related results may be found in Dynkin and Malyutov [1961]. There are close connections between classical moment problems and the Martin boundary for certain of these processes. A discussion of this point may be found in Watanabe [1960b]. Lamperti and Snell [1963] discussed the Martin boundary for the class of random walks introduced by Karlin and McGregor [1959]. This discussion was generalized by Kemeny [forthcoming]. Finally Blackwell and Kendall [1964] have given a discussion of the Martin boundary for the Polya urn scheme.

The result in Example 3 that the only positive regular functions for the symmetric random walk in three dimensions are the constants was proved by Murdoch [1954] by other methods. Murdoch obtained a better estimate of N_{0j} than that given here. The short proof of the estimate for N_{0j} that we give was supplied by E. Stein.

The results in Problems 30 to 34 were discovered by Harris [1957] and Veech [1963].

A point x of S^* is regular for the Dirichlet problem if for each continuous function $f \geq 0$ on S^* the superregular function h with f as boundary values has $\lim_{j \to x} h(j) = f(x)$. An equivalent condition on x is that for each open neighborhood U of x, $\lim_{j \to x} \Pr_j[x_v \in S^* - U] = 0$. Knapp [1966] showed that the set of regular points is a Borel set and gave an example of a chain P with $P\mathbf{1} = \mathbf{1}$ for which the set of regular points was empty.

Chapter 11:

The main results of this chapter were presented in a paper by Kemeny and Snell [1963a]. As noted in this paper, these authors are indebted to W. A. Veech for the important Lemma 11-11. The recurrent boundary was introduced independently by Orey [1964].

ADDITIONAL NOTES

The discussion below deals with some of the developments in the theory since the publication of the first edition. Some papers that predate that publication are mentioned to put matters in context. Citations point to the Additional References except when the bracketed date is followed by "R." The latter citations point to the References section.

Chapter 12:

The foundations for the theory of Markov fields and random fields in general were developed by Dobruschin (e.g., [1968]) in a series of papers. Our treatment of finite random fields and the equivalence theorem for Markov and neighbor Gibbs fields, as presented in Sections 2 and 3, is based on Griffeath [1973]. K. L. Chung and D. Dawson made helpful improvements in the presentation. Theorem 12-16 is due in essence to Averintsev [1970], though he considered only the case $T = \mathbb{Z}^d$. A series of papers, culminating in Grimmett [1973], exploited the Möbius inversion formula to obtain simpler proofs in a more general context. The Martin boundary approach to infinite Gibbs fields is due to Föllmer [1975a], and is based on the work of Dynkin [1971]. Their setting is far more general than the one presented here, so their arguments are not as elementary. The detailed study of countable Markov fields on $\mathbb{Z}$ was initiated by Spitzer [1975a]; much of Section 5 is based on his paper. Föllmer [1975b] has also treated this subject. The proof of Proposition 12-32 was supplied by H. Kesten. Theorems 12-36 and 12-37 were obtained by Spitzer using tail fields rather than the Martin boundary approach. The ratio-limit representation of Theorem 12-37, which does not appear in Spitzer's paper, is cited by Cox [1976]. Example 12-38, due to Spitzer, makes use of Doob, Snell, and Williamson [1960R]. The important Theorem 12-40 was discovered by Dobruschin [1968]. Example 12-41 is a special case of a family of phase transition examples discussed by Cox [1976]. The remarkable Theorem 12-42 is due to Kesten [1976]; the reader is referred to his paper for the proof. Spitzer [1974] gives a very nice exposition of many aspects of random field theory not discussed here. Another useful reference is Dawson [1974]. Tree processes were first studied by Preston [1974], whose book contains a wealth of information on random fields. A more recent reference is Spitzer [1975b]. A lucid exposition of the Ising model may be found in Griffiths [1972]. Problems 7 and 8 are derived from Spitzer [1975a]; Problem 9 is based on a construction of S. Kalikow [1976].

Corrections to the first edition:

Pitman [1974] pointed out that Theorem 9-53 was stated incorrectly in the first edition. It had stated incorrectly that $g = -Gf$, overlooking the possible failure of the relation $A({}^0Nf) = (A\,{}^0N)f$. This theorem and its consequences have been corrected in the second edition.

For general ergodic chains we can give no useful condition under which this associativity holds. However, for a strong ergodic chain associativity holds if f is bounded (and in particular if the potential g is bounded, since $f = (I - P)g$). In fact, it is enough to observe that $\alpha\,{}^0N1 \leq \alpha\,{}^0\bar{N}1 = M_{\alpha 0} < \infty$. Thus for a strong ergodic chain, $g = -Gf$ if the charge f is bounded. This conclusion was obtained in a more direct way in Proposition 9-73.

Martingales:

Two books that develop the subject of martingales are those by Meyer [1972] and Neveu [1972a]. Both books begin with the basic material on martingales. Neveu's contains a short chapter on the optimal stopping problem that is discussed later in these notes. The latter part of each book begins to reflect the explosion in the subject of martingales that has centered around integral inequalities.

Early developments were by Burkholder [1966] and Gundy [1967]. Suppose $(f_n, \mathscr{F}_n)$ is a martingale and d_n is the sequence of differences $d_0 = f_0$, $d_n = f_n - f_{n-1}$ for $n \geq 1$, so that $f_n = d_0 + \cdots + d_n$. If v_n is measurable with respect to $\mathscr{F}_{n-1}$, then the sequence g_n with

$$g_n = \sum_{k=0}^{n} v_k d_k$$

is called a *transform* of f_n. It is a martingale if $M[|g_n|] < \infty$ for all n, by imitation of the proof of Proposition 3-7. (The case that the v_n are characteristic functions arises in the proof of the Upcrossing Lemma and is the case of optional sampling.) Burkholder and Gundy deal with questions of convergence and integral boundedness of such transforms. The gambling interpretation is as follows: A gambler playing a sequence of rounds in a fair game can win d_n dollars in round n, and his fortune is then f_n. If $\sup M[|f_n|] < \infty$, his fortune converges to a finite limit f_∞ a.e. and $M[|f_\infty|] \leq \sup M[|f_n|]$. In the transformed game he is allowed to vary the stakes according to his past experience; at time n he can win $v_n d_n$ dollars. How can he improve his circumstances by choosing v_n suitably? Burkholder's first theorem is that if $\sup_n |v_n| < \infty$ a.e., then g_n converges a.e. to a finite limit g_∞; however, $M[|g_\infty|]$ may be infinite.

In studying a martingale $\{f_n\}$, Burkholder and Gundy work with the function $(\sum_{n=1}^{\infty} |f_n - f_{n-1}|^2)^{1/2}$ and generalizations. This is called the S-function; some of its properties of convergence and average size are comparable with those of $\{f_n\}$. The prototype for such conclusions is the Khintchine–Kolmogorov Theorem: Let $\{y_n\}$ be independent random variables with $M[y_n] = 0$ and $M[|y_n|^2] = \sigma_n^2$. If $\sum_{n=1}^{\infty} \sigma_n^2 = \sigma^2 < \infty$, then $\sum_{n=1}^{\infty} y_n$ is convergent a.e. and in L^2. Conversely if $\Sigma\, y_n$ is convergent in L^2, then $\Sigma\, \sigma_n^2 = \sigma^2 < \infty$ and $M[|\Sigma\, y_n|^2] = \sigma^2$. There is a corresponding martingale result with $f_n - f_{n-1}$ in place of y_n.

There is a parallel between the theory of martingales and some of the developments in Euclidean Fourier analysis, and some of the theorems in each of the areas motivate theorems in the other. Let Q_0 be the unit cube in n dimensional space $\mathbb{R}^n$, and let $\mathscr{R}_k$ be the partition of Q_0 consisting of cubes of side 2^{-k} with all coordinates of vertices at integral multiples of 2^{-k}. Now let $(f_k, \mathscr{R}_k^*)$ be a martingale; for example, $f_k = M[f\,|\,\mathscr{R}_k^*]$ is a martingale if f

is integrable on Q_0, and there are other examples. Meanwhile, consider harmonic functions u in $\mathbb{R}_+^{n+1} = \{(x, t) \mid x \in \mathbb{R}^n, t > 0\}$; for example, the Poisson integral of an integrable function on Q_0 (or in all of $\mathbb{R}^n$) is an example, and there are other examples. The parallel is obtained by comparing properties of the function f_k in the special martingale with the function $u(\cdot, 2^{-k})$, where u is harmonic.

This parallel was handled rigorously in one case when the martingale theorem of Burkholder and Gundy [1970] was generalized by Burkholder, Gundy, and Silverstein [1971] to deal with Brownian motion and then to prove a Euclidean theorem.

Except in this one case, however, the idea has been to make the comparison and to proceed by analogy. Under the parallel, the martingale S-function corresponds to the "Lusin area function" of Fourier analysis if the martingale $(f_k, \mathscr{R}_k^*)$ is general, and to the "Littlewood-Paley g-function" of Fourier analysis if the martingale has $f_k = M[f \mid \mathscr{R}_k^*]$. The parallel also gives useful information in dealing with functions of bounded mean oscillation. For an account of the martingale results, see Garsia [1973]. Fefferman [1975] has given a thorough exposition of the Euclidean results and described the parallel in more detail.

Strong ratio limit property:

A noncyclic recurrent chain P has the strong ratio limit property (SRLP) if there are positive numbers $\pi_j > 0$ such that

$$\lim_{n \to \infty} \frac{P_{ij}^{(n+m)}}{P_{kl}^{(n)}} = \frac{\pi_j}{\pi_l}$$

for all i, j, k, l, m. Chung and Erdos [1951 R] showed the SRLP holds for sums of independent random variables on the integers, and an example reproduced in Chung [1960 R] shows the SRLP fails for a certain non-cyclic recurrent P.

Orey [1961] proved the SRLP holds if $P_{00}^{(n+1)}/P_{00}^{(n)}$ tends to 1 and it holds if

$$\limsup_{n \to \infty} \frac{P_{00}^{(m(n+1))}}{P_{00}^{(mn)}} \leq 1.$$

The latter condition holds for a reversible chain since $P_{00}^{(n)}$ is non-increasing in n, and hence the SRLP holds for reversible chains.

The result of Chung and Erdos can be interpreted as showing the SRLP holds if there is spatial homogeneity of the right kind. Kingman and Orey [1964 R] proved the SRLP under a much weaker assumption of spatial homogeneity—that $N_{ii}^{(n)} \geq 1 + \epsilon$ for all i for some $\epsilon > 0$ and some n. Sums of independent random variables clearly have this property.

More recent work has concentrated on transient chains. The SRLP requires reformulation in these cases. Pruitt [1965] gives a definition and proves theorems analogous to those of Orey [1961]. The book by Orey [1971] treats these matters in some detail. See also Freedman [1971].

Applied uses of Markov chains:

Probabilistic functions of finite Markov chains: Let P be a finite Markov chain with state space S and starting vector π. Suppose, for each i in S, that $F_{i\cdot}$ is a probability measure on a finite set Y. We imagine a process in which P takes place unseen in the background and the matrix F is used at each time to produce an outcome in Y. Calling the outcomes y_n, we have

$$(*) \quad \Pr[y_1 = k_1 \wedge \cdots \wedge y_N = k_N] = \sum_{i_0,\ldots,i_N \in S} \pi_{i_0} P_{i_0 i_1} F_{i_1 k_1} P_{i_1 i_2} F_{i_2 k_2} \cdots P_{i_{N-1} i_N} F_{i_N k_N}.$$

For example, a subject in a psychological experiment may have different probabilities for making responses according to his state of mind. If we imagine his frame of mind as the outcome of a Markov chain, then S is the set of frames of mind and Y is the set of responses.

If each row vector $F_{i\cdot}$ has all its mass in one entry, then the situation is that in lumping. The states in S are lumped in some fashion and the lumped states are those in Y. The lumped process need not be a Markov chain. Such processes have been studied extensively for a long time. See Rosenblatt [1971], Chapter III.

The opposite extreme occurs when all F_{ij}, P_{ij}, and π_i are >0. The typical practical problem that arises is to estimate the parameters π, P, and F if a finite sequence of outcomes is all that is known. Specifically one wants values of π, P, and F that make (*) a maximum, given $k_1, \ldots, k_N$. Baum *et al.* [1970] give an iterative procedure in the last paragraph of their paper for passing from one set of values of π, P, and F to another with the property that (*) increases to a critical point. Their theorem that (*) increases to a critical point has been used in modeling letter patterns in English words, in predicting sunspot behavior, and in anticipating the stock market. As indicated above, it also has applications to psychology and sociology.

Optimal stopping problems: Let $(y_n, \mathscr{F}_n)$ be a denumerable stochastic process, and let $x_n = x_n(y_0, y_1, \ldots, y_n)$ be real-valued and measurable. Suppose the x_n are integrable. The problem is to find

$$V = \sup_t M[x_t],$$

where the supremum is taken over all random times t. V is the *value* of the x_n process. If the supremum is attained for some t, t is an optimal strategy, and a further problem is to describe t.

We are to regard the y_n's as some observable outcomes and the x_n's as rewards. We are allowed to choose the time of obtaining our reward, without clairvoyance, and the problem is to maximize the payoff. There is an extensive theory in this generality. See Chow, Robbins, and Siegmund [1971]. Neveu [1972a] treats the martingale case, beginning with "Le problème de Snell," solved in Snell [1952 R].

In many applications the $\{y_n\}$ are the outcome of a Markov chain, and the nth reward function is $x_n(y_0, \ldots, y_n) = f(y_n)$, with f a function on the state space that is independent of n. The value is simply $v_i = \sup_t M_i[f(y_t)]$. If f is bounded, then v is the least nonnegative superregular function $\geq f$. See Dynkin and Yushkevich [1969], Chapter 3, for a treatment of the problem and discussion of strategy. A deeper investigation is the book by Širjaev [1973].

Recurrent potential theory:

Orey [1964 R] developed a recurrent potential theory that avoids the notion of a normal chain and proceeds from axioms for a potential operator.

Neveu [1972b] developed recurrent potential theory from a different perspective, proceeding as follows. For a function h on S with $0 \le h \le 1$, define

$$U_h = \sum_{n=0}^{\infty} (PD_{1-h})^n P = \sum_{n=0}^{\infty} P(D_{1-h}P)^n,$$

where D_h is the diagonal matrix with h_i as ith diagonal entry.

When $h = 1$, $U_h = P$. When $h = 0$, $U_h = P + P^2 + P^3 + \cdots$. In the general case with $f \ge 0$,

$$(U_h f)_i = M_i\left[\sum_{n=0}^{\infty} (1 - h(x_1))(1 - h(x_2))\cdot \ldots \cdot(1 - h(x_{n-1}))f(x_n)\right].$$

If we write U_E for U_h when h is the characteristic function of E, we have

$$(U_E f)_i = M_i\left[\sum_{1 \le n \le t_E} f(x_n)\right].$$

When f is 1 on E and 0 elsewhere, the right side reduces to the probability that the chain started in state i ever returns to the set E.

Let P be recurrent with positive regular measure α. Neveu proves that there is some h on S with $0 < h \le 1$ such that $U_h h = 1$ and $U_h \ge 1\alpha$. Fix such an h, put $V = U_h - 1\alpha$, and define

$$W = \sum_{n=0}^{\infty} (VD_h)^n V.$$

Clearly W is ≥ 0. The finiteness of W is settled by the facts that $Wh = c1$ and $\alpha D_h W = c\alpha$, where c is the constant $(1 - \alpha h)/(\alpha h)$. The operator $I + W$ is the potential kernel, and Neveu develops an appropriate theory for it.

See also Revuz [1975].

Transient boundary theory:

Transient boundary theory for general denumerable Markov chains stands about where it was in 1966.

Dynkin [1969] gave an account of the theory that does not use extended chains. For the theory of the exit boundary the idea is to use a martingale-upcrossing argument to deal with a superregular measure μ. If $f(i) = \mu_i/(\pi N)_i$, the key result is that $\lim_{n\to\infty} f(x_n(\omega))$ exists a.e. on infinite paths, provided f satisfies a suitable integrability condition. From this result, the result we call Theorem 10-18 follows without reference to any extended chains, and the rest of the theory requires no change.

Athreya and Ney [1972] apply transient boundary theory to branching processes in Chapter II of their book.

Sums of independent random variables:

Sums of independent random variables for a countable group that is not necessarily abelian can be defined just as in the abelian case (see Chapter 4), as long as left and right are distinguished carefully. Let P be the transition matrix for such a chain, and suppose that the states form a single class.

Regarding the question of transience *vs.* recurrence, Kesten [1959] considered a symmetric P as a linear operator on square-summable sequences. He proved that the group admits such a P with spectral radius one if and only if the group is amenable, i.e., there is a non-zero left-invariant positive linear functional on the bounded functions on the group. If the spectral radius is less than one, then not only does N have finite entries but also N is a bounded operator on square-summable sequences; hence recurrence implies spectral radius one. Day [1964] removed the hypothesis of symmetry in Kesten's theorem and found further equivalent conditions. His theorem has been reproved several times by other authors.

For the exact question of transience *vs.* recurrence the results are less decisive. In his book Spitzer [1976] settles the case of processes on the lattice points in Euclidean space. Dudley [1962] proved that a countable *abelian* group has a recurrent P (with one class of states) if and only if the maximum number of linearly independent elements is at most 2. In the non-abelian case Kesten [1967] conjectured that the existence of a recurrent P for a group is related to the growth in n of the number of elements of the group expressible as a product of n generators. Milnor [1968a] showed that this growth function is approximately independent of the set of generators; he pointed out that the existence of a symmetric P with spectral radius less than 1 implies exponential growth, and he gave an example of a solvable group and a symmetric P with spectral radius 1 and with exponential growth. Milnor [1968b] and Wolf [1968] considered classes of countable groups and gave conditions under which the growth function is of polynomial size or of exponential size.

Ney and Spitzer [1966] compute the Martin boundary for transient sums of independent random variables on a lattice with nonzero mean. Kesten and Spitzer [1965] prove the existence of the potential kernel for recurrent sums of independent random variables on countable abelian groups, and they consider the Martin boundary. Kesten [1967] generalized this work to general countable groups, although the extent to which non-trivial non-abelian groups can admit recurrent processes is still not known.

Derriennic [1975] deals with sums of independent random variables on a free group with $n > 1$ generators. It is assumed that the transition matrix has only finitely many nonzero entries in each row and that all states communicate. He shows that such a process is transient and that the boundary consists of all "reduced infinite words." The special case in which the process in one step can move from a word only to the product of that word by a generator or its inverse was considered earlier by other authors. When in the special case all the $2n$ one-step probabilities are equally likely, the resulting example is one that arises in the theory of algebraic groups.

REFERENCES

Blackwell, D., and Kendall, D., "The Martin boundary for Pólya's urn scheme and an application to stochastic population growth," *J. Appl. Probability*, **1**, 284–296 (1964).

Blumenthal, R. M., "An extended Markov property," *Trans. Amer. Math. Soc.*, **85**, 52–72 (1957).

Brelot, M., "Le problème de Dirichlet, Axiomatique et frontière de Martin," *J. Math. Pures Appl.*, **35**, 297–335 (1956).

Brelot, M., *Éléments de la Théorie Classique du Potentiel*, Sorbonne, Paris, 1959.

Choquet, G., "Theory of capacities," *Ann. Inst. Fourier, Grenoble*, **5**, 131–295 (1953–54).

Choquet, G., and Deny, J., "Modèles finis en théorie du potentiel," *J. Anal. Math.*, **5**, 77–135 (1956–57).

Chung, K. L., "An ergodic theorem for stationary Markov chains with a countable number of states," *Proc. Intern. Congs. Math.*, Cambridge, Mass., **1**, 568 (1950).

Chung, K. L., "Contributions to the theory of Markov chains, II," *Trans. Amer. Math. Soc.*, **76**, 397–419 (1954).

Chung, K. L., *Markov Chains with Stationary Transition Probabilities*, Berlin–Göttingen–Heidelberg, Springer-Verlag, 1960.

Chung, K. L., "Some remarks on taboo probabilities," *Illinois J. Math.*, **5**, 431–435 (1961).

Chung, K. L., and Erdös, P., "Probability limit theorems assuming only the first moment I," *Mem. Amer. Math. Soc.*, **6**, 1–19 (1951).

Chung, K. L., and Ornstein, D., "On the recurrence of sums of random variables," *Bull. Amer. Math. Soc.*, **68**, 30–32 (1962).

Courant, R., and Hilbert, D., *Methods of Mathematical Physics*, Vol. I, New York, Interscience Publishers, Inc., 1953.

Derman, C., "A solution to a set of fundamental equations in Markov chains," *Proc. Amer. Math. Soc.*, **5**, 332–334 (1954).

Doeblin, W., "Sur deux problèmes de M. Kolmogoroff concernant les chaines dénombrables," *Bull. Soc. Math. France*, **66**, 210–220 (1938).

Doob, J. L., "Regularity properties of certain families of chance variables," *Trans. Amer. Math. Soc.*, **47**, 455–486 (1940).

Doob, J. L., *Stochastic Processes*, New York, John Wiley & Sons, Inc., 1953.

Doob, J. L., "Semimartingales and subharmonic functions," *Trans. Amer. Math. Soc.*, **77**, 86–121 (1954).

Doob, J. L., "Conditional Brownian motion and the boundary limits of harmonic functions," *Bull. Soc. Math. France*, **85**, 431–458 (1957).

Doob, J. L., "Probability theory and the first boundary value problem," *Illinois J. Math.*, **2**, 19–36 (1958).

Doob, J. L., "Discrete potential theory and boundaries," *J. Math. and Mech.*, **8**, 433–458 (1959).

Doob, J. L., "Notes on martingale theory," *Fourth Berkeley Symposium on Mathematical Statistics and Probability*, Vol. II, Berkeley, Calif., University of California Press, pp. 95–102, 1961.

Doob, J. L., Snell, J. L., and Williamson, R. E., "Application of boundary theory to sums of independent random variables," *Contributions to Probability and Statistics*, Stanford, Calif., Stanford University Press, pp. 182–197, 1960.

Dunford, N., and Schwartz, J. T., *Linear Operators*, Part I, New York, Interscience Publishers, Inc., 1958.

Dynkin, E. B., and Malyutov, M. B., "Random walk on groups with a finite number of generators," *Translation of Dokl. Acad. Sci. of USSR*, **2**, 399–402 (1961) (Original **137**, 1042–1045).

Feller, W., "Boundaries induced by non-negative matrices," *Trans. Amer. Math. Soc.*, **83**, 19–54 (1956).

Feller, W., *An Introduction to Probability Theory and Its Applications*, Vol. I, 2nd ed., New York, John Wiley & Sons, Inc., 1957.

Fréchet, M., "Méthode des fonctions arbitraires," *Théorie des évenements en chaine dans le cas d'un nombre fini d'états possibles*, 2nd livre, Paris, Gauthiers-Villars, 1938.

Gorden, P., *Théorie des chaînes de Markov finies et ses applications*, Paris, Dunod, 1965.

Halmos, P., *Measure Theory*, Princeton, N.J., D. Van Nostrand Co., Inc., 1950.

Harris, T. E., "Transient Markov chains with stationary measures," *Proc. Amer. Math. Soc.*, **8**, 937–942 (1957).

Harris, T. E., *The Theory of Branching Processes*, Berlin-Göttingen-Heidelberg, Springer-Verlag, 1963.

Hennequin, P.-L., and Tortrat, A., *Théorie des probabilités et quelques applications*, Paris, Masson, 1965.

Hildebrand, F. B., *Introduction to Numerical Analysis*, New York, McGraw-Hill Book Co., 1956.

Hunt, G. A., "Markoff processes and potentials I, II, III," *Illinois J. Math.*, **1**, 44–93, 316–369 (1957); *ibid.*, **2**, 151–213 (1958).

Hunt, G. A., "Markoff chains and Martin boundaries," *Illinois J. Math.*, **4**, 313–340 (1960).

Itô, K., and McKean, H. P., Jr., "Potentials and the random walk," *Illinois J. Math.*, **4**, 119–132 (1960).

Kac, M., "Random walk and the theory of Brownian motion," *Amer. Math. Monthly*, **54**, 369–391 (1947a).

Kac, M., "On the notion of recurrence in discrete stochastic processes," *Bull. Amer. Math. Soc.*, **53**, 1002–1010 (1947b).

Kac, M., "On some connections between probability theory and differential and integral equations," *Proceedings of the Second Berkeley Conference on Mathematical Statistics and Probability*, Berkeley, Calif., University of California Press, pp. 189–215, 1951.

Kakutani, S., "Two-dimensional Brownian motion and harmonic functions," *Proc. Imp. Acad. Tokyo*, **20**, 706–714 (1944).

Kakutani, S., "Markov processes and the Dirichlet problem," *Proc. Jap. Acad.*, **21**, 227–233 (1945).

Karlin, S., and McGregor, J., "Random walks," *Illinois J. Math.*, **3**, 66–81 (1959).

Kellogg, O. D., *Foundations of Potential Theory*, New York, Dover Publications, Inc., 1929.

Kemeny, J. G., "Doeblin's ratio limit theorem," *Notices Amer. Math. Soc.*, **9**, 390–391 (1962).

Kemeny, J. G., "A further note on discrete potential theory," *J. Math. Anal. and Appl.*, **6**, 55–57 (1963).

Kemeny, J. G., "Representation theory for denumerable Markov chains," *Trans. Amer. Math. Soc.*, **125**, 47–62 (1966).

Kemeny, J. G., Mirkil, H., Snell, J. L., and Thompson, G. L., *Finite Mathematical Structures*, Englewood Cliffs, N.J., Prentice-Hall, Inc., 1959.

Kemeny, J. G., and Snell, J. L., *Finite Markov Chains*, Princeton, N.J., D. Van Nostrand Co., Inc., 1960.

Kemeny, J. G., and Snell, J. L., "On Markov chain potentials," *Annals of Math. Stat.*, **32**, 709–715 (1961a).

Kemeny, J. G., and Snell, J. L., "Potentials for denumerable Markov chains," *J. Math. Anal. and Appl.*, **3**, 196–260 (1961b).

Kemeny, J. G., and Snell, J. L., "Notes on discrete potential theory," *J. Math. Anal. and Appl.*, **3**, 117–121 (1961c).

Kemeny, J. G., and Snell, J. L., "Boundary theory for recurrent Markov chains," *Trans. Amer. Math. Soc.*, **106**, 495–520 (1963a).

Kemeny, J. G., and Snell, J. L., "A new potential operator for recurrent Markov chains," *J. London Math. Soc.*, **38**, 359–371 (1963b).

Kemperman, J. H. B., *The Passage Problem for a Stationary Markov Chain*, Chicago, University of Chicago Press, 1961.

Kesten, H., "Ratio theorems for random walks, II," *J. d'Analyse Math.*, **11**, 323–379 (1963).

Kesten, H., and Spitzer, F., "Ratio theorems for random walks, I," *J. d'Analyse Math.*, **11**, 285–322 (1963).

Kingman, J. F. C., and Orey, S., "Ratio limit theorems for Markov chains," *Proc. Amer. Math. Soc.*, **15**, 907–910 (1964).

Knapp, A. W., "Connection between Brownian motion and potential theory," *J. Math. Anal. and Appl.*, **12**, 328-349 (1965).

Knapp, A. W., "Regular boundary points in Markov chains," *Proc. Amer. Math. Soc.*, **17**, 435–440 (1966).

Kolmogorov, A. N., "Anfangsgründe der Theorie der Markoffschen Ketten mit unendlichen vielen möglichen Zuständen," *Mat. Sbornik N.S.*, 607–610 (1936); *Bull. Univ. Moscow*, **1** (1937) (Russian).

Kolmogorov, A. N., *Foundations of the Theory of Probability*, 2nd English ed., New York, Chelsea Publishing Co., 1956.

Lahres, H., *Einführung in die diskreten Markoff-Prozesse und ihre Anwendungen*, Braunschweig, Germany, Friedr. Viewig und Sohn, 1964.

Lamperti, J., "Criteria for the recurrence or transience of stochastic process, I," *J. Math. Anal. and Appl.*, **1**, 314–330 (1960a).

Lamperti, J., "The first-passage moments and the invariant measure of a Markov chain," *Annals of Math. Stat.*, **31**, 515–517 (1960b).

Lamperti, J., "Criteria for stochastic processes II: Passage-time moments," *J. Math. Anal. and Appl.*, **7**, 127–145 (1963a).

Lamperti, J., "Wiener's Test and Markov chains," *J. Math. Anal. and Appl.*, **6**, 58–66 (1963b).

Lamperti, J., and Snell, J. L., "Martin boundaries for certain Markov chains," *J. Math. Soc. Japan*, **15**, 113–128 (1963).

Lévy, P., *Théorie de l'addition des variables aléatoires*, Paris, Gauthier-Villars, 1937.

Loève, Michel, *Probability Theory*, 3rd ed., Princeton, N.J., D. Van Nostrand, Co., Inc., 1963.

Markov, A. A., "Investigation of an important case of dependent trials," *Izvestia Acad. Nauk SPB*, VI, ser. I (Russian), **61** (1907).

Martin, R. S., "Minimal positive harmonic functions," *Trans. Amer. Math. Soc.*, **49**, 137–172 (1941).

McKean, H. P., Jr., "A problem about prime numbers and the random walk I," *Illinois J. Math.*, **5**, 351 (1961).

Murdoch, B. H., *Preharmonic Functions* (Thesis), Princeton, N.J., Princeton Univ. Library, May 1954.

Naïm, L., "Sur le rôle de la frontière de R. S. Martin dans la théorie du potentiel," *Ann. Inst. Fourier, Grenoble*, **7**, 183–281 (1957).

Nash-Williams, C. St J. A., "Random walk and electric currents in networks," *Proc. Cambridge Phil. Soc.*, **55**, 181–194 (1959).

Neveu, J., "Chaînes de Markov et théorie du potentiel," *Ann. Faculté des Sciences, Clermont*, **3**, 37–65 (1964).

Orey, S., "An ergodic theorem for Markov chains," *Z. Wahrescheinlichkeits-theorie*, **1**, 174–176 (1962).

Orey, S., "Potential kernels for recurrent Markov chains," *J. Math. Anal. and Appl.*, **8**, 104–132 (1964).

Polya, G., "Über eine Aufgabe der Wahrscheinlichkeitsrechnung betriffend die Irrfahrt im Strassennetz," *Math. Ann.*, **84**, 149–160 (1921).

Romanovskii, V. I., *Discrete Markov Chains*, Moscow–Leningrad (Russian), 1949.

Rudin, W., *Principles of Mathematical Analysis*, New York, McGraw-Hill Book Co., 1953.

Snell, J. L., "Applications of martingale system theorems," *Trans. Amer. Math. Soc.*, **73**, 293–312 (1952).

Spitzer, F. L., "Hitting probabilities," *J. Math. and Mech.*, **11**, 593–614 (1962).

Spitzer, F. L., *Principles of Random Walk*, Princeton, N.J., D. Van Nostrand Co., Inc., 1964.

Veech, W., "The necessity of Harris' condition for the existence of a stationary measure," *Proc. Amer. Math. Soc.*, **14**, 856–860 (1963).

Ville, J., *Étude Critique de la notion de collectif*, Paris, Gauthier-Villars, 1939.

Watanabe, T., "On the theory of Martin boundaries induced by countable Markov processes," *Mem. Coll. Science, University Kyoto Series* A, **33**, 39–108 (1960a).

Watanabe, T., "A probabilistic method in Hausdorff moment problem and Laplace–Stieltjes transform," *J. Math. Soc. of Japan*, **12**, 192–206 (1960b).

Yosida, K., and Kakutani, S., "Markov process with an enumerable infinite number of possible states," *Jap. J. Math.*, **16**, 47–55 (1940).

ADDITIONAL REFERENCES

Athreya, K. B., and Ney, P. E., *Branching Processes*, New York, Springer-Verlag, 1972.

Averintsev, M. B., "One method of describing random fields with a discrete argument," *Problems of Information Transmission*, **6**, 169–175 (1970). (Original: *Problemy Peredachi Informatsii*, **6**, 100–108.)

Baum, L. E., Petrie, T., Soules, G., and Weiss, N., "A maximization technique occurring in the statistical analysis of probabilistic functions of Markov chains," *Ann. Math. Stat.*, **41**, 164–171 (1970).

Blackwell, D., and Freedman, D., "The tail σ-field of a Markov chain and a theorem of Orey," *Ann. Math. Stat.*, **35**, 1291–1295 (1964).

Burkholder, D. L., "Martingale transforms," *Ann. Math. Stat.*, **37**, 1494–1504 (1966).

Burkholder, D. L., "Martingale inequalities," *Martingales*, Lecture Notes in Mathematics, **190**, New York, Springer-Verlag, 1–8, 1971.

Burkholder, D. L., and Gundy, R. F., "Extrapolation and interpolation of quasi-linear operators on martingales," *Acta Math.*, **124**, 249–304 (1970).

Burkholder, D. L., Gundy, R. F., and Silverstein, M. L., "A maximal function characterization of H^p," *Trans. Amer. Math. Soc.*, **157**, 137–153 (1971).

Chow, Y.-s., Robbins, H., and Siegmund, D., *Great Expectations: the Theory of Optimal Stopping*, Boston, Houghton Mifflin Company, 1971.

Cox, T., "An example of phase transition in countable one-dimensional Markov random fields," (forthcoming, 1976).

Dawson, D., *Discrete Markov Systems*, Carleton Mathematical Lecture Notes No. 10, Ottawa, Carleton University (1974).

Day, M. M., "Convolutions, means, and spectra," *Illinois J. Math.*, **8**, 100–111 (1964).

Derriennic, Y., "Marche aléatoire sur le groupe libre et frontière de Martin," *Z. Wahrscheinlichkeitstheorie*, **32**, 261–276 (1975).

Dobruschin, P. L., "The description of a random field by means of conditional probabilities and conditions of its regularity," *Th. Prob. and Its Appl.*, **13**, 197–224 (1968). (Original: *Teoriya Veroy. i ee Prim.*, **13**, 201–229.)

Dudley, R. M., "Random walks on abelian groups," *Proc. Amer. Math. Soc.*, **13**, 447–450 (1962).

Dynkin, E. B., *Markov Processes*, 2 vol., Springer-Verlag, New York, 1965.

Dynkin, E. B., "Boundary theory of Markov processes (the discrete case)," *Russian Math. Surveys*, **24**, no. 2, 1–42 (1969). (Original: *Uspekhi Mat. Nauk*, **24**, no. 2, 3–42.)

Dynkin, E. B., "Entrance and exit spaces for a Markov process," *Actes Congrès int. Math., 1970*, **2**, 507–512 (1971).

Dynkin, E. B., and Yushkevich, A. A., *Markov Processes, Theorems and Problems*, New York, Plenum Press, 1969.

Fefferman, C., "Recent progress in classical Fourier analysis," *Proc. International Congress Mathematicians, 1974* vol. 1, Canadian Mathematical Congress, 95–118 (1975).
Flatto, L., and Pitt, J., "Recurrence criteria for random walks on countable abelian groups," *Illinois J. Math.*, **18**, 1–19 (1974).
Föllmer, H., "Phase transition and Martin boundary," *Séminaire de Probabilités IX, Université de Strasbourg*, Lecture Notes in Mathematics, **465**, New York, Springer-Verlag, 305–317 (1975a).
Föllmer, H., "On the potential theory of stochastic fields," *Inter. Stat. Institute*, Warsaw (1975b).
Freedman, D., *Markov Chains*, San Francisco, Holden-Day, 1971.
Garsia, A. M., *Martingale Inequalities*, Reading, Mass., W. A. Benjamin, Inc., 1973.
Griffeath, D., "Markov and Gibbs fields with finite state space on graphs," unpublished manuscript (1973).
Griffeath, D., and Snell, J. L., "Optimal stopping in the stock market," *Ann. of Prob.*, **2**, 1–13 (1974).
Griffiths, R. B., "The Peierls argument for the existence of phase transitions," *Mathematical Aspects of Statistical Mechanics*, J. C. T. Pool (ed.), *SIAM-AMS Proceedings*, Providence, Amer. Math. Soc., **5**, 13–26, 1972.
Grimmett, G. R., "A theorem about random fields," *Bull. London Math. Soc.*, **5**, 81–84 (1973).
Gundy, R. F., "The martingale version of a theorem of Marcinkiewicz and Zygmund," *Ann. Math. Stat.*, **38**, 725–734 (1967).
Kalikow, S., "An entrance law which reaches equilibrium," (forthcoming, 1976).
Kemeny, J. G., "Slowly spreading chains of the first kind," *J. Math. Anal. and Appl.*, **15**, 295–310 (1966).
Kemeny, J. G., and Snell, J. L., "Markov chains and summability methods," *Z. Wahrscheinlichkeitstheorie*, **18**, 17–33 (1971).
Kesten, H., "Full Banach mean values on countable groups," *Math. Scand.*, **7**, 146–156 (1959).
Kesten, H., "The Martin boundary of recurrent random walks on countable groups," *Proc. Fifth Berkeley Symposium on Mathematical Statistics and Probability*, Berkeley, University of California Press, vol. 2, part 2, 51–74, 1967.
Kesten, H., "Existence and uniqueness of countable one-dimensional Markov random fields," (forthcoming, 1976).
Kesten, H., and Spitzer, F., "Random walk on countably infinite abelian groups," *Acta Math.*, **114**, 237–265 (1965).
Meyer, P.-A., *Martingales and Stochastic Integrals I*, Lecture Notes in Mathematics, **284**, New York, Springer-Verlag, 1972.
Milnor, J., "A note on curvature and fundamental group," *J. Diff. Geom.*, **2**, 1–7 (1968a).
Milnor, J., "Growth of finitely generated solvable groups," *J. Diff. Geom.*, **2**, 447–449 (1968b).
Neveu, J., *Martingales à temps discret*, Paris, Masson et Cie, 1972a.
Neveu, J., "Potentiel Markovien récurrent des chaines de Harris," *Ann. Inst. Fourier, Grenoble*, **22**, 2, 85–130 (1972b).
Ney, P., and Spitzer, F., "The Martin boundary for random walk," *Trans. Amer. Math. Soc.*, **121**, 116–132 (1966).

Orey, S., "Strong ratio limit property," *Bull. Amer. Math. Soc.*, **67**, 571–574 (1961).
Orey, S., *Limit Theorems for Markov Chain Transition Probabilities*, London, Van Nostrand Reinhold Company, 1971.
Pitman, J. W., "Uniform rates of convergence for Markov chain transition probabilities," *Z. Wahrscheinlichkeitstheorie*, **29**, 193–227 (1974).
Preston, C. J., *Gibbs States on Countable Sets*, Cambridge, Cambridge University Press, 1974.
Pruitt, W. E., "Strong ratio limit property for R-recurrent Markov chains," *Proc. Amer. Math. Soc.*, **16**, 196–200 (1965).
Revuz, D., *Markov Chains*, New York, American Elsevier Publishing Company, Inc., 1975.
Rosenblatt, M., *Markov Processes, Structure and Asymptotic Behavior*, New York, Springer-Verlag, 1971.
Ruelle, D., *Statistical Mechanics, Rigorous Results*, New York, W. A. Benjamin, Inc., 1969.
Seneta, E., *Non-Negative Matrices*, New York, John Wiley and Sons, 1973.
Širjaev, A. N., *Statistical Sequential Analysis, Optimal Stopping Rules*, Providence, American Mathematical Society, vol. 38, Translations of Mathematical Monographs, 1973.
Spitzer, F., "Introduction aux processus de Markov à paramètre dans Z_ν," *Ecole d'Eté de Probabilités de Saint-Flour III*, Lecture Notes in Math., **390**, New York, Springer-Verlag, 114–189, 1974.
Spitzer, F., "Phase transition in one-dimensional nearest-neighbor systems," *J. Functional Anal.*, **20**, 240–255 (1975a).
Spitzer, F., "Markov random fields on an infinite tree." *Ann. of Prob.*, **3**, 387–398 (1975b).
Spitzer, F., *Principles of Random Walk*, 2nd Edition, New York, Springer-Verlag, 1976.
Wolf, J. A., "Growth of finitely generated solvable groups and curvature of Riemannian manifolds," *J. Diff. Geom.*, **2**, 421–446 (1968).

INDEX OF NOTATION

General Notations

Specific Notations

Specific Notations—(*continued*)

INDEX